6G Key Technologies

6G Key Technologies

A Comprehensive Guide

Wei Jiang and Fa-Long Luo

IEEE PRESS

WILEY

Published by John Wiley & Sons, Inc., Hoboken, New Jersey.
Published simultaneously in Canada.

For general information on our other products and services or for technical support, please contact our Customer Care Department within the United States at (800) 762-2974, outside the United States at (317) 572-3993 or fax (317) 572-4002.

Wiley also publishes its books in a variety of electronic formats. Some content that appears in print may not be available in electronic formats. For more information about Wiley products, visit our web site at www.wiley.com.

Library of Congress Cataloging-in-Publication Data Applied for:

Hardback ISBN: 9781119847472

Cover Design: Wiley
Cover Image: © metamorworks/Shutterstock

Set in 9.5/12.5pt STIXTwoText by Straive, Chennai, India

Contents

Preface

Wireless technologies and applications are now being developed in a very rapid speed and massively large scale. In the early 2019, when South Korea's three mobile operators and US Verizon were arguing with each other about who would be the world's first provider of the fifth generation (5G) communication services, we stepped into the era of 5G. In the past three years, the term 5G has been remaining one of the hottest buzzwords in the media, attracting unprecedented attention from the whole society. It even went beyond the sphere of technology and economy, becoming the focal point of geopolitical tension. In addition to further improving network capacities such as the previous generations, 5G expands mobile communication services from human to things and from consumers to vertical industries. The potential scale of mobile subscriptions is substantially enlarged from merely billions of the world's populations to almost countless inter-connectivity among humans, machines, and things. It enables a wide variety of services from traditional mobile broadband to Industry 4.0, virtual reality, Internet of things, and automatic driving. In 2020, the outbreak of the COVID-19 pandemic leads to a dramatic loss of human life worldwide and imposes unprecedented challenges on societal and economic activities. However, this public health crisis highlights the unique role of telecommunication networks and digital infrastructure in keeping society running and families connected, especially the values of 5G services and applications, such as remote surgeons, online education, remote working, driver-less vehicles, unmanned delivery, robots, smart healthcare, and autonomous manufacturing.

As of the writing of this book, the 5G mobile networks are still on its way to being deployed across the world, but it is already time for the academia and industry to shift their attention to beyond 5G or the sixth generation (6G) system to satisfy the future demands for information and communications technology in 2030. Even though discussions are ongoing within the wireless community as to whether there is any need for 6G or whether counting the generations should be stopped at 5, many research groups, standardization and regulatory organizations, and countries across the world have kicked off their initiatives toward 6G. A focus group called *Technologies for Network 2030* within the International Telecommunication Union Telecommunication (ITU-T) standardization sector was established in July 2018. This group intends to study the capabilities of networks for 2030 and beyond, when it is expected to support novel forward-looking scenarios, such as holographic-type communications, pervasive intelligence, tactile internet, multi-sense experience, and digital twin. The European Commission initiated to sponsor beyond 5G research activities, as its recent Horizon 2020 calls – ICT-20 5G Long Term Evolution and ICT-52 Smart Connectivity beyond 5G – where a batch of pioneer research projects for key 6G technologies was kicked off at the early beginning of 2020. The European Commission has also announced its strategy to accelerate investments in Europe's "Gigabit Connectivity" including 5G and 6G to shape Europe's digital future. In October 2020, the Next Generation Mobile Networks (NGMN) has launched its new "6G Vision and Drivers" project, intending to provide early and timely direction for global 6G activities. At its meeting in February 2020, the International Telecommunication Union Radiocommunication (ITU-R) sector decided to start studying future technology trends for the future evolution of International Mobile Telecommunications (IMT-2030).

Inspired by the tremendous impact of 5G technology, governments and the public further recognized the significance of mobile systems for economic prosperity and national security. In the past two years, many countries have announced ambitious plans on the development of 6G or already launched research initiatives officially. In Finland, the University of Oulu began ground-breaking 6G research as part of Academy of Finland's flagship program called "6G-Enabled Wireless Smart Society and Ecosystem (6Genesis)", which focuses on several challenging research areas including reliable near-instant unlimited wireless connectivity, distributed computing and intelligence, as well as materials and antennas to be utilized in future for circuits and devices. In March 2019, the Federal Communications Commission (FCC) of the United States announced that it opens experimental license for the use of frequencies between 95 GHz and 3 THz for 6G and beyond, fostering the test of THz communications. In October 2020, the Alliance for Telecommunications Industry Solutions (ATIS) launched Next G Alliance with founding members including AT&T, T-Mobile, Verizon, Qualcomm, Ericsson, Nokia, Apple, Google, Facebook, Microsoft, etc. It is an industry initiative that intends to advance North American mobile technology leadership in 6G over the next decade through private sector-led efforts. Another US company, SpaceX, which is famous for its revolutionary innovation of reusable rockets, announced the Starlink project in 2015. Starlink is a very large-scale LEO communication satellite constellation aiming to offer ubiquitous Internet access services across the whole planet. It is envisioned that such as a space communication infrastructure will reshape the architecture of the next-generation mobile communications. In November 2019, the Ministry of Science and Technology of China has officially kicked off the 6G technology research and development works coordinated by the ministry, together with five other ministries or national institutions. A promotion working group from the government that is in charge of management and coordination, and an overall expert group that is composed of 37 experts from universities, research institutes, and industry were established at this event. In late 2017, a working group was established by the Japanese Ministry of Internal Affairs and Communications to study next-generation wireless technologies. South Korea announced a plan to set up the first 6G trial in 2026 and is expected to spend around 169 million US dollars over five years to develop key 6G technologies. Germany has also entered this crowded field. Beginning in August 2021, the German Federal Ministry of Education and Research (BMBF) funded the establishment of four hubs for research into the future technology 6G with up to 250 million euros in the first four year term.

At this historical crossroad, it is strongly believed that a book on 6G will serve as an enlightening guideline to spur interests and further investigations related to 6G communication systems. Such a book will also attract a broad audience in both academia and industry of all related fields. By far, a book to deal in a systematic and unified manner to cover the state-of-the-art vision of 6G and elaborate a complete list of identified key technologies is still lacking but highly desirable. From technology research and development point of views and authored by two experts in the field, this book aims to be the world's first book to provide a comprehensive and highly consistent treatment on the enabling techniques for 6G radio transmission and signal processing by covering New Vision, New Spectrum, New Propagation, as well as New Air Interface and Smart Radio Systems. This book is organized into 10 chapters in three parts.

Part 1: Vision of 6G and Technical Evolution

In terms of the technology development and practice applications, the first part of this book will present the vision of 6G by reviewing the evolution from 1G standards to 5G standards and put the emphasis on the driving factors, requirements, use cases, key performance indicators (KPI), and roadmap toward 6G. Consisting of three chapters, the first part will also drive a detailed list of potential 6G technologies to support the targeted use cases and KPIs.

Chapter 1: Standards History of Cellular Systems Toward 6G

To well understand the complex cellular systems of today, it is vital to have a complete view of how cellular systems have evolved. To this end, the motivation of the first chapter is to provide the readers a brief review of the whole history of mobile cellular systems, from pre-cellular to 5G. Then, the readers should be able to well prepare for getting insights into the 6G technology.

Chapter 2: Pre-6G Technology and System Evolution

The second chapter will provide an in-depth study of the previous generations from the technological perspective. A symbolic standard, which achieved a dominant position commercially in the market worldwide and adopted mainstreaming technologies at that time, is elaborated. In this regard, the readers will understand the overall architecture of the representative cellular system for each generation, including major network elements, functionality split, interconnection, interaction, and operational flows.

Chapter 3: Vision of 6G: Drivers, Enablers, Uses, and Roadmap

It is necessary to clarify the significance and the fundamental motivation of developing 6G and convince the readers that 6G will definitely come like the previous generations of mobile communications before jumping into 6G technical details. This chapter will provide a comprehensive vision concerning the driving forces, use cases, and usage scenarios and demonstrate essential performance requirements and identify key technological enablers.

Part 2: Full-Spectra Wireless Communications in 6G

Organized into three chapters, the second part of this book focuses on the full-spectra wireless communications for 6G by including new spectrum opportunities related to millimeter wave, terahertz communications, visible light communications, and optical wireless communications.

Chapter 4: Enhanced Millimeter-Wave Wireless Communications in 6G

Millimeter-wave (mmWave) has been adopted in 5G systems, but its utilization is infancy. Hence, enhanced mmWave wireless communications is highly probably a key enabler of 6G. Regardless of the abundant spectral resources and contiguous large-volume bandwidths, the characteristics of mmWave signal propagation are distinct from that of low-frequency bands at UHF and microwave bands. This chapter will focus on the characteristics of mmWave signal propagation and key mmWave transmission technologies to unleash the potential of high frequencies.

Chapter 5: Terahertz Technologies and Systems for 6G

To satisfy the demand of extreme transmission rate on the magnitude order of tera-bits-per-second envisioned in the 6G system, wireless communications need to exploit the abundant spectrum in the Terahertz (THz) band. In addition to THz communications, the THz band is also applied for other particular applications, such as imaging, sensing, and positioning, which are expected to achieve synergy with THz communications in 6G. In addition to shed light on its high potential, this chapter will analyze major challenges such as high free-space path loss, atmospheric absorption, rainfall attenuation, blockage, and high Doppler fluctuation and will introduce enabling THz technologies such as array-of-subarrays beamforming in an ultra-massive MIMO system and lens antenna arrays. Meanwhile, the world's first THz communications standard, i.e. IEEE 802.15.3d, will be presented to give an insight.

Chapter 6: Optical and Visible Light Wireless Communications in 6G

Optical Wireless Communications points to wireless communications that use the optical spectrum, including infrared, visible light, and ultraviolet, as the transmission medium. Optical wireless communications show great advantages in deployment scenarios such as home networking, vehicular communications in intelligent transportation systems, airplane passenger lighting, and electronic medical equipment that is sensitive to RF interference. This chapter will introduce the main technical features, the physical fundamentals of optical devices, and the major challenges in optical communications systems.

Part 3: Smart Radio Networks and Air Interface Technologies for 6G

Moving to radio network and air interface level of 6G, the third part of this book consists of four chapters by mainly addressing the related technologies including intelligent reflecting surface (IRS)-based systems, multiple dimensional and antenna techniques, cellular and cell-free massive MIMO, and adaptive and non-orthogonal multiple access.

Chapter 7: Intelligent Reflecting Surface-Aided Communications in 6G

Using a large number of small, passive, and low-cost reflecting elements, IRS can proactively form a smart and programmable wireless environment. Thus, it provides a new degree of freedom to the design of 6G wireless systems, enabling sustainable capacity and performance growth with affordable cost, low complexity, and low energy consumption. This chapter will introduce the system model and the signal transmission of IRS-aided systems with either single-antenna or multi-antenna base stations in both frequency-flat and frequency-selective fading channels. It can serve as a tutorial for the readers as a starting point to further carry out research works on this promising topic.

Chapter 8: Multiple Dimensional and Antenna Techniques for 6G

This chapter will explore the fundamentals of multi-antenna transmission, including spatial diversity, beamforming, and spatial multiplexing. Although some of these multi-antenna techniques have been adopted in pre-6G standards, they will also play a critical role in the 6G systems, especially when combining with emerging technologies, such as intelligent reflecting surface, cell-free structure, non-orthogonal multiple access and terahertz as well as optical, and visible-light communications. Therefore, this chapter will provide comprehensive insights into these techniques aiming to unleash their full potentials in 6G.

Chapter 9: Cellular and Cell-Free Massive MIMO Techniques for 6G

This chapter will first introduce the critical issues of multi-user MIMO techniques, including the principle of the well-known dirty-paper coding. Then, a revolutionary technique called massive MIMO that breaks the scalability barrier by not attempting to achieve the full Shannon limit and paradoxically by increasing the size of the system is shown. This chapter will emphasize a cutting-edge technology called cell-free massive MIMO. The cell-free architecture is particularly attractive for the upcoming 6G deployment scenarios, such as a campus or a private network dedicated to an industrial vertical, and it is envisioned as a key technology enabler of 6G.

Chapter 10: Adaptive and Non-orthogonal Multiple Access Systems in 6G

A cellular network needs to accommodate a lot of active subscribers simultaneously over a finite amount of time frequency resources. Both orthogonal and non-orthogonal multiple accesses are envisioned to be evolved further and play a critical role in the upcoming 6G system design. Although OFDM, OFDMA, SC-FDMA, and NOMA have been adopted in pre-6G standards, these techniques will play a fundamental role in 6G, in the conventional sub-6 GHz band and high-frequency bands. There exists a big room for further exploiting these techniques when combining with intelligent reflecting surface, cell-free structure, mmWave, terahertz, and optical transmission techniques.

For Whom Is this Book Written?

It is hoped that this book serves not only as a complete and invaluable reference for professional engineers, researchers, manufacturers, network operators, software developers, content providers, service providers, broadcasters, and regulatory bodies aiming at development, standardization, deployment, and applications of 6G system, but also as a textbook for graduate students in circuits, signal processing, wireless communications, artificial intelligence, microwave technology, information theory, antenna and propagation, system-on-chip implementation, and computer networks.

Wei Jiang, Ph.D
Kaiserslautern, Germany
March, 2022

Fa-Long Luo, Ph.D
Silicon Valley, California, USA
March, 2022

List of Abbreviations

0G	Zeroth Generation
1G	First Generation
2G	Second Generation
3G	Third Generation
3GPP	Third Generation Partnership Project
3GPP2	Third Generation Partnership Project 2
4G	Fourth Generation
5G	Fifth Generation
5GC	5G Core network
6G	Sixth Generation
AAS	Active Antenna System
ACI	Adjacent Channel Interference
AF	Application Function
AGV	Automated Guided Vehicle
AI	Artificial Intelligence
AMF	Access and Mobility Management Function
AMPS	Advanced Mobile Phone System
AOA	Angle of Arrival
AOD	Angle of Departure
AP	Access Point
APP	Application
ARIB	Association of Radio Industries and Businesses
AUSF	Authentication Server Function
AWGN	Additive White Gaussian Noise
BBU	Baseband Unit
BCI	Brain–Computer Interface
BD	Block Diagonalization

BER	Bit Error Ratio
BPSK	Binary Phase-Shift Keying
BPTT	Back-Propagation Through Time
BS	Base Station
BSC	Base Station Controller
BSS	Base Station Subsystem
BTS	Base Transceiver Station
CAPEX	Capital Expenditure
CDD	Cyclic Delay Diversity
CDF	Cumulative Distribution Function
CDMA	Code-Division Multiple Access
CDPD	Cellular Digital Packet Data
CEPT	Conference of European Postal and Telecommunications
CFR	Channel Frequency Response
CIR	Channel Impulse Response
CoMP	Coordinated Multi-Point transmission and reception
CP	Cyclic Prefix
CPRI	Common Public Radio Interface
CPU	Central Processing Unit
CR	Cognitive Radio
C-RAN	Cloud Radio Access Network
CRC	Cyclic Redundancy Code
CS	Circuit-Switched
CSI	Channel State Information
CTN	Cellular Telephone Number
CU	Centralized Unit
D2D	Device-to-Device
DAC	Digital-to-Analog Convertor
D-AMPS	Digital Advanced Mobile Phone System
DFT	Discrete Fourier Transform
DHCP	Dynamic Host Configuration Protocol
DL	Downlink
DPC	Dirty Paper Coding
DPSK	Differential Phase Shift Keying
DSA	Dynamic Spectrum Allocation
DSL	Digital Subscriber Line
DSP	Digital Signal Processing (DSP)
DSS	Dynamic Spectrum Sharing
DU	Distributed Unit
DVB	Digital Video Broadcasting

E2E	End-to-End
EDGE	Enhanced Data Rates for GSM Evolution
eICIC	Enhanced Inter-Cell Interference Coordination
EIRP	Effective Isotropic Radiated Power
eMBB	Enhanced Mobile Broadband
EPC	Evolved Packet Core
ESN	Electronic Serial Number
ESPRIT	Estimation of Signal Parameters via Rational Invariance Techniques
ETSI	European Telecommunications Standards Institute
ETU	Extended Typical Urban
E-UTRA	Evolved Universal Terrestrial Radio Access
FBMC	Filter-Bank Multi-Carrier
FCC	Federal Communications Commission
FD	Full-Duplex
FDD	Frequency Division Duplexing
FDMA	Frequency-Division Multiple Access
FEC	Forward Error Correction
FFR	Fractional Frequency Reuse
FFSK	Fast Frequency-Shift Keying
FFT	Fast Fourier Transform
FIR	Finite Impulse Response
FM	Frequency Modulation
FOMA	Freedom of Mobile Multimedia Access
FOV	Fields of View
FPGA	Field-Programmable Gate Array
FSK	Frequency-Shift Keying
FSO	Free-Space Optical
FSPL	Free Space Pass Loss
FSTD	Frequency-Switched Transmit Diversity
FTN	Faster-Than-Nyquist
GDPR	General Data Protection Regulation
GERAN	GSM EDGE Radio Access Network
GFDM	Generalized Frequency Division Multiplexing
GGSN	Gateway GPRS Support Node
GMSK	Gaussian Minimum Shift Keying
gNB	Next-generation Node B
GNSS	Global Navigation Satellite System
GPRS	General Packet Radio Service

GPS	Global Positioning System
GRU	Gated Recurrent Unit
GSM	Global System for Mobile communications
HAP	High Altitude Platform
HARQ	Hybrid Automatic Repeat Request
HDTV	High-Definition Television
HetNet	Heterogeneous Network
HLR	Home Location Register
HSDPA	High Speed Downlink Packet Access
HSPA	High Speed Packet Access
HSUPA	High Speed Uplink Packet Access
IAB	Integrated Access and Backhaul
ICI	Inter-Cell Interference
ICIC	Inter-Cell Interference Coordination
ICNIRP	International Commission on Non-Ionizing Radiation Protection
ICT	Information and Communications Technology
IFFT	Inverse Fast Fourier Transform
IIoT	Industrial Internet of Things
IMEI	International Mobile Equipment Identity
IMSI	International Mobile Subscriber Identity
IMT	International Mobile Telecommunications
IMTS	Improved Mobile Telephone Service
IoT	Internet of Things
IR	Infra-Red
IRC	Interference Rejection Combining
IRS	Intelligent Reflecting Surface
ISD	Inter-Site Distance
ISDN	Integrated Services Digital Network
ISI	Inter-Symbol Interference
JT	Joint Transmission
KPI	Key Performance Indicator
LAA	License-Assisted Access
LD	Laser Diode
LDPC	Low Density Parity Check

LED	Light-Emitting Diode
LEO	Low Earth Orbit
LIDAR	Light Detection and Ranging
Li-Fi	Light Fidelity
LOS	Line-Of-Sight
LPF	Low Pass Filter
LSTM	Long-Short Term Memory
LTE	Long-Term Evolution
LTE-Advanced	Long-Term Evolution-Advanced
MAC	Medium Access Control
MAP	Maximum a Posteriori
MC-CDMA	Multi-Carrier Code-Division Multiple Access
MCM	Multi-Carrier Modulation
MCS	Modulation and Coding Scheme
MEC	Mobile Edge Computing
MF	Matched Filtering
MIB	Master Information Block
MIMO	Multiple Input Multiple Output
ML	Maximum-Likelihood
MME	Mobility Management Entity
MMS	Multimedia Messaging Service
MMSE	Minimum Mean Square Error
mMTC	Massive Machine-Type Communications
mmWave	Millimeter Wave
MRC	Maximum-Ratio Combining
MRT	Maximum-Ratio Transmission
MSS	Mobile Station Subsystem
MTC	Machine-Type Communications
MTS	Mobile Telephone Service
MTSO	Mobile Telephone Switching Office
MUD	Multi-User Detection
MU-MIMO	Multi-User Multiple-Input Multiple-Output
MUSA	Multi-User Shared Access
MUSIC	Multiple Signal Classification
MUST	Multi-User Superposition Transmission
NaaS	Network-as-a-Service
NB-IoT	Narrow-Band Internet of Things
NEF	Network Exposure Function

NFV	Network Function Virtualization
NGMA	Next Generation Mobile Networks
NLoS	Non-Line-of-Sight
NOMA	Non-Orthogonal Multiple Access
NR	New Radio
NRF	NF Repository Function
NSA	Non-Standalone
NSS	Network and Switching Subsystem
NTIA	National Telecommunications and Information Administration
NTT	Nippon Telegraph and Telephone
NTT	Non-Terrestrial Networks
OAM	Orbital Angular Momentum
OFDM	Orthogonal Frequency-Division Multiplexing
OFDMA	Orthogonal Frequency-Division Multiple Access
OMA	Orthogonal Multiple Access
OOB	Out-Of-Band
OOK	On-Off Keying
OPEX	Operational Expenditure
O-RAN	Open Radio Access Network
OSS	Operation and Support Subsystem
OTT	Over-The-Top
OWC	Optical Wireless Communications
QAM	Quadrature Amplitude Modulation
QoE	Quality of Experience
QoS	Quality of Service
QPSK	Quadrature Phase Shift Keying
P2P	Peer-to-Peer
PA	Power Amplifier
PAM	Pulse Amplitude Modulation
PAPR	Peak-to-Average-Power Ratio
PAS	Power Angular Spectrum
PBCH	Physical Broadcast Channel
PCEF	Policy and Charging Enforcement Function
PCF	Policy Control Function
PCM	Pulse Coded Modulation
PCRF	Policy and Charging Rules Function
PD	Photo Diode
PDC	Personal Digital Cellular

PDN	Packet Data Network
PER	Packet Error Rate
P-GW	Packet Data Network Gateway
PHICH	Physical Hybrid ARQ Indicator Channel
PHY	Physical Layer
PIC	Parallel Interference Cancelation
PMI	Precoding Matrix Indicator
PPN	Polyphase Network
PRACH	Physical Random Access Channel
PS	Packet-Switched
PSD	Power Spectral Density
PSS	Primary Synchronization Signal
PSTN	Public Switched Telephone Network
RAN	Radio Access Network
RAT	Radio Access Technology
RB	Resource Block
RF	Radio Frequency
RFU	Radio Frequency Unit
RIS	Reconfigurable Intelligent Surface
RMS	Root Mean Square
RNC	Radio Network Controller
RNN	Recurrent Neural Network
RRC	Radio Resource Control
RRH	Remote Radio Head
RRM	Radio Resource Management
RU	Radio Unit
SAR	Synthetic Aperture Radar
SBA	Service-Based Architecture
SC	Single Carrier
SC-FDMA	Single Carrier-Frequency-Division Multiple Access
SCM	Spatial Channel Model
SCMA	Sparse Code Multiple Access
SDMA	Space-Division Multiple Access
SDR	Software-Defined Radio
SDSF	Structured Data Storage network Function
SFBC	Space-Frequency Block Coding
SFR	Soft Frequency Reuse
SGD	Stochastic Gradient Descent
SGSN	Serving GPRS Support Node

S-GW	Serving Gateway
SIC	Successive Interference Cancellation
SIM	Subscriber Identity Module
SIMO	Single-Input Multiple-Output
SINR	Signal-to-Interference-plus-Noise Ratio
SISO	Single-Input Single-Output
SLA	Service Level Agreement
SLAM	Simultaneous Localization And Mapping
SMF	Session Management Function
SMS	Short Messaging Service
SNR	Signal-to-Noise Ratio
SON	Self-Organizing Network
SSS	Secondary Synchronization Signal
STBC	Space-Time Block Coding
STTC	Space-Time Trellis Code
SU-MIMO	Single-User Multiple-Input Multiple-Output
SVD	Singular Value Decomposition
SWIPT	Simultaneous Wireless Information and Power Transfer
TAS	Transmit Antenna Selection
TDD	Time Division Duplexing
TDMA	Time-Division Multiple Access
TD-SCDMA	Time Division-Synchronous Code Division Multiple Access
THP	Tomlinson–Harashima Precoding
THz	Terahertz
TIA	Telecommunications Industry Association
TRI	Transmit Rank Indicator
TTI	Transmission Time Interval
UAV	Unmanned Aerial Vehicle
UCA	Uniform Circular Array
UDM	Unified Data Management
UDN	Ultra-Dense Network
UDSF	Unstructured Data Storage network Function
UE	User Equipment
UHF	Ultra High Frequency
UL	Uplink
ULA	Uniform Linear Array
UMTS	Universal Mobile Telecommunications System
UPF	User Plane Function
URLLC	Ultra-Reliable and Low-Latency Communications

UTRA	Universal Terrestrial Radio Access
UV	Ultraviolet
V2I	Vehicle-to-Infrastructure
V2P	Vehicle-to-Pedestrian
V2V	Vehicle-to-Vehicle
V-BLAST	Vertical Bell Laboratories Layered Space-Time Architecture
VLC	Visible Light Communications
VLR	Visitor Location Register
VR	Virtual Reality
WAP	Wireless Application Protocol
WCDMA	Wideband Code-Division Multiple Access
Wi-Fi	Wireless Fidelity
WiMAX	Worldwide Interoperability for Microwave Access
WLAN	Wireless Local Area Network
WRAN	Wireless Regional Area Networks
WRC	World Radiocommunication Conference
ZFP	Zero-Forcing Precoding

Part I

The Vision of 6G and Technical Evolution

1

Standards History of Cellular Systems Toward 6G

In the summer of 1895, a few decades after the invention of the wire telephone, Guglielmo Marconi successfully demonstrated the feasibility of radio transmission. Since then, a wide variety of radio communications and broadcasting services were adopted throughout the world. Around half of a century later, the world-renowned research institution, Bell Labs, accomplished two historic innovations in the same year of 1947 – the Transistor and the Cellular concept. In the early 1980s, with tens of years' technical development, the first-generation cellular networks were finally rolled out to offer commercial mobile telephony service to the public. With its ease of deployment, economic efficiency, portability, flexibility, and scalability compared with wire-line networks, mobile cellular networks experienced explosive growth in the last decades. It became one of the critical infrastructures to empower modern society and drastically reshaped human behaviors in business, education, entertainment, and personal life. Recently, the term *5G* has been remained as one of the hottest buzzwords in the media, attracting unprecedented attention from the public. The whole society has got a consensus that the fifth-generation (5G) cellular system is one of the greatest innovations in the 2020s and will bring tremendous economic and societal benefits. At the moment of starting the writing of this book, more than 400 mobile operators in approximately 130 countries are deploying 5G networks, and the number of 5G subscribers in either the consumer market or vertical industries has already reached an enormous scale in many regions. Now, the attention of academia and industry is increasingly shifting toward the next generation. Since the first experiments with radio communications in the 1890s, it was quite a long journey to reach cutting-edge mobile communications. To well understand the complex cellular systems of today, it is vital to have a complete view of how cellular systems have evolved. To this end, the motivation of the first chapter in association with the following chapter is to provide the readers a brief review of the whole history of mobile cellular systems, from pre-cellular to the

6G Key Technologies: A Comprehensive Guide, First Edition. Wei Jiang and Fa-Long Luo.

fifth generation. Then, the readers should be able to well prepared for getting insights into the forthcoming six-generation (6G) system. This chapter will be organized chronologically in terms of the generations. Each section dedicates to one generation of cellular systems, where the main content generally consists of three main parts:

– The underlying motivation of evolution.
– The milestones of development, standardization, and deployment.
– The review of various competing standards with their major technical features.

1.1   0G: Pre-Cellular Systems

Wireless communications had been exploited already in early ancient times when people tried to transfer critical messages such as the invasion of enemies by means of smoke, torches, flashing mirrors, signal flares, or semaphore flags. Long-range transmission was realized through the signal relaying over a network of observation stations built on beacon towers or mountain peaks. These infant communication systems were replaced by the electric telegraph (invented by Samuel Morse in 1837) that transferred text messages over landlines, and later the wire telephone (invented by Alexander Graham Bell in 1876), carrying information-rich voice signals. In the summer of 1895, a few decades after the invention of the telephone, Guglielmo Marconi successfully carried out the first experiment to illustrate the ability of radio communications. Since then, a wide variety of radio services such as wireless telegraph, mobile telephony, radio broadcasting, television broadcasting, satellite communications, wireless local area networks, and Bluetooth were adopted worldwide and sharply reshaped modern society. As the most successful form of radio technologies, mobile communications have experienced explosive growth in the last decades. Nowadays, cellular networks serve as the critical infrastructure and the basis for the mobile Internet that is an industry worth trillions of dollars per year.

The first cellular system originated in a portable radiophone known as Walkie-Talkie during World War II, symbolized by SCR-536 developed by Motorola for the US military. This handheld radio transceiver operated in a push-to-talk manner, allowing one radio to transmit while others in its range to listen (i.e. the half-duplex operation). It was primitive but gained much experience for the later development of pre-cellular mobile telephone systems. One of the earliest mobile telephone systems was known as Mobile Telephone Service (MTS), which was connected to the public telephone network as an extension of the wire telephone service and operated commercially in the United States in 1946 by Motorola in conjunction with the Bell System. On 17 June 1946, the Bell System demonstrated the world's first mobile call in the City of St. Louis through a car phone weighed around 36 kg. Initially, only 3 channels were available for all the subscribers in the metropolitan area but increased to 32 channels later. Within three years, this service had been expanded to 100 cities across the United States, attracting a total of 5000 users. In 1964, an enhanced system named Improved Mobile Telephone Service (IMTS) was rolled out to replace the previous MTS system. It achieved two major advances: direct dialing allowing a phone call without manual connection by a human operator, and the full-duplex transmission, by which two communicating parties can talk simultaneously.

Such pre-cellular systems were the forerunners of the first generation of cellular networks, sometimes referred to as the zeroth generation (0G). These initial systems utilized a central transmission station to serve an entire metropolitan area. An IMTS base station generally covered a wide area with a diameter of 60–100 kilometers (km) using a transmit power of 100 Watts (100 W), in comparison with less than 1 W on modern base stations. Each voice conversation exclusively occupied a radio channel, but even a large cite was licensed with only a few channels, leading to very limited system capacity. In the 1970s, before the deployment of cellular networks, a customer wishing to subscribe to mobile telephone service had to wait for up to three years until an incumbent subscriber terminated his or her mobile subscription.

Cellular Network

The constraint of network capacity was the main driver for a more elegant network design known as the cellular system.

In 1947, William R. Young, an engineer who worked at AT&T Bell Labs, reported his idea about the hexagonal layout throughout each city so that every mobile telephone can connect to at least one cell. Douglas H. Ring, also at Bell Labs, expanded on Young's concept. He sketched out the basic design for a standard cellular network and published the intellectual groundwork as a technical memorandum entitled *Mobile Telephony – Wide Area Coverage* on Bell Labs' internal journal on 11 December 1947 [Ring, 1947]. In a cellular network, a wide area can be divided into small geographical areas called cells, each covered by a radio station. It allowed efficient reuse of precious spectral resources at spatially separated sites taking advantage of the fact that the power of a transmitted signal decays dramatically with distance.

Nevertheless, the development process of the cellular system from an initial concept to a practical network was quite a long journey due to technological barriers. AT&T requested a spectrum license for cellular service from the US Federal Communications Commission (FCC) as early as 1947, and the system design had been mostly completed in the 1960s. The first trial network consisting of 10 cells was eventually installed until 1977, when many of the original technologies were outdated [Goldsmith, 2005]. Based on this trial network, Bell Labs worked out the first cellular network standard in the United States called Advanced Mobile Phone System (AMPS) [Young, 1979], which was successfully deployed in many countries and smoothly evolved into a second-generation cellular standard known as IS-54 (where IS stands for Interim Standard).

1.2 1G: The Birth of Cellular Network

In December 1979, the Japanese network operator Nippon Telegraph and Telephone (NTT) launched the first commercial cellular system in the world. The initial network comprised 88 cells covering all metropolitan area districts in Tokyo, and inter-cell handover was supported. It operated in the frequency band around 900 MHz and offered a total of 600 pairs of channels for Frequency-Division Duplexing (FDD) operation. The voice signal of each mobile user was transmitted over an analog channel with a bandwidth of 25 kHz. Within five years, the network was expanded to cover the entire population of Japan, making it the first country to provide a nationwide cellular communications service.

However, the early mobile stations in the NTT network were still car phones, which had to be fitted into automobiles and were first commercialized in the 1940s. Motorola demonstrated the world's first car call in October 1946, but the phone was too heavy (the original equipment weighs around 36 kg) and consumed too much power. In 1985, NTT released shoulder phones that were still bulky but at least can be carried freely by a human. The gifted engineer Martin Cooper led a Motorola team to develop the first cellphone prototype and demonstrated the first cellphone call at the New York City Hilton in midtown Manhattan on 3 April 1973. Ten years later, Motorola introduced its historic product - DynaTAC 8000X – the first commercial cellphone that was lightweight and small enough to carry. Owning a cellphone at that time was a symbol of affluence and social status since, for example, the Motorola DynaTAC 8000X was priced at $3 995 in 1984 with, in addition, an expensive subscription cost. Motorola played an exceptionally influential role in the early days of the development of cellphones. Followed its iconic DynaTAC 8000 series, the company released the world's first flip phone Motorola MicroTAC and then the first clamshell phone Motorola StarTAC, which was not only the world's smallest at the time but also most lightweight with an extreme weight of 105 g. These early days also witnessed the rise of Nokia to become the world's second-largest cellphone maker with the launch of their Cityman series followed by the Nokia 101 candy bar design as opposed to the previous "bricks" [Linge and Sutton, 2014].

While Motorola was developing the cellphone, Bell Labs worked out the AMPS system, which became the first cellular network standard in the United States [Frenkiel and Schwartz, 2010]. In October 1983, the United States eventually had got its first commercial cellular network launched by Ameritech in Chicago. Although it was later than other regions, the cellular service in the United States was offered through cellphones rather than car phones. In Europe, the Scandinavian countries pioneered the development of the first European cellular standard called Nordic Mobile Telephone (NMT). The first NMT network was rolled out in the Nordic countries of Norway and Sweden in 1981, followed by Denmark and Finland in the subsequent year. It was the first mobile network that can support international roaming. In 1985, the number of subscribers had grown to 110 000 in Scandinavia and Finland, made it the world's largest mobile network then. The initial NMT network was operated in 450 MHz (hence also known as NMT-450) and adopted a channel bandwidth of 25 kHz. Additional frequency bands, i.e. 890–915 MHz for the uplink and 935–960 MHz for the downlink, were allocated in 1986, and the system operating in these bands became known as NMT-900. As of 2021, according to Wikipedia, a limited NMT-450 network is still in operation in some remote regions of Russia to offer basic communication services in sparsely populated areas with long distances. In addition to NMT, European countries developed several different cellular standards, including Total Access Communication System (TACS) first implemented by the United Kingdom in 1983, C-450 in Germany (1985), and Radiocom 2000 in France (1986). However, the first-generation European standards were incompatible due to the adoption of different frequency bands, air interfaces, and communication protocols, as summarized in Table 1.1.

Among all first-generation analog standards, NMT and AMPS are regarded as two good representatives that achieved great success at that time, which are briefly introduced as follows:

1.2.1 Nordic Mobile Telephone (NMT)

Nordic Telecommunications Administrations developed the NMT standard to meet the heavy demand of voice service, which cannot be accommodated by the overcrowding mobile telephone networks then: Auto Radio Phone (ARP) in Finland, Mobile Telephony System (MTD) in Sweden and Denmark, and Public Land Mobile Telephony (OLT) Telephony in Norway. The principle technologies were ready by 1973, and the specifications for base stations were completed in 1977. In 1981, the first NMT system was launched in Norway and Sweden, followed by Denmark and Finland in the subsequent year. Using the FDD operation mode, the uplink transmission was assigned to the frequency band of 453–458 MHz while 463–468 MHz for the downlink. In 1986, another pair

Table 1.1 First-generation cellular standards.

Feature	AMPS	NMT	NTT	TACS	C-450	RC2000
Launch time	1983	1981	1979	1983	1985	1986
DL band (MHz)	869–894	463–468[a]	870–885[b]	935–960	460–465.74	424.8–428[c]
UL band (MHz)	824–849	453–458	925–940	890–915	450–455.74	414.8–418
Bandwidth (kHz)	30	25	25	25	10	12.5
No. of channel	832	180	600	1 000	573	256
Multiple access	FDMA					
Duplexing	FDD					
Modulation	FM					

a) NMT also operated in the frequency bands around 900 MHz, known as NMT-900.
b) NTT also operated in several other frequency bands around 900 MHz.
c) Radiocom2000 also operated in several other frequency bands around 200 MHz.
Source: Adapted from Goldsmith [2005].

of frequency bands, i.e. 890–915 MHz and 935–960 MHz for the uplink and downlink, respectively, were allocated. The system employed Frequency-Division Multiple Access (FDMA) to accommodate a large number of mobile users. As a consequence, the spectrum was subdivided into a magnitude of narrow-band channels with a bandwidth of 25 kHz. The voice channels were analog, where the speech signals were modulated through Frequency Modulation (FM). Nevertheless, the control signaling between the base station and the mobile station was transmitted digitally, using Fast Frequency-Shift Keying (FFSK) modulation with a rate of up to 1200 bps. The cell sizes in an NMT network ranged from 2 to 30 km. To serve car phones, the system utilized a transmission power of up to 15 W (NMT-450) and 6 W (NMT-900), while the power was lower (up to 1 W) for mobile handsets. NMT was the first cellular system with fully automatic switching (dialing), and supported the handover among cells from the beginning. It was also the first cellular system to realize international roaming. The NMT specifications were free and open, allowing many companies such as Nokia and Ericsson to produce network equipment and pushing the deployment cost down.

1.2.2 Advanced Mobile Phone System (AMPS)

AMPS was developed in the United States primarily by Bell Labs, inspired by the heavily congested mobile telephone system. Originated in the cellular concept proposed in 1947, it underwent quite a long journey to become a practical network. The system design had been almost completed in the 1960s, followed by an extensive trial (technical and commercial) to optimize the system parameters and verify the basic planning rules for a cellular layout. In 1978, Bell Labs set up a large-scale and fully operational trial network, working in cooperation with Illinois Bell Telephone Co., the American Telephone and Telegraph Co. (AT&T), and the Western Electric Co. The trial network consisted of 10 cells to cover approximately 3 000 square miles in the Chicago, IL area, aiming to provide a capacity for more than 2 000 users [Ehrlich, 1979]. It was not until 1983 that commercial operation licenses were issued when the FCC allocated an initial spectrum of 40 MHz (later increased to 50 MHz) for analog cellular networks. The downlink and uplink transmission were separated using FDD, where a pair of frequency bands on 824–849 and 869–894 MHz were assigned. The spectrum was subdivided into a total of 416 paired channels consisting of 21 control channels and 395 voice channels. The speech signal of a mobile user was tuned to the carrier frequency using the FM analog modulation and transmitted over a 30 kHz channel. Although AMPS was an analog cellular system, its control channels were already digitized. Control signaling was exchanged between a base station and mobile stations at a data rate of 10 kbps. The data was modulated using Frequency-Shift Keying (FSK) and the Manchester encoding was used for error correction.

1.3 2G: From Analog to Digital

Similar to the first generation of anything, the first-generation cellular system was not called 1G until the term 3G was adopted to name the third-generation system around 20 years later. Although 1G opened the era of mobile cellular communications, it was considered primitive and exhibited many deficiencies such as

- Worse voice quality
- Limited system capacity
- No security protection
- Limited international roaming
- Poor handover reliability
- Bulky and expensive handsets
- Short battery life

Consequently, the mobile industry initiated the development of second-generation digital systems in the early 1980s, and it gradually replaced the first-generation analog systems in the 1990s. The digital cellular standards in this generation included Global System for Mobile (GSM) in Europe, Digital Advanced Mobile Phone System (D-AMPS) and IS-95 in the United States, and Personal Digital Cellular (PDC) in Japan [Mishra, 2005].

Digitization

The transition from the first-generation to second-generation cellular system was driven by digital technology. A digital system can achieve a higher capacity than an analog system since digital communications can apply more spectral-efficient digital modulation and more efficient multiple-access techniques. Digitization facilitates the compression of voice signals, the encryption of information against eavesdropping, and the support for data services. In addition, digital components are more powerful, more lightweight, smaller, cheaper, and more power-efficient than analog components.

1.3.1 Global System for Mobile Communications (GSM)

The incompatibility among various European systems made it different for the travelers among European countries to get continuous communication service with a single analog phone. It motivated the necessity for a uniform European standard and unified frequency allocation throughout Europe. As early as 1982, the Conference of European Postal and Telecommunications (CEPT) set up a working group called the Global System for Mobile (the initial meaning of GSM) to coordinate the development work. From 1982 to 1985, discussions were held in the GSM group to select between an analog and a digital system. After multiple field trials, it was decided to develop a digital cellular system based on narrow-band Time-Division Multiple Access (TDMA). The requirements for this pan-European system, such as good subjective voice quality, low terminal and service cost, and international roaming, were defined. In 1988, the CEPT formed the European Telecommunications Standards Institute (ETSI), and then the responsibility for specifying the standard was transferred to ETSI. In 1990, the Phase I recommendations for the Global System for Mobile communications (GSM) was released. Meanwhile, a variant of GSM operating in a higher frequency band, known as Digital Cellular System at 1800 MHz (DCS-1800), was standardized within ETSI and approved in February 1991 [Mouly and Pautet, 1995]. In addition to the basic voice service, GSM terminals can connect with the Integrated Services Digital Network (ISDN) for various data services with a rate of 9.6 kbps. The commercial operation of the first GSM network started in Finland. On 1 July 1991, the Finnish Prime Minister Harri Holkeri made the world's first GSM call on the Radiolinja mobile network built with the equipment from Nokia and Siemens. Since then, GSM has rapidly gained acceptance and became the dominant 2G digital cellular standard [Vriendt et al., 2002]. It achieved remarkable commercial success, with a global market share of more than 90%. By early 2004, more than 1 billion population in more than 200 countries and territories enjoyed their mobile telephony services thanks to GSM.

1.3.2 Digital Advanced Mobile Phone System (D-AMPS)

Due to its economy of scale, the AMPS standard achieved a relatively better position over sporadic and competitive European standards in the era of 1G. Nevertheless, the United States did not continue the same success in the second round. The development of the second-generation digital cellular fell into a raged debate on the selection of spectrum sharing techniques between TDMA and Code-Division Multiple Access (CDMA). The result of this debate was two incompatible systems: IS-54 (and its evolution known as IS-136) versus IS-95. IS-54 and IS-136 constituted D-AMPS, which was the digital advancement of the existing AMPS systems in the United States. D-AMPS inherited the basic architecture and signaling protocols from its predecessor, allowing for a smooth transition from analog to digital. The D-AMPS network was deployed in the same frequency bands of AMPS, i.e. 869–894 MHz for the downlink and 824–849 MHz for the uplink. But each 30 kHz channel was further subdivided into three time slots using TDMA. Capacity was tripled by multiplexing compressed voice signals from three users over a single analog channel. The specification of IS-54 was completed in 1992, and was deployed in the United States and Canada ever since its first commercial launch in 1993. It was enhanced over time and these enhancements evolved into the IS-136 standard. IS-136 introduced several new features to the original IS-54 standard, including circuit-switched data, text messaging, and the support of operating in 1900 MHz. IS-136 opened the possibility for an all-digital TDMA system instead of the dual-mode operation adopted by its predecessor.

1.3.3 Interim Standard 95 (IS-95)

CDMA has some unique technical advantages over TDMA for cellular systems, e.g. higher system capacity and simple frequency planning due to universal frequency reuse, high quality of service during soft handover, no hard limit on the number of users (soft capacity), the ability to exploit voice activity to reduce the aggregated interference automatically, and improved robustness using noise-like spread-spectrum signals. Qualcomm developed the first CDMA-based cellular system, and the initial specifications of the system were finalized in 1993. The Telecommunications Industry Association (TIA) and Electronic Industries Alliance (EIA) of the United States approved it as a digital standard in 1995, hence named IS-95 or IS-95A. In October 1995, Hutchison Telephone launched the world's first commercial CDMA cellular network in Hong Kong, under the name of cdmaOne. In contrast to previous narrow-band mobile communications, it was a wideband system that spreads information bits over a signal bandwidth of 1.25 MHz using the direct-sequence spread-spectrum technique. The CDMA system required complicated air interfaces and communication protocols, e.g. the

Rake receiver was adopted to mitigate the effect of multi-path transmission. Due to multi-user interference and inter-cell interference, the system performance depended heavily on accurate power control, especially in the uplink, to compensate for the near-far effect. A power-control bit was transmitted 800 times per second on the forward link to instruct a mobile station to adjust its transmit power with a granularity of 1 dB. The enhanced version was IS-95B, also called 2.5G of CDMA technology, which combined the standards IS-95, ANSI-J-STD-008, and TSB-74. The standardization of IS-95B was completed in 1997 and the world's first IS-95B commercial network was launched by a South Korean operator in 1998. It provided a higher data rate through code aggregation, where a base station can assign up to eight code channels to a single mobile station, increasing the achievable rate from 11.4 kbps in IS-95A to 115 kbps. There was much debate about the relative merits of the IS-54 and IS-95 standards throughout the early 1990s, claiming that IS-95 could achieve 20 times the capacity of AMPS, whereas IS-54 could only achieve three times this capacity. In the end, both systems turned out to achieve approximately the same capacity increase over AMPS [Goldsmith, 2005].

1.3.4 Personal Digital Cellular (PDC)

Japan independently developed its digital cellular standard known as PDC, exclusively deployed in Japan. Similar to D-AMPS and GSM, PDC adopted TDMA as the multiple-access technique. To be compatible with the Japanese analog systems, it selected a signal bandwidth of 25 kHz for voice channels. Each channel was divided into three time slots for full-rate (11.2 kbps) or six time slots for half-rate (5.6 kbps) voice codecs. The Research and Development Center for Radio System (RCR), later became the Association of Radio Industries and Businesses (ARIB), completed the specifications in April 1991. Using the network equipment manufactured by NEC, Motorola, and Ericsson, NTT DoCoMo launched its digital service in March 1993. After a peak of nearly 80 million subscribers, it was slowly phased out in favor of 3G technologies and was shut down on April 1, 2012. The PDC network offered mobile voice services (full- and half-rate), supplementary services (call waiting, voice mail, three-way calling, call forwarding, etc.), circuit-switched data service (up to 9.6 kbps), and packet-switched data service (up to 28.8 kbps). Regardless of its isolation from the rest of the world, the Japanese 2G network fostered an eye-catching innovation known as i-mode, which was regarded as a pioneer of mobile Internet (Tables 1.2–1.4).

1.3.5 General Packet Radio Service (GPRS)

In addition to the improved security due to digital encryption and significantly increased system capacity over their predecessors, the milestone progress of 2G

Table 1.2   Second-generation cellular standards.

	GSM	D-AMPS	PDC	IS-95
Launch year	1991	1993	1993	1995
DL band (MHz)	935–960	869–894	940–960, 1 477–1 501	869–894
UL band (MHz)	890–915	824–849	810–830, 1 429–1 453	824–849
Bandwidth (kHz)	200	30	25	1 250
User capacity	1 000	2 500	3 000	~2 500 (soft)
Multiple access	TDMA			CDMA
Receiver	Equalizer			RAKE
Duplexing	FDD			
Modulation	GMSK	π/4-DPSK	π/4-DPSK	BPSK/QPSK
Speech (kbps)	13	7.95	11.2 (full)/5.6 (half)	1.2–9.6 (variable)

cellular was the introduction of data service into the mobile network. In 1992, for the first time, Short Messaging Service (SMS), which supports a data rate of 9.6 kbps, was born. Neil Papworth, a 22-year-old software engineer working at Vodafone, sent the world's first text message on 3 December 1992 when he typed "Merry Christmas" from a computer to Richard Jarvis on an Orbitel 901 handset. With the phenomenal success of SMS and the rising demand for accessing the Internet via mobile phones and laptop computers, the 2G cellular standards evolved to enhance the capability of carrying High-Rate Packet Data (HRPD) services.

ETSI developed General Packet Radio Service (GPRS) in response to the earlier Cellular Digital Packet Data (CDPD), overlaying the AMPS system to provide a rate of 19.2 kbps, and Japanese i-mode services. The Cellular Packet Radio (CELLPAC) protocol that introduced packet-switching in GSM was the root for the specification of GPRS starting from 1993 [Walke, 2003]. In June 2000, British Telecom Cellnet launched the world's first commercial GPRS network in the United Kingdom. GPRS is an overlaying packet-switched data network on the circuit-switched GSM network. Relying on the legacy air interface, an operator only needs to install some network nodes to upgrade a voice-only GSM network to a voice-plus-data GPRS network. Base station controllers separate the data and voice traffic, and direct the data to GPRS support nodes connected to the data network. Operating in a best-effort style, GPRS typically reached a data rate of 40 kbps in the downlink and 14 kbps in the uplink by aggregating multiple time slots into one bearer. Enhancement in later specifications can theoretically achieve a peak rate of 171.2 kbps by aggregating eight time slots at the same time for a single user.

1.3.6 Enhanced Data Rates for GSM Evolution (EDGE)

On the one hand, GPRS exhibited some limitations, such as low practical data rates much lower than the theoretical values. On the other hand, the mobile operators, who failed to win a 3G license, needed a further-enhanced GPRS standard to offer data services at speeds near those available on 3G networks. Enhanced Data Rates for GSM Evolution (EDGE) was first developed by the ETSI, as an evolution of GPRS in 1997. Although EDGE reused the GSM carrier bandwidth and time slot structure, it was not restricted to GSM cellular systems. Instead, it aimed to become a generic technology facilitating an evolution of existing cellular systems toward third-generation capabilities [Furuskar et al., 1999]. After evaluating several different proposals, the Universal Wireless Communications Consortium (UWCC) approved EDGE in January 1998 as the outdoor component of IS-136HS to provide 384 kbps data services. The first commercial EDGE network was launched in 2003 by AT&T in the United States. With the introduction of a higher-order modulation scheme named eight phase-shift keying (8PSK), opposed to Gaussian Minimum-Shift Keying (GMSK) in its predecessors, it can support maximal data rate up to 470 kbps. Several new techniques, including link adaptation, Hybrid Automatic Repeat Request (HARQ) with soft combining, and advanced scheduling, were first applied in EDGE, followed by Wideband Code-Division Multiple Access (WCDMA), CDMA2000, and other standards.

Meanwhile, the IS-54 and IS-136 systems provided a data rate of up to 60 kbps by aggregating time slots and using high-order modulation. The further evolution of the IS-136 standard was called IS-136HS (high-speed), based on EDGE. It increased the data throughput of IS-136 systems to over 470 kbps per carrier. The initial IS-95 system supported circuit-mode and packet-mode data services at a data rate of 14.4 kbps. Without breaking the legacy air interface design to maintain strict backward-compatibility, it was upgraded to IS-95B that offered an increased data rate of 115 kbps [Knisely et al., 1998].

The transition from 1G analog to 2G digital also facilitated the innovation of mobile terminals, where much smaller, lighter, cheaper, and power-efficient cell phones were popularized at an extraordinary pace. Mobile phones moved from being commercial users to general users as an essential part of modern living. Nokia successfully recognized the desire of people to turn mobile phones into personalized devices. In 1994, Nokia released the first cell phone to use the iconic ring-tone – Nokia 2110, the first cell phone allowing the user to change the phone's covers to reflect their mood or style – Nokia 5110, and the first to feature the mobile game Snake – Nokia 6110. As a consequence, Nokia achieved the dominant market share in the world's cell phone market. In 2002, the world had completed the transition to digital cellular networks, and the number of mobile subscribers surpassed that of the fixed-line telephone subscribers for the first time, making cellular networks the dominant technique to provide communication service.

1.4 3G: From Voice to Data-Centric

Second-generation digital systems were designed to address the weaknesses, such as limited system capacity, easy eavesdropping, and worse voice quality, in the first-generation systems. However, standards like GSM, IS-95, and IS-136 were still designed for voice communications but performed not well for data service.

> **Data-Centric Cellular Network**
>
> *The proliferation of Internet-based services, such as web browsing, multimedia messaging, email, interactive gaming, and high-fidelity audio and video streaming, and the expansion of these services from wired networks to mobile networks, imposed a need for data-optimized cellular systems to replace previous voice-centric cellular systems.*

Incompatible mobile environment and fragmented spectrum usage that annoyed the previous generations motivated the International Telecommunications Union (ITU) to begin the effort on a global standard with full interoperability and interworking in the 1980s. In 1990, the ITU released the first recommendation on the Future Public Land Mobile Telecommunications System (FPLMTS). FPLMTS was renamed to the International Mobile Telecommunications-2000 (IMT-2000) [ITU-R M.1225, 1997] in the late 1990s since the old acronym was hard to pronounce. Meanwhile, the World Radio Conference held in February 1992 identified 230 MHz spectrum in the bands 1885–2025 MHz and 2110–2200 MHz for IMT-2000 on a worldwide basis. IMT-2000 recommendations defined the minimal technical requirements of the 3G system, including high data rate, asymmetric data transmission, global roaming, multiple simultaneous services, improved voice quality, security, and greater capacity. The evaluation criteria specified in International Telecommunication Union Radiocommunication (ITU-R) M.1225 [ITU-R M.1225, 1997] set the target data rates for the 3G circuit-switched and packet-switched data services:

- Up to 2 Mbps for an indoor environment.
- Up to 144 kbps for outdoor-to-indoor and pedestrian environments.
- Up to 64 kbps for a vehicular environment.

Nevertheless, the ITU did not specify technological solutions to meet these requirements but only solicited proposals from interested organizations. Based on the successful standardization of GSM, the ETSI initiated a new organization called the Third Generation Partnership Project (3GPP) together with other standard development organizations across the world, including ARIB (Japan), ATIS (USA), CCSA (China), TTA (South Korea), and TTC (Japan). At the same time, another group in the United States formed the Third Generation Partnership Project 2 (3GPP2), intended to the specifications for a 3G system based on the evolution of IS-95. Although some differences remained, both 3GPP and 3GPP2 selected CDMA as the underlying baseline technology. The standard developed by 3GPP was WCDMA or Universal Mobile Telecommunications System (UMTS). 3GPP2 focused on the development of CDMA2000, which reused the spectrum bands of IS-95 and inherited the bandwidth set of 1.25 MHz. In 1998, the ITU received numerous technical proposals, among which five standards were approved for terrestrial service, i.e. UTRA FDD, UTRA Time-Division Duplex (TDD), CDMA2000, TDMA Single-Carrier, and FDMA/TDMA. In 2007, Worldwide Interoperability for Microwave Access (WiMAX) specified by IEEE 802.16 specifications was approved by the ITU as the sixth IMT-2000 standard, also known as IMT2000 OFDMA TDD WMAN. Unlike other 3G standards based on CDMA, WiMax adopted more pre-4G technologies such as orthogonal frequency-division multiplexing (OFDM), multiple-input multiple-output multi-input multi-output (MIMO), and low-density parity-check (LDPC) coding. IMT-2000 TDMA Single-Carrier also called Universal Wireless Communications 136 (UWC-136), developed by a consortium consisting of more than 85 wireless network operators and vendors, is based on TDMA for backward compatible with the IS-136 standard. Digital Enhanced Cordless Telecommunications (DECT) was also called IMT-2000 FDMA/TDMA, which was developed under DECT Forum and ETSI. Although UWC-136 and DECT were also approved 3G standards by the ITU, they received fewer supports from the industry and were not widely deployed. The ITU-R family of IMT-2000 terrestrial radio interface standards are illustrated in Figure 1.1.

1.4.1 Wideband Code-Division Multiple Access (WCDMA)

In the late 1990s, NTT DoCoMo developed Wideband CDMA technology for their 3G system known as Freedom of Mobile Multimedia Access (FOMA). WCDMA only defined the air-interface part, therefore also named as Universal Terrestrial Radio Access (UTRA). WCDMA was selected as the air interface of UMTS, as the 3G successor to GSM. Different systems, including FOMA, UMTS, and J-Phone, shared WCDMA air interface but have different protocols for a complete stack of communication standards. 3GPP submitted it as an IMT-2000 proposal, and the ITU-R approved it as part of the IMT-2000 family standards. It employed direct-sequence code-division multiple access with a chip rate of 3.84 Mcps. The radio access specifications provided both FDD and TDD variants, utilizing a 5 MHz channel to achieve peak rates of up to 5 Mbps. In October 2001, NTT DoCoMo launched the first commercial FOMA network in Japan as the successor to i-mode. In many European countries, mobile operators acquired a license for the 3G spectrum and had to pay enormous fees in auctions. For example, the

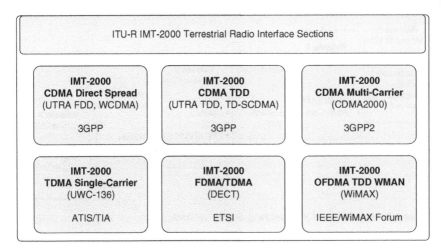

Figure 1.1 IMT-2000 standards approved by ITU-R [ITU-R M.1457, 2000]. Source: Data from ITU-R M.1457 [2000].

mobile operators of the United Kingdom spent 33 billion USD in the auction of April 2000, and 47.5 billion USD was recorded in German auction later that year. Such a high financial pressure of mobile operators raised by the high licensing cost resulted in a delay of the commercial roll-out of European 3G networks. As an example, the United Kingdom's first commercial 3G network was deployed by Hutchison Telecom as late as March 2003.

WCDMA specified in Release 99 and Release 4 of the specifications contains all technical features to satisfy the IMT-2000 requirements, but further enhancement had never stopped, as shown in Figure 1.2. High Speed Packet Access (HSPA) appeared in 2002 as the first significant evolution of the WCDMA radio interface.

- *Release 5* increased the downlink capability with a rate of up to 14 MHz, known as High Speed Downlink Packet Access (HSDPA). To achieve this, it introduced a set of technical features, including shared-channel transmission, channel-dependent scheduling, higher-order modulation (i.e. 16QAM), H-Automatic Repeat Request (ARQ) with soft combining, and link adaptation.
- *Release 6* was finalized in March 2005 added the enhancement known as High Speed Uplink Packet Access (HSUPA), offering a rate of 5.74 MHz in the uplink.
- *Release 7* published in September 2007, as a further evolution of HSPA called HSPA Evolution or HSPA+. It utilized multiple antennas technique (2x2 MIMO) and higher-level modulation (i.e. 16QAM in the uplink and 64QAM in the downlink), to achieve 28 Mbps in the downlink and 11 Mbps in the uplink over a bandwidth of 5 MHz.
- *Release 8* enabled the simultaneous usage of two-layer spatial multiplexing and 64QAM modulation in the downlink. It employed carrier aggregation in a similar way as later done for long-term evolution (LTE), thereby increasing

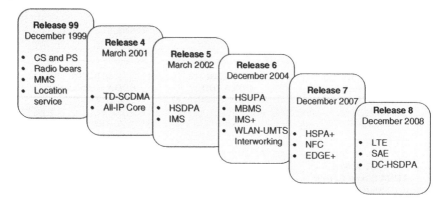

Figure 1.2   Releases of 3GPP specifications for WCDMA.

the maximal bandwidth 10 MHz. Dual-carrier HSDPA can double data rates to 56 Mbps in the downlink by aggregating two carrier channels.

• *Release 9* enhanced the uplink by introducing two aggregated carriers, leading to a rate of 22 Mbps in the uplink.

• *Release 10* can achieve the downlink peak data rate of 168 Mbps by adding the support of aggregating four component carriers for the maximal bandwidth of 20 MHz [Dahlman et al., 2011].

1.4.2   Code-Division Multiple Access 2000 (CDMA2000)

IS-95 was the first cellular system that employed CDMA technology, and therefore it was easier to evolve into a CDMA-based 3G standard. When it became a global IMT-2000 standard, the name was changed to CDMA2000, and the standardization work was transferred from the US TIA to 3GPP2. Being a sister organization of 3GPP, 3GPP2 pushed forward the CDMA2000 technology with an evolution path similar to that of WCMDA. The focus was shifted from circuit-switched voice communications to packet-switched data services. Two parallel evolutionary paths, as demonstrated in Figure 1.3, were initiated to improve the support of data transmission further. The primary path was *Evolution-Data Only (EV-DO)*, or also interpreted as *Evolution – Data Optimized*. In contrast, another path dedicated to the simultaneous support of both circuit-switched and packet-switched services on the same carrier, hence referred to as *Evolution for integrated Data and Voice (EV-DV)* [Attar et al., 2006].

• *CDMA2000 1x*: The initial version of IMT-2000 CDMA Multi-Carrier approved by the ITU-R supported two operation modes: single carrier (CDMA2000 1x) and multiple carriers (CDMA2000 3x). Although the 3x mode was an essential component of the submission of CDMA2000 to the ITU-R, it was

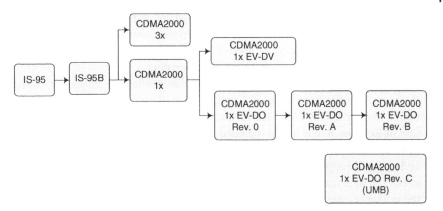

Figure 1.3 The evolution path of CDMA2000.

never commercially deployed in a large-scale network. CDMA2000 1x was a fully backward-compatible advance of IS-95, inheriting the basic design of direct-sequence spread spectrum and the channel bandwidth of 1.25 MHz. It added several enhancements over the earlier versions of IS-95 to improve spectral efficiency and offer higher data rates. Most importantly, it provided a structure opening the possibility for further evolution of packet-switched data services. CDMA2000 1x can be deployed on the IS-95 frequency bands so that an IS-95 network operator can smoothly upgrade from 2G to 3G without the need to acquire a license for the 3G spectrum. In October 2000, SK Telecom rolled out the world's first commercial CDMA2000 1X network in South Korea. As of 2014, the CDMA Development Group stated that 314 operators in 118 countries had offered CDMA2000 1X or 1xEV-DO services.

- *CDMA2000 1x EV-DO Revision 0*: The 1x version of CDMA2000 was evolved along different paths, resulting in two options: CDMA2000 1xEV-DV and CDMA2000 1xEV-DO. The former focused on the improvement of voice capacity and received limited development under 3GPP2. In contrast, EV-DO played as a main evolution track and went through several evolutionary steps in Revision 0, Revision A, Revision B, and Revision C. CDMA2000 1x EV-DO has later also been named HRPD. EV-DO Revision 0 re-designed the uplink and downlink structure of CDMA2000 1x, optimized for packet-switched data transmission while removed the constraint of supporting circuit-switched voice communications. An operator would deploy an additional carrier for EV-DO, separating voice and packet-data connections on different carriers. 3GPP2 added a set of data-optimized technologies into the Revision 0 of CDMA2000 EV-DO, including shared-channel transmission, channel-dependent scheduling, short Transmission Time Interval (TTI), link adaptation, higher-order modulation (i.e. 16QAM in the downlink), HARQ, virtual soft handover, and receive diversity. These techniques were also adopted by 3GPP for the evolution of HSPA. Thanks to a new air interface and a separate channel used only for data transmission, a data rate of 2.4 Mbps in the forward link was achieved while 153 kbps on the reverse link over a 1.25 MHz carrier.

- *CDMA2000 1x EV-DO Revision A*: The next evolution after Revision 0 was named Revision A, rather than Revision 1, which focused on the enhancement of the uplink similar to HSUPA in 3GPP. The forward link of Revision A is similar to that of Revision 0, but it also included some updates, increasing the data rate from 2.4 to 3.1 Mbps. In the reverse link, higher-order modulation (i.e. QPSK and optional support of 8PSK), in contrast to BPSK used in Revision 0, as well as HARQ, were introduced, achieving the uplink rate of up to 1.8 Mbps. Besides, the utilize of smaller packet sizes and a shorter TTI enabled a lower latency of up to 50% compared with its predecessor for well supporting of voice over IP (VoIP) and delay-sensitive data services.

- *CDMA2000 1x EV-DO Revision B*: The further enhanced version was Revision B, which supports higher data rates by using multiple carriers. Up to 16 carriers can be aggregated to form a 20 MHz bandwidth, achieving a theoretical rate of up to 46.5 Mbps. Due to the constraint of cost, hardware size, and battery life, the mobile terminal in a Revision B network supports up to three carriers, resulting in a peak rate of 9.3 Mbps. The radio interface of Revision B was backward compatible with Revision 0 and Revision A, making it possible for a multi-carrier network to further support legacy single-carrier terminals. It supported an asymmetric operation where carriers do not have to be symmetrically allocated between the downlink and the uplink. For asymmetric applications such as file download and video streaming, more carriers can be used for the forward link. One reverse link can carry control signaling and feedback information for multiple forward links, reducing the amount of uplink signaling overhead.

- *CDMA2000 1x EV-DO Revision C*: The next step was 1x EV-DO Revision C, also known as Ultra Mobile Broadband (UMB). Adopting Loosely Backward Compatible (LBC) option, Revision C is not compatible with the previous revisions of CDMA2000 specifications. The objectives for designing a disruptive air interface were to achieve higher peak rates, improved spectral efficiency, lower latency, and enhanced user experiences for delay-sensitive data applications, like the development of LTE in 3GPP. The significant new features in Revision C are the introduction of typical 4G technologies, namely, OFDM and MIMO. OFDM multi-carrier transmission selected a subcarrier spacing of 9.6 kHz with different Fast Fourier Transform (FFT) sizes (128, 256, 512, 1024, and 2048) to flexibly support various transmission bandwidths. Spatial multiplexing supported up to four transmission layers in the forward link but it was only supported in conjunction with OFDM. In the reverse link, up to two spatial layers were

Table 1.3 Third-generation cellular standards.

Parameter	WCDMA	CDMA2000	TD-SCDMA	WiMAX
Version	Release 99	1x	Release 4	IEEE802.16e
Launch time	2001	2000	2009	2006
Bandwidth (MHz)	5	1.25	1.6	1.25/5/10/20
Multi-access		CDMA		OFDMA
Duplexing	FDD	FDD	TDD	TDD
Chip rate (Mcps)	3.84	1.2288	1.28	N/A
Power control	1500 Hz	800 Hz	200 Hz	N/A
Frame length (ms)	10	20	10	5
Modulation	QPSK (DL)/BPSK (UL)		~8 PSK	~64 QAM
Channel coding		Turbo codes		Turbo/LDPC

specified with codebook-based precoding under the control of a base station. Using a bandwidth of 20 MHz, the maximal rates are 260 Mbps in the forward link and 70 Mbps in the reverse link. CDMA2000 and WCDMA competed for the 3G market worldwide in the first several years of 3G deployment. Regardless of the incompatibility with legacy GSM standards, late introduction, and the high upgrade cost of deploying an all-new air-interface technology, WCDMA has won this competition and eventually became the dominant 3G standard. As a consequence, the supports for the evolution of CDMA2000 gradually became weak. In the 3GPP2 camp, UMB was the target to the planned 4G successor of CDMA2000, competed with LTE in 3GPP. However, Qualcomm, the main stakeholder of UMB, announced to cease the further evolution of CDMA2000 technologies in November 2008. 3GPP2 had its last activity in 2013, and the group has been dormant ever since.

1.4.3 Time Division-Synchronous Code-Division Multiple Access (TD-SCDMA)

In parallel to the development of UTRA FDD/WCDMA and its evolution to HSPA, 3GPP also worked on the TDD version of UTRA. Regardless of the similar high-layer protocols between FDD and TDD, the physical-layer designs were quite different. For historical reasons, there were three variants with different chip rates. The initial version of UTRA TDD adopted the chip rate of 3.84 Mcps, and 7.68 and 1.28 Mcps were added later. UTRA low chip rate (1.28 Mcps) TDD, also called Time Division-Synchronous Code Division Multiple Access (TD-SCDMA), is substantially different from the other two. TD-SCDMA was developed as an industry standard under the lead of the Chinese Academy of Telecommunications Technology (CATT). In March 2001, it was approved to merge into Release 4 of the 3GPP specifications as an alternative UTRA TDD version. The significant difference between TD-SCDMA and the other two 3G standards (WCDMA and CDMA2000) is its use of TDD operation instead of FDD for duplex signaling. It adopted a signal bandwidth of 1.6 MHz, 8PSK modulation, and a shorter TTI of 5 ms. Some technical features such as multi-frequency operation and smart antenna/beamforming support with eight antennas were introduced by this system. Of the three versions, TD-SCDMA was the only UTRA TDD standard deployed on a large scale, with the other two being limited to niche deployment. China Mobile, the world's biggest mobile operator in terms of the number of subscribers, was granted a 3G license in early 2009 for operating a TD-SCDMA network. The unique TD-SCDMA deployment worldwide finally become a network consisting of around 500 000 base stations and the peak number of subscribers reached approximately 250 million. Although this technology was exclusively applied in China, it promoted TDD systems' advantages and pushed forward the development of a TDD version of 4G known as TD-LTE or LTE TDD. The HSPA enhancements of TD-SCDMA were similar to those applied to UTRA FDD, such as the application of high-order modulation (16QAM) and hybrid ARQ.

1.4.4 Worldwide Interoperability for Microwave Access (WiMAX)

The IEEE 802.16 specifications were developed by the Institute of Electrical and Electronics Engineers (IEEE) under the umbrella of Wireless Metropolitan Area Network (WMAN). The initial version in 2001 was designed for line-of-sight communications in the millimeter-wave frequency range of 10–60 GHz, targeting fixed Wireless Broadband (WiBro) access. In 2003, the enhanced version called IEEE 802.16a introduced the support for non-line-of-sight operation over the low-frequency bands of 2–11 GHz, but still limited to fixed-wireless-access applications. The monumental milestone was the IEEE 802.16e-2005 specification released in 2005 as the first mobile WiMAX system [Etemad, 2008]. Empowered by cutting-edge technologies at that time, it offered the peak data rates of 128 Mbps in the downlink and 56 Mbps in the uplink over a 20 MHz channel. The first commercial network was deployed in South Korea in 2006 (branded as WiBro) and then deployed in many parts of the world.

The IEEE 802.16 specifications usually provide the specifications of the physical and Medium Access Control (MAC) layers instead of the overall communication protocol stack. Furthermore, IEEE 802.16 specifications contain multiple alternatives for the basic physical-layer transmission scheme. Implementing all these options and alternative features in a mobile system was not necessary. The WiMAX Forum is an industry-led, non-profit alliance formed to promote and certify compatibility and interoperability of IEEE 802.16-based products. Its responsibility was to select technical features from the full set of features defined by IEEE 802.16 specifications to form a complete and implementable standard called WiMAX System Profile. The first such profile, WiMAX Release 1.0, was published in 2007, with the second profile, Release 1.5, finalized in 2009. IEEE 802.16e, also referred to as Mobile WiMAX, was submitted to the ITU-R as a proposal for IMT-2000. It was approved in 2007 by the ITU as IMT-2000 OFDMA TDD WMAN, in parallel with WCDMA, CDMA2000, and TD-SCDMA.

Although IEEE 802.16 provided several alternatives for the basic physical-layer transmission scheme, including both OFDM and single-carrier transmission, Mobile WiMAX is based on OFDM transmission. Like LTE, IEEE 802.16e improve spectrum flexibility by adopting variable bandwidths, i.e. 1.25, 5, 10, and 20 MHz. With a common subcarrier spacing of 10.94 kHz, only scaling the number of subcarriers (128, 512, 1024 and 2048) within the transmission bandwidth. IEEE 802.16e specified both TDD and FDD, including the possibility for half-duplex FDD, whereas the first version of Mobile WiMAX merely support the TDD operation, where a 5 ms frame is divided into downlink and uplink part, together consisting of 48 OFDM symbols. Similar to LTE, Mobile WiMAX supports QPSK, 16QAM, and 64QAM modulation, and link adaptation (adaptive modulation and coding) in terms of instantaneous channel conditions. The 802.16e specifications support various channel coding schemes, including Turbo codes, similar to HSPA and LTE, and LDPC codes. However, Mobile WiMAX only supports Turbo codes.

Video content and web browsing raised the need for larger and better screen displays, fostering the revolution of the mobile terminal. The first smartphone was the IBM Simon released in 1994, followed by Nokia 9000 released in 1996, combining email, word processor, diary, and QWERTY keyboard. On 7 January 2007, Steve Jobs announced that Apple was entering the mobile phone market and, by considering the mobile as a computer first and phone second, brought a whole new insight to handset design. The Apple iPhone proved to be a disruptive piece of technology that redefined mobile phone design and introduced the world to the APPlication (APP) – even though the first model was only a 2G device [Linge and Sutton, 2014].

1.5 4G: Mobile Internet

The number of mobile subscribers grew tremendously in the first decade of the twenty-first century. The landmark for the first billion achieved in 2002, but the number boomed quickly to over 5 billion in 2010. Another driving force for the fourth-generation cellular system was the explosive growth of traffic brought by mobile broadband. In cellular networks, for the first time, the data traffic exceeded the voice traffic and was expected to saturate 3G networks soon. In addition, the proliferation of Internet-based services on mobile devices imposed a significant challenge on cellular networks that were optimized for voice communications. Since the 2.5G system, cellular networks had to simultaneously operate two parallel infrastructures: a packet-switched network for data services and a circuit-switched network for voice calls.

All-IP Cellular Network

The 4G cellular system was driven toward end-to-end all-IP architecture such that Internet-based services can be well supported. For the first time in the history of cellular networks, the circuit-switched network was thoroughly abandoned, and only a packet-switched network was provided for more flexible and efficient operation.

The 3G standardization bodies such as 3GPP, 3GPP2, WiMAX, and IEEE were working on the enhancements of 3G standards toward 4G, using advanced air-interface technologies and all-IP network infrastructure. To ensure the competitiveness of UMTS, 3GPP initiated the study item of *LTE* air interface, also known as Evolved Universal Terrestrial Radio Access (E-UTRA), in 2004. In December 2008, the first release (Release 8) of the LTE air interface and its core network called the Evolved Packet Core (EPC) was completed, followed by an enhanced version (Release 9) frozen in December 2009. Similar to the development of IMT-2000, the ITU-R WP5D defined the minimal technical requirements for the 4G system, named the International Mobile Telecommunications Advanced (IMT-Advanced) in 2008. However, since LTE cannot fully comply with the requirements of IMT-Advanced, e.g. the peak data rate of 1 Gbps for low mobility and 100 Mbps for high mobility, 3GPP then continued to work on an enhanced version known as long-term evolution-advanced (LTE-Advanced) from 2009. Meanwhile, the WiMAX standard continuously evolved under the development of IEEE and WiMAX Forum. The specifications of WirelessMAN-Advanced (also known as Mobile WiMAX Release 2.0) were completed in 2011 to comply with IMT-Advanced. Significant enhancements were introduced on IEEE 802.16e-2005 to form a new standard named IEEE 802.16m-2011. IEEE and WiMAX Forum submitted IEEE 802.16m-2011 to the ITU as one of the IMT-Advanced proposals, competing with LTE-Advanced. In addition, 3GPP2 worked on the development of UMB as a fourth-generation successor to CDMA2000. However, Qualcomm, the leading sponsor of UMB, announced the end of developing this technology in November 2008, favoring LTE instead. As a result, 3GPP2 had its last activity in 2013, and the group has been dormant ever since.

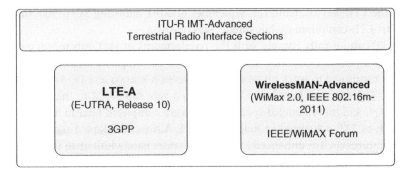

Figure 1.4 IMT-Advanced standards approved by ITR-R.

In October 2010, the ITU-R had completed the evaluation of six candidate submissions and approved two industry-developed technologies as the global 4G standards, as shown in Figure 1.4. Recommendation M.2012 [ITU-R M.2012, 2012] identifies the terrestrial radio interface technologies of IMT-Advanced and provides the detailed radio interface specifications.

1.5.1 Long-Term Evolution-Advanced (LTE-Advanced)

3GPP initiated the study item on LTE in late 2004, aiming at a disruptive radio-access technology dedicated to packet-switched data transmission. The study focused on technical requirements for LTE. The significant outcomes approved in June 2005 include low latency, high data rates at the cell edge, and spectrum flexibility. The 3GPP Radio Access Network (RAN) plenary meeting in December 2005 made a decision that the downlink of LTE should be based on Orthogonal Frequency-Division Multiple Access (OFDMA) and Single-Carrier Frequency-Division Multiple Access (SC-FDMA) in the uplink. In December 2008, the first release (Release 8) of the LTE air interface and its core network called the EPC was completed, followed by an enhanced version (Release 9) frozen in December 2009. LTE is a disruptive standard that was designed without the restriction of backward compatibility so that it is flexible to adopt new technical features. It can be operated in either FDD or TDD mode, referred to as LTE FDD and TD-LTE, respectively, with low latency and flat system architecture. The peak data rates reached 300 Mbps in the downlink and 75 Mbps in the uplink over a signal bandwidth of 20 MHz. In December 2009, TeliaSonera launched the world's first commercial LTE mobile services in the Scandinavian capitals Stockholm and Oslo, with network equipment provided by Ericsson and Huawei [Astely et al., 2009]. No LTE-compliant mobile phone was commercially available at that time, and the subscribers used computers with a USB wireless network adapter to access the LTE service. Until the September of 2010, Samsung SCH-r900, the world's first LTE-compliant mobile phone, was released.

Since LTE cannot fully comply with the requirements of IMT-Advanced, e.g. the peak data rate of 1 Gbps for low mobility and 100 Mbps for high mobility, 3GPP then continued to work on an enhanced version known as LTE-Advanced from 2009. The proposal based on LTE-Advanced was submitted to the ITU in October 2009, and more detailed specifications were completed later to form the first version of LTE-Advanced in Release 10. LTE-Advanced adopted significant technical features such as enhanced MIMO and a wider bandwidth up to 100 MHz to achieve high-speed transmission of 1 Gbps in the downlink and 500 Mbps in the uplink. In 2012, the Russian operator YOTA Networks announced the launch of the world's first LTE-Advanced network in Moscow using equipment from Huawei. Release 13 of the 3GPP specifications completed in early 2016 was the first version of *LTE-Advanced Pro*. The amount of new technical features was considered sufficient to merit a new LTE marker, but neither of *LTE-Advanced* and *LTE-Advanced Pro* implies a break of backward compatibility. As of this writing, 3GPP has completed Release 16 and is working on Release 17, which also includes further enhancements of LTE in addition to the specifications of fifth-generation (5G) [Dahlman et al., 2021].

The releases of the 3GPP specifications focusing on LTE are briefly summarized in the following paragraphs, as also shown in Figure 1.5.

- *Release 8* is the first definition of the LTE radio-access technology and the all-IP EPC network, forming the foundation for the following evolution. Spectrum

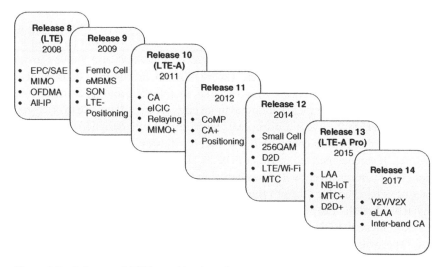

Figure 1.5 Releases of 3GPP specifications for LTE.

flexibility was emphasized by supporting paired and unpaired spectrum using FDD and TDD. It also supports flexible bandwidths (1.4, 3, 5, 10, 15, and 20 MHz) by scaling the number of OFDM subcarriers. Over a bandwidth of 20 MHz, the peak data rate can reach 150 Mbps in the uplink with 2×2 MIMO and 300 Mbps in the downlink with 4×4 MIMO.

- *Release 9* is the first evolution of LTE. It provided some improvements left behind from Release 8 and some minor enhancements, with the technical features including femtocell, MIMO beamforming, self-organized networks (SON), Enhanced Multimedia Broadcast Multicast Services (eMBMS), LTE positioning, and public warning system.

- *Release 10* specified the LTE-Advanced standard to ensure that the LTE radio-access technology can fully compliant with the IMT-Advanced requirements. It was frozen in March 2011 and introduced the technical features including carrier aggregation, enhanced uplink multiple access, MIMO enhancement, relaying, enhanced inter-cell interference coordination (eICIC), heterogeneous network deployment, and SON improvement.

- *Release 11* was finalized in September 2012, which further enhanced the performance and capabilities of LTE-Advanced. One of the most significant features of LTE Release 11 was the introduction of coordinated multi-point (CoMP) transmission and reception. Other improvements were carrier aggregation enhancement, a new control-channel structure, network-based positioning, RAN overload control for machine-type communication, and smartphone battery-saving technique.

- *Release 12* was completed in June 2014 and focused on the optimization and enhancements for small cells, including dual connectivity, dense small-cell deployment, small-cell on/off, and semi-dynamic TDD. Higher-order modulation (256QAM) was introduced to make use of high signal strength in small-cell environment. Another priority of this release was applying LTE technology for emergency events and public safety, with technical specifications for mission-critical application layer functional elements. Other features included device-to-device (D2D) communications, LTE TDD-FDD joint operation including Carrier Aggregation, Security Assurance Methodology, and LTE/WiFi integration.

- *Release 13* marked the start of LTE-Advanced Pro, which was sometimes in marketing dubbed 4.5G and seen as an intermediate step between the first release of LTE and the advent of 5G. Release 13 was a significant step, with many exciting features, such as license-assisted access (LAA) to support unlicensed spectra, improved support for machine-type communications, and further enhancements in MIMO, D2D communications, and carrier aggregation. Other efforts for expanding it to a set of new services and new verticals included the introduction of narrow-band Internet of Things (NB-IoT) and the initial studies on Vehicle-to-Vehicle (V2V) communications.

- *Release 14* was frozen in 2017. Apart from enhancements to some of the features introduced in earlier releases, such as enhanced license-assisted access (eLAA) and inter-band carrier aggregation, it brought the support for V2V and Vehicle-to-Everything (V2X) communications, as well as wide-area broadcast with a reduced subcarrier spacing.

1.5.2 WirelessMAN-Advanced

As already mentioned, WiMAX Release 1.0 based on IEEE 802.16e-2005 802.16m was approved by the ITU in 2007 as the sixth global 3G standard under the name IMT-2000 OFDMA TDD WMAN. The subsequent step taken within the IEEE WMAN community was the development of IEEE 802.16m, aiming to extend the performance and capabilities of 802.16 radio-access technologies to ensure compliance with the IMT-Advanced requirements. In contrast to LTE-Advanced being a fully backward-compatible evolution of LTE, IEEE 802.16m was not a smooth evolution of IEEE 802.16e with some disruptive features [Dahlman et al., 2011]. Instead, IEEE 802.16m was recognized as a new standard although holding some essential characteristics of IEEE 802.16e, including the basic OFDM numerology. With time multiplexing, these two radio-access technologies can coexist on the same carrier within the IEEE 802.16e 5 ms frame structure. IEEE 802.16m adopted many features similar to LTE-Advanced, such as the use of carrier aggregation for bandwidths beyond 20 MHz and the support for relaying functionality. It also introduced much shorter subframes of length roughly 0.6 ms to reduce hybrid ARQ round-trip time and, in general, allow for reduced latency over the radio interface. Instead of inheriting the resource-mapping schemes defined for IEEE 802.16e, IEEE 802.16m introduced physical resource units consisting of a magnitude of frequency-contiguous subcarriers during one subframe, similar to the resource blocks in LTE. Each resource unit comprises 18 subcarriers with the subcarrier spacing of 10.94 kHz, resulting in a bandwidth close to the LTE resource-block bandwidth of 180 kHz. Due to many similarities between LTE-Advanced and IEEE 802.16m, it is not surprising that performance evaluations indicate the comparable performance of the two radio-interface technologies. Thus, similar to LTE-Advanced, IEEE 802.16m can also fulfill all the requirements for IMT-Advanced as defined by the ITU. In October 2010, WiMAX Release 2.0 based on IEEE 802.16m, along with LTE-Advanced, was approved by the ITU-R as one of the two IMT-Advanced standards under the name WirelessMAN-Advanced.

Table 1.4 Comparison of the main system parameters between LTE/LTE-Advanced and WiMAX.

Parameter	LTE	LTE-Advanced	WiMAX 1.0	WiMAX 2.0
Standard	3GPP Release 8	3GPP Release 10	IEEE 802.16e-2005	IEEE 802.16m-2011
Launch time	2009	2012	2006	2012
Bandwidth (MHz)	1.4, 3, 5, 10, 15, 20	(Aggregated) up to 100	1.25, 5, 10, 20	5, 10, 20, 40
Multi-access	DL: OFDMA UL: SC-FDMA	DL: OFDMA UL: SC-FDMA	DL: OFDMA UL: OFDMA	DL: OFDMA UL: OFDMA
OFDM subcarrier	15 kHz	15 kHz	10.94 kHz	10.94 kHz
Duplex mode	FDD/TDD	FDD/TDD	TDD	TDD/FDD
Multi-Antenna	DL: 2x2,4x2,4x4 UL: 1x2,1x4	DL: up to 8x8 UL: up to 4x4	DL: 2x2 UL: 2x1	DL: up to 8x8 UL: up to 4x4
Modulation	QPSK, 16QAM, 64QAM	Up to 256QAM[a]	QPSK, 16QAM, 64QAM	QPSK, 16QAM, 64QAM
Channel coding	Turbo codes	Turbo codes	Turbo codes/LDPC	Turbo codes/LDPC
Frame length	10 ms	10 ms	5 ms	5 ms
Mobility (km/h)	350	350	120	350
Data rate	DL: 300 Mbps UL: 75 Mbps	DL: 1 Gbps UL: 500 Mbps	DL: 75 Mbps UL: 20 Mbps	DL: 1 Gbps UL: 200 Mbps
Latency	UP: 5 ms, CP: 50 ms	UP: 5 ms, CP: 50 ms	UP: 20 ms, CP: 50 ms	UP: 10 ms, CP: 30 ms

a) 3GPP Release 12 added the support of 256QAM in LTE-Advanced.
UP, user plane; CP, control plane.

WiMAX was pushed to the market much earlier than LTE and was a superior technology in terms of data throughput for a few years (2005–2009). Besides, it pioneered to adopt pre-4G technologies such as MIMO and OFDM and support new features such as variable transmission bandwidths. As early as 2006, two South Korean telecom operators launched the world's first mobile WiMAX service based on IEEE 802.16e standard under the brand of WiBro. As of October 2010, the WiMAX Forum claimed over 592 WiMAX (fixed and mobile) networks deployed in over 148 countries, covering over 621 million people. However, LTE was the evolution of dominant standards (i.e. GSM and WCDMA), whereas WiMAX was a relatively disruptive technology without a large user base. As a result, major mobile operators such as Verizon, Vodafone, China Mobile, NTT, and Deutsche Telekom chose to upgrade their legacy infrastructure from 3G to LTE smoothly rather than adopt a new technology standard. Ultimately, LTE/LTE-Advanced won the competition to become the dominant 4G standard. With LTE/LTE-Advanced, the world has thus converged into a universal global standard for mobile communications, deployed by essentially all mobile-network operators worldwide and applicable to both paired and unpaired spectra.

1.6 5G: From Human to Machine

5G was built on the success of 4G LTE and the exploration of 5G started before the full deployment of 4G. Some of the earliest efforts toward 5G technology commenced in the early 2010s, which demonstrated the feasibility of techniques that could be adopted. In August 2012, the New York University founded a multi-disciplinary academic research center known as NYU Wireless to develop the fundamental theories and pioneering work for 5G wireless communications. A key focus was on millimeter wave (mmWave) communications operating in the high frequency bands above 10 GHz. It got many research accomplishments such as the world's first radio channel measurements proving that the potential of mmWave spectrum and demonstrated the safety of mmWave radiation for the human body. Just two months later than the foundation of NYU Wireless, the University of Surrey in the United Kingdom announced to establish a new 5G research center jointly funded by the British government and a consortium of key mobile operators and vendors such as Huawei, Samsung, Telefonica, Fujitsu, and Rohde&Schwarz. In November 2012, a research project entitled *Mobile and wireless communications Enablers for the Twenty-twenty Information Society (METIS)* funded by European Commission was kicked off [Osseiran et al., 2014]. METIS achieved an early global consensus on the picture of what would be 5G, prior to global standardization activities such as ITU-R and 3GPP.

Cellular Network for Human and Machine

In contrast to the previous generation cellular systems that focused merely on human-centric communication services, 5G needs to expand the sphere of mobile communications from human to things, from consumers to vertical industries, and from public to private networks. The potential scale of mobile subscription is substantially enlarged from merely billions of the world's population to almost countless inter-connectivity among humans, machines, and things. It enables a wide variety of disruptive use cases such as Industry 4.0, virtual reality, Internet of Things, and automatic driving.

In February 2013, the ITU-R Working Party 5D initiated two study items to analyze the IMT Vision for 2020 and future technology trends for the terrestrial IMT systems, known as IMT-2020. IMT-2020 was envisaged to support diverse usage scenarios and applications beyond the previous IMT systems. Furthermore, a broad variety of capabilities would be tightly coupled with these intended different usage scenarios and applications, as shown in Figure 1.6. Part of the study outcomes was transferred to ITU-R Recommendation M.2083 released in 2015 [ITU-R M.2083, 2015], where three usage scenarios were firstly defined:

- *enhanced Mobile Broadband (eMBB):* Mobile broadband addresses the human-centric use cases for access to multi-media content, services, cloud, and data. With the proliferation of smart devices (smartphones, tablets, and wearable electronics) and the rising demand for video streaming, the need for mobile broadband continues to grow, setting new requirements for what ITU-R calls eMBB.

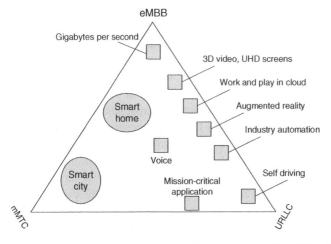

Figure 1.6 Usage scenarios of IMT-2020 defined in ITU-R M.2083. Source: Adapted from ITU-R M.2083 [2015].

This usage scenario comes with new use cases and requirements for improved capabilities and an increasingly seamless user experience. eMBB covers various cases, including wide-area coverage and hot spots, which have different requirements. For the hot-spot case, i.e. for an area with high user density, very high traffic capacity is needed, while the requirement for mobility is low and user data rate is higher than that of wide-area coverage. Seamless coverage and medium to high mobility are desired for the wide-area coverage case, with a substantially improved data rate than existing data rates. However, the data rate requirement may be relaxed compared with hot spots.

- *Ultra-Reliable Low-Latency Communications (URLLC):* This scenario aims to support both human-centric and critical machine-type communications. It is a disruptive promotion over the previous generations of cellular systems that focused merely on the services for mobile subscribers. It opens the possibility for offering mission-critical wireless applications such as automatic driving, vehicle-to-vehicle communication involving safety, wireless control of industrial manufacturing or production processes, remote medical surgery, distribution automation in a smart grid, and transportation safety. It is characterized by stringent requirements such as ultra-low latency, ultra-reliability, and availability.

- *massive Machine-Type Communications (mMTC):* This scenario supports massive connectivity with a vast number of connected devices that typically have very sparse transmissions of delay-tolerant data. Such devices, e.g. remote sensors, actuators, and monitoring equipment, are required to be low cost and low power consumption, allowing for a very long battery life of up to 10 years due to the possibility of remote Internet of Things (IoT) deployment.

IMT-2020 was expected to provide far more enhanced capabilities than those of IMT-Advanced. In addition, IMT-2000 can be considered from multiple perspectives, including the users, manufacturers, application developers, network operators, and service and content providers. Therefore, it is recognized that technologies for IMT-2020 can be applied in a variety of deployment scenarios and can support a range of environments, service capabilities, and technology options [Andrews et al., 2014]. Based on the usage scenarios and applications described as Recommendation M.2083, the ITU-R defined a set of technical performance requirements. In November 2017, ITU released Recommendation M.2410 *Minimum requirements related to technical performance for IMT-2020 radio interface* [ITU-R M.2410, 2017], as the baseline for the evaluation of IMT-2020 candidate technologies. In addition to the peak data rates of 20 Gbps in the downlink and 10 Gbps in the uplink following the tradition of offering higher transmission rates, a number of novel Key Performance Indicators (KPIs) such as reliability, energy efficiency, and connection density were set up. The key

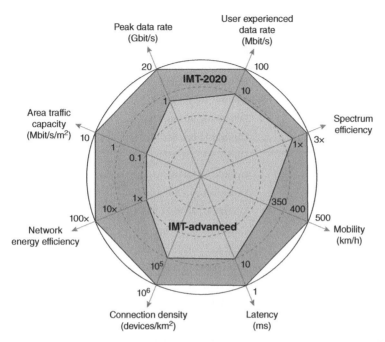

Figure 1.7 Improvement of key performance indicators from IMT-Advanced to IMT-2020. Source: ITU-R M.2083 [2015]/ITU.

capabilities of IMT-2020 are shown in Figure 1.7, compared with those of IMT Advanced. Table 1.5 provides a summary of these performance requirements.

Another milestone for 5G development was the identification of its spectrum discussed in the World Radio Conference. At WRC-15, a new set of frequency bands below 6 GHz (e.g. 470–694, 694–790, and 3300–3400 MHz) were identified for IMT on a global basis. This conference also appointed an agenda item for the following WRC-19 toward the identification of higher spectrum above 24 GHz for IMT-2020 mobile services. Based on the studies conducted by the ITU-R after WRC-15, WRC-19 noted that the ultra-low latency and very-high data-rate applications require larger contiguous blocks of spectrum. As a consequence, a total of 13.5 GHz spectrum consisting of a set of high-frequency bands were assigned for the deployment of 5G mmWave communications:

- 24.25–27.5 GHz
- 37–43.5 GHz
- 45.5–47 GHz
- 47.2–48.2 GHz
- 66–71 GHz

Table 1.5 Minimum technical performance requirements for IMT-2020.

KPI	Minimum performance requirement
Peak data rate	Downlink: 20 Gbps
	Uplink: 10 Gbps
Peak spectral efficiency	Downlink: 30 bps/Hz
	Uplink: 15 bps/Hz
User-experienced rate	Downlink: 100 Mbps
	Uplink: 50 Mbps
5th-percentile user	Downlink: 0.12–0.3 bps/Hz
Spectral efficiency	Uplink: 0.045–0.21 bps/Hz
Average spectral efficiency	Downlink: 3.3–9 bps/Hz
	Uplink: 1.6–6.75 bps/Hz
Area traffic capacity	10 Mbps/m^2 (indoor hot spot)
User plane latency	4 ms – eMBB
	1 ms – URLLC
Control plane latency	20 ms
Connection density	1 000 000 devices per km^2
Energy efficiency	The support for two aspects:
	(1) Efficient data transmission in a loaded case
	(2) Low energy consumption when there is no data
Reliability	1–10^{-5} (99.999%)
Mobility	up to 500 km/h
Mobility interruption time	0 ms
Maximal bandwidth	100 MHz for sub-6 GHz
	1 GHz for mmWave

With the IMT-2020 framework defined by the ITU-R and the frequency bands identified by the WRC, the task of specifying detailed technologies fell into the standardization bodies. Unlike conflicting technical paths and multiple standardization authorities in the previous generations, 3GPP played a dominant role during the development of 5G technologies. Its technical specifications that were organized as releases acted actually as *de facto* standards. As early as 2015, 3GPP RAN group decided to set up a study item in Release 14 for 5G New Radio (NR) and initiated the task on channel modeling for frequency bands above 6 GHz. The specification of the initial 5G NR was carried out through a work item in Release 15. To meet commercial requirements on early large-scale trials and deployments in 2018, earlier than the initially envisaged timeline of around 2020, 3GPP committed to accelerating the process, agreeing that a Non-Standalone (NSA) variant would be finalized earlier. At the end of 2017, the first version of 5G specifications was available. In such an NSA deployment, the NR air interface is connected to the existing EPC core network, making the capacities offered by the NR (lower latency, etc.) available without network replacement. The world's first 5G NSA call was jointly completed in Spain by Vodafone and Huawei, just ahead of Mobile World Congress, which started on 26 February 2018. After the initial delivery of NSA, much effort of 3GPP was shifted on timely completion of Release 15 to form the first complete set of 5G standards. Therefore, 3GPP also developed a new core network referred to as the 5G Core (5GC) network in parallel to the NR radio-access technology. In June 2018, the final version of Release 15 that can support the Standalone (SA) operation of 5G NR was available, marking the completion of 5G Phase 1.

The focus of Release 15 was primarily on eMBB and (to some extent) URLLC, while mMTC was still supported by using LTE-based machine-type communication technologies such as eMTC and NB-IoT. Release 15 provided the foundation on which 3GPP continues its work to evolve the capability and functionality of 5G so as to support new spectrum and new applications, and further enhance existing core features. The evolution of 5G NR continued in Release 16, often referred to informally as "5G Phase 2", which was completed in June 2020. Technical features were added into 5G NR for the support of Industrial Internet of Things (IIoT) and the enhancement of URLLC applications. This release aimed to meet the IMT-2020 requirements and, along with Release 15, acted as the initial complete 3GPP 5G specifications submitted to the ITU-R. The 3GPP final proposal included two separate and independent submissions, defined as the single Radio Interface Technology (RIT) and the combined Sets of Radio Interface Technologies (SRIT). In November 2020, the ITU-R announced that 3GPP 5G-SRIT and 3GPP 5G-RIT conform with the IMT-2020 vision and stringent performance requirements. As of this writing, 3GPP works on Release 17, the third version of 5G, with a rolling schedule of completing Stage 2 in 2021 and stage 3 in 2022. Release 17 is probably the most versatile release in the history of 3GPP in terms of the number of technical features, as shown in Figure 1.8. 3GPP also announced its evolution of 5G toward 5.5G with the official new name of *5G-Advanced*, which will be standardized in Release 18 and beyond.

In April 2019, when South Korea's three mobile operators – SK Telecom, LG U+, and KT – and US Verizon were arguing with each other about who is the world's first provider of the 5G communication services, we stepped into the era of 5G. In the past two years, we have witnessed a strong expansion of 5G networks across the world and a great growth of 5G subscriptions in major countries. At the end of 2020, for instance, the penetration rate of 5G usage in South Korea

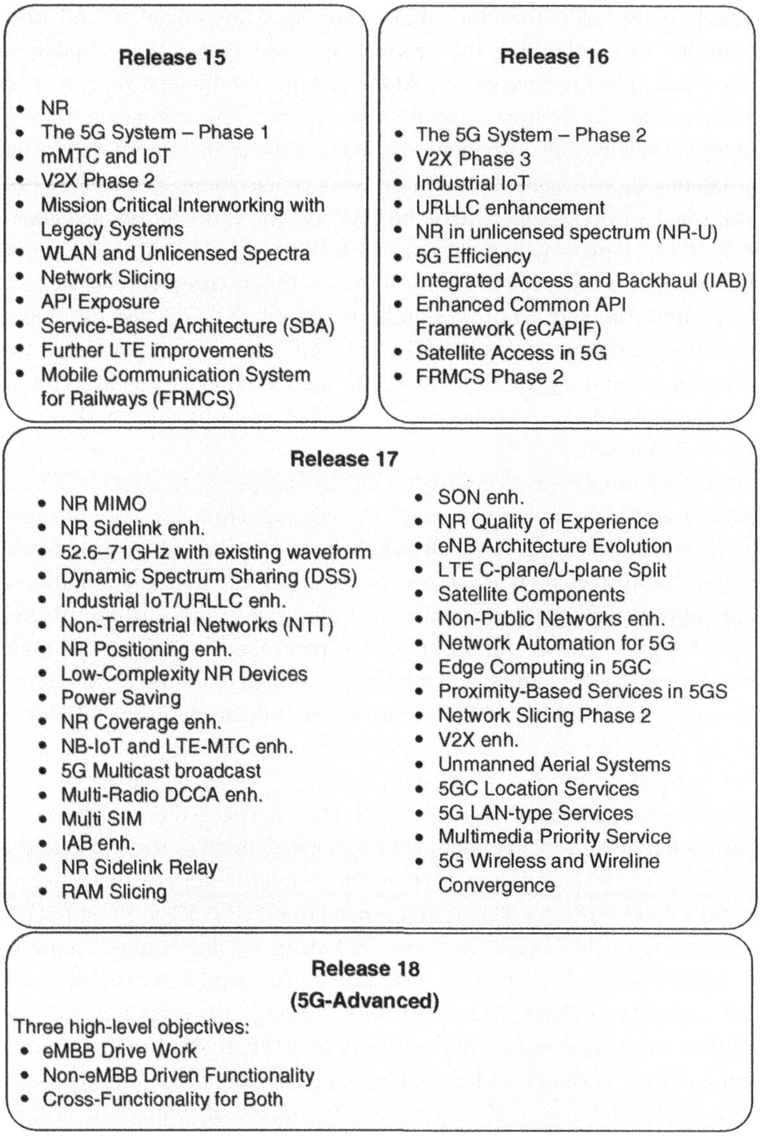

Release 15

- NR
- The 5G System – Phase 1
- mMTC and IoT
- V2X Phase 2
- Mission Critical Interworking with Legacy Systems
- WLAN and Unlicensed Spectra
- Network Slicing
- API Exposure
- Service-Based Architecture (SBA)
- Further LTE improvements
- Mobile Communication System for Railways (FRMCS)

Release 16

- The 5G System – Phase 2
- V2X Phase 3
- Industrial IoT
- URLLC enhancement
- NR in unlicensed spectrum (NR-U)
- 5G Efficiency
- Integrated Access and Backhaul (IAB)
- Enhanced Common API Framework (eCAPIF)
- Satellite Access in 5G
- FRMCS Phase 2

Release 17

- NR MIMO
- NR Sidelink enh.
- 52.6–71GHz with existing waveform
- Dynamic Spectrum Sharing (DSS)
- Industrial IoT/URLLC enh.
- Non-Terrestrial Networks (NTT)
- NR Positioning enh.
- Low-Complexity NR Devices
- Power Saving
- NR Coverage enh.
- NB-IoT and LTE-MTC enh.
- 5G Multicast broadcast
- Multi-Radio DCCA enh.
- Multi SIM
- IAB enh.
- NR Sidelink Relay
- RAM Slicing

- SON enh.
- NR Quality of Experience
- eNB Architecture Evolution
- LTE C-plane/U-plane Split
- Satellite Components
- Non-Public Networks enh.
- Network Automation for 5G
- Edge Computing in 5GC
- Proximity-Based Services in 5GS
- Network Slicing Phase 2
- V2X enh.
- Unmanned Aerial Systems
- 5GC Location Services
- 5G LAN-type Services
- Multimedia Priority Service
- 5G Wireless and Wireline Convergence

Release 18
(5G-Advanced)

Three high-level objectives:
- eMBB Drive Work
- Non-eMBB Driven Functionality
- Cross-Functionality for Both

Figure 1.8 Releases of the 3GPP NR-5GC specifications.

had surpassed 15.5% while China had deployed more than 700 000 base stations to serve around 200 million 5G subscribers. Meanwhile, the term *5G* has been remaining one of the hottest buzzwords in the media, attracting unprecedented attention from the whole society. It even went beyond the sphere of technology and economy, becoming the focal point of geopolitical tension. When we start the writing of this book, more than 400 mobile operators in approximately 130 countries are investing in 5G networks and the number of 5G subscribers already reaches a very large scale in many regions. In 2020, the outbreak of the COVID-19 pandemic leads to a dramatic loss of human life worldwide and imposes unprecedented challenges on societal and economic activities. But this public health crisis highlights the unique role of networks and digital infrastructure in keeping society running and families connected, especially the values of 5G services and applications, such as remote surgeon, online education, remote working, driver-less vehicles, unmanned delivery, robots, smart healthcare, and autonomous manufacturing.

1.7 Beyond 5G

Currently, 5G is still on its way being deployed across the world, but it is already the time for academia and industry to shift their attention to beyond 5G or the sixth generation (6G) systems, in order to satisfy the future demands for information and communications technology (ICT) in 2030 (Figure 1.9). Even though discussions are ongoing within the wireless community as to whether there is any need for 6G or whether counting the generations should be stopped at 5, and even there is an opposition to talking about 6G [Fitzek and Seeling, 2020], several pioneering works on the next-generation wireless networks have been initiated [Jiang and Schotten, 2021]. A focus group called *Technologies for Network 2030* within the International Telecommunication Union Telecommunication (ITU-T) standardization sector was established in July 2018. The group intends to study the capabilities of networks for 2030 and beyond [ITU-T NET-2030, 2019], when it is expected to support novel forward-looking scenarios, such as holographic-type communications, ubiquitous intelligence, Tactile Internet, multi-sense experience, and digital twin. The European Commission initiated to sponsor beyond 5G research activities, as its recent Horizon 2020 calls – ICT-20 *5G Long Term Evolution* and ICT-52 *Smart Connectivity beyond 5G* – where a batch of pioneer research projects for key 6G technologies were kicked off at the early beginning of 2020. The European Commission has also announced its strategy to accelerate investments in Europe's "Gigabit Connectivity" including 5G and 6G to shape Europe's digital future [EU Gigabit Connectivity, 2020]. In October 2020, the Next Generation Mobile Networks (NGMN) has launched its new "6G Vision and Drivers" project, intending to provide early and timely direction for global 6G activities. At its meeting in February 2020, the ITU-R sector decided to start study on future technology trends for the future evolution of International Mobile Telecommunications (IMT) [ITU-R WP5D, 2020]. In Finland, the University

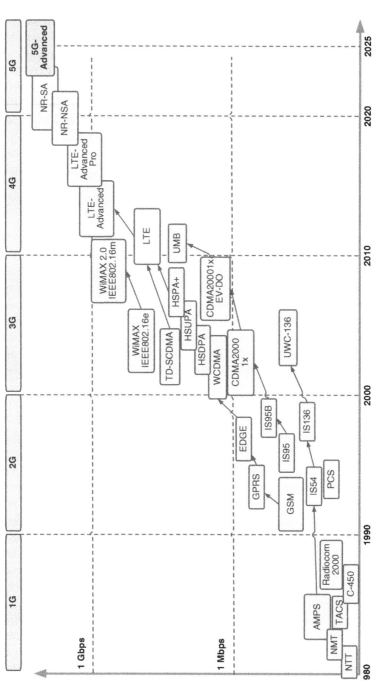

Figure 1.9 The evolution of cellular systems.

of Oulu began ground-breaking 6G research as part of Academy of Finland's flagship program [Latva-aho et al., 2019] called 6G-Enabled Wireless Smart Society and Ecosystem (6Genesis), which focuses on several challenging research areas including reliable near-instant unlimited wireless connectivity, distributed computing and intelligence, as well as materials and antennas to be utilized in future for circuits and devices. Besides, other traditional main players in mobile communications, such as the United States, China, Germany, Japan, and South Korea, already initiated their 6G research officially or at least announced their ambitions and tentative roadmap [Jiang et al., 2021].

1.8 Conclusions

As the starting point of this book, this chapter reviewed the evolution of mobile systems from the pre-cellular systems to the latest 5G cellular networks. The content was organized in terms of the generation of mobile systems. Each section dedicates to one generation of cellular systems, where the main content generally consists of three main parts: (i) The underlying motivation of evolution; (ii) The milestones of development and deployment; and (iii) Comparative review of different technical standards. The purpose of this chapter is to provide the reader an overall view of all previous generations and the evolution path, so that they can well understand the state-of-the-art advances for the upcoming 6G system. Next chapter we will provide an in-depth introduction of the previous generations consisting of their system architecture and key technologies.

References

Andrews, J. G., Buzzi, S., Choi, W., Hanly, S. V., Lozano, A., Soong, A. C. K. and Zhang, J. C. [2014], 'What will 5G be?', *IEEE Journal on Selected Areas in Communications* **32**(6), 1065–1082.

Astely, D., Dahlman, E., Furuskär, A., Jading, Y., Lindström, M. and Parkvall, S. [2009], 'LTE: The evolution of mobile broadband', *IEEE Communications Magazine* **47**(4), 44–51.

Attar, R., Ghosh, D., Lott, C., Fan, M., Black, P., Rezaiifar, R. and Agashe, P. [2006], 'Evolution of CDMA2000 cellular networks: Multicarrier EV-DO', *IEEE Communications Magazine* **44**(3), 46–53.

Dahlman, E., Parkvall, S. and Sköld, J. [2011], *4G LTE/LTE-Advanced for Mobile Broadband*, Academic Press, Elsevier, Oxford, The United Kingdom.

Dahlman, E., Parkvall, S. and Sköld, J. [2021], *5G NR - The Next Generation Wireless Access Technology*, Academic Press, Elsevier, London, The United Kingdom.

Ehrlich, N. [1979], 'The advanced mobile phone service', *IEEE Communications Magazine* **17**(2), 9–16.

Etemad, K. [2008], 'Overview of mobile WiMAX technology and evolution', *IEEE Communications Magazine* **46**(10), 31–40.

EU Gigabit Connectivity [2020], Shaping Europe's digital future, Communication COM(2020)67, European Commission, Brussels, Belgium.

Fitzek, F. H. P. and Seeling, P. [2020], 'Why we should not talk about 6G', *arXiv*.

Frenkiel, R. and Schwartz, M. [2010], 'Creating cellular: A history of the AMPS project (1971–1983)', *IEEE Communications Magazine* **48**(9), 14–24.

Furuskar, A., Mazur, S., Muller, F. and Olofsson, H. [1999], 'EDGE: Enhanced data rates for GSM and TDMA/136 evolution', *IEEE Personal Communications* **6**(3), 56–66.

Goldsmith, A. [2005], *Wireless Communications*, Cambridge University Press, Stanford University, California.

ITU-R M.1225 [1997], Guidelines for evaluation of radio transmission technologies for IMT-2000, Recommendation M.1225-0, ITU-R.

ITU-R M.1457 [2000], Detailed specifications of the terrestrial radio interfaces of International Mobile Telecommunications-2000 (IMT-2000), Recommendation M.1457-0, ITU-R.

ITU-R M.2012 [2012], Detailed specifications of the terrestrial radio interfaces of International Mobile Telecommunications-Advanced (IMT-Advanced), Recommendation M.2012-0, ITU-R.

ITU-R M.2083 [2015], IMT Vision-Framework and overall objectives of the future development of IMT for 2020 and beyond, Recommendation M.2083-0, ITU-R.

ITU-R M.2410 [2017], Minimum requirements related to technical performance for IMT-2020 radio interface(s), Recommendation M.2410-0, ITU-R.

ITU-R WP5D [2020], Future technology trends for the evolution of IMT towards 2030 and beyond, Liaison Statement, ITU-R Working Party 5D.

ITU-T NET-2030 [2019], A blueprint of technology, applications and market drivers towards the year 2030 and beyond, White Paper, ITU-T Focus Group NET-2030.

Jiang, W. and Schotten, H. D. [2021], The kick-off of 6G research worldwide: An overview, *in* 'Proceedings of 2021 Seventh IEEE International Conference on Computer and Communications (ICCC)', Chengdu, China.

Jiang, W., Han, B., Habibi, M. A. and Schotten, H. D. [2021], 'The road towards 6G: A comprehensive survey', *IEEE Open Journal on the Communications Society* **2**, 334–366.

Knisely, D., Kumar, S., Laha, S. and Nanda, S. [1998], 'Evolution of wireless data services: IS-95 to CDMA2000', *IEEE Communications Magazine* **36**(10), 140–149.

Latva-aho, M., Leppänen, K., Clazzer, F. and Munari, A. [2019], Key drivers and research challenges for 6G ubiquitous wireless intelligence, White paper, University of Oulu.

Linge, N. and Sutton, A. [2014], 'The road to 4G', *The Journal of the Institute of Telecommunications Professionals* **8**(1), 10–16.

Mishra, A. R. [2005], *Advanced Cellular Network Planning and Optimisation*, John Wiley & Sons, West Sussex, England.

Mouly, M. and Pautet, M.-B. [1995], 'Current evolution of the GSM systems', *IEEE Personal Communications* **2**(5), 9–19.

Osseiran, A., Boccardi, F., Braun, V., Kusume, K., Marsch, P., Maternia, M., Queseth, O., Schellmann, M., Schotten, H., Taoka, H., Tullberg, H., Uusitalo, M. A., Timus, B., and Fallgren, M. [2014], 'Scenarios for 5G mobile and wireless communications: The vision of the METIS project', *IEEE Communications Magazine* **52**(5), 26–35.

Ring, D. H. [1947], 'Mobile telephony - Wide area coverage', Bell Labs, Murray Hill, NJ, USA, Internal Tech. Memo.

Vriendt, J. D., Laine, P., Lerouge, C. and Xu, X. [2002], 'Mobile network evolution: A revolution on the move', *IEEE Communications Magazine* **40**(4), 104–111.

Walke, B. H. [2003], 'The roots of GPRS: The first system for mobile packet-based global internet access', *IEEE Wireless Communications* **20**(5), 12–23.

Young, W. R. [1979], 'Advanced mobile phone service: Introduction, background, and objectives', *Bell System Technical Journal* **58**(1), 1–14.

2

Pre-6G Technology and System Evolution

The previous chapter reviewed the history of mobile communications, from the pre-cellular systems to the latest fifth-generation networks. Thus, the readers have already got a complete view of the driving forces for the development of each generation, the competition among conflicting technologies, representative applications, and services per generation, competing standards and their respective evolution paths, standardization activities, spectrum regulation, and commercial deployment. However, these pieces of knowledge are not adequate to get insights into the future technological trend toward the six-generation mobile communications. In order to bridge this gap, this chapter will provide an in-depth study of the previous generations from the technological perspective. A symbolic standard, which achieved a dominant position commercially in the market worldwide and adopted mainstreaming technologies at that time, is taken to represent each generation. As a result, Advanced Mobile Phone System for the first generation, GSM for the second generation, WCDMA for the third generation, LTE for the fourth generation, and New Radio for the fifth generation will be introduced in this chapter. This chapter will be organized chronologically in terms of the generations and each section dedicates to one generation of cellular systems. After reading this chapter you should be able to:

- Understand the overall architecture of the representative cellular system for each generation, including major network elements, functionality split, interconnection, interaction, and operational flows.
- Identify principal technologies for each generation, and know the basic principles, benefits, and challenges of these technologies.
- Get an insight into the identification of essential technologies and the synergy of these technologies to build a cellular system.

6G Key Technologies: A Comprehensive Guide, First Edition. Wei Jiang and Fa-Long Luo.

2.1 1G – AMPS

As the most influential 1G standard, Advanced Mobile Phone System (AMPS) was developed in the USA primarily by Bell Labs, inspired by the heavily congested IMTS system. Originated from the initial concept of cellular networking proposed in 1947, it underwent quite a long journey to become a practical network. The system design had been almost completed in the 1960s, followed by an extensive trial (technical and commercial) to optimize the system parameters and verify the basic planning rules for a cellular network layout. In 1978, Bell Labs set up a large-scale and fully operational trial, working in cooperation with Illinois Bell Telephone Co., the American Telephone and Telegraph Co., and the Western Electric Co. The trial system consisted of 10 cells to cover approximately 3000 square miles in the Chicago, IL area, aiming to provide a capacity for more than 2000 users [Ehrlich, 1979]. It was not until 1983 that commercial operation licenses were issued when the Federal Communications Commission (FCC) allocated an initial spectrum of 40 MHz (later increased to 50 MHz) for analog cellular networks. In AMPS, the frequency range of 824–849 MHz was applied for the reverse transmission, while the 869–894 MHz range for the forward transmission. Spectral sharing, particularly termed multiple accesses in wireless communications, is implemented by multiplexing the signaling dimensions along the time, frequency, code, or space domain. The AMPS system adopted the frequency-division multiple access (FDMA) technique to divide the whole spectrum into a set of orthogonal frequency channels with a bandwidth of 30 kHz. Two licensees knew as A and B carriers with different channels were issued in a geographical region to encourage competition. In other words, a total of 416 paired channels consisting of 21 control channels and 395 voice channels were available for each carrier. The speech signals are modulated analogously by employing Frequency Modulation (FM). Each control channel can be associated with a group of voice channels. Thus each set of voice channels can be split into groups of 16 channels, controlled by a different control channel. Although AMPS is an analog cellular system, control channels then were already digitized. Signaling was transmitted between a base station and mobile stations at a data rate of 10 kbps. The signaling data was digitally modulated using Frequency-Shift Keying (FSK) and the Manchester encoding for error correction.

2.1.1 System Architecture

In addition to mobile stations, a typical AMPS network consists of two main components: Base Transceiver Station (BTS) and Mobile Telephone Switching Office (MTSO). Its architecture is illustrated in Figure 2.1. A BTS is generally placed in the center of a cell and contains the transceivers for transmitting

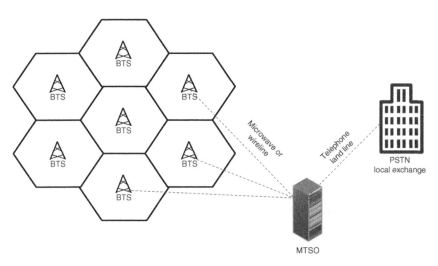

Figure 2.1 Architecture of a typical AMPS system.

and receiving radio signals and the transmission equipment connected to the MTSO. Local processing such as call setup, call monitoring, mobile station locating, and call terminating is performed at the BTS. All base stations within a geographical area are connected to an MTSO via high-speed wireline or microwave links. The main role of MTSO is to perform the switching function to provide connection toward the public switched telephone network (PSTN). It also takes responsibility for handling the overall control of the network, allocating channels within each cell, coordinating inter-cell handover when a mobile station traverses a cell boundary, routing calls to and from mobile users, and detecting system faults. An MTSO generally controls all mobile stations within a metropolitan service area.

The deployment of AMPS temporarily alleviated the capacity constraint of mobile phone service, but it also exhibited many drawbacks. For example, the system suffered from spectrum inefficiency. It applied a frequency reuse factor of 7, i.e. the total channels are divided into seven groups to avoid inter-cell interference (ICI) among adjacent cells. As a result, the capacity of a cell was too limited to accommodate a large number of active users. Moreover, AMPS did not provide sufficient security protection, where a handset's identification information such as ESN (Electronic Serial Number) or CTN (Cellular Telephone Number) can be cloned by using a fake base station and illegally reused in other sites. These triggered the need for a more capable, efficient, and robust system. The successor of AMPS was a digital system named Digital Advanced Mobile Phone System (D-AMPS). The capacity was increased by adopting the time-division multiple access (TDMA) technique to simultaneously support three users on each 30 kHz channel and digitally compressing the voice data. D-AMPS inherited the same basic architecture and signaling protocol from its predecessor, facilitating smooth transition from analog to digital systems. D-AMPS mobile stations can initially access the network via the traditional AMPS control channels and then request a digital voice channel if such a channel is available or keep in the analog mode. D-AMPS was also named IS-54 and later IS-136 from the Electronics Industries Association and Telecommunication Industries Association (EIA/TIA), as the first American 2G digital cellular standard.

2.1.2 Key Technologies

In early mobile telephone systems, a base station generally covered a wide area with a diameter of tens of kilometers by mounting its antenna at a high elevation and transmitting radio signals with high power. All mobile users within this coverage shared the allocated spectrum, leading to a limited capacity. With the increasing demand for mobile subscriptions, the need for a capable, economical, and portable system fostered the birth of the cellular concept. In 1947, William Rae Young, an engineer who worked at AT&T Bell Labs, reported his idea about the hexagonal cell layout throughout each city so that every mobile telephone can connect to at least one cell. Douglas H. Ring, also at AT&T Bell Labs, expanded on Young's concept. He sketched out the basic design for a standard cellular network and published the intellectual groundwork as a technical memorandum entitled *Mobile Telephony – Wide Area Coverage* on 11 December 1947 [Ring, 1947].

Two essential features behind the cellular system is *frequency reuse* and *cell splitting* [Donald, 1979].

2.1.2.1 Frequency Reuse

Because signal power decays drastically with the propagation distance, the same frequency spectrum is possible to reuse at spatially separated locations. If the distance is sufficiently large, co-channel interference is not objectionable. Base stations with moderate power distribute throughout the coverage area, and each base station covers its nearby zone called a cell. Radio and television broadcasting had already adopted frequency reuse before the advent of the cellular network. Nevertheless, the latter requires two-way communication between mobile stations and the network conveying personalized messages, rather than common information, imposing a fundamental difference in system design. Divide the available spectrum into N narrow-band channels with a bandwidth of W/N Hz per channel. Each cell gets n channels, and the channels allocated to a cell are not necessarily continuous. The same channel is not reused in neighboring cells to lower the co-channel interference. The ratio N/n denotes how often

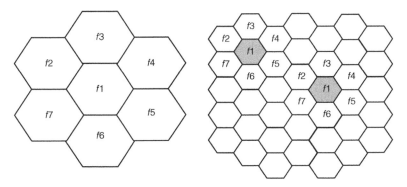

Figure 2.2 Cellular network layout illustrating frequency reuse and cell splitting.

a channel can be reused and is termed the *frequency reuse factor* (some literature use n/N as the reuse factor). In a typical hexagonal layout, as shown in Figure 2.2, the reuse factor is 7 where the channels are divided into seven groups denoted by $\{f1, f2, f3, f4, f5, f6, f7\}$. Through frequency reuse, every seven neighboring cells, sometimes also named a cluster, use the whole spectrum to provide the mobile service. Depending on the geometry of the cellular arrangement and the interference avoidance pattern, the reuse factor can be different. The AMPS network selected a factor of 7 and GSM with a factor of 3, while the Code Division Multiple Access (CDMA) system implemented universal frequency reuse. With the aid of advanced interference suppression techniques, the entire spectrum was maximally reused in each cell with a factor of 1.

2.1.2.2 Cell Splitting

At the initial stage, a cellular network preferred to employ only a few large cells to cover an entire city or region. The coverage area of this large cell is at least several square kilometers, referred to as Macro-cell. The base stations were usually installed on tall buildings or mountains and transmitted with relatively high power. This arrangement was mainly determined by two factors: the high cost of base stations and the low density of mobile subscriptions. When the traffic demand in an area saturates, the direct solution is to get more bandwidth. However, the licensed spectrum is not only expensive but also thoroughly unavailable in some regions. An alternative solution to further increase the capacity without extra spectrum is to revise cell boundaries so that the area formerly covered by a single cell can now divide into multiple cells. This process is called cell splitting. As illustrated in Figure 2.2, an area covered by seven cells can split into several clusters. Deploying more base stations with smaller transmit power, the spectral resources $\{f1, f2, f3, f4, f5, f6, f7\}$ can be reused in each cluster.

2.1.2.3 Sectorization

The signal power of a base station equipped with an omnidirectional antenna was radiated uniformly at all angles. Inspired by dividing an area into cells to improve spectrum efficiency, a cell is further divided into smaller areas called sectors. In 1985, Philip T. Porter proposed to employ directional antennas at base stations [Porter, 1985]. His proposal can lower interference to allow a seven-cell reuse pattern. Sectorization provides an effective method to further increase the system capacity without building new sites or network infrastructure. Using three sets of 120° directional antennas, the capacity of a cell can be tripled in theory. Sectorization is effective when the base station is tall with few surrounding obstacles, whereas it becomes far less effective in the conditions of more scattering and reflectors due to inter-sector interference.

The techniques of frequency reuse and cell splitting allow a cellular network to cover a wide area and serve a large number of mobile users using limited spectrum allocation. Cell splitting also enables a scalable network adapting to the growth of mobile traffic and various spatial density. Low-demand areas can be served by large cells while high-demand areas can be served by small cells. However, the cellular network raises a new problem – When a mobile user moves from one cell to another cell, the communication quality decreases. In order to guarantee the user experience, the connection needs to hand over (also called hand off) between two cells seamlessly.

2.1.2.4 Handover

Each base station broadcasts a beacon signal with constant power. The signals from different base stations are differentiated by employing pseudo-random sequences. The carrier frequency, signal format, power, and pseudo-random sequences are specified beforehand and known by all mobile stations within the network. A mobile station periodically measures the signal strength of the surrounding base stations. If it locates at the center of a cell, the beacon signal of this cell is strong while others are weak. At the cell edge, the mobile station may receive several beacon signals with similar strength. But in either case, the mobile station can select the best cell with the strongest signal strength to access if the admission control of this cell allows it to associate. When the mobile station moves away this cell, the measured signal strength decreases. Once it is below the predefined threshold that is required for the minimum acceptable performance, a handover procedure is triggered. The involved base stations and the controller assist the mobile station to release the occupied channel in the outgoing cell and tune to a new channel in the incoming cell.

In addition to the key techniques for cellular arrangement, another critical technology empowering the 1G systems was FDMA.

2.1.2.5 Frequency-Division Multiple Access

Unlike the radio and television systems broadcasting common signals in a uni-directional manner, mobile communications are two-way with dedicated messages for different users. Such a multi-user system requires the allocation of resources to specific users, referring to as multiple access. Real-time applications like voice or video communications need dedicated channels to ensure that signal transmission is not interrupted. Orthogonal channelization techniques such as frequency-division, time-division, space-division, or hybrid combinations are applied to generate dedicated channels. In contrast, delay-tolerant services, e.g. bursty data delivery, generally adopt non-orthogonal multiple access known as random access. The 1G systems such as AMPS, Nordic Mobile Telephony (NMT), Total Access Communication System (TACS), and C-450 employed FDMA for multiple access, where the system bandwidth is divided along the frequency axis into orthogonal channels. Each user gets a dedicated channel without multi-user interference, and there is no frequency-selective fading if the individual channel is narrow-band. Suffering from the impairments, such as imperfect hardware, spectrum spreading due to the Doppler shift, and adjacent-channel spectral leakage, an FDMA channel has to use guard bands at both sides. It leads to the waste of spectral resources. For example, each AMPS user is assigned a 30 kHz channel, corresponding to 24 kHz for the FM signal transmission and 3 kHz guard bands on each side. A terminal also needs to frequency-agile radio frequency (RF) components that can tune to different channels.

2.2 2G – GSM

The incompatibility of the various European systems made it different for the travelers among European countries to get continuous communication service with a single analog phone. It motivated the necessity for a uniform European standard and unified frequency allocation throughout Europe. As early as 1982, the Conference of European Postal and Telecommunications (CEPT) set up a working group called the Groupe Special Mobile (the initial meaning of GSM) to coordinate the development work. From 1982 to 1985, discussions were held in the GSM group to select between an analog and a digital system. After multiple field trials, it was decided to develop a digital cellular system based on narrow-band TDMA. The requirements for this pan-European system, such as good subjective voice quality, low terminal and service cost, and international roaming, were defined. In 1988, the CEPT formed the European Telecommunications Standards Institute (ETSI), and then the responsibility for specifying the standard was transferred to ETSI. In 1990, the Phase I recommendations for the Global System for Mobile communications (GSM) was released. Meanwhile, a variant of GSM operating in a higher frequency band, known as Digital Cellular System at 1800 MHz (DCS-1800), was standardized within ETSI and approved in February 1991 [Mouly and Pautet, 1995]. In addition to the most basic voice service, GSM terminals can connect with the Integrated Services Digital Network (ISDN) for various data services with a rate of 9.6 kbps. The commercial operation of the first GSM network started in Finland. On 1 July 1991, the Finnish Prime Minister Harri Holkeri made the world's first GSM call on the Radiolinja mobile network built with the equipment from Nokia and Siemens. Since then, GSM has rapidly gained acceptance and became the dominant 2G digital cellular standard [Vriendt et al., 2002]. It achieved remarkable commercial success, with a global market share of more than 90%. By early 2004, more than one billion population in more than 200 countries and territories enjoyed their mobile telephony services thanks to GSM.

2.2.1 System Architecture

A generic GSM system is composed of four components: Mobile Station Subsystem (MSS), Base Station Subsystem (BSS), Network and Switching Subsystem (NSS), and Operation and Support Subsystem (OSS). Each subsystem contains various functional entities that are mutually inter-connected through specified interfaces [Rahnema, 1993].

2.2.1.1 Mobile Station Subsystem
The hardware of MSS is mobile equipment, which supports voice calls, SMS, and low-speed data access for subscribers. A serial number called the International Mobile Equipment Identity (IMEI) is applied to identify a mobile station uniquely. Besides, a smart card known as Subscriber Identity Module (SIM) stores the International Mobile Subscriber Identity (IMSI) to identify the subscriber, a secret key for authentication, and other subscriber information. The mobility of mobile users is supported since the IMEI and IMSI are independent.

2.2.1.2 Bases Station Subsystem
BTS and Base Station Controller (BSC) make up the BSS. The former contains antennas for radiating and receiving electromagnetic waves, transceivers for generating and detecting radio signals, and equipment for encrypting and decrypting communications with the BSC. A BTS is usually installed at the center of a cell using one and more transceivers in terms of the density of users. It is the most complicated part of the system, equipped with expensive and power-consuming RF components such as:

- Combiner that combines multiple feeds into a single antenna
- Power amplifier that amplifies the transmit signal
- Duplexer that allows bi-directional (duplex) communication.

The BSC manages the radio resources for one or more BTSs and controls user admission, radio channel setup, inter-BTS handover, and frequency hopping.

2.2.1.3 Network and Switching Subsystem

The main role of NSS is to perform the switching function to provide the connection for the serving mobile stations to other mobile users and fixed telephony users toward the PSTN. It also provides functionalities to allow the mobile user necessities to be supported like authentication, registration, location updating, inter-Mobile Switching Center (MSC) handover, and call routing. MSC acts as the central component of the NSS, with the assist of a set of databases storing subscriber and mobility information, which are described as follows:

- *Home Location Register (HLR)* is a critical database that holds the administrative details of all subscribers belonging to the service area. Stored information for each subscriber includes the IMSI, the phone number, the authentication key, the list of subscribed services, and some temporary data such as the current location where the mobile terminal was last registered. When a terminal switches on, it registers with the network and determines which BTS it communicates with so that incoming calls can be routed appropriately. Even when the terminal is not active (but switched on), it reports periodically to ensure that the network is aware of its latest location.
- *Visitor Location Register (VLR)* is a database that stores information of visiting subscribers. When an MSC detects a new mobile station in its service area, the VLR associated with this MSC requests necessary information from the HLR or the mobile station. The subscriber information such as the IMSI, the authentication key, the phone number, and the address of its HLR is extracted in order to assure the subscribed services and call routing.
- *Authentication Center (AuC)* stores a copy of the secret key of each subscriber needed for authentication and encryption over the radio channel. When a mobile station attempts to connect to the network, AuC needs to authenticate the validation of its SIM card. During the communication, it also provides an encryption key to encrypt the data between the mobile phone and the Core Network (CN).
- *Equipment Identity Register (EIR)* is also used for security purposes. It maintains a catalog of all valid terminals on the network identified by its IMEI. EIR can forbid calls from invalid terminals that are registered as stolen or unauthorized.

2.2.1.4 Operation and Support Subsystem

The OSS is the functional entity facilitating the network operator to monitor and manage the system. The operation and maintenance center (OMC) is the equipment that connects to different components in the switching system and the BSC. The OMC is in charge of the functions such as administration operation (subscription, termination, charging, and statistics), security management, network configuration, performance monitoring, and maintenance.

2.2.1.5 General Packet Radio Service

The support for General Packet Radio Services (GPRS) was specified in the Release'97 version of the GSM standard [3GPP TS23.060, 1999]. It was implemented by overlaying a packet-switched sub-network on the legacy circuit-switched network, as shown in Figure 2.3. A GSM network can smoothly upgrade to support data services such as Internet browsing, Wireless Application Protocol (WAP) access, and Multimedia Messaging Service (MMS) through adding some new nodes and software. Two kinds of nodes, i.e. Serving GPRS Support Node (SGSN) and Gateway GPRS Support Node (GGSN), were introduced to support routing and proper enhancement of the data packets.

SGSN The SGSN in the packet-switched sub-network takes the role as same as the MSC for the voice traffic in the circuit-switched network. The SGSN serves mobile stations through base stations and connects to the GGSN for the access of external networks. It provides functions such as

- *Mobility management*: When a mobile station attaches to the packet-switched network, the SGSN generates mobility management information according to the current location of mobile stations. The SGSN tracks the movement of a

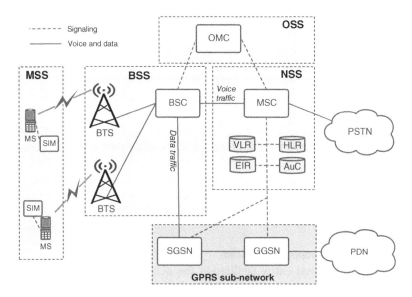

Figure 2.3 Architecture of the GSM system, overlapped with a GPRS network.

registered mobile station and forwards the incoming packets to the approximate address.

- *Session management*: The SGSN manages the initiating, maintaining, and terminating of real-time or non-real-time data sessions and provides mechanisms to guarantee the required quality of services (QoS) for a variety of data services.
- *Switching*: The SGSN forwards incoming and outgoing data packets from the BSC and GGSN. It also communicates with other areas of the network, e.g. MSC and other circuit-switched areas, to get necessary management information.
- *Charging*: The SGSN is also responsible for charging and statistics collection by monitoring the flow of user data across the GPRS network. The SGSN generates the record of call details for the charging entities.

2.2.1.6 Gateway GPRS Support Node

It handles inter-working with external packet data networks such as the Internet or X.25 networks and is connected with SGSNs via the GRPS backbone network. From the perspective of an external network, it is a router. This node supports gateway functionality such as publishing subscriber addresses, mapping addresses, routing and tunneling packets, screening messages, and counting packets. The GPRS packets from the SGSN are converted into the appropriate data format (e.g. IP or X.25) at the GGSN and then are forwarded to an external data network. The format of incoming data packets are converted at the GGSN and are forwarded to the SGSN associated with the target mobile station [Lin et al., 2001].

2.2.2 Key Technologies

Compared with the 1G analog system, the 2G system offered much larger system capacity, improved QoS, security provisioning, and efficient spectrum usage. The digitization of the cellular system facilitated the adoption of advanced techniques that were hard to be applied before. The major 2G technologies are briefly reviewed as follows:

2.2.2.1 Time-Division Multiple Access

TDMA divides the signaling dimensions along the time axis into orthogonal channels. Each user transmits over the entire bandwidth but cyclically accesses the assigned time slot. It implies non-continuous transmission, which simplifies the system design since some processing such as channel estimation can be performed during the time slots of other users. Another advantage is that the system is able to assign multiple time slots for a single user, facilitating the implementation of high data rates. One major challenge in a TDMA system is the synchronization of the uplink channel. The signals transmitted from spatially separated users induce different propagation delays. Moreover, the mobile users may move continuously, and the multi-path propagation environment also varies, making the synchronization of the uplink channel hard to achieve. The bandwidth of a TDMA channel is generally larger than that of an FDMA channel in the previous analog system. If the signal bandwidth exceeds the coherent bandwidth of wireless channel, inter-symbol interference (ISI) raises and therefore an equalizer is required at the receiver to compensate for the ISI.

The GSM system employed the TDMA technique combined with FDMA as the multiple access scheme. The original GSM system was assigned paired frequency bands with a bandwidth of 25 MHz each: 890–915 MHz for the uplink transmission and 935–960 MHz for the downlink direction. Setting guard bands between GSM and other systems operating in the neighboring band, the remaining bandwidth is divided, using an FDMA scheme, into a total of 124 channels with a width of 200 kHz. A TDMA channel multiplexes eight time slots using a TDMA scheme. Each time slot has a duration of approximately 0.577 ms, and the content carrying in a time slot is called a burst. To compensate for synchronization error and multi-path delay spread, a guard period is inserted at the tail of each burst. Several burst types were defined for different functions, including normal burst, frequency correction burst, synchronization burst, access burst, and dummy burst. Eight bursts with a total length of 4.615 ms form a TDMA frame, which cyclically repeats.

2.2.2.2 Frequency Hopping

To mitigate the effect of narrowband interference and ISI due to multi-path channels, the spread spectrum technique is an effective tool. Two methods, i.e. direct-sequence and frequency hopping, can be applied to spread the signal over a bandwidth much larger than the width of the original signal. The frequency-hopping technology was invented by the film star Hedy Lamarr and the composer George Antheil during World War II and released in their patent *Secret communication system* [Markey and Antheil, n.d.]. The transmitted signal hops over a wide bandwidth by continuously changing its carrier frequency. A frequency synthesizer at the transmitter generates the hopping frequency carriers according to a pseudorandom sequence termed the spreading code. At the receiver, the same spreading code is input into a frequency synthesizer to generate the frequency carriers to down-convert the received signals. If the hop time exceeds a symbol period, it is called slow frequency hopping. Otherwise, it is fast frequency hopping when the hop time is less than a symbol period. GSM adopted slow frequency hopping, taking advantage of the inherent frequency agility of the transceivers that can transmit and receive on different channels. The carrier frequency changes every TDMA frame at a prescribed rate of 217 times per second.

2.2.2.3 Speech Compression

The most basic service provided in early mobile systems was the voice transmission. In the 1G system, the speech signal was modulated analogously. The digitization of 2G also required to digitize analog speech signals before transmission. In the wireline telephone system, Pulse Coded Modulation (PCM) was employed to code speech signals for multiplexing over high-speed backbone or optical fiber lines. PCM has a coding rate of 64 kbps, which is too high for the transmission over air interface due to the constraint of radio resources. Hence, the GSM working group studied a variety of speech coding and synthesis algorithms to reduce the redundancy in the sounds of the voice. On the basis of subjective speech quality, processing delay, power consumption, and complexity, the final choice for the GSM speech codec is RPE-LTP (Regular Pulse Excitation Long-Term Prediction). It uses a number of past samples to predict the current sample. The speech signal is sampled every 20 ms to obtain 260-bit data blocks, equivalent to a coding rate of 13 kbps.

2.2.2.4 Channel Coding

The 1G system suffered from worse voice quality since the induced noise and interference on the analog transmitted signal cannot filter out. The 2G digital system enabled the adoption of channel coding to improve performance (The AMPS system applied channel coding but only for the control channel). Contrary to the speech coding that tries to compress the amount of data as lower as possible, channel coding intentionally adds redundancy bits to the original information to detect or correct errors incurred during the transmission. In GSM, for example, different levels of protection are provided for the speech bits, according to their importance. The speech codec of a GSM terminal generates a block of 260 bits every 20 ms. From the users' perspective, the perceived speech quality depends more on some part of this block rather than every bit equally. Consequently, the block is divided into three parts: Class Ia: 50 bits – most important, Class Ib: 132 bits – moderately important, and Class II: 78 bits – least important. First, Cyclic Redundancy Code (CRC) is added at the tail of Class Ia for error detection. These 53 bits, together with Class Ib and a 4-bit tail sequence, amounting to a total of 189 bits, are input into a convolutional encoder with a rate of 1/2 and a constraint length of 4. The remaining bits in Class II are added into the encoded sequence without any protection. Finally, every 20 ms speech signal is transformed into a transmission block with a length of 456 bits, resulting in a coding rate of 22.8 kbps. Moreover, to further protect against the burst errors common to the radio interface, the transmission block is interleaved.

2.2.2.5 Digital Modulation

Digital modulation holds many advantages over analog modulation, including high spectral efficiency, high power efficiency, robustness against channel impairments, improved security and privacy, and cheaper hardware implementation [Goldsmith, 2005]. To be specific, high-level digital modulation such as M-ary Quadrature Amplitude Modulation (MQAM) offers much higher transmission rates over the same signal bandwidth than analog modulation. A digital transceiver can apply advanced techniques such as channel coding, equalization, and spread spectrum to resist hardware impairments, channel fading, noise, and interference. The information bits carried on modulation constellations are much easier to encrypt than analog signals, resulting in a high level of security and privacy. The GSM system adopted the Gaussian Modulation Shift Keying (GMSK) to modulate the information bits. It was selected over other modulation schemes as a compromise among spectral efficiency, transmitter complexity, and limited spurious emissions. GMSK is a constant-modulus signal (constant envelope signal), which reduces problems caused by the non-linear distortion of a power amplifier. As an enhancement of GSM, Enhanced Data Rates for GSM Evolution (EDGE) employed higher-order phase-shift keying (8PSK) to improve the data transmission rate.

2.2.2.6 Discontinuous Transmission (DXT)

The principle of the DXT is to suspend the signal transmission during the interval called the silence period, taking the fact that a person speaks generally less than 50% during a conversation. The significant benefits include the reduction of co-channel interference among the cells reused the same channel, low power consumption, increased system capacity, and prolonged battery life. Two major components were required, i.e. Voice Activity Detection (VAD) and a generator for comfort noise, to implement this function. The VAD was employed to distinguish between noise and voice, even when it is trivial. If a voice signal is identified as noise, the transmitter is turned off, producing an unpleasant effect called clipping. There is an absolute silence due to the digital signaling during the turn-off period of the transmitter. The users' subjective perception might be very annoying on the reception side because it seems that the connection drops. In order to overcome this problem, the receiver needs to generate a small comfort signal to mimic the background noise.

2.3 3G – WCDMA

In the late 1990s, NTT DoCoMo developed Wideband CDMA technology for their 3G system known as Freedom of Mobile Multimedia Access (FOMA). Wideband

Code Division Multiple Access (WCDMA) only defined the air-interface part, therefore also named as Universal Terrestrial Radio Access (UTRA). WCDMA was selected as the air interface of UMTS, as the 3G successor to GSM. Different systems, including FOMA, UMTS, and J-Phone, shared WCDMA air interface but have different protocols for a complete stack of communication standards. 3GPP submitted it as an IMT-2000 proposal, and the International Telecommunication Union Radiocommunication Secto (ITU-R) approved it as part of the IMT-2000 family standards. It employed direct-sequence code-division multiple access with a chip rate of 3.84 Mcps. The radio access specifications provided both Frequency Division Duplex (FDD) and Time Division Duplex (TDD) variants, utilizing a 5 MHz channel to achieve peak rates of up to 5 Mbps. In October 2001, NTT DoCoMo launched the first commercial FOMA network in Japan as the successor to i-mode. In many European countries, mobile operators acquired a license for the 3G spectrum and had to pay enormous fees in auctions. For example, the mobile operators of the United Kingdom spent 33 billion USD in the auction of April 2000, and 47.5 billion USD was recorded in German auction later that year. Such a high financial pressure of mobile operators raised by the high licensing cost resulted in a delay of the commercial roll-out of European 3G networks. As an example, the United Kingdom's first commercial 3G network was deployed by Hutchison Telecom as late as March 2003.

WCDMA specified in Release 99 and Release 4 of the specifications contains all technical features to satisfy the IMT-2000 requirements, but further enhancement had never stopped. High Speed Packet Access (HSPA) appeared in 2002 as the first significant evolution of the WCDMA radio interface.

2.3.1 System Architecture

The network elements of WCDMA are categorized into three main groups: User Equipment (UE), UMTS Terrestrial Radio Access Network (UTRAN) that deals with the radio interfaces, and CN that is mainly in charge of switching voice calls and routing data packets to and from external networks. In the initial deployment of WCDMA, some CN entities were directly inherited from GSM, aiming to smooth transition. In contrast, the radio part of the legacy GSM network was therefore termed GSM EDGE Radio Access Network (GERAN) [Holma and Toskala, 2004].

2.3.1.1 User Equipment

Because of the considerably more number of applications and functionalities that it can support, a new term *UE* was made rather than what was previously called mobile phone or cell phone (Figure 2.4). It could be any device between a mobile phone for voice talking to a computer without voice capability. The UE consists of two parts: the mobile terminal used for radio communications and the UMTS

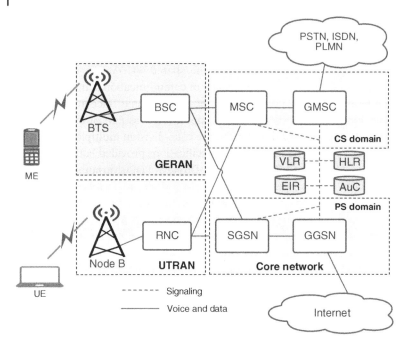

Figure 2.4 Architecture of the WCDMA system. UTRAN stands for new air interface of WCDMA while GERAN is the radio access of the legacy GSM system.

Subscriber Identity Module (USIM) that stores the subscriber identification, authentication and encryption keys, and subscription information.

2.3.1.2 UMTS Terrestrial Radio Access Network

UTRAN is composed of one or more Radio Network Subsystems (RNS). An RNS consists of a number of base stations termed Node B, and their corresponding controller, i.e. Radio Network Controller (RNC).

Node B is equipped with radio transceivers for air-interface processing and also takes part in some radio resource management. Its major functions include

- Radio Signal Transmission and Reception
- Modulation and Demodulation
- Spreading and Despreading
- Channel Coding and Decoding
- Frequency and Time Synchronization
- Closed-Loop Power Control

WCDMA is a wideband direct-sequence CDMA system where user information bits are spreaded over a bandwidth of 5 MHz by multiplying quasi-random chips

Table 2.1 The main WCDMA radio parameters, in comparison with GSM and IS-95.

Parameter	WCDMA	IS-95	GSM
Signal bandwidth	5 MHz	1.25 MHz	200 kHz
Multiple access	DS-CDMA	DS-CDMA	TDMA
Duplex mode	FDD/TDD	FDD	FDD
Chip rate	3.84 Mcps	1.2288 Mcps	N/A
Frame length	10 ms	20 ms	4.615 38 ms
Frequency reuse factor	1	1	3
Power control	Downlink: 1500 Hz	Downlink: 800 Hz	Optional, 2 Hz
	Uplink: 1500 Hz	Uplink: Slow	
Peak rate[a]	2 Mbps	14.4 kbps	9.6 kbps

a) The values are the rates of the initial version of the systems, i.e. Release 99 of WCDMA, IS-95A, and GSM without GPRS enhancement.

Source: Mcps, Mega chips per second; N/A, not applicable.

of spreading codes. WCDMA can operate in either FDD or TDD mode. It supports variable data rates by using different spreading codes, and the user rate can flexibly change per frame with a duration of 10 ms. The main parameters of WCDMA air interface are listed in Table 2.1, in comparison with that of GSM and IS-95.

RNC carries out the control of radio resources and some mobility management of UTRAN. The RNC connects to the MSC for voice communications and the SGSN for packet data delivery. The main functions of RNC are:

- Radio Resource Control
- Admission Control
- Channel Allocation
- Open-Loop Power Control
- Mobility Management
- Ciphering
- Macro Diversity

2.3.1.3 Core Network

The CN of UMTS is divided into Circuit-Switched (CS) and Packet-Switched (PS) domains. It is the equivalent to the NSS of the GSM network, providing switching, routing, and all central processing and management functions. It is the interface to external networks, including the PSTN, ISDN, Public Land Mobile Network (PLMN), and the Internet. The databases, i.e. HLR, VLR, AuC, and EIR, are shared by both domains and smoothly inherited from the GSM network.

CS Domain: It is primarily based on the GSM network elements optimized for voice transmission in a circuit-switched manner, i.e. a long-duration channel for the conversation of a call. The functions are carried out by the MSC (as same as the MSC in GSM), and Gateway Mobile Switching Center (GMSC), which is the interface to the external networks.

PS Domain: The network entities in this domain are optimized for packet data delivery. It includes SGSN and GGSN inherited from the GPRS network and operated similarly (Table 2.2).

2.3.2 Key Technologies

The transition from voice-only communications to the hybrid of voice and data services imposed many challenges on the design of air interface and CN. Accordingly, novel technologies that can support advanced multiple access, flexible resource sharing, high and variable data rates, and improved spectral efficiency were adopted. Among a dozen of 3G techniques, CDMA, the Rake receiver, and Turbo Coding, were recognized as the fundamental enablers, which are introduced as follows:

2.3.2.1 Code-Division Multiple Access

In narrow-band systems, users within a cell transmit their signals over orthogonal time-frequency resource units (through FDMA or TDMA), and users in adjacent cells are assigned different frequency blocks complying with the cellular layout. In a CDMA system, each user applies the direct-sequence spread-spectrum technique to spread its signal over the entire transmission bandwidth by multiplexing a pseudo-random sequence called spreading code. The transmission bandwidth is much higher than the original signal bandwidth. Their ratio is called the *processing gain*. The spread signals from different users share the same bandwidth simultaneously. Simply treating other users' signals as random noise, the desired signal can be detected by applying the same spreading code at the receiver. With orthogonal spreading codes such as Walsh-Hadamard codes, the information signals of different users can be well separated. Non-orthogonal spreading codes are also effective to differentiate multi-user signals, but mutual interference raises. That is the principle of *code-division multiple access* enabled by direct-sequence spread spectrum. In addition to allowing multiple access, the spread-spectrum technique provides frequency diversity to against multi-path fading and narrow-band interference.

Compared with narrow-band signal transmission, the CDMA technology has the following advantages:

- *Universal Frequency Reuse*: It implies the maximal frequency reuse, namely, a frequency-reuse factor of 1. In FDMA or TDMA systems, multi-user interference

Table 2.2 Comparison of major technical features in previous generations.

System	1G	2G	3G	4G	5G
Deployment	1979	1991	2000	2009	2019
Data rate	N/A	9.6–384 kbps	2–56 Mbps	1 Gbps	20 Gbps
Frequency band	400 MHz/800 MHz	900 MHz/1800 MHz	Up to 2.1 GHz	Sub-6 GHz	Sub-6 GHz/mmWave
Bandwidth	20 kHz, 30 kHz	200 kHz	5 MHz	20 MHz/100 MHz	400 MHz/1 GHz
Multi-access	FDMA	TDMA	CDMA	OFDMA	OFDMA/NOMA
Duplexing	FDD	FDD	FDD/TDD	FDD/TDD	FDD/TDD
Modulation	FM	GMSK	8PSK	64QAM/256QAM	1024QAM
Channel coding	N/A	Convolutional code	Turbo codes	Turbo/LDPC	LDPC/polar codes
Multi-antenna	N/A	Receive diversity	Beamforming	MIMO 8x8	MIMO 256x32
Networking	Circuit-switched	Circuit-/packet-switched	Circuit-/packet-switched	All-IP	All-IP
Service	Analog voice	Digital voice, SMS, low-rate data	VoIP, high-rate data, video	Mobile Internet	eMBB, URLLC, mMTC
Standards	AMPS, NMT, NTT, TACS, C-450, Radiocom2000	GSM, D-AMPS, IS-95	WCDMA, CDMA2000, TD-SCDMA, WiMAX	LTE, LTE-Advanced, WiMAX2.0	NR-NSA, NR-SA, 5G-Advanced
Technologies	Cellular, FDMA, FDD, FM	TDMA, FHSS, GMSK	CDMA, Rake receiver, Turbo Codes	MIMO, OFDM, carrier aggregation, CoMP, relaying, HetNet, D2D, LAA	Massive MIMO, mmWave, SDN/NFV, network slicing, LDPC, polar codes

is avoided by assigning orthogonal time-frequency slots to all users in the same cell and disjoint frequency bands to adjacent cells. The number of degrees of freedom per user is substantially reduced in terms of the number of users and the frequency-reuse factor. In a CDMA system, not only the users in the same cell but also the users in different cells, including adjacent ones, share the same time-frequency resources. In addition to increase the degrees of freedom per user, universal frequency reuse also simplifies the network planning. These benefits come at the price of a lower signal-to-interference-plus-noise ratio (SINR) of the individual links.

- *Soft Handover*: The neighboring cells in narrow-band systems assigned to different frequency blocks and the mobile terminals can only tune to a single carrier frequency at a time. Consequently, a terminal has to disconnect the outgoing cell before connecting to a new cell. Such a *hard handover* causes the drop of voice calls, which was one of the major problems worsening the user experience in the first- and second-generation systems. Since the neighboring cells use the same frequency in a CDMA system, a user at the edge of a cell can communicate with more than one base stations simultaneously. In the uplink, the user signal is modulated by multiple spreading codes associated with different cells and then is received by more than one base stations. The received signals are combined and detected at a central processing unit such as the RNC. In the downlink, the same signal can be transmitted to the user by more than one base station simultaneously. Soft handover is also a diversity technique sometimes called macro diversity.

- *Soft Capacity*: In a narrow-band system, the time-frequency resources are divided into a fixed number of orthogonal channels. There is a limit of hard capacity since new users cannot be admitted into a network once all channels are occupied. Using non-orthogonal spreading codes, a CDMA system has no hard limit on the number of serving users. The more users share the degrees of freedom using non-orthogonal codes, the higher the level of multi-user interference.

- *Interference Sharing*: An allocated channel is idle during the silence period of a voice conversation when he or she listens. But this channel cannot share with other users in FDMA or TDMA systems. The capacity of a CDMA system is interference limited and the number of channels has no hard limit. A CDMA system can automatically benefit from the source variability of users. When a user has no signal to transmit, the level of the aggregated interference goes down and the performance of other users improves.

The performance of a CDMA system depends heavily on accurate power control, especially in the uplink, to compensate for the near-far effect. If two users transmit at the same power, the signal strength of the nearby user is probably tens-of-decibel stronger than that of the user at the cell edge. The far user is hard to detect its desired signal from the overwhelming multi-user inference. As a consequence, power control is required such that the received signal strength of all users is roughly the same. But the frequent transmission of power-control signaling incurs a significant overhead. In contrast, power control is optional in narrow-band systems, used for lowering power consumption rather than interference.

2.3.2.2 Rake Receiver

Multipath fading is the major impairment to constraint the performance of the wireless transmission, where several delayed copies of the transmitted signal combine at the receiver. The combination is sometimes constructive but usually destructive. Since each component contains the original information, the ideal case is to add all components coherently to improve the power of the desired signal and lower the probability of deep fades. To this end, an advanced receiver structure called the Rake receiver can be employed. It has several branches, referred to as fingers, to deal with different signal paths. Each finger is synchronized to a signal path and equipped with an individual correlator to despread the received signal. The outputs of the Rake fingers are combined to detect the transmitted symbols, where different methods such as selective combination and maximal ratio combination can be applied. Direct-sequence CDMA is well suited for the application of Rake reception since the large signal bandwidth allows a high resolution of distinguishing multiple paths. It is particularly suitable for WCDMA transceiver with a signal bandwidth of 5 MHz. The Rake reception is in essence a diversity technique, which was proposed by Price and Green [1958]. It has been described as the historically most important adaptive receiver for multipath fading channels.

2.3.2.3 Turbo Codes

According to the Information Theory, a randomly chosen code of sufficiently large block length can approach the Shannon capacity. The complexity of the maximum likelihood decoding of such a code increases exponentially with the block length until decoding becomes physically unrealizable. In the mid-1990s, Berrou, Glavieux, and Thitimajshima invented a powerful coding scheme that can transmit data with a code rate within a fraction of a decibel of the Shannon capacity on a Gaussian channel [Berrou and Glavieux, 1996]. This coding scheme combines parallel concatenation of convolutional codes, large block length, interleaving, and iterative decoding approaches, relying on the exchange of soft-decision information. A typical encoder consists of two parallel convolutional encoders and an interleaver. The information bits m concatenate with two parity-bit sequences X_1 and X_2. One convolutional encoder generates X_1 in terms of m, while another convolutional encoder makes X_2 using the interleaved information bits. The concatenated data (m, X_1, X_2) is transmitted to the receiver. The iterative decoding of Turbo codes is implemented by two parallel decoders, an interleaver, and a deinterleaver. Berrou and Glavieux used a maximum a posteriori probability (MAP) algorithm to perform maximum-likelihood (ML) estimation, yielding reliability information (soft decision). To be specific, the first decoder generates a probability measure $p(m_1)$ on each information bits based on the codeword (m, X_1). This probability measure is passed to the second decoder. Then, another probability measure $p(m_2)$ is generated by the second decoder in terms of (m, X_2) and is passed to the first encoder. This process iterates until the convergence condition reaches when, ideally, the decoders eventually agree on the probability measure. Berrou and Glavieux exhibited that a Bit Error Rate (BER) of 10^{-5} can be achieved for an E_b/N_0 at 0.5 dB above the Shannon capacity for 1/2-rate quadrature phase-shift keying (QPSK) on a Gaussian channel, and this gap was narrowed to 0.35 dB later. Turbo codes received wide recognition quickly and triggered a wave of research and development addressing its design, implementation, performance evaluation, and application in digital communications systems.

The CDMA2000 system employed 1/5-rate Turbo coding that consists of two identical, eight-state, parallel, 1/3-rate Recursive Systematic Convolutional (RSC) encoders. This code is punctured to rates between 1/2 and 1/4, with the interleaver length ranges from 378 to 12 282. Turbo coding in WCDMA is similar to that of CDMA2000, except that two Turbo encoders use different pseudo-random interleavers.

2.4 4G – LTE

In contrast to the CDMA technology employed in the 3G system, Long-Term Evolution (LTE) adopted a more efficient multiple-access technique called Orthogonal Frequency-Division Multiplexing (OFDM), which affected not only the design of air interface but also the networking structure. Specifically, the downlink of LTE air interface used OFDMA, and the uplink employed its variant called Single-Carrier Frequency-Division Multiple Access (SC-FDMA) to allow a low-cost power amplifier on the terminal side. For the first time in a mobile system, multiple signal bandwidths are supported, ranging from 1.4 to 20 MHz, thanks to the flexibility enabled by the OFDM modulation. Over a bandwidth of 20 MHz, the peak data rate can reach 150 Mbps with 2×2 multiple-input multiple-output (MIMO) and 300 Mbps with 4×4 MIMO. The degrees of freedom are divided into small resource blocks of 180 kHz for flexible resource allocation and packet scheduling. The difference between TDD and FDD operation modes was minimized to achieve a large commonality. The CN adopted an end-to-end

IP architecture optimized for packet-switched data services while the traditional voice service was replaced with Voice over IP (VoIP).

2.4.1 System Architecture

In addition to advanced air-interface techniques, high network capacity needs efficient networking architecture. Therefore, a flat structure by reducing the number of network elements was employed to minimize end-to-end latency and improve network scalability. The LTE system comprises three subsystems: UE, Evolved Universal Terrestrial Radio Access Network (E-UTRAN), and Evolved Packet Core. The following provides a brief introduction of E-UTRAN and Evolved Packet Core (EPC).

2.4.1.1 Evolved Universal Terrestrial Radio Access Network

The radio access network of LTE is simplified with only one kind of network element called E-UTRAN Node B or Evolved Node B (eNodeB), in comparison with two elements – Node B and RNC – in UTRAN. All radio signal processing, radio resource management, and packet handling locate in eNodeB. An eNodeB acts as a switch, communicating with UEs, other eNodeBs, and the EPC through various interfaces. In addition to data packet processing and forwarding, it is also in charge of many control functions, which include:

1. *Radio Resource Management*: the setup, maintenance, and release of radio bearers, radio resource allocation, traffic scheduling and prioritizing, and resource utilization monitoring.
2. *Mobility Management*: the inter-cell handover and radio signal measurement
3. *Admission Control*: the acceptance or denial of access request
4. *Mobility Management Entity (MME) Selection*: selecting an available MME for serving a UE, enabling a UE shifted to a different MME, the establishment of the route toward an MME
5. *Packet Compression*: IP header compression for downlink packets for radio-efficient transmission over the air interface, and IP header decompression for uplink packets
6. *Ciphering*: encryption and decryption of packets through ciphering algorithms
7. *Messaging*: the transmission of paging messages, operation and maintenance messages, or broadcast information, the reception of broadcast information and paging messages from an MME.

2.4.1.2 Evolved Packet Core

The CN of LTE adopted a flat architecture with only a few network elements, allowing low end-to-end latency and cost-efficient packet data transmission.

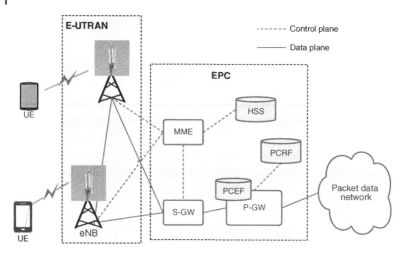

Figure 2.5 Architecture of the LTE system. Source: Secureroot/FreeImages.

The User Plane (UP) and Control Plane are decoupled to make scaling and independent upgrading possible. Thanks to this functional split, the operators can dimension and adapt their network easily. The middle part of Figure 2.5 shows a basic architecture of the Evolved Packet System (EPS), which is composed of the following network elements: Serving Gateway (S-GW), Packet Data Network Gateway (P-GW), MME, Home Subscriber Server (HSS), Policy and Charging Rules Function (PCRF), and Policy and Charging Enforcement Function (PCEF).

- *S-GW* is a UP entity used mainly for IP packet forwarding and tunneling, with minimal control functions. It relays all the incoming and outgoing IP packets belonging to a UE in connected mode between the eNodeB and the P-GW. Once received packets for a UE in idle mode, S-GW buffers the data and requests the MME to set up paging of the UE. The S-GW can replicate the monitored user's data to the authorities in case of lawful interception. To support mobility management, the S-GW acts like the local mobility anchor for handover between neighboring eNodeBs. The S-GW also handles mobility between LTE and other circuit-switched networks.
- *P-GW* is the edge router to interconnect the EPC and external packet data networks. It routes incoming and outgoing data packets and ensures connectivity to external networks. A UE can connect to more than one P-GW for multiple-Packet Data Network (PDN) accessing. When a UE communicates with other IP hosts in external networks, the P-GW allocates an IP address to the UE by performing Dynamic Host Configuration Protocol or queries an external Dynamic Host Configuration Protocol (DHCP) server. In addition, the P-GW handles policy enforcement such as packet filtering and screening for lawful

interception, and collects the related charging information, in cooperation with PCRF and PCEF.

- *MME* is the main element in the Control Plane of the EPC, which has the following functions:
  - *Authentication and Security*: When a UE registers to the network, the MME requests the authentication key from the HSS in this UE's home network and performs challenge-response authentication to assure that the UE is whom it claims to be. The MME generates a key for ciphering and integrity protection from the authentication key to prevent eavesdropping and unauthorized alteration. To protect users' privacy, the MME also provides each UE a temporary ID named the Global Unique Temporary Identity (GUTI) to hide the permanent ID.
  - *Mobility Management*: When a UE registers to the network for the first time, the MME creates a location entry and notifies the HSS in the UE's home network. The MME continuously tracks the location of the UE. The location information of an active UE is at the level of eNodeB, while at the level of a tracking area (i.e. a group of eNodeBs) for a UE in idle mode. The MME controls the resource allocation and release for the UE and controls the handover procedure.
  - *User Profile Management*: The MME is in charge of retrieving the user profile, such as subscription information from the home network. This information decides which PDN connections should be assigned to the UE at network attachment.
- *HSS* is a database that contains user profile and subscription information. The role of HSS is similar to a combination of HLR and Authentication Center in previous cellular networks. It stores information about the subscribed services, information about the allowed PDN connections, and information about the roaming permission to a particular network. It also stores the permanent key for the local UEs. The key is applied to generate the authentication key requested by the MME of a visiting network. In addition, the HSS records the locations of the users using the address of the MME in the visiting network.
- *PCRF* is a CP network element combining Charging Rules Function and Policy Decision Function. It is in charge of policy control decision-making and controlling the flow-based charging functionalities in the PCEF. The PCRF provides the QoS authorization that decides how a data flow will be served in the PCEF and ensures that this is in accordance with the user's subscription profile.
- *PCEF* is a UP network element locating in the P-GW. The PCEF enforces the policies that have been statically set or dynamically configured by the PCRF. It ensures that data flows are processed in accordance with network policy and services are charged accordingly.

2.4.2 Key Technologies

The essential requirements for the LTE development were high spectral efficiency and spectrum flexibility. As the foundation of LTE, MIMO and OFDM, also jointly referred to as MIMO-OFDM, were adopted as the basic transmission scheme. To meet new expectations, 3GPP continuously added enhancement into the initial LTE standard under the restriction of backward compatibility. To ensure that the LTE system fully complies with the IMT-Advanced requirements, LTE-Advanced improved spectrum flexibility through Carrier Aggregation (CA), enhanced multi-antenna transmission, added the support for relaying and Heterogeneous Network (HetNet) deployment. Release 11, finalized in late 2012, further enhanced the capabilities of LTE. The most notable feature of Release 11 was the enhancements of radio-interface functionalities for Coordinated Multi-Point (CoMP) transmission and reception. The focus of Release 12 froze in 2014 was on new scenarios with the introduction of direct Device-to-Device (D2D) communications and low-complexity devices for machine-type communications. Release 13, which marked the start of a significant evolution called *LTE-Advanced Pro*, adopted License-Assisted Access (LAA) to support unlicensed spectra as a complement to licensed spectra. The critical technological enablers are briefly reviewed as follows:

- *Multi-Input Multi-Output*: The great potential demonstrated by the vertical Bell laboratories layered space-time architecture (V-BLAST) multi-antenna wireless system in the late 1990s attracted tremendous interest in MIMO technologies. The use of multiple antennas allows the exploitation of spatial domain as another degree of freedom for higher spectral efficiency. In rich-scattering propagation environments, the theoretical spectral efficiency scales linearly with the minimum of the number of transmit and receive antennas. MIMO can bring three benefits:
- *Spatial Multiplexing*: Transmission of multiple data streams simultaneously at the same frequency by means of multiple spatial layers created by multiple antennas. As a result, higher spectral efficiency is obtained.
- *Spatial Diversity*: Use of independent propagation paths enabled by multiple antennas to improve the reliability of the signal transmission against the effect of multiple fading.
- *Array Gain*: Concentrate the transmit energy of an antenna array in particular directions to improve the received power of desired signals and suppress co-channel interference.

The MIMO technologies acted as one of the cornerstones of the LTE radio transmission. Due to the constraints on hardware, cost, and energy consumption, the deployment of multiple antennas at the base station is more attractive and practical. There are a variety of multi-antenna transmission schemes implemented in LTE, i.e.

- *Transmit Diversity*: This MIMO scheme transmits a single layer stream over multiple low-correlation signal paths. Spatial diversity is obtained by means of sufficiently large inter-antenna spacing or different antenna polarization. It is valuable for high-reliable scenarios such as control and broadcast channels. Transmit diversity in LTE was only specified for two or four transmit antennas. The transmit symbols can be encoded flexibly using Space-Time Block Coding (STBC), Space-Frequency Block Coding (SFBC), Frequency-Switched Transmit Diversity (FSTD), or Cyclic Delay Diversity (CDD).
- *Beamforming*: If the knowledge of the downlink channels is available, the transmitter can provide, in addition to transmit diversity, beamforming to concentrate the transmit energy toward particular directions. It can increase the strength of the received signal with a factor in proportion to the number of transmit antennas and accordingly suppress the interference in other directions. This scheme can be implemented on either high-correlation antenna arrays in the form of *classical beamforming* or low-correlation arrays as applying *transmit precoding*.
- *Open-Loop Spatial Multiplexing*: This MIMO scheme used within the LTE system transmits two data streams over two or more antennas. There is no channel information from the UE except a Transmit Rank Indicator (TRI) used to determine the number of spatial layers.
- *Closed-Loop Spatial Multiplexing*: Implicit Channel State Information (CSI) such as Precoding Matrix Indicator (PMI) is fed back from a UE to an eNodeB to select the most desirable precoding matrix. A codebook consists of a set of predefined precoding matrices is known by both the transmitter and receiver. It maximizes the transmission capacity by adapting the precoded symbols to the condition of the channel and enables the receiver to differentiate the data streams efficiently.
- *Multi-User Multiple-Input Multiple-Output (MU-MIMO)*: This scheme can improve the network capacity and system flexibility by serving multiple users simultaneously at the same frequency, at the cost of more complex signal processing. The early release of LTE focused on transmit diversity and Single-User Multiple-Input Multiple-Output (SU-MIMO). The basic support for MU-MIMO was also provided in Release 8, but it adopted the same codebook-based scheme for SU-MIMO using implicit CSI. The enhanced support for MU-MIMO was added in later releases by the introduction of UE-specific reference signals. Explicit CSI feedback removes the limit of the codebook and provides flexibility to apply advanced transmission schemes such as zero-forcing precoding and conjugate beamforming.

2.4.2.1 Orthogonal Frequency-Division Multiplexing

ISI raised by the time dispersion of multi-path channels is the most critical impairment to limit high-speed transmission. The complexity of the traditional equalization technique in the narrow-band signal transmission becomes unacceptable because the number of taps of frequency-selective fading channels can be several hundred. Another primary feature of LTE is the adoption of multi-carrier transmission as the basic modulation technique. Dividing a high-rate data stream into a large number of parallel low-rate streams, the symbol duration on each subcarrier significantly increases to far exceed the channel delay spread.

The major advantages of OFDM include:

- *High Transmission Rate* by dividing the wide bandwidth into a multitude of narrow-band subcarriers and adding a guard interval referred to as Cyclic Prefix (CP) at the beginning of each symbol, the effect of ISI is effectively suppressed.
- *Low-Complexity Receiver* by applying frequency-domain equalization to compensate for channel impairments individually on each subcarrier.
- *Flexibility* on simply combining with spatial signal processing resulting in the ease of MIMO-OFDM transmission.

However, an OFDM signal suffers from high Peak-to-Average Power Ratio (PAPR), leading to a need for a highly linear power amplifier. The uplink transmission is hard to tolerate the high PAPR of OFDM since the mobile terminal needs to compromise between the output power required for good outdoor coverage, the power consumption, and the cost of the power amplifier. Therefore, a variant of OFDM, referred to as Discrete Fourier Transform (DFT) spread Orthogonal Frequency Division Multiplexing (DFT-s-OFDM), was adopted to provide a low-PAPR option for the uplink transmission since it is in essence a single-carrier signal.

A multitude of orthogonal subcarriers, which carry independent data streams either individually or in groups, enables a very efficient multiple-access scheme. OFDMA extends the multi-carrier technology of OFDM to provide orthogonal time-frequency resources among multiple users. This scheme exploits the frequency domain as another degree of freedom, which improves the system flexibility in various ways:

- Scalable system bandwidths can be supported without changing the fundamental system parameters or equipment design. It substantially improves the deployment flexibility in smaller or fragmented bands and the smooth expansion of system capacity.
- Agile allocation of time-frequency resources to different users and frequency-domain scheduling to exploit the frequency-diversity gain.
- Facilitating fractional or soft frequency reuse and inter-cell interference coordination (ICIC).

After the initial evaluation of proposals, the candidate schemes for the downlink of the LTE air interface were OFDMA and Multi-Carrier CDMA (MC-CDMA), while the candidate schemes for the uplink were SC-FDMA, OFDMA, and MC-CDMA. In December 2005, 3GPP RAN plenary meeting made the final choice of multiple-access schemes, with OFDMA for the downlink and SC-FDMA for the uplink.

2.4.2.2 Carrier Aggregation

The development of LTE-Advanced aimed to meet the IMT-Advanced require-ments to support the maximal bandwidth of at least 40 MHz and a peak data rate of 1 Gbps. However, such large portions of the continuous spectrum are not available in most cases due to the high competition of spectrum utilization and the fragmen-tation of the existing spectrum allocation. Therefore, LTE-Advanced makes use of CA to implement such large bandwidths. Up to five component carriers, possibly each of different bandwidths, can be aggregated to support a maximal transmission bandwidth of 100 MHz for a single terminal. Each aggregated carrier was designed based on an LTE Release 8 structure so that LTE-Advanced can be configured in a backward-compatibility way to support legacy UEs. An LTE-Advanced termi-nal can simultaneously utilize multiple component carriers to achieve higher data rates, while an LTE terminal, on the other hand, can be served transparently on a single component carrier.

The aggregation approaches can be classified into three categories:

- *Intra-Band Contiguous*: The simplest approach to arrange aggregation is to use contiguous component carriers within the same frequency band. Contiguous spectrum can save the spectral resources used as guard bands and can apply a single base-band processing chain if the bandwidth of the RF transceiver is wide enough. However, this case is usually not possible due to the fragmentation of frequency allocation.
- *Intra-Band Non-Contiguous*: If contiguous component carriers are not manda-tory, fragmented frequency bands can be exploited. LTE-Advanced supports intra-band non-contiguous CA, where component carriers belong to the same frequency band but have a gap, or gaps, in between.
- *Inter-Band Non-Contiguous*: The component carriers are not only fragmented but also belonging to different operating frequency bands. Discontinuous aggre-gation has the advantage of frequency diversity since different frequency bands correspond to different channel fading. But it requires several independent RF and base-band processing chains.

2.4.2.3 Relaying

The high possibility of a terminal transmitting with higher data rates requires a rel-atively high signal-to-noise ratio. The most influential factor for link performance is path loss, which depends heavily on the propagation distance. Unless the link budget can be increased, for example, using higher transmit power or deploying antenna arrays for beamforming, a denser infrastructure is required to reduce the propagation distance. One of the effective approaches is relaying, which can reduce the distance between the terminal and the infrastructure and thereby improve the link budget. A basic requirement for the relaying deployment is that the relay nodes should be transparent to the terminals. That is, the relay node acts as an ordinary low-power base station from a terminal perspective. It is an important feature to simplify the terminal implementation and making the relay node backward compatible. On the other hand, the relay node can access the infrastructure to get LTE-based wireless backhaul like a common LTE terminal. Accordingly, the connection between eNodeB-relay and relay-UE are termed *backhaul link* and *access link*, respectively. The relaying can be classified into *inband* and *outband* in terms of the frequency bands used for backhaul and access links. Release 10 added the support for *Decode-and-forward* relaying in LTE-Advanced networks while another simple scheme called *Amplify-and-forward* relaying does not require additional standardization.

- *Amplify-and-forward* relays, commonly referred to as repeaters, amplify and forward the received analog signals and are, on some markets, relatively common as a tool for handling coverage holes. Repeaters are transparent to both the terminal and the base station and can therefore be introduced in existing networks. The basic principle of a repeater is to amplify whatever it receives, including noise and interference and the useful signal, implying that repeaters are mainly useful in high-SNR environments.
- *Decode-and-forward* relays decode and re-encode the received signal before forwarding it to the served users. The decode-and-re-encode process results in this class of relays not amplifying noise and interference. They are therefore also useful in low-SNR environments. However, the decode-and-forward operation induces a considerable delay compared with an amplify-and-forward repeater.

2.4.2.4 Heterogeneous Network

The demand of providing high data rates to a large number of mobile users shifted the network architecture. In cellular networks, high system capacity can be achieved by employing advanced radio transmission techniques to improve spectral efficiency, acquiring more spectral resources, or deploying denser network nodes. In a traditional homogeneous network, cell splitting is applied to reduce the cell size for a higher capacity, but network re-planning and system re-configuration are required. Meanwhile, the traffic demand is not expected to be uniform. Therefore, LTE-Advanced introduced HetNet, a mixture of the cells with different sizes, transmit powers, coverage, and hardware capabilities, i.e. macro-, micro-, and micro-, pico-, and femto-cell. Low-power, low-cost base stations are added to the macro-cell networks as underlying nodes to provide high capacity in hot spots or fill coverage gaps – both indoors and outdoors. If the system has enough radio resources to allocate different frequency bands for different cell types, there is no mutual interference. However, with the increased traffic demand, there is no room to allocate a dedicated carrier for small cells. Heterogeneous cells have to mutually overlap in a geographical area, operating on the same set of frequencies. The most challenging aspect in such heterogeneous deployment is to deal with ICI. As a consequence, 3GPP set up a work item on the ICIC and later enhanced inter-cell interference coordination (eICIC) to address this issue. LTE-Advance also makes use of CA to support the deployment of a HetNet, where *cross-carrier scheduling* is applied to enable control signaling to be transmitted on one common component carrier so that the interference on control channels between the layer of macro-cells and the layer of small cells is avoided.

2.4.2.5 Coordinated Multi-Point Transmission and Reception

Mobile users expect to get high QoS anywhere at any time. However, the users at the cell edge suffer from not only high signal attenuation due to longer propagation distances but also strong ICI. Over the years, several different techniques such as interference averaging, frequency hopping, and interference coordination have been proposed to mitigate ICI. An example is ICIC or its enhanced version eICIC, which was introduced by LTE-Advanced as a technical feature. A tighter interference coordination referred to as CoMP transmission and reception was also considered by 3GPP as a tool to improve cell-edge spectral efficiency. The coordination among multiple geographically separated sites can be performed either in the downlink or in the uplink, resulting in different level of coordination:

- *Joint Processing*: Multiple sites simultaneously transmit or receive the signals to and from a single UE located in the coordination area. It improves the received signal quality, suppresses or avoids ICI, and gets macro-diversity gain. However, this method imposes high requirements on the backhaul network since the transmitted or received data, channel knowledge, and computed transmission weights are required to be exchanged among coordinated sites.
- *Coordinated Scheduling or Beamforming*: Only a single site is selected to communicate with a typical user. The user data do not need to be made available at multiple coordinated sites, whereas only control signaling on scheduling decisions or generated beams are needed. It simplifies the implementation since the requirements on the backhaul network are much lowered.

2.4.2.6 Device-to-Device Communications

In a traditional mobile network, the information is first sent to a base station in the uplink and then forwarded to the target terminal in the downlink, even if two communicating parties are close to each other. As we know, the performance of the transmission link heavily depends on the propagation distance; namely, a shorter distance gains a high signal-to-noise ratio. The emergence of new application scenarios such as content distribution and Proximity-based Service (ProSe) defined in 3GPP Release 12 fostered the application of D2D communication in cellular networks. Direct communication in such a scenario has great potential, such as high spectral efficiency, high system capacity, high energy efficiency, low latency, and fairness. D2D communication is usually not transparent to the cellular network, and it can operate on the licensed spectrum (i.e. inband) or license-exempt spectrum (i.e. outband).

- *In-Band*: It refers to both D2D and cellular links utilize the cellular spectrum. Inband D2D communication is further categorized into two types – underlay and overlay – in terms of the used frequencies in D2D and cellular links. The underlay type can improve the spectral efficiency of a cellular network by reusing the same spectrum in both links. The overlay type needs dedicated spectral resources to D2D links that directly connect the transmitter and the receiver. The key disadvantage of inband D2D communications is the interference caused by D2D users to cellular communications and vice versa.
- *Out-Band*: The motivation behind it is to exploit the unlicensed spectrum and eliminate the interference between D2D and cellular links. Using an unlicensed spectrum requires a different interface and usually adopts other wireless technologies such as Wi-Fi and Bluetooth. Although the use of unlicensed spectrum avoids the in-band interference, it may suffer from the uncontrolled interference on unlicensed spectra.

2.4.2.7 License-Assisted Access

LAA is a technical feature of LTE, introduced in Release 13 as part of LTE Advanced Pro. The basic idea is to make use of unlicensed spectra in combination with licensed spectra to deliver higher data rates to UEs and better user experiences. With CA, a secondary carrier on the unlicensed spectrum of 5 GHz is applied to offload the traffic of the primary carrier in the licensed band. Consumers can leverage the combination of licensed and unlicensed bands to achieve higher peak rates indoor and outdoor. A major challenge of LAA is the interoperability between 3GPP and non-3GPP technologies sharing the same unlicensed spectrum. LAA incorporates some mechanisms, such as channel sensing and Listen-before-talk (LBT) protocol, to manage co-channel and adjacent channel interference. Since Wi-Fi is a popular system operating in the 5 GHz unlicensed band, LAA needs to select clear channels that Wi-Fi does not occupy dynamically. If there is no white space available in a geographical area, LTE LAA can still contend with other unlicensed users fairly to share a channel.

2.5 5G – New Radio

As early as 2015, 3GPP RAN group decided to set up a study item for New Radio (NR) in Release 14 and initiated the task on channel modeling for frequency bands above 6 GHz. The initial specification was carried out through a work item in Release 15. To meet commercial requirements on early large-scale trials and deployments in 2018, earlier than the initially envisaged timeline of around 2020, 3GPP committed to accelerating the process, agreeing that a Non-Standalone (NSA) variant would be finalized earlier. At the end of 2017, the first version of NR specifications was available. The world's first NSA NR call was jointly completed in Spain by Vodafone and Huawei, just ahead of the Mobile World Congress, which started on 26 February 2018. After the initial delivery of NSA, much effort of 3GPP was shifted on timely completion of Release 15 to form the first complete set of NR standards. Therefore, 3GPP also developed a new CN referred to as the 5G Core (5GC) network in parallel to the NR radio-access technology. In June 2018, the final version of Release 15 that can support the Standalone (SA) operation of NR was available. The focus of Release 15 was primarily on eMBB (enhanced Mobile Broadband) and (to some extent) URLLC (Ultra Reliable Low Latency Communications), while mMTC (massive Machine Type Communications) was still supported by using LTE-based machine-type communication technologies such as eMTC and NB-IoT. Release 15 provided the foundation on which 3GPP continues its work to evolve the capability and functionality of 5G so as to support new spectrum and new applications and further enhance existing core features. The evolution of NR continued in Release 16, which was completed in June 2020. Technical features were added into NR to support Industrial Internet of Things (IIoT) and enhance URLLC applications. This release aimed to meet the IMT-2020 requirements and, along with Release 15, acted as the initial complete 3GPP specifications submitted to the ITU-R. The 3GPP final proposal included two separate and independent submissions, defined as the single Radio Interface Technology (RIT) and the combined Sets of Radio Interface Technologies (SRIT). In November 2020, the ITU-R announced that 3GPP 5G-SRIT and 3GPP 5G-RIT conform with the IMT-2020 vision and stringent performance requirements.

Compared with LTE, NR offers many advantages:

- Exploitation of millimeter-wave frequency bands to obtain sufficient spectral resources and wide transmission bandwidth for extremely high data rates. NR can operate in the frequency range up to 100 GHz with heterogeneous deployment: macro base stations at lower carrier frequencies and small base stations at higher carrier frequencies.
- Ultra-lean design to significantly improve network energy efficiency and reduce interference in high traffic load conditions. Synchronization signals, system

broadcast information, and reference signals are only transmitted when necessary, instead of *always on* transmissions in the previous generations.

- Forward compatibility to prepare for further enhancements to support future yet unknown use cases and applications. It is achieved by means of self-containment and well-confined transmission. Self-containment refers to that data in a slot and a beam is detectable without dependency on other slots and beams. Well-confined transmission means keeping transmissions confined in the frequency and time domains to allow future inclusion of new types of transmissions in parallel with legacy transmission [Zaidi et al., 2017].
- System flexibility adapting to a wide range of carrier frequencies, heterogeneous deployment (macro, micro, and pico-cells), and diverse usage scenarios (eMBB, URLLC, and mMTC) with stringent and sometimes contradictory requirements. The physical-layer design of NR is flexible and scalable, including highly adaptive modulation schemes (from Binary Phase-Shift Keying [BPSK] in the UL of mMTC to 1024QAM in the downlink of eMBB), a scalable OFDM numerology, Low-Density Parity-Check (LDPC) codes with rate-compatible structure, and flexible frame structure.
- Beam-centric design to enable the extensive usage of Massive Multiple-Input Multiple-output (massive MIMO) for not only data transmission but also control signaling. Reference signals can be beamformed with configurable granularity in the time domain and frequency domain.

2.5.1 System Architecture

In contrast to previous generations, the 3GPP 5G system adopted the service-based architecture (SBA) that organizes the architectural elements into a set of service-oriented network functions (NFs). The interaction between NFs is represented in two ways:

- *Service-based representation*, where NFs enables other authorized NFs to access their services via the interfaces of a common framework.
- *Reference point representation*, where the interaction between any two NFs is described by point-to-point reference point.

The 5G System architecture is designed to support a wide variety of use cases with stringent but sometimes conflicting performance requirements. It enabling deployments to use techniques such as Network Function Virtualization (NFV), Software-Defined Networking (SDN), and Network Slicing. Some key principles and concept [3GPP TS23.501, 2021] are to:

- Separate the UP functions from the Control Plane functions, allowing independent scalability, evolution and flexible deployments.

- Modularize the function design to enable flexible and efficient network slicing. Wherever applicable, defining procedures (i.e. the set of interactions among NFs) as services, to maximize the re-usability.
- Enable each NF and its NF services to interact with other NF directly or indirectly via a service communication proxy if required.
- Minimize dependencies between the Access Network (AN) and the CN. The architecture is defined with a converged CN with a common AN – CN interface integrating both 3GPP and non-3GPP access technologies.
- Support a unified authentication framework.
- Support stateless NFs, where the computing resource is decoupled from the storage resource.
- Support capability exposure.
- Support concurrent access to local and centralized services. To support low-latency services and access to local data networks, UP functions can be deployed close to the access network.
- Support roaming with both home routed traffic as well as local breakout traffic in the visited PLMN.

2.5.1.1 5G Core Network

Some of the functionalities for the 5G CN look similar to that of the previous generations. It is not surprising, as the network always has to carry out some basic functions, such as authentication, charging, resource allocation, and mobility management. However, we also recognize some significant functions not present before, which are needed to enable new network paradigms, like network slicing, service-based networking, and user/control plane split. To facilitate the enablement of different data services and requirements, the NFs have been further simplified, with most of them being software-based so that they can run on generic computer hardware. The service-based representation of the 5G System architecture is shown in Figure 2.6, consisting mainly of the following NFs:

- *Application Function (AF)*: application influence on traffic routing, and interaction with policy framework for policy control.
- *Access and Mobility Management Function (AMF)*: Termination of Non-Access Stratum (NAS) signaling, NAS ciphering and integrity protection, registration management, connection management, mobility management, access authentication and authorization, and security context management.
- Authentication Server Function (AUSF) acts as an authentication server.
- *Network Exposure Function (NEF)*: exposure of capabilities and events, secure provision of information from external application to 3GPP network, translation of internal or external information.

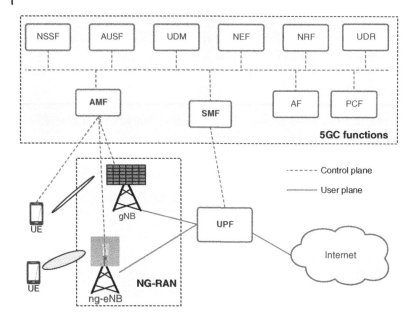

Figure 2.6 Architecture of the 5G SA system. 5GC, 5GC network; AF, Application Function; AUSF, Authentication Server Function; AMF, Access and Mobility Management Function; gNB, next-generation Node B; IoT, Internet of Things; SDSF, Structured Data Storage network Function; UDSF, Unstructured Data Storage network Function; NEF, Network Exposure Function; NRF, NF Repository Function; PCF, Policy Control Function; SMF, Session Management Function; UDM, Unified Data Management; UPF, User Plane Function. This figure only shows some typical network functions rather than an exhaustive list of 5GC network functions. Source: Secureroot/FreeImages.

- *NF Repository Function (NRF)*: service discovery function, maintains NF profile and available NF instances.
- Network Slice Admission Control Function (NSACF).
- Network Slice Specific Authentication and Authorization Function (NSSAAF).
- *Network Slice Selection Function (NSSF)*: selecting of the network slice instances to serve the UE, determining the allowed Network Slice Selection Assistance Information (NSSAI), determining the AMF set to be used to serve the UE.
- Network Data Analytics Function (NWDAF).
- *Policy Control Function (PCF)*: unified policy framework, providing policy rules to CP functions, access subscription information for policy decisions.
- *Session Management Function (SMF)*: session management (session establishment, modification, and release), IP address allocation and management, DHCP functions, termination of NAS signaling related to session management, downlink data notification, traffic steering configuration for User Plane Function (UPF) for proper traffic routing.

- *Unified Data Management (UDM)*: generation of Authentication and Key Agreement (AKA) credentials, user identification handling, access authorization, and subscription management.
- Unified Data Repository (UDR)
- *UPF*: packet routing and forwarding, packet inspection, QoS handling, acts as external Protocol Data Unit (PDU) session point of interconnect to data network, and is an anchor point for intra- and inter-Radio Access Technology (RAT) mobility.
- UE radio Capability Management Function (UCMF)
- Unstructured Data Storage Function (UDSF)

2.5.1.2   Next Generation Radio Access Network

Next Generation Radio Access Network (NG-RAN) is the radio-access part of the 5G system. The NG-RAN consists of a set of NG-RAN nodes connected to the 5GC. The node can be either a gNodeB (gNB) or a next-generation eNodeB (ng-eNB):

- gNB providing NR user-plane and control-plane protocols
- ng-eNB providing E-UTRA user-plane and control-plane protocols

The gNB (or ng-eNB) is responsible for all radio signal processing and some radio-related control in one or several cells. The gNBs and ng-eNBs are connected to the 5GC, more specifically to the AMF for control functions and to the UPF for the user data transmission. According to the specification of 3GPP TS 23.501 [3GPP TS23.501, 2021], the functionalities for the nodes or NFs related to NG-RAN are summarized as follows.

The gNB and ng-eNB host the following functions:

- *Functions for Radio Resource Management*: radio bearer control, radio admission control, connection mobility control, dynamic allocation of resources to UEs in both uplink and downlink (scheduling);
- IP and Ethernet header compression, encryption, and integrity protection of data;
- Selection of an AMF at UE attachment when no routing to an AMF can be determined from the information provided by the UE;
- Routing of UP data toward UPF(s);
- Routing of Control Plane information toward AMF;
- Connection setup and release;
- Scheduling and transmission of paging messages;
- Scheduling and transmission of system broadcast information (originated from the AMF or OAM);
- Measurement and measurement reporting configuration for mobility and scheduling;

- Transport level packet marking in the uplink;
- Session management;
- Support of network slicing;
- QoS flow management and mapping to data radio bearers;
- Support of UEs in RRC_INACTIVE state;
- Distribution function for NAS messages;
- Radio access network sharing;
- Dual connectivity;
- Tight interworking between NR and E-UTRA;
- Maintain security and radio configuration for UP CIoT 5GS optimization.

The AMF hosts the following main functions:

- NAS signaling termination;
- NAS signaling security;
- AS security control;
- Inter CN node signaling for mobility between 3GPP access networks;
- Idle mode UE reachability (including control and execution of paging retransmission);
- Registration area management;
- Support of intra-system and inter-system mobility;
- Access authentication;
- Access authorization including check of roaming rights;
- Mobility management control (subscription and policies);
- Support of network slicing;
- SMF selection.
- Selection of CIoT 5GS optimizations;

The UPF hosts the following main functions:

- Anchor point for Intra-/Inter-RAT mobility (when applicable);
- External PDU session point of interconnect to data network;
- Packet routing and forwarding;
- Packet inspection and user plane part of policy rule enforcement;
- Traffic usage reporting;
- Uplink classifier to support routing traffic flows to a data network;
- Branching point to support multi-homed PDU session;
- QoS handling for user plane, e.g. packet filtering, gating, UL/DL rate enforcement;
- Uplink traffic verification (SDF to QoS flow mapping);
- Downlink packet buffering and downlink data notification triggering.

The SMF hosts the following main functions:

- Session management;
- UE IP address allocation and management;

- Selection and control of UP function;
- Configures traffic steering at UPF to route traffic to proper destination;
- Control part of policy enforcement and QoS;
- Downlink data notification.

2.5.2 Key Technologies

To meet the stringent performance requirements defined in IMT-2020, revolutionary technologies in both radio access and networking were applied for the 5G system. The major technological breakthroughs include massive MIMO, millimeter wave (mmWave) communications, non-orthogonal multiple access, polar codes, NF virtualization, software-defined networking, and network slicing.

2.5.2.1 Massive MIMO

Massive MIMO, also known as large-scale antenna array, very large MIMO, hyper MIMO, full-dimension MIMO, is a key component of NR. With a large number of antennas, the transmission energy can be directed with extremely sharpness into a very small area. Directivity can bring huge improvements in spectral efficiency and energy efficiency. Massive MIMO is cost-efficient by leveraging low-cost, low-precision RF components, where expensive, high-linear power amplifiers used for conventional systems can be replaced by hundreds of cheap power amplifiers with output power in the magnitude order of milli-Watt. Other benefits of massive MIMO include reduced latency on the air interface, simplification of the multiple-access layer, and increased robustness against both unintended interference and intentional jamming [Larsson et al., 2014].

For lower frequencies, NR employs a low to moderate number of antennas (up to 64 transmit and receive antennas at the base station side around 700 MHz) [3GPP TR38.913, 2020]. FDD operation can be supported in this case, where the acquisition of CSI requires transmission of Channel State Information Reference Signal (CSI-RS) in the downlink and CSI reporting in the uplink. The limited bandwidths available in this frequency region require high spectral efficiency enabled by MU-MIMO and higher order spatial multiplexing, which is achieved via higher resolution CSI reporting compared with LTE. For higher frequencies, a larger number of antennas (NR supports up to 256 transmit and receive antennas around 4 GHz) can be employed with the same hardware size, which increases the capability for beamforming and MU-MIMO. Since the number of reference signals is proportional with the number of transmit antennas, massive MIMO has to operate in the TDD mode by exploiting channel reciprocity. In this case, the base station acquire the CSI of the downlink by estimating the channel sounding reference signals in the uplink. In the downlink data transmission, there is no reference signals and some forms of precoding schemes such as conjugate beamforming and zero-forcing precoding are applied to simplify the signal reception at the UE side. For even higher frequencies (in the mmWave range) an analog beamforming implementation is typically required currently, which limits the transmission to a single beam direction per time unit and radio chain. Since an isotropic antenna element is very small in this frequency region owing to the short carrier wavelength, a great number of antenna elements is required to maintain coverage. Beamforming needs to be applied at both the transmitter and receiver ends to combat the increased path loss, even for control channel transmission. A new type of beam management process for CSI acquisition is required, in which the BS needs to sweep radio transmitter beam candidates sequentially in time, and the UE needs to maintain a proper radio receiver beam to enable reception of the selected transmitter beam. In the frequency bands around 30 GHz and 70 GHz, NR supports up to 256 antenna elements at the base station side and up to 32 antenna elements at the UE side.

2.5.2.2 Millimeter Wave

Previous generations of cellular systems usually operated in low-frequency bands ranging from several hundred megahertz to several gigahertz. On the one hand, spectral resources over these bands are pretty limited compared with the demand for mobile broadband. On the other hand, there are a wide variety of applications operating in these bands, such as television broadcasting, satellite communications, radar, radio astronomy, and maritime navigation, leading to a global bandwidth shortage. It motivated mobile network operators to explore the underutilized mmWave spectrum to offer mobile broadband services. Millimeter wave, also known as millimeter band, refers to electromagnetic spectrum with wavelengths between 10 mm (equivalent to 30 GHz) and 1 mm (300 GHz). It was envisaged that mobile communications could benefit from mmWave by a set of use cases, e.g. low-cost fiber replacement for mobile backhaul, dense mmWave small cells, wireless broadband access, and low-latency uncompressed high-definition media transfers [Rappaport et al., 2013]. 3GPP defined the relevant spectrum for NR, which was divided into two frequency ranges: the First Frequency Range (FR1), which includes sub-6 GHz frequency bands extending from 450 MHz to 6 GHz, and the Second Frequency Range (FR2) covering 24.25–52.6 GHz. Initial mmWave deployments are expected in 28 GHz (3GPP NR band n257 and n261) and 39 GHz (3GPP n260), followed by 26 GHz (3GPP n258). More mmWave bands are expected to be opened with the rising demand for NR services. However, its practical rollout in mobile networks imposes significant technical challenges for designing and developing RF and antenna components. An mmWave signal suffers from atmospheric losses due to water vapor and oxygen absorption that can easily exceed the usual free-space losses. In addition, mmWave signals cannot generally penetrate solid materials such as reinforced concrete walls. In order to compensate for such significantly large propagation losses, multi-element antenna arrays are required at both the base station and UE sides to focus the transmission energy into a small region. Moreover, extremely high data throughput puts the necessity of applying broader transmission bandwidths. The requirements of supporting up to 400 MHz in a single carrier and more than 1 GHz in CA make the implementation of RF and antenna elements much challenging.

2.5.2.3 Non-Orthogonal Multiple Access

Multiple access refers to a technique that allows multiple users to share radio resources as a fundamental component of a cellular communication system. Over the past few decades, cellular systems have witnessed a radical evolution in their multiple access schemes. In particular, FDMA, TDMA, code-division multiple access, and OFDMA have been adopted for 1G–4G cellular networks. These schemes belong to the category of Orthogonal Multiple Access (OMA), where each user transmits or receives a user-specific signal over an orthogonal radio-resource unit in the frequency-, time-, code-domain, or their combinations. OMA has been the primary choice for previous generations of cellular communications since it simplifies the transceiver design and alleviates multi-user interference. However, the pool of radio resources restricts the system capacity in terms of the maximal number of active users.

In contrast to OMA, Non-Orthogonal Multiple Access (NOMA) allows multiple users to share the same radio-resource unit, improving system capacity and connection densities. It can be implemented by means of multi-user interference cancelation with the price of higher computational complexity at the receiver. As a special case of the NOMA technique, Multi-User Superposed Transmission (MUST) has been studied in LTE Release 13, mainly focusing on the downlink transmission [Chen et al., 2018]. According to adaptive power control and bit-labeling at the transmitter side, the MUST schemes are divided into three categories. In particular, Category 1 independently maps the coded bits of two or more co-scheduled users to component constellation symbols without Gray mapping, Category 2 jointly maps the coded bits of two or more co-scheduled users to component constellations with Gray mapping, and Category 3 directly maps the coded bits onto the symbols of a composite constellation. During a study item in Release 14, different NOMA schemes, such as Sparse Code Multiple Access (SCMA), Multi-User Shared Access (MUSA), Pattern Division Multiple Access (PDMA), and Resource Spread Multiple Access (RSMA), have been proposed. Grant-based NOMA, which typically operates in Radio Resource Control (RRC) connected state, was specified in Release 14 LTE to support downlink eMBB. In 3GPP Release 15, a study item was established to continue studying signal processing at the transmitter side, multi-user receiver design, complexity analysis, and NOMA-related procedures such as hybrid Automatic Repeat Request

(ARQ), link adaptation, and power allocation. Since the enormous performance gain brought by massive MIMO in NR, applying NOMA in the NR downlink can only bring a marginal gain. Therefore, the focus of NOMA study item in Release 16 was shifted on uplink grant-free transmission, which is expected to reduce the control signaling overhead, transmission latency, and devices' power consumption.

2.5.2.4 SDN/NFV

SDN is a network paradigm that decouples the control plane from the data forwarding function, under the development of a non-profit operator-led consortium called Open Networking Foundation (ONF). Network control can then be centralized in the SDN controller, and the underlying infrastructure is abstracted to become a pool of forwarding elements. The SDN controller contains a global view of the whole network and makes the network control such as routing, congestion control, traffic engineering, and security inspection directly programmable. SDN does not directly address any technical challenges of network control, but it opens new opportunities to create and deploy innovative solutions to these problems by exposing the network as a service to SDN applications over the northbound interface. The SDN controller interprets the instructions from SDN applications into specific configuration commands for the underlying infrastructure through the southbound interface using the OpenFlow protocol. SDN brings several technical benefits, including

- Centralized control of multi-vendor network elements;
- Network automation and programmability by abstracting the underlying infrastructure and exposing it through a standard interface;
- Rapid innovation through deploying new network applications and services without the need to configure specific elements or wait for vendor releases;
- Increased network reliability and security due to centralized and automated management of network elements, uniform policy enforcement, and fewer configuration errors.

NFV is another network paradigm that decouples software from hardware, aiming to transform the way of network deployment. NFV allows network operators to deploy NFs as virtualized software instances to replace dedicated hardware appliances. Virtualized NFs can run on standard general-purpose high-volume servers and migrate among various locations on-demand without the need to install new equipment. NFV offers many benefits, including:

- Low equipment costs and low power consumption through exploiting the economy of scale of the IT industry;
- Faster time-to-market of new services;

- Flexibility with elastic scale up and scale down of network capacity;
- Multi-tenancy, which allows the sharing of a single platform for different applications, users, and tenants;
- Enables boarder independent software eco-systems and encourages openness.

Regardless of their independent development under different organizations (ETSI and ONF), SDN and NFV complement each other. On the one hand, SDN controller and SDN applications can be implemented as typical Virtual Network Functions (VNFs) and deployed in standard IT platforms. Under the unified control of NFV Management and orchestration (MANO), SDN-related software instances can flexibly instantiate, scale, migrate, update, and deploy in the virtualized infrastructure. On the other hand, NFV can take advantage of network programmability enabled by SDN to implement various NFs.

2.5.2.5 Network Slicing

Network slicing refers to a set of technologies to create specialized, dedicated logical networks over a shared physical network. Through the customized design of functions, isolation mechanisms, and management tools, network slicing is capable of providing Network-as-a-Service (NaaS) to meet diversified requirements from vertical industries. Following the Service Level Agreement (SLA) between a mobile operator and a customer, a self-contained virtual network, referred to as a network slice, is instantiated. A network slice is an independent end-to-end logical network with its virtual resources, topology, traffic flow, and provisioning rules, but running on shared physical infrastructure. With scalable resource allocation and flexible configuration, a network slice can offer customized network capabilities such as data throughput, coverage, QoS, latency, reliability, security, and availability. There are various types of network slices to meet the specific communication needs of different users. Major concepts for network slicing are as follows:

- *Network Slice Instance*: A network slice instance is a set of shared or dedicated NFs and physical or virtual resources to run these NFs. It forms a complete instantiated logical network to meet specific network characteristics such as ultra-reliability and ultra-low latency. An instance typically covers multiple technical domains, including terminal, access network, transport network, CN, and data center that hosts third-party applications from vertical industries.
- *Network Slice Type*: Network slice types are high-level categories for network slice instances, reflecting the distinguished demands for network solutions. Three basic types have been identified for 5G, namely, eMBB, URLLC, and mMTC. This shortlist could be further extended according to the demand or the evolution of 5G.

- *Tenant*: Tenants are customers of network slices (e.g. vertical industries) or network operators themselves. They utilize network slice instances to provide services to their users. As a result, tenants typically have independent operation and management policies, which are uniquely applicable to their network slice instances.

2.5.2.6 Polar Codes

Approaching the limit of Shannon capacity with a practical complexity is the major challenge in the coding theory of digital communications over the past several decades. With the introduction of coding randomness by interleaving the information bits at the encoder, Turbo codes can achieve near-optimal performance with reasonable complexity. It has been widely applied in 3G and 4G cellular systems such as WCDMA, CDMA2000, and LTE. Similarly, LDPC codes realized coding randomness employing pseudo-random connections between the variable and check nodes. Given its excellent performance, LDPC was also successfully adopted by WiMAX in IEEE 802.16e and IEEE 802.16m specifications. In 2009, Arikan invented a new coding scheme called polar codes, which opened a new frontier of constructing error-correcting codes to achieve the Shannon capacity [Arikan, 2009]. It relies on an elegant phenomenon, called channel polarization, which can be regarded as a Matthew effect in the digital world (*the rich gets richer and the poor gets poorer*). Channel polarization can be recursively implemented by transforming multiple independent uses of a given Binary-input Discrete Memoryless Channel (B-DMC) into a set of successive uses of synthesized binary-input channels. At first, independent channels are transformed into two kinds of synthesized channels: the good and bad channels. These channels are *polarized*, which transmit a single bit with slightly different reliability. By recursively applying such polarization transformation over the resulting channels, the mutual information of the synthetic channels tend to two extremes: either close to 0 (the noisy channels) or close to 1 (the noiseless channels). Then, transmitting information bits over the noiseless channels while assigning frozen bits to the noisy ones, resulting in extreme channel capacities. In October 2016, Huawei announced that it had successfully reached a downlink rate of 27 Gbps using polar codes. It demonstrated that polar codes could simultaneously meet all three usage scenarios of eMBB (up to 20 Gbps), URLLC (1 ms latency), and mMTC (massive connections) as per ITU IMT-2020 definition. Polar codes can provide an efficient channel coding technique for 5G, allowing significantly higher spectrum efficiency and the practical decoding ability of linear complexity to minimize the implementation cost. In November 2016, 3GPP approved to employ polar codes for the control channel while LDPC codes for the data channel in 5G NR.

2.6 Conclusions

On the basis of the first chapter, the second chapter provided an in-depth introduction of key technologies used for the previous generations. Each section consisted of two main parts: (i) the system architecture of a representative standard for each generation, including its major network elements, functionality split, interconnection, interaction, and operational flows; and (ii) the basic principles, benefits, and challenges of essential technologies for each generation, and the synergy of these technologies to build a cellular system. The purpose of this chapter is to offer an overview of the technological evolution in mobile communications, so that the readers can well understand the state-of-the-art advances toward the upcoming 6G system. The next chapter will provide a vision of 6G, including potential use cases, usage scenarios, performance requirements, roadmap, as well as the identification of potential 6G technologies.

References

3GPP TR38.913 [2020], Study on scenarios and requirements for next generation access technologies (Release 16), Report TR38.913, The 3rd Generation Partnership Project.

3GPP TS23.060 [1999], General Packet Radio Service (GPRS): Service description, Specification TS23.060, The 3rd Generation Partnership Project.

3GPP TS23.501 [2021], System architecture for the 5G system (5GS); stage 2 (Release 17), Specification, The 3rd Generation Partnership Project.

Arikan, E. [2009], 'Channel polarization: A method for constructing capacity-achieving codes for symmetric binary-input memoryless channels', *IEEE Transactions on Information Theory* **55**(7), 3051–3073.

Berrou, C. and Glavieux, A. [1996], 'Near optimum error correcting coding and decoding: Turbo-codes', *IEEE Transactions on Communications* **44**(10), 1261–1271.

Chen, Y., Bayesteh, A., Wu, Y., Ren, B., Kang, S., Sun, S., Xiong, Q., Qian, C., Yu, B., Ding, Z., Wang, S., Han, S., Hou, X., Lin, H., Visoz, R. and Razavi, R. [2018], 'Toward the standardization of non-orthogonal multiple access for next generation wireless networks', *IEEE Communications Magazine* **56**(3), 19–27.

Donald, V. H. M. [1979], 'Advanced mobile phone service: The cellular concept', *The Bell System Technical Journal* **58**(1), 15–41.

Ehrlich, N. [1979], 'The advanced mobile phone service', *IEEE Communications Magazine* **17**(2), 9–16.

Goldsmith, A. [2005], *Wireless Communications*, Cambridge University Press, Stanford University, California.

Holma, H. and Toskala, A. [2004], *WCDMA for UMTS-Radio Access for Third Generation Mobile Communications (Third Edition)*, John Wiley & Sons Inc., England.

Larsson, E. G., Edfors, O., Tufvesson, F. and Marzetta, T. L. [2014], 'Massive MIMO for next generation wireless systems', *IEEE Communications Magazine* **52**(2), 186–195.

Lin, Y.-B., Rao, H. C.-H. and Chlamtac, I. (2001), 'General packet radio service (GPRS): Architecture, interfaces, and deployment', *Wiley - Wireless Communications and Mobile Computing* **1**(1), 77–92.

Markey, H. K. and Antheil, G. (n.d.) , 'US2292387A: Secret communication system', *US Patent*.

Mouly, M. and Pautet, M.-B. [1995], 'Current evolution of the GSM systems', *IEEE Personal Communications* **2**(5), 9–19.

Porter, P. [1985], 'Relationships for three-dimensional modeling of co-channel reuse', *IEEE Transactions on Vehicular Technology* **34**(2), 63–68.

Price, R. and Green, P. E. [1958], 'A communication technique for multipath channels', *Proceedings of the IRE* **46**, 555–570.

Rahnema, M. [1993], 'Overview of the GSM system and protocol architecture', *IEEE Communications Magazine* **31**(4), 92–100.

Rappaport, T. S., Sun, S., Mayzus, R., Zhao, H., Azar, Y., Wang, K., Wong, G. N., Schulz, J. K., Samimi, M. and Felix Gutierrez, J. [2013], 'Millimeter wave mobile communications for 5G cellular: It will work!', *IEEE Access* **1**, 335–349.

Ring, D. H. [1947], 'Mobile telephony - wide area coverage', *Bell Telephone Laboratories*.

Vriendt, J. D., Laine, P., Lerouge, C. and Xu, X. [2002], 'Mobile network evolution: A revolution on the move', *IEEE Communications Magazine* **40**(4), 104–111.

Zaidi, A. A., Baldemair, R., Andersson, M., Faxér, S., Molés-Cases, V. and Wang, Z. [2017], 'Designing for the future: The 5G NR physical layer', *Ericsson Technology Review* **7**, 1–13.

3

The Vision of 6G: Drivers, Enablers, Uses, and Roadmap

As of this writing, more than 400 mobile operators in approximately 130 countries and territories are investing in 5G technology, while the number of 5G subscribers reaches a vast scale in many regions. On the one hand, academia and industry have already shifted their attention toward the next-generation technology known as the sixth generation (6G). The major players in the mobile communications industry have initiated many activities to explore potential 6G technologies. On the other hand, there still exist different voices like *Is there any need for 6G?* or *Do we really need 6G?*. Consequently, it is necessary to clarify the significance and the fundamental motivation of developing 6G and convince the readers that 6G will definitely come like the previous generations of mobile communications before jumping into 6G technical details. This chapter will provide a comprehensive vision concerning the driving forces, use cases, and usage scenarios and demonstrate essential performance requirements and identify key technological enablers. This chapter consists of the following sections:

- Section 3.1 introduces the current background of the evolution toward 6G.
- Section 3.2 reports the prediction of explosive traffic growth by 2030, which will drive the further evolution of cellular communications systems.
- Section 3.3 envisions high-potential use cases and applications for 6G.
- Section 3.4 reviews three 5G usage scenarios and provides possible definitions of 6G usage scenarios.
- Section 3.5 gives key performance requirements for 6G compared with that of 5G.
- Section 3.6 summarizes the state-of-the-art research initiatives worldwide and foresees the potential roadmap toward 6G.
- Section 3.7 lists a number of potential 6G technologies.

6G Key Technologies: A Comprehensive Guide, First Edition. Wei Jiang and Fa-Long Luo.
© 2023 The Institute of Electrical and Electronics Engineers, Inc. Published 2023 by John Wiley & Sons, Inc.

3.1 Background

The era of the fifth-generation (5G) mobile communications arrived in April 2019, when South Korea's three mobile operators – SK Telecom, LG U+, and KT – were competing with the United States' carrier Verizon to launch the world's first 5G commercial network. In the past two years, we have witnessed a substantial expansion of 5G coverage globally and the tremendous growth of 5G subscriptions in major countries. For instance, at the end of 2020, the penetration rate of 5G usage in South Korea had surpassed 15.5%, while China had deployed more than 700 000 base stations to serve around 200 million 5G subscribers. Unlike previous generations of cellular networks that focused on human-centric communication services, 5G aims at not only mobile broadband but also massive machine-type communications and ultra-reliability low-latency communications for mission-critical Applications (APPs). The advent of 5G expands the sphere of mobile communications from humans to things, from consumers to vertical industries, and from public service to public–private hybrid APPs. The scale of mobile subscriptions is substantially enlarged from merely billions of the world's population to countless inter-connectivity among humans, machines, and things. The deployment of 5G networks will facilitate a wide variety of new APPs such as Industry4.0, eHealth, Virtual Reality (VR), Internet of Things (IoT), and automatic driving [Andrews et al., 2014]. In 2020, the outbreak of the COVID-19 pandemic led to a dramatic loss of human life across the world and brought substantial challenges to societal and economic activities. However, this public health crisis highlighted the importance of networks and digital infrastructure in keeping society running and families connected. In particular, 5G-empowered APPs such as remote surgeons, online education, remote working, high-definition video conferencing, driver-less vehicles, unmanned delivery, robots, contactless healthcare, and autonomous manufacturing demonstrated their utilities during the battle against the pandemic.

As of this writing, 5G is still on its way for wide deployment worldwide, but academia and industry have already shifted their attention to beyond 5G or sixth generation (6G) technologies to satisfy the future demands for Information and Communications Technology (ICT) in 2030. Even though discussions are ongoing within the wireless community as to whether there is any need for 6G or whether counting the generations should be stopped at 5, and even there exists opposition to talking about 6G at this time [Fitzek and Seeling, 2020], several pioneering works on new mobile communication technologies have been initiated. A focus group called *Technologies for Network 2030* within the International Telecommunication Union Telecommunication (ITU-T) standardization sector was established in July 2018. The group intends to study the capabilities of networks for 2030 and beyond [ITU-T NET-2030, 2019] when it is expected to support novel forward-looking scenarios, such as holographic type communications, pervasive intelligence, Tactile Internet, multi-sense experience, and digital twin. Furthermore, the European Commission initiated to sponsor beyond 5G research activities, as its recent Horizon 2020 calls – ICT-20 *5G Long Term Evolution* and ICT-52 *Smart Connectivity beyond 5G* – where a batch of pioneer research projects for potential 6G technologies was kicked off at the early beginning of 2020. The European Commission has also announced its strategy to accelerate investments in Europe's *Gigabit Connectivity* including 5G and 6G to shape Europe's digital future [EU Gigabit Connectivity, 2020]. In October 2020, the Next Generation Mobile Networks (NGMN) alliance had launched its new *6G Vision and Drivers* project, intending to provide early and timely direction for global 6G activities. At its meeting in February 2020, the International Telecommunication Union Radiocommunication Sector (ITU-R) decided to start the study on future technology trends for the future evolution of International Mobile Telecommunications (IMT) [ITU-R WP5D, 2020].

Traditional leading players in mobile communications, such as the United States, China, Finland, Germany, Japan, and South Korea, already initiated their national 6G research program officially. In Finland, the University of Oulu began ground-breaking 6G research as part of Academy of Finland's flagship program [Latva-aho et al., 2019] called 6G-Enabled Wireless Smart Society and Ecosystem (6Genesis), which focuses on several challenging research areas, including reliable near-instant unlimited wireless connectivity, distributed computing, and intelligence, as well as materials and antennas to be utilized in future for circuits and devices. In October 2020, the Alliance for Telecommunications Industry Solutions (ATIS) launched "Next G Alliance" with founding members including AT&T, T-Mobile, Verizon, Qualcomm, Ericsson, Nokia, Apple, Google, Facebook, Microsoft, etc. It is an industry initiative that intends to advance North American mobile technology leadership in 6G over the next decade through private sector-led efforts. With a strong emphasis on technology commercialization, its ambition is to encompass the full lifecycle of 6G research and development, manufacturing, standardization, and market readiness. As early as 2018, the 5G working group at the Ministry of Industry and Information Technology of China started a concept study of potential 6G technologies, making China one of the first countries to explore 6G technology. In November 2019, the Ministry of Science and Technology of China had officially kicked off the 6G technology research and development work coordinated by the ministry, together with five other ministries or national institutions. A promotion working group from the government is in charge of management and coordination. Meanwhile, an overall expert group that is composed of 37 experts from universities, research institutes, and industry was established at this event.

3.2 Explosive Mobile Traffic

Since the middle of 2019, 5G networks have been commercially deployed globally, and the number of 5G subscriptions has already reached a massive scale in some regions. A new generation of mobile communications usually appears every decade, so both academia and industry have initiated the exploration of the successor of 5G. On the way toward 6G, however, the first problem we encounter is that there are many concerns like "Do we really need 6G?" or "Is 5G already enough?." To move this barrier, the wireless community first needs to clarify the critical driving forces for the development of 6G.

A new generation system is driven by the exponential growth of mobile traffic and mobile subscriptions and new disruptive services and APPs on the horizon. In addition, it is also driven by the intrinsic need of mobile communication society to continuously improve network efficiencies, namely cost efficiency, energy efficiency, spectrum efficiency, and operational efficiency. Furthermore, with the advent of advanced technologies such as Artificial Intelligence (AI), Terahertz (THz) communications, and large-scale satellite constellation, the communication network can evolve toward a more powerful and more efficient system to better fulfill the requirements of current services and open the possibility for offering disruptive services that have hitherto never been seen. This section reports the trend on mobile traffic that is expected to continue to grow explosively by 2030. The following two sections will demonstrate potential use cases and usage scenarios.

We are in an unprecedented era where many interactive, intelligent products, services, and APPs emerge and evolve promptly, imposing a huge demand on mobile communications. It can be foreseen that the 5G system will be hard to accommodate the tremendous volume of mobile traffic in 2030 and beyond. In 2015, the ITU-R released a report [ITU-R M.2370, 2015] to analyze IMT traffic growth for the years 2020–2030. According to this report, the main drivers behind the anticipated traffic growth are:

- *Video Usage*: Usage of video-on-demand services will continue to grow, and the resolution of these videos will continue to increase. People want to watch high-resolution visual content, regardless of how the content is delivered. A study by Bell Labs revealed that video streaming has accounted for almost two-thirds of all mobile traffic in 2016.
- *Device Proliferation*: There are over five billion people in the world who own smartphones, and more than one billion new smartphones are sold in the market per year. That is 66.5% of the world's population. In addition, various new smart devices such as wearable electronics, VR glasses, and smart automobiles are connected to mobile networks.

- *Application Uptake*: The speed of the spreading of mobile APPs is accelerating. The annual global downloading of APPs was 102 billion in 2013 and grew to 270 billion in 2017. Most APPs are not used more than once after being downloaded. This APP uptake and the usage of those will contribute to increased mobile broadband traffic, and, in addition, the number of regular updates to those hundreds of billions of APPs will also increase mobile broadband traffic.

In addition to these major drivers of data traffic growth, there are some other characteristics and trends impacting the overall traffic demand by 2030.

- *Deployment of IMT-2020*: New technologies will improve the perceived Quality of Experience (QoE) and decreases the cost per bit, which in turn creates more traffic demand.
- *Machine-to-Machine (M2M)* APPs and devices are also one of the fastest-growing segments for mobile services and, eventually, increased mobile data demand. The amount of M2M connections could be several orders of magnitude larger than the world population. Billions of machines will potentially utilize mobile networks to access online services and interconnect with each other.
- *Enhanced Screen Resolution*: Continuous improvement in the screen capabilities, e.g. 4K Ultra-High-Definition (UHD), and increasing demand for video downloading and streaming will bring more traffic on mobile networks.
- *Proliferation of Ambient Screens* or info-bearing surfaces to internet-connected devices for up-to-date information, such as screens in elevators and buses, will increase traffic.
- *Cloud Computing*: The demand for mobile cloud services is expected to grow because users are increasingly adopting more services that are required to be ubiquitously accessible. With the increasing number of users connecting through the mobile network to the cloud, the mobile data traffic between mobile terminals, cloud servers, and cloud storage will continue to grow.
- *Fixed Broadband (FBB) Replacement by MBB*: In areas and contexts where MBB is used as an alternative to wired broadband, such as copper, cable, and optic fiber, this would contribute to an increase in IMT traffic.
- *Multimedia Streaming*: People use their mobile devices more often for multimedia streaming entertainment, with increased unicast media consumption due to time shift (expansion of cloud), space shift (availability of content anywhere), and device shift (multi-screen, switch between mobile and portable devices). The live TV still represents 90% of the world's audio-visual service in 2014, growing 4.2% from the previous year; while over-the-top (OTT) video transmission represents 4.4%, growing 37%. Most of the audio-visual traffic is currently delivered through non-IMT networks, which will be transferred into IMT networks.

In other words, the traffic over mobile networks will continuously grow in an explosive manner due to the proliferation of rich-video APPs, enhanced screen resolution, M2M communications, mobile cloud services, etc. According to the estimation by ITU-R [ITU-R M.2370, 2015] in 2015, the global mobile traffic will reach up to 5016 EB[1] per month in 2030 compared with 62 EB in 2020. A report from Ericsson [Ericsson Report, 2020] reveals that the global mobile traffic has reached 33 EB per month at the end of 2019, which justifies the correctness of ITU-R's estimation.

In the last decade, the number of smartphones and tablets has experienced exponential growth due to the proliferation of mobile broadband. This trend will continue in the 2020s since the penetration of smartphones and tablets is still far from being saturated, especially in developing countries. Meanwhile, new-style user terminals, such as wearable electronics and VR glasses, emerge in the market quickly and are adopted by consumers at an unprecedented pace. As a result, it is expected that the total number of MBB subscribers worldwide will reach 17.1 billion by 2030, as shown in Figure 3.1. On the other hand, the traffic demand per MBB user continuously rises, in addition to the rising number of MBB users. That is mainly because of the popularity of mobile video services such as YouTube, Netflix, and more recently, short-video service Tik-Tok, and the stable improvement of screen resolution on mobile devices. The traffic coming from mobile video services already accounts for two-thirds of all mobile traffic nowadays [Ericsson Report, 2020] and is estimated to be more dominant in the future. In some developed countries, a strong traffic growth before 2025 will be driven by rich-video services, and a long-term growth wave will continue due to the penetration of augmented reality (AR) and VR APPs. The average data consumption for every mobile user per month, as illustrated in Figure 3.1, will increase from around 5 GB in 2020 to over 250 GB in 2030. In addition to human-centric communications, the scale of M2M terminals will increase more rapidly and become saturated no earlier than 2030. The number of M2M subscriptions will reach 97 billion, around 14 times over 2020 [ITU-R M.2370, 2015], serving as another driving force for the explosive growth of mobile traffic.

3.3 Use Cases

With the advent of new technologies and continuous evolution of existing technologies, e.g. holography, robotics, microelectronics, new energy, photo-electronics, AI, and space technology, many unprecedented APPs can be fostered in mobile networks. To explicitly highlight the unique characteristics and define

---

1 1 exabyte (EB)=1 000 000 terabytes (TB), 1 TB = 1000 gigabytes (GB).

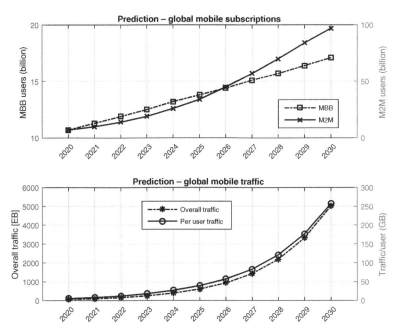

Figure 3.1 The trend of global mobile subscriptions and mobile traffic from 2020 to 2030 estimated by the ITU. Source: Jiang et al. [2021]/With permission of IEEE.

the technical requirements of 6G, the researchers in wireless community tried to foresee disruptive use cases, e.g. as defined in Jiang et al. [2021]:

Holographic-Type Communication (HTC): In contrast to traditional 3D videos using binocular parallax, genuine holograms can satisfy all visual cues of observing 3D objects by the naked eye as naturally as possible. With a significant advance of holographic display technology in recent years, such as Microsoft's HoloLens, it is envisioned that its application will become a reality in the next decade. Remote rendering high-definition holograms through a mobile network will bring a truly immersive experience. For example, holographic telepresence will allow remote participants to be projected as holograms into a meeting room or allow the attendee of online training or education to interact with ultra-realistic objects. However, HTC leads to a demand for huge bandwidths on the order of terabits per second, even with image compression. In addition to the frame rate, resolution, and color depth in two-dimensional (2D) video, the hologram quality also involves the volumetric data such as tilt, angle, and position. If representing an object with images every 0.3°, an image-based hologram with 30° field of view and a tilt of 10° needs a 2D array of 3300 separate images [Clemm et al., 2020]. HTC also requires ultra-low latency for true immersiveness and high-precision synchronization across massive bundles of interrelated streams for reconstructing holograms.

Extended Reality (ER): Combining augmented, virtual, and mixed realities, ER starts stepping into practical APPs at the age of 5G, but it is still in its infancy analog to the video service at the beginning of mobile Internet. To achieve the same level of image quality, ER devices with 360° field of view need much higher data throughput in comparison to 2D video streaming. For an ideal immersion experience, the quality of video with higher resolution, higher frame rate, more color depth, and high dynamic range is required, leading to a bandwidth demand of over 1.6 Gbps per device [Huawei VR Report, 2018]. Similar to video traffic that saturates the 4G networks, the proliferation of ER devices will be restricted by the limited capacity of 5G with the peak rate of 20 Gbps, especially at the cell edge. Interactive ER APPs such as immersive gaming, remote surgery, and mission-critical remote manipulation require low latency and high reliability in addition to high data throughput.

Tactile Internet: It provides extremely low End-to-End (E2E) latency to satisfy the 1-millisecond (ms) or less reaction time reaching the limit of human sense [Fettweis et al., 2014]. In combination with high reliability, high availability, high security, and sometimes high throughput, many disruptive real-time APPs emerge. It will play a critical role in real-time monitoring and remote industrial management for Industry 4.0 and Smart Grid. For example, with immersive audio-visual feeds provided by ER or HTC streaming, together with haptic sensing data, a human operator can remotely control the machinery in a place surrounded by biological or chemical hazards, as well as remote robotic surgery carried out by doctors from hundreds of miles away. The typical closed-loop controlling, especially for devices or machinery rotating rapidly, is very time-sensitive, where an E2E latency below 1 ms is expected.

Multi-Sense Experience: Human has five senses (sight, hearing, touch, smell, and taste) to perceive the external environment, whereas current communications focus only on optical (text, image, and video) and acoustic (audio, voice, and music) media. The involvement of the senses of taste and smell can create a fully immersive experience, which may bring some new services in food and texture industries [ITU-T NET-2030, 2019]. Furthermore, the application of haptic communication will play a more critical role and raise a wide range of APPs such as remote surgery, remote controlling, and immersive gaming. However, this use case brings a stringent requirement on low latency.

Digital Twin is used to create a complete and detailed virtual copy of a physical (a.k.a. real) object. The software copy is equipped with a wide range of characteristics, information, and properties related to the original object. Such a twin is then used to manufacture multiple copies of an object with full automation and intelligence. The early rollouts of the digital twin have attracted significant attention from many vertical industries and manufacturers. However, its full deployment is expected to be realized with the development of 6G networks.

Pervasive Intelligence: With the proliferation of mobile smart devices and the emergence of new-style connected equipment such as robots, smart cars, drones, and VR glasses, over-the-air intelligent services are envisioned to boom. These intelligent tasks mainly rely on traditional computation-intensive AI technologies: computer vision, Simultaneous Localization And Mapping (SLAM), face and speech recognition, natural language processing, motion control, to name a few. Since mobile devices have limited computing, storage, and connectivity resources, 6G networks will offer pervasive intelligence in an AI-as-a-Service manner [Letaief et al., 2019] by utilizing distributed computing resources across the cloud, mobile edge, and end-devices, and cultivating communication-efficient Machine Learning (ML) training and interference mechanisms. For example, a humanoid robot such as Atlas from Boston Dynamics can off-load its computational load for SLAM toward edge computing resources to improve motion accuracy, prolong battery life, and become more lightweight by removing some embedded computing components. In addition to computation-intensive tasks, pervasive intelligence also facilitates time-sensitive AI tasks to avoid the latency constraint of cloud computing when fast decisions or responses to conditions are required.

Intelligent Transport and Logistics: In 2030 and beyond, millions of autonomous vehicles and drones provide a safe, efficient, and green movement of people and goods. Connected autonomous vehicles have stringent requirements on reliability and latency to guarantee the safety of passengers and pedestrians. Unmanned aerial vehicles, especially the swarm of drones, open the possibility for a wide variety of unprecedented APPs while also bringing disruptive requirements for mobile networks.

Enhanced On-Board Communications: With the development of the economy, the activity sphere of humans and the frequency of their movement will rapidly increase in the next decade. The number of passengers traveling by commercial planes, helicopters, high-speed trains, cruise ships, and other vehicles will be huge, bringing skyrocketing demands on high-quality communication services on board. Despite the efforts in the previous generations until 5G, it is undeniable that on-board connectivity is far from satisfactory in most cases due to high mobility, frequent handover, sparse coverage of terrestrial networks, limited bandwidth, and high cost of satellite communications. Relying on reusable space launching technologies and massive production of satellites, the deployment of large-scale satellite constellations such as SpaceX's Starlink becomes a reality, enabling cost-efficient and high-throughput global coverage. Keep this in mind, 6G is expected to be an integrated system of terrestrial networks, satellite constellation, and other aerial platforms to provide seamless 3D coverage, which offers high-quality, low-cost, and global-roaming on-broad communication services.

Global Ubiquitous Connectability: The previous generations of mobile communications focused mainly on the dense metropolitan areas, primarily indoor scenarios. However, a large population in remote, sparse, and rural areas has no access to essential ICT services, digging a significant digital divide among humans across the world. Besides, more than 70% of the Earth's surface is covered by water, where the growth of maritime APPs requires network coverage for both water surface and underwater. However, ubiquitous coverage across the whole planet with sufficient capacity, acceptable Quality of Service (QoS), and affordable cost are far from a reality. On the one hand, it is technically impossible for terrestrial networks to cover remote areas and extreme topographies such as the ocean, desert, and high mountain areas. At the same time, it is too costly to offer terrestrial communication services for sparsely populated areas. On the other hand, Geostationary Earth Orbit (GEO) satellites are expensive to deploy, and their capacity is currently limited to several Gbps per satellite [Qu et al., 2017], which is dedicated only for high-end users such as maritime and aeronautic industries. As mentioned earlier, the deployment of a large-scale Low Earth Orbit (LEO) satellite constellation will enable low-cost and high-throughput global communication services. The 6G system is envisioned to make use of the synergy of terrestrial networks, satellite constellation, and other aerial platforms to realize ubiquitous connectivity for global MBB users and wide-area IoT APPs.

3.4 Usage Scenarios

In February 2013, the ITU-R Working Party 5D initiated two study items to analyze the IMT Vision for 2020 and future technology trends for the terrestrial IMT systems, known as IMT-2020. IMT-2020 was envisaged to meet more diverse QoS requirements arising from a wide variety of vertical APPs and services, which have never been encountered by mobile subscribers in the previous generations. Furthermore, a broad variety of capabilities would be tightly coupled with these intended different usage scenarios and APPs. Part of the study outcomes was transferred to ITU-R Recommendation M.2083 released in 2015 [ITU-R M.2083, 2015], where three usage scenarios were firstly defined:

- *Enhanced Mobile Broadband (eMBB)*: Mobile broadband addresses the human-centric use cases for access to multi-media content, services, cloud, and data. With the proliferation of smart devices (smartphones, tablets, and wearable electronics) and the rising demand for video streaming, the need for mobile broadband continues to grow, setting new requirements for what ITU-R calls enhanced mobile broadband. This usage scenario comes with new use cases and requirements for improved capabilities and an increasingly seamless user experience. eMBB covers various cases, including wide-area coverage and hot spots, which have different requirements. For the hot-spot case, i.e. for an area with high user density, very high traffic capacity is needed, while the requirement for mobility is low and user data rate is higher than that of wide-area coverage. Seamless coverage and medium-to-high mobility are desired for the wide-area coverage case, with a substantially improved data rate than existing data rates. However, the data rate requirement may be relaxed compared with hot spots.

- *Ultra-Reliable Low-Latency Communications (URLLC):* This scenario aims to support both human-centric and critical machine-type communications. It is a disruptive promotion over the previous generations of cellular systems that focused merely on the services for mobile subscribers. It opens the possibility for offering mission-critical wireless APPs such as automatic driving, vehicle-to-vehicle communication involving safety, wireless control of industrial manufacturing or production processes, remote medical surgery, distribution automation in a smart grid, and transportation safety. It is characterized by stringent requirements such as ultra-low latency, ultra-reliability, and availability.

- *Massive Machine-Type Communications (mMTC):* This scenario supports massive connectivity with a vast number of connected devices that typically have very sparse transmissions of delay-tolerant data. Such devices, e.g. remote sensors, actuators, and monitoring equipment, are required to be low cost and low power consumption, allowing for a very long battery life of up to 10 years due to the possibility of remote IoT deployment.

It can be seen that these 5G usage scenarios cannot satisfy the technical requirements of the aforementioned 6G use cases. For instance, a user that wears a lightweight VR glass to play interactively immersive games requires not only ultra-high bandwidth but also low latency. Autonomous vehicles on the road or flying drones need ubiquitous connectivity with high throughput, high reliability, and low latency. Some discussions have been carried out in wireless community with respect to potential usage scenarios for 6G. For example, Jiang et al. [2021] applied a holistic methodology to define 6G scenarios through extending the scope of current usage scenarios, as shown in Figure 3.2. Three new scenarios are proposed to meet the requirement of aforementioned use cases, which cover the overlapping areas of 5G scenarios so as to form a complete set.

- *Ubiquitous Mobile Broadband (uMBB)*: To support high-quality on-board communications and global ubiquitous connectability, the MBB service should be available across the whole surface of the Earth in the era of 6G, called *ubiquitous MBB* or uMBB. In addition to its ubiquitousness, another enhancement of uMBB is a remarkable boost of network capacity and transmission rate for hot spots so

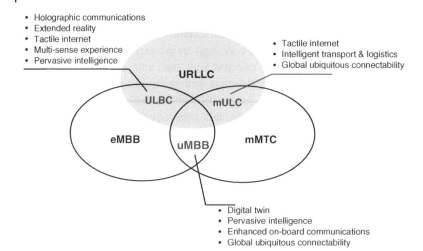

- Holographic communications
- Extended reality
- Tactile internet
- Multi-sense experience
- Pervasive intelligence

- Tactile internet
- Intelligent transport & logistics
- Global ubiquitous connectability

- Digital twin
- Pervasive intelligence
- Enhanced on-board communications
- Global ubiquitous connectability

Figure 3.2 An example of 6G usage scenarios: three new scenarios of uMBB, ULBC, and mULC added on the top of three typical 5G usage scenarios (i.e. eMBB, ULRRC, and mMTC). Source: Jiang et al. [2021]/With permission of IEEE.

as to support disruptive services, e.g. a group of users wearing lightweight VR glasses gathering in a small room where a data rate of several Gbps per user is needed. The uMBB scenario will be the foundation of digital twin, pervasive intelligence, enhanced on-board communications, and global ubiquitous connectability, as the mapping relationship shown in Figure 3.2. In addition to Key Performance Indicators (KPIs) that are applied to evaluate eMBB (such as peak data rate and user-experienced data rate), other KPIs become as same critical as the others in uMBB, i.e. mobility, coverage, and positioning.

- *Ultra-reliable low-latency broadband communication (ULBC)* supports the APPs requiring not only URLLC but also extreme high throughput, e.g. HTC-based immersive gaming. It is expected that the use cases of HTC, ER, Tactile Internet, multi-sense experience, and pervasive intelligence will benefit from this scenario.
- *Massive Ultra-Reliable Low-Latency Communication (mULC)* combines the characteristics of both mMTC and URLLC, which will facilitate the deployment of massive sensors and actuators in vertical industries. Together with eMBB, URLLC, and mMTC, three new scenarios fill the gaps in-between and then a complete set of usage scenarios is formed to support all kinds of use cases and APPs in 6G, as shown in Figure 3.2.

There are also other definitions for possible usage scenarios. For example, Huawei announced its vision toward beyond 5G system [Huawei NetX2025, 2021], which can improve real-time interaction experience for individual users, enhance cellular IoT capabilities, and explore new scenarios, including Uplink-Centric Broadband Communication (UCBC), Real-Time Broadband Communication (RTBC), and Harmonized Communication and Sensing (HCS), for a better, intelligent world. 5.5G aims to evolve from the Internet of everything to the intelligent Internet of everything, and is expected to create entirely new value as a result.

- *UCBC* accelerates the intelligent upgrade of industries. UCBC delivers ultra-broadband uplink experience (Figure 3.3). Relying on 5G capabilities, it will enable a 10-fold increase in uplink bandwidth. This is a perfect fit for manufacturers who need to upload videos in machine vision and massive broadband IoT, accelerating their intelligent upgrade. UCBC also significantly improves user experience of mobile phones in indoor areas requiring intensive coverage. Through multi-band uplink aggregation and uplink massive antenna array technologies, the uplink capacity and user experience in these scenarios can also witness a considerable improvement.
- *RTBC* delivers an immersive, true-to-life experience. RTBC supports large bandwidth and low communication latency. It aims to provide a 10-fold increase in bandwidth at a specified latency, thereby creating an immersive experience for physical-virtual interactions such as XR Pro and holographs. It leverages

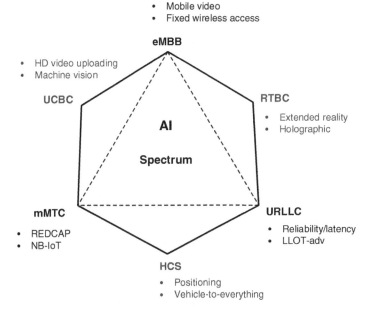

Figure 3.3 Beyond 5G vision of Huawei. Source: Adapted from Huawei NetX2025 [2021].

typical carriers to quickly expand network capabilities, and the E2E cross-layer XR experience mechanism to build real-time interaction capabilities featuring large bandwidth.

- *HCS* is designed to enable connected cars and connected drones, scenarios in which autonomous driving is the key requirement. HCS predominantly enables V2X and Unmanned Aerial Vehicle (UAV) scenarios, which place autonomous driving as a key requirement. These scenarios need wireless cellular networks with both communications and perception capabilities. On the basis of 5G capabilities, UCBC will enable a fivefold increase in uplink bandwidth. This is a perfect fit for manufacturers who need to upload videos in machine vision and massive broadband IoT, accelerating their intelligent upgrade.

- *New Usage Mode of Sub-100 GHz* will maximize spectral efficiency. Spectrum is the most important resource in the wireless industry, and to achieve the industry vision, 5.5G needs more spectra in the sub100 GHz segment. Different spectrum types have different features. For example, FDD symmetric spectrum features low latency, TDD spectrum features high bandwidth, and mmWave can achieve ultra-large bandwidth and low latency. Within this context, one of the main goals involves tapping into the full potential of spectrum. The hope is to achieve maximum spectral efficiency by remodeling the sub-100 GHz spectrum usage through uplink and downlink decoupling and on-demand, flexible aggregation on all frequency bands.

- *Intelligence* will make 5G connections smarter. In the 5G era, operators have to deal with considerably more frequency bands and types of terminals, services, and customers. Given this, 5.5G needs intelligent penetration from multiple perspectives to advance the L4 High Autonomous Driving Network (High ADN) and L5 Full ADN.

3.5 Performance Requirements

To well support disruptive use cases and APPs in 2030 and beyond, the 6G system will provide extreme capacities and performance. Like the minimal requirements related to technical performance for IMT-2020 radio interfaces, specified in ITU-R M2410 [ITU-R M.2410, 2017], a number of quantitative or qualitative KPIs are utilized to specify the technical requirements for 6G. Most of the KPIs that are applied for evaluating 5G are still valid for 6G while some new KPIs would be introduced for the assessment of new technological features, which are briefly introduced [Jiang et al., 2021] as follows:

- *Peak data rate* is the highest data rate under ideal conditions, e.g. all assignable radio resources (excluding radio resources for physical layer synchronization, reference signals, guard bands, and guard intervals) for the corresponding link direction are utilized for a single mobile station. It is proportional to the channel bandwidth and the peak spectral efficiency in that band. This requirement is defined for the purpose of evaluation in the eMBB usage scenarios. Traditionally, it is the most symbolic parameter to differentiate different generations of mobile systems. Driven by both user demand and technological advances such as THz communications, it is expected to reach up to 1 Tbps, tens of times that of 5G, which has the peak rate of 20 Gbps for downlink and 10 Gbps for uplink.

- *Peak spectral efficiency* is an important KPI to measure the advance of air-interface technologies. The minimum requirement in IMT-2020 for peak spectral efficiencies are 30 bps/Hz in the downlink and 15 bps/Hz in the uplink assuming an antenna configuration to enable eight spatial streams in the downlink and four spatial streams in the uplink. Following the empirical data, it is expected that advanced 6G radio technologies with higher spatial degree of freedom could achieve three times higher spectral efficiency over the 5G system.

- *Signal bandwidth* is the maximum aggregated system bandwidth. The bandwidth may be supported by single or multiple RF carriers. The 5G system is required to support a minimum bandwidth of 100 MHz, and up to 1 GHz in higher frequency bands above 6 GHz. To reach a peak rate of Tbps, 6G shall support bandwidths up to 10 GHz for operation in higher frequency bands or even higher bandwidth in THz communications or Optical wireless communications (OWC). Like 5G, it shall operate in scalable bandwidths to support spectral flexibility.

- *User-experienced data rate* is defined as the 5th percentile point (5%) of the Cumulative Distribution Function (CDF) of user throughput. The user throughput is defined as the number of correctly received bits in the Service Data Units (SDUs) delivered to Layer 3, over a certain period of time during the active mode. In other words, a mobile user can get at least this data rate at any time or location with a possibility of 95%. It is more meaningful to measure the perceived performance, especially at the cell edge, and reflect the quality of network design such as site density, architecture, and inter-cell optimization. In the dense-urban deployment scenario of 5G, the target of user-perceived rate is defined as 100 Mbps in the downlink and 50 Mbps in the uplink. It is expected that 6G can offer ten times user-experienced rates up to 1 Gbps even higher.

- *5th percentile user spectral efficiency* is defined as the 5th percentile point (5%) of the CDF of the normalized user throughput. The normalized user throughput is defined as the number of correctly received bits in the SDUs delivered to Layer 3 over a certain period of time divided by the channel bandwidth. The channel bandwidth in this context is defined as the effective bandwidth times the frequency reuse factor, where the effective bandwidth is the operating

Table 3.1 Performance requirement of 5th percentile user spectral efficiency.

Deployment environment	Downlink (bps/Hz)	Uplink (bps/Hz)
Indoor hotspot	3	2.1
Dense urban	2.25	1.5
Rural	1.2	0.45

bandwidth normalized appropriately considering the ratio of downlink and uplink. This KPI is expected to promote ten times over IMT-2020 corresponding to the improvement of user-experienced data rate, as given in Table 3.1:

- *Average spectral efficiency* is the aggregate throughput of all mobile users (the number of correctly received bits, i.e. the number of bits contained in the SDUs delivered to Layer 3, over a certain period of time) divided by the channel bandwidth of a specific band divided by the number of Transmission and Reception Points (TRxPs). Average spectral efficiency is expected to improve two to three times over IMT-2020, as given in Table 3.2:
- *Latency* can be differentiated into two categories: user plane latency and control plane latency. The former is the time delay induced in a radio network from a packet being sending out at the source until the destination receives it, assuming a mobile station is in the active state. To be specific, it is defined as the one-way time it takes to successfully deliver a small application-layer packet (e.g. 0 byte payload with an IP header) from the radio protocol layer 2/3 SDU ingress point to the radio protocol layer 2/3 SDU egress point of the radio interface in either downlink or uplink in the network for a given service in unloaded conditions. In 5G, the minimum requirement for user plane latency is 4 ms for eMBB and 1 ms for URLLC. This value is envisioned to be further reduced to 100 μs or even 10 μs. Control plane latency refers to the transition time from a most "battery efficient" state (e.g. the idle state) to the start of continuous data transfer (e.g. the active state). The minimum latency for control plane should be 20 ms in 5G and is expected to be also remarkably improved in 6G. In addition to over-the-air

Table 3.2 Performance requirement of average spectral efficiency.

Deployment environment	Downlink (bps/Hz/TRxP)	Uplink (bps/Hz/TRxP)
Indoor hotspot	25	15
Dense urban	20	10
Rural	10	5

delay, round-trip or E2E delay is more meaningful but also complicated due to the large number of network entities involved. In 6G, the E2E latency may be considered as a whole.

- *Mobility* means the maximal velocity of a mobile station supported by a network with the provisioning of acceptable QoS and QoE. To support the deployment scenario of high-speed trains, the highest mobility supported by 5G is 500 km/h. Different classes of mobility are defined:
  - Stationary: 0 km/h
  - Pedestrian: 0–10 km/h
  - Vehicular: 10–120 km/h
  - High Speed Vehicular: 120–500 km/h

  In 6G, the maximal speed of 1000 km/h is targeted if commercial airline systems are considered.

- *Connection density* is the KPI applied for the purpose of evaluation in the usage scenario of mMTC. Given a limited number of radio resources, the minimal number of devices with a relaxed QoS per square kilometer (km 2) is 10^6 in 5G, which is envisioned to be further improved 10 times to 10^7 per km 2.

- *Energy efficiency* is important to realize cost-efficient mobile networks and reduce the total Carbon Dioxide (CO_2) emission for green ICT, playing a critical role from the societal-economic respective. Energy efficiency of the network and the device can relate to the support for the following two features:
  (a) Efficient data transmission in a loaded case;
  (b) Low energy consumption when there is no data.

  The network and mobile devices should have the capability to support a high sleep ratio and long sleep duration for low energy consumption. The sleep ratio is the fraction of unoccupied time resources (for the network) or sleeping time (for the device) in a period of time corresponding to the cycle of the control signaling (for the network) or the cycle of discontinuous reception (for the device) when no user data transfer takes place. Furthermore, the sleep duration, i.e. the continuous period of time with no transmission (for network and device) and reception (for the device), should be sufficiently long. After the early deployment of 5G networks, there are already some complaints about its high energy consumption although the energy efficiency per bit has been substantially improved in comparison with the previous generations. In 6G networks, this KPI would be 10–100 times better over that of 5G so as to improve the energy efficiency per bit while reducing the overall power consumption of the mobile industry.

- *Area traffic capacity* is a measurement of the total mobile traffic that a network can accommodate per unit area, relating to the available bandwidth, spectrum efficiency, and network densification. The minimal requirement for 5G is

10 Mbps per square meter (m 2), which is expected to reach 1 Gbps/m^2 in some deployment scenarios such as indoor hot spots.

- *Reliability* relates to the capability of transmitting a given amount of traffic within a predetermined time duration with high success probability. This requirement is defined for the purpose of evaluation in the usage scenario of URLLC. In 5G networks, the minimum requirement for the reliability is measured by a success probability of $1-10^{-5}$ when transmitting a data packet of 32 bytes within 1 ms given the channel quality of coverage edge for the deployment scenario of urban macro environment. It is expected to improve at least two orders of magnitude, i.e. $1-10^{-7}$ or 99.999 99 % in the next-generation system.

- *Positioning accuracy* of the 5G positioning service is better than 10 m. Higher accuracy of positioning has a strong demand in many vertical and industrial APPs, especially in indoor environment that cannot be covered by satellite-based positioning systems. With the application of THz radio station, which has a strong potential in high-accuracy positioning, the accuracy supported by 6G networks is expected to reach centimeter (cm) level.

- *Coverage* in the definition of 5G requirement mainly focuses on the received quality of radio signal within a single base station. The coupling loss, which is defined as the total long-term channel loss over the link between a terminal and a base station and includes antenna gains, path loss, and shadowing, is utilized to measure the area served by a base station. In 6G networks, the connotation of coverage should be substantially extended considering that the coverage will be globally ubiquitous and will be shifted from only 2D in terrestrial networks to 3D in a terrestrial-satellite-aerial integrated system.

- *Security and privacy* are necessary for assessing whether the operation of a network is secure enough to protect infrastructure, devices, data, and assets. The main security tasks for mobile networks are *confidentiality* that prevents sensitive information from being exposing to unauthorized entities, *integrity* guaranteeing that information is not modified illegally, and *authentication* ensuring that the communicating parties are who they say they are. On the other hand, privacy becomes a high priority to address growing concern and privacy legislation such as the General Data Protection Regulation (GDPR) in Europe. Some KPIs can be applied to quantitatively measure security and privacy, e.g. percentage of security threats that are identified by threat identification algorithms, with which the effectiveness of anomaly detection can be evaluated.

- *Capital and operational expenditure (OPEX)* is a critical factor to measure the affordability of mobile services, influencing substantially the commercial success of a mobile system. The expenditure of a mobile operator can be divided into two main aspects: capital expenditure (CAPEX) that is the cost spent to

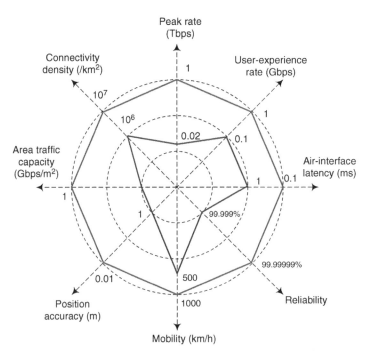

Figure 3.4 Improvement of key performance indicators from IMT-2020 to 6G. Source: Jiang et al. [2021]/With permission of IEEE.

build communication infrastructure and OPEX used for maintenance and operation. Due to the network densification, mobile operators suffer from a pressure of high CAPEX. Meanwhile, mobile networks' troubleshooting (systems failures, cyber-attacks, and performance degradations, etc.) still cannot avoid manual operations. A mobile operator has to keep an operational group with a large number of network administrators with high expertise, leading to a costly OPEX that is currently three times that of CAPEX and keeps rising [Jiang et al., 2017a]. During the design of 6G, the expenditure will be a key factor to consider. To provide a quantitative performance comparison between 5G and 6G, eight representative KPIs are visualized, as shown in Figure 3.4.

3.6 Research Initiatives and Roadmap

Even though discussions are ongoing within the wireless community regarding whether there is any need for 6G and whether counting the generations should be stopped at 5, a few pioneering research works on the next-generation wireless technologies have been initiated. This section provides the readers an up-to-date advance of 6G explorations from representative institutions, i.e. the ITU and Third Generation Partnership Project (3GPP), major countries, and companies. In addition, the possible roadmap of definition, specification, standardization, and regulation is demonstrated in Figure 3.5.

3.6.1 ITU

As a pointer to the new horizon for the future digital society and networks, ITU-T Focus Group "Technologies for Network 2030" (FG NET-2030) [ITU-T NET-2030, 2019] was established in July 2018. It aims to identify the gaps and technological challenges toward the capabilities of networks for the year 2030 and beyond when it is expected to satisfy extreme performance requirements to support disruptive use cases such as holographic-type communications, Tactile Internet, multi-sense networks, and digital twin. Although it mainly focuses on fixed communication networks, the future network architecture, requirements, use cases, and capabilities of the networks identified in this group will be a guideline for the definition of the 6G mobile system. Furthermore, the ITU-R Sector has recently published Recommendation M.2150 titled "Detailed specifications of the radio interfaces of International Mobile Telecommunications (IMT)-2020," which can be regarded as the finalization of 5G specification. With the great success achieved by ITU for the evolution of IMT-2000 (the third generation (3G)), IMT-Advanced (the 4G), and IMT-2020 (the 5G), a similar process will be applied once again for the development of *IMT toward 2030 and beyond*. According to the IMT process, ITU-R starts studying ITU Vision on 6G as the first step and then publishes the minimum requirements and evaluation criteria for *IMT toward 2030 and beyond* in the middle of the 2020s, and will step into an invitation for proposals and the evaluation phase afterward. At its meeting in February 2020, ITU-R working party 5D decided to start the study on future technology trends [ITU-R WP5D, 2020] and plans to complete this study at the meeting in June 2022. It invited organizations within and external to the ITU-R to provide inputs for its June and October meetings in 2021, which will help develop the first draft "Future Technology Trends towards 2030 and beyond." ITU-R is also responsible for organizing the world radiocommunication conference (WRC) that governs the frequency assignment, being held every three to four years. For example, the spectrum allocation issue for the 5G system was approved in WRC-19. It is expected that the WRC probably scheduled in 2023 (WRC-23) will discuss the spectrum issues for 6G, and the spectrum allocation for 6G communications may be formally decided in 2027 (WRC-27).

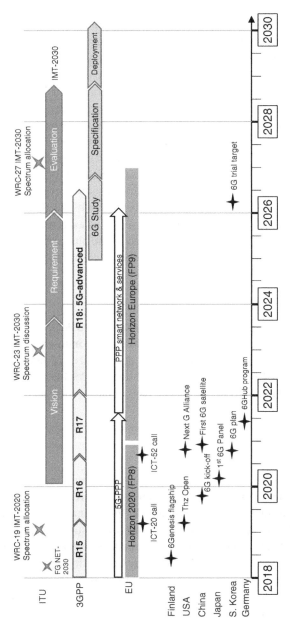

Figure 3.5 Estimation of a roadmap of research, definition, specification, spectrum regulation, development, and deployment for the 6G system. Source: Jiang et al. [2021]/With permission of IEEE.

3.6.2 Third Generation Partnership Project

In early 2019, the 3GPP has frozen the Release 15 specifications as the initial version of 5G standards. In July 2020, the subsequent release (i.e. Release 16) had completed as the enhancement of initial 5G standards [Ghosh et al., 2019]. Currently, a more advanced version (Release 17) is being specified by 3GPP and is expected to be completed in 2021 despite a delay due to the COVID-19 pandemic. Driven by a multitude of key stakeholders from the traditional mobile industry, a wide range of verticals, and the non-terrestrial industry, it is envisioned as the most versatile release in 3GPP history in terms of features content, including NR over non-terrestrial networks, NR beyond 52.6 MHz, NR sidelink enhancement, network automation, etc. In addition, 3GPP announced its evolution of 5G toward 5.5G with the official new name of *5G-Advanced*, which will be standardized in Release 18 and beyond. According to the experiences got in previous generations, 6G will be a disruptive system that will be developed without the restriction of backward compatibility. In parallel, therefore, 3GPP is expected to initiate the study item for 6G around the year 2025, followed by the phase of specification, to guarantee the first commercial deployment roll-out of 6G by 2030.

3.6.3 Industry

In the mobile industry, the vendors, mobile operators, and device manufacturers are the real driving forces. Hence, the viewpoints of the industry, especially the major vendors play an important role in the development of new generation. In January 2020, NTT Docomo published its white paper "5G Evolution and 6G," followed by other major players. We provide a summary of their main viewpoints in Table 3.3.

3.6.4 Europe

We emphasize the 6G research activities in Europe because their programs are not only the earliest worldwide but also the most transparent and open with rich public-available information. In the past decade, Europe has successfully carried out its 5G research and development under the 5G Infrastructure Public–Private Partnership (5G-PPP). Some key 5G concepts and technologies such as Network Function Virtualization (NFV) and Mobile Edge Computing (MEC), also known as Multi-access Edge Computing, were proposed and standardized by European Telecommunications Standards Institute (ETSI). In April 2018, the University of Oulu in Finland announced the world's first 6G research project called 6G-Enabled Wireless Smart Society and Ecosystem (6Genesis) as part of the Academy of Finland's flagship program. It focuses on ground-breaking 6G

Table 3.3 The 6G vision of mobile industry.

Company	White Paper	Release	Main viewpoints
NTT DOCOMO	5G Evolution and 6G	Jan. 2020	5G is the first generation mobile system utilizing the mmWave band and further improvement on coverage and mobility of mmWave technology is needed for 5G evolution. Extreme-high capacity and extreme-low latency in uplink from physical to cyber space, while extreme-high reliability and extreme-low latency in downlink from cyber to physical space, are needed to support the cyber-physical fusion, requiring substantial enhancement of uplink transmission in 5G evolution and 6G. The requirements of 6G will be new combinations of 5G requirements for new use cases or extreme requirements for specific use cases.
Nokia	Communications in the 6G era	Mar. 2020	5G will continue to be evolved to improve the performance by adopting new technologies in a backward-compatible manner. Modifications that are not compatible with the 5G framework or can only be implemented with high cost will be part of 6G, which provides the inter-connection of physical, biological, and digital worlds. Key 6G technologies are likely to be high spectrum bands and cognitive spectrum sharing, AI-driven air interface, new networking paradigms involving private networks and RAN-Core convergence, the integration of communication, sensing, and localization, extreme latency and reliability, and new security and privacy.
Rohde & Schwarz	5G evolution – on the path to 6G	Mar. 2020	It is important to follow standardization activities to get insight into what vendors, mobile operators, and device makers are attempting to accomplish and fix. Weaknesses in the 5G standard are the driver on the path toward 6G, and this is especially true with respect to the weaknesses in security and privacy. What will distinguish 6G development is that nations now recognize the significance of wireless standards for economic prosperity and national security.

Table 3.3 (Continued)

Company	White Paper	Release	Main viewpoints
Samsung	The next hyper-connected experience for all	Jul. 2020	The mega-trends driving the mobile industry toward 6G are massive connected machines, AI, openness of networks, and increasing societal impact of mobile communications. 6G will support three key services: truly immersive extended reality (XR), high-fidelity mobile hologram, and digital twin. Key technological enablers are THz technologies, advanced antenna techniques, evolved duplex technology, dynamic network topology, spectrum sharing, comprehensive AI, split computing, and high-precision network.
Ericsson	Ever-present intelligent communication: a research outlook toward 6G	Nov. 2020	Despite the built-in flexibility of 5G, it will evolve to 6G following the pull from society needs and the push from more advanced technologies. The main driving forces behind 6G are trustworthiness, sustainability, automatization, digitalization, and limitless connectivity anywhere, anytime, and for anything. Going beyond connectivity, 6G should become a trusted compute and data platform encouraging innovation and serving as the information backbone of society.
Huawei	Defining 5.5G for a Better, Intelligent World	Nov. 2020	Continuous evolution is necessary to fully unleash the potential of the technology, and the industry should work together to drive a thriving 5.5G ecosystem by making the most of sub-100 GHz spectrum. Three new scenarios, i.e. Uplink Centric Broadband Communication (UCBC) facilitating high-definition video uploading and machine vision, Real-Time Broadband Communication (RTBC) offering an immersive experience, and Harmonized Communication and Sensing (HCS) for autonomous driving and connected drones, are envisaged.

research with four interrelated strategic areas: wireless connectivity, distributed computing, devices and circuit technology, and services and APPs. After publishing the world's first 6G White Paper [Latva-aho et al., 2019] in September 2019 as an outcome of the first 6G Wireless Summit, a series of white papers covering 12 particular areas of interest such as ML, edge intelligence, localization, sensing, and security have been published. Within its eighth Framework Programme (FP8) for Research and Technological Development, also known as *Horizon2020*, the European Commission started beyond 5G research through ICT-20-2019 call "5G Long Term Evolution." Eight pioneering projects were kicked off in early 2020 after a competitive selection process with a total of 66 high-quality proposals submitted by consortia consisting of vendors, mobile operators, academia, research institutions, small companies, and verticals. At the beginning of 2021, another batch of research projects focusing on 6G sponsored through its call ICT-52-2020 "Smart Connectivity beyond 5G" were kicked off. The details of ICT-20 and ICT-52 research projects are summarized in Table 3.4. Following the success of the Horizon2020 5G-PPP program, the research and development of 6G in Europe will continue with the upcoming Public–Private Partnership (PPP) "Smart Network & Services" under the ninth Framework Program (FP9), also called Horizon Europe. In February 2020, the European Commission had also announced to accelerate investments in Europe's "Gigabit Connectivity" including 5G and 6G to shape Europe's digital future [EU Gigabit Connectivity, 2020].

3.6.5 The United States

In 2016, the U.S. Defense Advanced Research Projects Agency (DARPA), along with companies from the semiconductor and defense industries like Intel, Micron, and Analog Devices, have set up the joint university microelectronic project (JUMP) with six research centers to address existing and emerging challenges in microelectronic technologies. The Center for Converged TeraHertz Communications and Sensing (ComSecTer) aims to develop technologies for future cellular infrastructure. The Federal Communications Commission (FCC) announced in March 2019 that it opens up experimental licenses for the use of frequencies between 95 GHz and 3 THz for 6G and beyond, fostering the test of THz communications. In October 2020, the ATIS launched "Next G Alliance" with founding members including AT&T, T-Mobile, Verizon, Qualcomm, Ericsson, Nokia, Apple, Google, Facebook, Microsoft, etc. It is an industry initiative that intends to advance North American mobile technology leadership in 6G over the next decade through private sector-led efforts. With a strong emphasis on technology commercialization, its ambition is to encompass the entire lifecycle of 6G research and development, manufacturing, standardization, and market readiness. Another U.S. company, SpaceX, which is famous for its revolutionary

Table 3.4 European Commission Horizon 2020 5G-PPP beyond-5G and 6G research projects.

	Acronym	Project title	Major research topics
ICT-20	5G-CLARITY	Beyond 5G multi-tenant private networks integrating Cellular, Wi-Fi, and LiFi, Powered by ARtificial Intelligence and Intent Based PolicY	Private networks, AI-driven network automation, intent-based network
	5G-COMPLETE	A unified network, Computational and stOrage resource Management framework targeting end-to-end Performance optimization for secure 5G muLti-tEchnology and multi-Tenancy Environments	Computing-storage-network convergence, architecture, post-quantum cryptosystem, fiber-wireless fronthaul
	5G-ZORRO	Zero-touch security and trust for ubiquitous computing and connectivity in 5G networks	Security, privacy, distributed ledge technology (DLT), zero-touch automation, E2E network slicing
	ARIADNE	Artificial Intelligence Aided D-band Network for 5G Long Term Evolution	D-band, metasurfaces, AI-based management
	INSPIRE-5G+	INtelligent Security and PervasIve tRust for 5G and Beyond	Trusted multi-tenancy, security, AI, blockchain
	LOCUS	LOCalization and analytics on-demand embedded in the 5G ecosystem, for Ubiquitous vertical applicationS	Localization and analytics, location-as-a-service
	MonB5G	Distributed management of Network Slices in beyond 5G	Network slice management, zero-touch automation, AI-assisted security
	TERAWAY	Terahertz technology for ultra-broadband and ultra-wideband operation of backhaul and fronthaul links in systems with SDN management of network and radio resources	THz, photonics-defined transceiver, backhaul and fronthaul, network and resource management

Table 3.4 (Continued)

	Acronym	Project title	Major research topics
ICT-52	6G BRAINS	Bring Reinforcement-learning Into Radio Light Network for Massive Connections	THz, OWC, AI, 3D SLAM, D2D cell-free network, reinforcement learning
	AI@EDGE	A secure and reusable Artificial Intelligence platform for Edge computing in beyond 5G Networks	AI for network automation, AI-enabled network applications, edge computing, security
	DAEMON	Network intelligence for aDAptive and sElf-Learning MObile Networks	Network intelligence, AI, E2E architecture
	DEDICAT 6G	Dynamic coverage Extension and Distributed Intelligence for human Centric applications with assured security, privacy and trust: from 5G to 6G	Distributed intelligence, security and privacy, AI, blockchain, smart connectivity
	Hexa-X	A flagship for B5G/6G vision and intelligent fabric of technology enablers connecting human, physical, and digital worlds	High frequency, localization and sensing, connected intelligence, AI-driven air interface, 6G architecture
	MARSAL	Machine learning-based, networking and computing infrastructure resource management of 5G and beyond intelligent networks	Optical-wireless convergence, fixed-mobile convergence, distributed cell-free, O-RAN, AI, blockchain, secured multi-tenant slicing
	REINDEER	REsilient INteractive applications through hyper Diversity in Energy Efficient RadioWeaves technology	Intelligent surfaces, cell-free wireless access, distributed radio, computing, and storage, channel measurement
	RISE-6G	Reconfigurable Intelligent Sustainable Environments for 6G Wireless Networks	RIS, architecture and operation for multiple RISs, radio propagation modeling
	TeraFlow	Secured autonomic traffic management for a Tera of SDN flows	SDN, DLT, ML-based security, cloud-native arctecture

innovation of reusable rockets, announced the Starlink project [Foust, 2019] in 2015. Starlink is a very large-scale LEO communication satellite constellation aiming to offer ubiquitous Internet access services across the whole planet. The FCC has approved its first-stage plan to launch 12 000 satellites, and another application for 30 000 additional satellites is under consideration. Since its first launch in May 2019, more than 1100 satellites have been successfully deployed through 19 times space launching by far. In the first half of 2021, hundreds of Starlink satellites have been deployed through a dozen launches, achieving its previous plan of deploying 120 satellites per month through two launch times. As of this writing, the Starlink service has offered commercial services using more than 1500 satellites while the number of subscribers surpassed 60 000. It is too exaggerated when somebody claims that Starlink will replace 5G or it stands for 6G, but the impact of such a very large-scale LEO satellite constellation on 6G should be seriously taken into account in the mobile industry.

3.6.6 China

As early as 2018, the 5G working group at the Ministry of Industry and Information Technology of China started a concept study of potential 6G technologies, making China one of the first countries to explore 6G technology. In November 2019, the Ministry of Science and Technology of China had officially kicked off the 6G technology research and development work coordinated by the ministry, together with five other ministries or national institutions. A promotion working group from the government that is in charge of management and coordination. Besides, an overall expert group composed of 37 experts from universities, research institutes, and industry was established at this event. In November 2020, the first 6G experimental satellite developed by the University of Electronic Science and Technology of China was successfully launched. Its task is to test communications from space using a high-frequency terahertz spectrum. The leading Chinese ICT company, Huawei, said the company is now at the initial stage of 6G research. It provided a roadmap of 6G development with major milestones, including the vision for 6G by around 2023, standardization by 2026, rolling out relevant technologies by 2028, and preliminary commercial deployment by 2030. In 2020, another telecommunication equipment giant ZTE and one of the three Chinese mobile operators China Unicom initiated their cooperation on 6G technological innovation and standards. Recently, a consortium called the *6G Alliance of Network AI (6GANA)* was established under the lead of Huawei and China Mobile, focusing on integrating AI technologies with the mobile network to support AI as a Service (AIaaS) in the next-generation system.

3.6.7 Japan

In late 2017, a working group was established by the Japanese Ministry of Internal Affairs and Communications to study next-generation wireless technologies.

Their findings are that 6G should provide transmission rates at least ten times faster than 5G, near-instant connection, and massive connection of 10 million devices per square kilometer. In early 2020, the Japanese government set up a dedicated panel including representatives from the private sector and academia to discuss technological development, potential use cases, and policy. Japan reportedly intends to dedicate around $2 billion to encourage private-sector research and development for 6G technology.

3.6.8 South Korea

South Korea announced a plan to set up the first 6G trial in 2026 and is expected to spend around $169 million over five years to develop key 6G technologies. The trial aims to realize the peak data rate of 1 Tbps and extreme-low latency that is ten times lower than that of 5G. The government of South Korea will push the research programs in six key areas (hyper-performance, hyper-bandwidth, hyper-precision, hyper-space, hyper-intelligence, and hyper-trust) to preemptively secure next-generation technology. As Huawei's 5G technology is banned in the United States, Australia, and the United Kingdom, Samsung Electronics speeds up its global ambition to become a major telecommunication vendor. Samsung Research has established the Next-Generation Communication Research Center on the basis of Samsung Research's Standard Research Team that researches 6G, as the largest among Samsung Research's units. LG Electronics announced its ambition to lead global standardization on 6G and create new business opportunities. As a follow-up, it launched a 6G Research Center in January 2019 and the Korea Advanced Institute of Science and Technology (KAIST).

3.7 Key Technologies

To satisfy extreme performance requirements of disruptive use cases and APPs, disruptive technologies on radio transmission and networking would be developed and then applied to build the 6G system. By far, the wireless community has provided some views of potential technologies through academic publications, white papers, the topics of research projects, etc. This section provides the readers a complete view of potential 6G technological enablers, which can be categorized into several groups:

- *New Spectrum* consisting of mmWave, THz communications, and optical wireless communications
- *New Air Interface* including massive MIMO, intelligent reflecting surface, and next-generation multiple access
- *New Networking* comprising open Radio Access Network (RAN) and non-terrestrial networks
- *New Paradigm* empowered by the convergence of communication, computing, and sensing, as well as the integration with artificial-intelligence technology.

3.7.1 Millimeter Wave

The mmWave technology was introduced by the 5G new radio and is believed to remain an essential component in future 6G networks. Compared with legacy RF technologies working below 6 GHz, it significantly broadens the available bandwidth with new carrier frequencies up to 300 GHz. As Shannon's theorem revealed, such a huge bandwidth will inflate the channel capacity and quench the imminent thirst for higher data rates. Meanwhile, the shorter wavelength also leads to a smaller antenna size. Thus, it not only improves the portability and integration level of devices but also allows to increase the dimension of antenna arrays and, in addition to that, narrow the beams [Wang et al., 2018], which is beneficial to specific APPs such as detection radars and physical layer security. Furthermore, the atmospheric and molecular absorption exhibits highly variant characteristics at different frequencies across the mmWave band, providing potential for diverse use cases. On the one hand, low attenuation can be observed at particular bands such as 35, 94, 140, and 220 GHz, making long-distance transmission possible at these frequencies. On the other hand, severe propagation loss is experienced at some "attenuation peaks" such as 60, 120, and 180 GHz, which can benefit short-range covert networks with stringent safety requirements.

The benefits of mmWave technologies come with the price of distinct technical challenges. First of all, the broad bandwidth in the mmWave band and high transmission power can lead to severe non-linear signal distortions, imposing higher technical requirements for the integrated circuits than those for RF devices. Meanwhile, since the effective transmission range of mmWave, especially in the 60 GHz band, is severely limited by the atmospheric and molecular absorption, mmWave channels are commonly dominated by the Line-of-Sight (LOS) path. It becomes a major drawback that is further magnified by the poor diffraction at this short wavelength, which causes substantial blockage in scenarios with the dense presence of small-scale obstacles such as vehicles, pedestrians, or even the human body. High propagation loss and LOS-dependency also significantly raise the channel state sensitivity to the mobility, i.e. the impact of fading is much more substantial than that in the RF bands.

3.7.2 Terahertz Communications

Despite its current abundance in spectral redundancy, mmWave is hardly adequate to tackle the increasing cravenness on bandwidth for another decade. Therefore, looking forward to the 6G era, wireless technologies operating at even higher frequencies, such as THz or optical frequency bands, are expected to play an essential role in the next generation RAN, providing huge bandwidths.

Similar to mmWave, THz waves also suffer from high path loss and therefore highly rely on directive antennas and LOS channels while providing a very small coverage. However, when a strong LOS link is available, the high frequency brings significantly larger bandwidths than any legacy system, making it possible to simultaneously provide ultra-high performance in throughput, latency, and reliability. Moreover, compared with both mmWave systems working at lower frequencies and wireless optical systems working in higher frequency bands, THz communication systems are not so sensitive to atmospheric effects, which eases the tasks of beamforming and beam tracking. This shapes THz communication into a good supplementary solution in addition to the mainstream RF technologies for specific use cases, such as indoor communications and wireless backhaul; and a competitive option for future cyber-physical APPs with extreme QoS requirements.

Furthermore, the high frequency also allows smaller antenna size for a higher integration level. Thus, it is envisaged that more than 1000 antennas can be embedded into a single THz Base Station (BS) to provide hundreds of super-narrow beams simultaneously and overcome the high propagation loss [Zhang et al., 2019]. Thus, the support of extreme traffic capacity and massive connectivity unlocks its application in ultra-massive machine-type communications such as Internet-of-Everything (IoE). Nevertheless, while THz outperforms mmWave in many ways, it also faces severe technical challenges, especially from the aspect of implementing essential hardware circuits, including antennas [Vettikalladi et al., 2019], amplifiers [Tucek et al., 2016], and modulators. Primarily, it has been the most critical challenge for practical deployment of THz technologies to efficiently modulate baseband signals onto such high-frequency carriers with integrated circuits.

3.7.3 Optical Wireless Communications

Optical Wireless Communications points to wireless communications [Elgala et al., 2011] that use Infrared (IR), visible-light, or Ultraviolet (UV) frequency bands as the transmission medium. It is a promising complementary technology for traditional wireless communications operating over RF bands. The optical band can provide almost unlimited bandwidth without permission from spectrum regulators worldwide. It can be applied to realize high-speed, low-cost access thanks to the availability of optical emitters and detectors. Since the IR and UV waves have similar behavior as visible light, the security risks and interference can be significantly confined, and the concern about the radio radiation to human health can be eliminated. It is expected to have unique advantages in deployment scenarios that are sensitive to electromagnetic interference, such as vehicular communications in intelligent transportation systems, airplane passenger lighting, and medical machines. Despite its advantage, OWC suffers from the impairments such as ambient light noise, atmospheric loss, the nonlinearity of light-emitting diodes (LEDs), multi-path dispersion, and pointing errors.

OWC systems operating in the visible-light band are commonly referred to as Visible Light Communications (VLC), attracted much attention from both academia and industry recently. VLC operates in the frequency range of 400–800 THz. Differing from the RF technologies in lower THz range that use antennas, VLC relies on illumination sources – especially LEDs – and image-sensor or photodiode arrays to implement the transceivers. With these transceivers, a large bandwidth can easily achieve with low power consumption (100 mW for 10–100 Mbps) without generating electromagnetic or radio interference. The high power efficiency, long lifetime (up to 10 years), and low cost of mainstream LEDs, in addition to the unlicensed access to spectrum, make VLC an attractive solution for use cases sensitive to battery life and access costs, such as massive IoT and Wireless Sensor Networks (WSNs). Moreover, VLC also exhibits better propagation performance than RF technologies in some non-terrestrial scenarios, such as aerospace and underwater, which can be an important aspect of the future 6G ecosystem.

Terrestrial point-to-point OWC, also known as Free-Space Optical (FSO) communications [Juarez et al., 2006], takes place at the near IR band. Using a high-power, high-concentrated Laser beam at the transmitter, the FSO system can achieve a high data rate, i.e. 10 Gbps per wavelength, over long distances (up to 10 000 km). Thus, it offers a cost-effective solution for the backhaul bottleneck in terrestrial networks, enables crosslinks among space, air, and ground platforms, and facilitates high-capacity inter-satellite links for the emerging LEO satellite constellation. Furthermore, there has also been growing interest in UV communication [Xu and Sadler, 2008] as a result of recent progress in solid-state optical transmitters and detectors for non-LOS UV communications that offer broad coverage and high security.

3.7.4 Massive MIMO

Massive Multiple-Input Multiple-output (Massive MIMO) has already been applied as a critical component of 5G NR. With a large number of antennas, the transmission energy can focus with extreme sharpness on a tiny area. Directivity can bring significant improvement in spectral efficiency and energy efficiency. Massive MIMO is cost-efficient by leveraging low-cost, low-precision RF components, where hundreds of cheap power amplifiers in the magnitude order of milli-Watt output power can replace expensive, high-linear power amplifiers used for conventional systems. For lower frequencies, NR employs a moderate number of antennas (up to 64 transmit and receive antennas at the base station side around 700 MHz). For higher frequencies, a more significant number of antennas (NR supports up to 256 transmit and receive antennas around 4 GHz) can be employed with the same hardware size, which increases the capability for beam-forming and MU-MIMO. Since the number of reference signals is proportional to the number of transmit antennas, massive MIMO has to operate in the TDD mode by exploiting channel reciprocity. Analog beamforming is typically wanted for even higher frequencies (in the mmWave range), limiting the transmission to a single beam direction per time unit and radio chain. It is expected that massive MIMO will also play a critical role in the 6G air interface, especially for deploying higher mmWave frequency bands and THz bands. The number of antenna elements will be further increased to an extremely large number, imposing technical challenges on precoding schemes, detection algorithms, RF implementation, hardware impairments, and interference management. For the success of 6G, more spectral-efficient, energy-efficient, and cost-efficient massive MIMO schemes are expected. In addition to the collocated massive MIMO, a kind of distributed massive MIMO scheme referred to as cell-free massive MIMO has recently received much attention from both academia and industry [Ngo et al., 2017]. There are no cells or cell boundaries. A large number of distributed access points (APs) serve a smaller number of users at the same time-frequency resource, offering a uniform QoS for all users to eliminate the under-served problem at the cell edge of conventional cellular networks. Different aspects of cell-free massive MIMO such as resource allocation, power control, pilot assignment, energy efficiency, backhaul constraint, and scalability have been studied. Cell-free massive MIMO provides a promising solution for the deployment of private networks for industrial APPs. The synergy between cell-free and cellular networks is expected to bring a driving force for the next-generation mobile networks.

3.7.5 Intelligent Reflecting Surfaces

While releasing a significant bandwidth to support high throughput, the use of high-frequency bands over 6 GHz also introduces new challenges, such as higher propagation loss, lower diffraction, and severe blockage. In the frequency range of mmWave, massive MIMO has been proven effective in realizing active beamforming to provide high antenna gain to compensate for propagation loss. Nevertheless, its capability can be insufficient for the future 6G new spectrum. Among all potential solutions to enhance current beamforming approaches, the technology of Intelligent Reflecting Surfaces (IRS) has been widely considered promising for 6G mobile networks.

The so-called IRS, a.k.a. Reconfigurable Intelligent Surfaces (RIS) [ElMossallamy et al., 2020], is assembled by a category of programmable and reconfigurable material sheets that are capable of adaptively modifying their radio reflecting characteristics. When attached to environmental surfaces such as walls, glasses, and ceilings, IRS enables to transform wireless environment into smart reconfigurable reflectors, known as Smart Radio Environment (SRE). Consequently, it forms a kind of passive beamforming that can significantly improve the channel gain at low implementation cost and low power consumption compared with active massive MIMO antenna arrays. Moreover, unlike antenna arrays that must be compact enough for integration, SREs are implemented on large-size surfaces, making it easier to realize accurate beamforming with ultra-narrow beams. Furthermore, unlike active mMIMO antenna arrays that implement specifically for every individual Radio Access Technology (RAT), the passive reflection mechanism that IRS is relying on works almost universal for all RF and optical frequencies, which is especially cost beneficial for the 6G systems that work in an ultra-broad spectrum. Though IRS is showing great technical competitiveness in the context of the 6G new spectrum, it still lacks mature techniques for accurate modeling and estimation of the channels and the surface themselves, especially in the near-field range. Moreover, commercial deployment is only possible after addressing the business concern that IRS relies on external assessments, such as buildings that do not belong to the mobile operators. Therefore, it calls for a thoughtful design and standardization of framework providing virtual interfaces, agreements, and signaling protocols, so that 6G operators become capable of widely accessing and exploiting IRS-equipped objects in public and private domains.

3.7.6 Next-Generation Multiple Access

Both LTE and NR adopted OFDMA [Jiang and Kaiser, 2016] as the orthogonal multiple-access technique, which is a typical instance of OMA technologies prohibiting the same physical resource from supporting multiple users simultaneously. Compared with CDMA deployed in 3G systems, OFDMA shows a conspicuous superiority in combating multi-path fading by simple channel equalization. Furthermore, when combined with MIMO, OFDMA is capable of outperforming CDMA in spectral efficiency overwhelmingly. Nevertheless, the full performance of MIMO-OFDM highly relies on the MIMO precoding and resource mapping, which have to be precisely adapted to the channel condition to achieve the optimum. As the dimension of MIMO increases, from up to 8×4 in LTE-A gradually to over 256×32 massive MIMO, eventually to the future ultra-massive MIMO (e.g. 1024×64), the complexity of MIMO-OFDM adaptation is dramatically increasing. Meanwhile, in response to the demand for higher mobility, which implies higher fading dynamics, the computation latency constraint to this online adaptation procedure is also becoming more strict. To cope with these emerging challenges, a new architecture of AI-driven MIMO-OFDM transceivers has been proposed toward future 6G systems, which relies on AI techniques to efficiently solve the problems of online MIMO precoding and resource mapping.

Compared with orthogonal multiple access in most traditional cellular networks, Non-Orthogonal Multiple Access (NOMA) can offer higher system throughput and connection densities by allowing multiple users to share the same radio-resource unit. As a special case of the NOMA technique, Multi-User Superposed Transmission (MUST) has been studied in LTE Release 13, mainly focusing on the downlink transmission [Chen et al., 2018]. Grant-based NOMA, which typically operates in Radio Resource Control (RRC) connected state, was specified in Release 14 LTE to support downlink eMBB. The study item of NOMA in Release 16 was on uplink grant-free transmission, which can reduce the control signaling overhead, transmission latency, and devices' power consumption. Since the beyond 5G and the 6G networks are expected to support massive connectivity, NOMA will play an important role in next-generation networks. Recent studies have also demonstrated that NOMA can be effectively exploited in the new spectrum, including mmWave, THz, and optical frequencies. Additionally, when deployed together with CoMP, NOMA has been proven to outperform CoMP-OMA in both power efficiency and spectral efficiency. Being entirely based on successive interference cancellation, NOMA has a significantly higher complexity in its receiver design than OMA, which increases in polynomial or even exponential order along with the number of users. Especially in some scenarios that require cooperative decoding across different UEs, specific D2D interfaces must be reserved for this functionality, and security/trust concerns shall be taken into account to enable the deployment of NOMA in 6G.

3.7.7 Open Radio Access Network

To support disruptive use cases, the network infrastructure is expected to be flexible, intelligent, and open to multi-vendor equipment and multi-tenancy. To that aim, softwarization, cloudification, virtualization, and network slicing will be further tailored to the 6G network. In addition, a new network paradigm referred to as Open Radio Access Network (O-RAN) has recently received much attention from both academia and industry. The critical concepts of O-RAN, including its vision, architecture, interfaces, technologies, objectives, and other vital aspects, were introduced by the O-RAN alliance in a white paper [Lin et al., 2018]. The O-RAN alliance has then further studied the use cases leveraging the O-RAN architecture to demonstrate its capability in real-time behavior. The main objective of the openness and intelligence in RAN architecture is to build a radio network that is resource-efficient, cost-effective, software-driven, virtualized, slicing-aware, centralized, open-source, open hardware, intelligent, and therefore more flexible and dynamic than any previous generation of mobile networks. To do so, the research community has introduced the utilization of AI and ML techniques on every single layer of the RAN architecture to fulfill the requirements of dense network edge in beyond 5G and 6G mobile communication systems.

Opening up the RAN from a single vendor environment to a standardized, open, multi-vendor, and AI-powered hierarchical structure allows third parties and mobile operators to deploy innovative APPs and emerging services that cannot be deployed or supported in legacy RAN architectures. In addition, the O-RAN is built upon the NFV Management and Orchestration (NFV-MANO) reference architecture proposed by the ETSI, which deploys commercial off-the-shelf hardware components, virtualization techniques, and software pieces. The virtual machines abstracted (or virtualized) from the underlying physical resources are easily created, deployed, configured, and decommissioned. Therefore, such virtualized environment brings flexibility to the O-RAN architecture and lowers the CAPEX, OPEX, and energy consumption. Despite the flexibility and interoperability that O-RAN offers, it also has several key problems that require further research efforts for its full realization in future mobile networks, including the convergence of multi-vendor technologies on the same platform, the harmonization of various management and orchestration frameworks, and the validation and troubleshooting of performance issues related to the network. Researchers from industry and academia are also expected to take part in theoretical analysis and practical roll-outs of this technology toward an open and intelligent RAN for 6G mobile networks to overcome these challenges.

3.7.8 Non-Terrestrial Networks

The focus of legacy cellular systems was on terrestrial infrastructure, leading to a problem of wide-area coverage. In marine, oceanic, and wild terrestrial areas, which are impossible or economically challenging to be covered by terrestrial cellular networks, satellites have been since long the most common communication solution. Aiming at a better coverage rate, deploying non-terrestrial infrastructures as part of the 6G network is being treated as an emerging topic, known as the Integrated Space and Terrestrial Network (ISTN). An ISTN is expected to consist of three layers:

- The ground layer built by terrestrial base stations
- The airborne layer empowered by High Altitude Platform (HAP) and UAV
- The spaceborne layer implemented by a constellation of satellites

Terrestrial networks cover only a small portion of the whole surface of the Earth. First, it is technically impossible to install terrestrial base stations to offer large-scale coverage in ocean and desert. Second, it is difficult to cover extreme topographies, e.g. high mountain areas, valleys, and cliffs, while it is not cost-efficient to use a terrestrial network to provide services for sparsely populated areas. Third, terrestrial networks are vulnerable to natural disasters such as earthquake, flood, hurricane, and tsunami, where there is a vital demand of communications but the infrastructure is destroyed or in an outage of service. With the expansion of human activity, e.g. the passengers in commercial planes and cruise ships, the demand for MBB services in uncovered areas increasingly grows. Also, the connectivity demand of IoT deployment scenarios like wild environmental monitoring, offshore wind farm, and smart grid requires wide-area ubiquitous coverage. Satellite communications have been the most common solution for wide coverage but currently the mobile communication service offered by GEO satellites is costly, low-data rate, and high latency due to the expensive cost for launching and its wide-area coverage (1/3 surface of Earth per GEO satellite).

The satellites in LEO [Hu and Li, 2001] have some advantages over GEO satellites for providing communication services. An LEO satellite operates in orbit generally lower than 1000 km, which can substantially lower the latency due to the signal propagation compared with the GEO satellite in the orbit of 36 000 km. Meanwhile, the propagation loss of LEO is much smaller, facilitating the direct connectivity to mobile and IoT devices that are strictly constraint by the battery supply. Moreover, a stationary ground terminal like an IoT device mounted in a monitoring position may suffer from an obstacle in the line of sight from GEO. An early attempt to implement a global satellite mobile communication system, i.e. the Iridium constellation, became commercially available in November 1998. It consists of 66 LEO satellites at the altitude of approximately 781 km and provides mobile phone and data services over the entire Earth surface. Even though it fails due to expensive costs and lack of demands, it is a significant technological breakthrough. It still operates today, and the second-generation Iridium system was successfully deployed in 2019. In recent years, the high-tech company SpaceX gains much attention due to its revolutionary development of space launching technologies. Its reusable rockets, namely Falcon 9, dramatically lower the cost of space launching, opening the possibility for deploying large space infrastructures. Looking forward to a future global ubiquitous converge available anywhere and anytime, it is strongly suggested to integrate satellite networks into the 6G network as part of it.

3.7.9 Artificial Intelligence

On the list of 6G enabling technologies, AI is recognized as the most potential one. As mobile networks are increasingly complex and heterogeneous, many optimization tasks become intractable, offering an opportunity for advanced

ML techniques. Categorized typically into supervised, unsupervised, and reinforcement learning, ML is considered as a promising data-driven tool to provide computational radio and network intelligence from the physical layer [Jiang and Schotten, 2018] to network management [Jiang et al., 2017b]. As a sub-branch of ML, deep learning [Jiang and Schotten, 2020] can mimic biological nervous systems and automatically extract features, extending across all three mentioned learning paradigms. It has a wide variety of APPs in wireless communications, where it can be applied to form more adaptive transmission (power, precoder, coding rate, and modulation constellation) in massive MIMO, enable more accurate estimation and prediction of fading channels, provide a more efficient RF design (pre-distortion for power amplifier compensation, beam-forming, and crest-factor reduction), deliver a better solution for intelligent network management, and offer more efficient orchestration for MEC, networking slicing, and virtual resources management.

In addition to deep learning, a few cutting-edge ML techniques represented by federated learning and transfer learning start showing strong potential in wireless communications. Data-driven methods always have to take into account the issue of data privacy, which limits the manner of processing collected data. In some scenarios, distributing data is strictly prohibited, and only local processing on the device where the data was collected is allowed. Federated learning is a method to fulfill this requirement by processing the raw data locally and distributing the processed data in a masked form. The mask is designed such that each of the individual data processing exposes no information, whereas their cooperation allows for meaningful parameter adjustments toward a universal model. While federated learning gives a method of training ML models from many data sources without ever exposing sensitive data, it creates only one shared model for universal applicability. When individual adjustments of models are required for their deployment to be successful, transfer learning can be used to enable these adjustments and do so in a manner requiring a much lower amount of data. By reusing the major part of pre-trained models in a different environment and only adjusting some of the parameters, transfer learning is able to provide quick adaptations using only a low amount of local data.

In addition to using AI to assist with the operation of the networks (i.e. AI for Networking), it is also essential to use the ubiquitous computing, connectivity, storage resources to provide mobile AI services to end-users in an AI-as-a-Service paradigm (i.e. Networking for AI). Principally, this provides deep edge resources to enable AI-based computation for new-style terminals such as robots, smart cars, drones, and VR glasses, which demand a large number of computing resources but are limited by embedded computing components and power supply. Such AI tasks mainly mean the traditional computation-intensive AI tasks, e.g. computer vision, SLAM, speech and facial recognition, natural language processing, and motion control.

3.7.10 Communication-Computing-Sensing Convergence

Mobile edge networks provide computing and caching capabilities at the network edge, making low-latency, high-bandwidth, location-aware pervasive computing services a reality. With the proliferation of the IoT and Tactile Internet, a huge number of sensors and actuators are connected to mobile networks. The next-generation system is envisioned to become a giant computer that would converge ubiquitous communication, computation, storage, sensing, and controlling as a whole to provide disruptive APPs. Due to their superiority in integration and mobility over wired connections, wireless connectivity is gradually applied in modern and future controlling systems to close the signal loop, giving birth to the usage scenario of URLLC. The spirit of system design behind this concept follows the traditional methodology of independently designing control and communication. It starts from designing the controlling component without concerning the characteristics of the communication system, where a set of communication requirements will be raised regarding the expected controlling performance. There follows the design of a wireless system, aiming at achieving the target performance proposed by the last stage. The KPI requirements of URLLC, such as 99.999% reliability and 10 ms, were formulated for a generic controlling scenario in this fashion. Nevertheless, recent studies have revealed a necessity of in-loop co-designing of communication and controlling systems tightly coupled. For instance, the close-loop reliability of the controlling system has been proven exponentially decreasing along with the Age of Information (AoI) over the feedback channel. Meanwhile, the AoI over control/feedback channels, as a communication metric, is *convex* about the arrival rates of controlling command and feedback information, which correspond to the sampling rate of sensors and decision rate in the control system. The performance of communication systems is therefore limited by the design of sensing and controlling systems.

Similar issues are also to be addressed in cloud computing. While it occasionally happens that some computing task occupies the cloud server for a long time, blocking all other pending tasks in the waiting queue and causing severe congestion, its source may have re-issued the same task with a more up-to-date status, making the previous task outdated and lack of utility. Preemption of the server, i.e. terminating the ongoing task in advance of its completion, will help reduce the aging of the task and improve the quality of cloud computing service. Furthermore, it is also critical for reducing the AoI to schedule the order of computing tasks from multiple APPs to be offloaded to the cloud. However, optimal decisions of such preemption and scheduling cannot be solely solved by the computing server, nor by the communication system, but only achievable in a collaboration among the terminal devices, the network controller, and the cloud computing server.

3.8 Conclusions

This chapter provided a comprehensive vision of the driving forces, use cases, usage scenarios, technical performance requirements, research initiatives, and technological enablers for the sixth-generation mobile system. The traditional evolution of a new generation every decade will not terminate at 5G, and the first 6G network is expected to deploy in 2030 or even earlier. 6G will enable unprecedented use cases and APPs that cannot be supported by 5G, e.g. holographic-type communications, pervasive intelligence, and ubiquitous global coverage, as well as others that we are unable to imagine yet. The 6G system is required to meet extremely stringent requirements on throughput, latency, reliability, coverage, mobility, and security, which will be implemented by adopting a magnitude of disruptive technologies. The potential 6G key technologies probably include millimeter wave, terahertz, optical wireless communications, massive MIMO, intelligent reflecting surfaces, next-generation multiple access, O-RANs, non-terrestrial networks, AI, and communication-computing-sensing convergence. The rest of this book will introduce some of these technologies in detail. We will mainly focus on radio transmission and signal processing of the next-generation wireless communications.

References

Andrews, J. G., Buzzi, S., Choi, W., Hanly, S. V., Lozano, A., Soong, A. C. K. and Zhang, J. C. [2014], 'What will 5G be?', *IEEE Journal on Selected Areas in Communications* **32**(6), 1065–1082.

Chen, Y., Bayesteh, A., Wu, Y., Ren, B., Kang, S., Sun, S., Xiong, Q., Qian, C., Yu, B., Ding, Z., Wang, S., Han, S., Hou, X., Lin, H., Visoz, R. and Razavi, R. [2018], 'Toward the standardization of non-orthogonal multiple access for next generation wireless networks', *IEEE Communications Magazine* **56**(3), 19–27.

Clemm, A., Vega, M. T., Ravuri, H. K., Wauters, T. and Turck, F. D. [2020], 'Toward truly immersive holographic-type communication: Challenges and solutions', *IEEE Communications Magazine* **58**(1), 93–99.

Elgala, H., Mesleh, R. and Haas, H. [2011], 'Indoor optical wireless communication: Potential and state-of-the-art', *IEEE Communications Magazine* **49**(9), 56–62.

ElMossallamy, M. A., Zhang, H., Song, L., Seddik, K. G., Han, Z. and Li, G. Y. [2020], 'Reconfigurable intelligent surfaces for wireless communications: Principles, challenges, and opportunities', *IEEE Transactions on Cognitive Communications and Networking* **2**(3), 990–1002.

Ericsson Report [2020], Mobile data traffic outlook, Report, Ericsson.

EU Gigabit Connectivity [2020], Shaping Europe's digital future, Communication COM(2020)67, European Commission, Brussels, Belgium.

Fettweis, G., Boche, H., Wiegand, T., Zielinski, E., Schotten, H. D., Merz, P., Hirche, S., Festag, A., Häffner, W., Meyer, M., Steinbach, E., Kraemer, R., Steinmetz, R., Hofmann, F., Eisert, P., Scholl, R., Ellinger, F., Weiß, E. and Riedel, I. [2014], The Tactile Internet, Technology Watch Report, ITU-T.

Fitzek, F. H. P. and Seeling, P. [2020], 'Why we should not talk about 6G', *arXiv*.

Foust, J. [2019], 'SpaceX's space-internet woes: Despite technical glitches, the company plans to launch the first of nearly 12,000 satellites in 2019', *IEEE Spectrum* **56**(1), 50–51.

Ghosh, A., Maeder, A., Baker, M. and Chandramouli, D. [2019], '5G evolution: A view on 5G cellular technology beyond 3GPP release 15', *IEEE Access* **7**, 127639–127651.

Hu, Y. and Li, V. O. K. [2001], 'Satellite-based internet: A tutorial', *IEEE Communications Magazine* **39**(3), 154–162.

Huawei NetX2025 [2021], Netx2025 target network technical white paper, White Paper, Huawei.

Huawei VR Report [2018], Cloud VR network solution white paper, White Paper, Huawei.

ITU-R M.2083 [2015], IMT Vision-Framework and overall objectives of the future development of IMT for 2020 and beyond, Recommendation M.2083-0, ITU-R.

ITU-R M.2370 [2015], IMT traffic estimates for the years 2020 to 2030, Recommendation M.2370-0, ITU-R.

ITU-R M.2410 [2017], Minimum requirements related to technical performance for IMT-2020 radio interface(s), Recommendation M.2410-0, ITU-R.

ITU-R WP5D [2020], Future technology trends for the evolution of IMT towards 2030 and beyond, Liaison Statement, ITU-R Working Party 5D.

ITU-T NET-2030 [2019], A blueprint of technology, applications and market drivers towards the year 2030 and beyond, White Paper, ITU-T Focus Group NET-2030.

Jiang, W. and Kaiser, T. [2016], From OFDM to FBMC: Principles and comparisons, *in* F. L. Luo and C. Zhang, eds, '*Signal Processing for 5G: Algorithms and Implementations*', John Wiley&Sons and IEEE Press, United Kindom, chapter 3.

Jiang, W. and Schotten, H. D. [2018], Multi-antenna fading channel prediction empowered by artificial intelligence, *in* 'Proceedings of the 2018 IEEE Vehicular Technology Conference (VTC)', Chicago, USA.

Jiang, W. and Schotten, H. D. [2020], 'Deep learning for fading channel prediction', *IEEE Open Journal of the Communications Society* **1**, 320–332.

Jiang, W., Strufe, M. and Schotten, H. D. [2017a], Experimental results for artificial intelligence-based self-organized 5G networks, *in* 'Proceedings of the 2017 IEEE

International Symposium on Personal, Indoor and Mobile Radio Communications (IEEE PIMRC)', Montreal, QC, Canada.

Jiang, W., Strufe, M. and Schotten, H. D. [2017b], A SON decision-making framework for intelligent management in 5G mobile networks, *in* 'Proceedings of the 2017 IEEE International Conference on Computer and Communication (ICCC)', Chengdu, China.

Jiang, W., Han, B., Habibi, M. A. and Schotten, H. D. [2021], 'The road towards 6G: A comprehensive survey', *IEEE Open Journal of the Communications Society* **2**, 334–366.

Juarez, J. C., Dwivedi, A., Hammons, A. R., Jones, S. D., Weerackody, V. and Nichols, R. A. [2006], 'Free-space optical communications for next-generation military networks', *IEEE Communications Magazine* **44**(11), 46–51.

Latva-aho, M., Leppänen, K., Clazzer, F. and Munari, A. [2019], Key drivers and research challenges for 6G ubiquitous wireless intelligence, White paper, University of Oulu.

Letaief, K. B., Chen, W., Shi, Y., Zhang, J. and Zhang, Y.-J. A. [2019], 'The roadmap to 6G: AI empowered wireless networks', *IEEE Communications Magazine* **57**(8), 84–90.

Lin, C., Katti, S., Coletti, C., Diego, W., Duan, R., Ghassemzadeh, S., Gupta, D., Huang, J., Joshi, K., Matsukawa, R., Suciu, L., Sun, J., Sun, Q., Umesh, A. and Yan, K. [2018], O-RAN: Towards an Open and Smart RAN, White Paper, O-RAN Alliance.

Ngo, H. Q., Ashikhmin, A., Yang, H., Larsson, E. G. and Marzetta, T. L. [2017], 'Cell-free massive MIMO versus small cells', *IEEE Transactions on Wireless Communications* **16**(3), 1834–1850.

Qu, Z., Zhang, G., Cao, H. and Xie, J. [2017], 'LEO satellite constellation for internet of things', *IEEE Access* **5**, 18391–18401.

Tucek, J. C., Basten, M. A., Gallagher, D. A. and Kreischer, K. E. [2016], Operation of a compact 1.03 THz power amplifier, *in* 'Proceedings of the 2016 IEEE International Vacuum Electronics Conference (IVEC)', Monterey, CA, USA, pp. 1–2.

Vettikalladi, H., Sethi, W. T., Abas, A. F. B., Ko, W., Alkanhal, M. A. and Himdi, M. [2019], 'Sub-THz antenna for high-speed wireless communication systems', *International Journal of Antennas and Propagation* **2019**.

Wang, X., Kong, L., Kong, F., Qiu, F., Xia, M., Arnon, S. and Chen, G. [2018], 'Millimeter wave communication: A comprehensive survey', *IEEE Communications Surveys & Tutorials* **20**(3), 1616–1653.

Xu, Z. and Sadler, B. M. [2008], 'Ultraviolet communications: Potential and state-of-the-art', *IEEE Communications Magazine* **46**(5), 67–73.

Zhang, Z., Xiao, Y., Ma, Z., Xiao, M., Ding, Z., Lei, X., Karagiannidis, G. K. and Fan, P. [2019], '6G wireless networks: Vision, requirements, architecture, and key technologies', *IEEE Vehicular Technology Magazine* **14**(3), 28–41.

Part II

Full-Spectra Wireless Communications in 6G

4

Enhanced Millimeter-Wave Wireless Communications in 6G

Traditional mobile cellular systems were built upon low-frequency bands, which have favorable propagation characteristics, resulting in a large coverage area for outdoor communications and good penetration of buildings for indoor communications. However, the spectrum below 6 GHz has overcrowded for a long time, and available spectral resources for the IMT services are highly constrained. Millimeter wave (mmWave) refers to the spectrum bands from 30 to 300 GHz, corresponding to the range of signal wavelengths from 1 to 10 mm. On the one hand, mmWave provides high potential for next-generation mobile communications due to abundant spectral resources and contiguous large-volume bandwidths. On the other hand, the characteristics of mmWave signal propagation are distinct from that of low-frequency bands at Ultra High Frequency and microwave bands. High propagation and penetration losses impose a lot of technical challenges on the design of mmWave communications systems.

This chapter will focus on the characteristics of mmWave propagation and key mmWave transmission technologies, consisting of

- Large-scale fading of mmWave signal propagation, including free-space path loss, atmospheric attenuation, weather effects, and shadowing.
- Small-scale fading of mmWave signal propagation occurs over the distances on the order of the carrier wavelength, caused by the constructive and destructive superposition of multi-path signal components.
- Large-scale and small-scale channel models for mmWave frequencies up to 100 GHz.
- The potential and challenges for beamforming at mmWave frequencies, and the principles of digital beamforming, analog beamforming, and hybrid beamforming.

6G Key Technologies: A Comprehensive Guide, First Edition. Wei Jiang and Fa-Long Luo.
© 2023 The Institute of Electrical and Electronics Engineers, Inc. Published 2023 by John Wiley & Sons, Inc.

- The technical challenges of initial access for a mmWave communication system, multi-beam synchronization and broadcasting schemes, and the standardized initial-access procedures for LTE and NR.
- Omnidirectional coverage of broadcast channels in multi-antenna systems, including random beamforming and its enhanced schemes.

4.1 Spectrum Shortage

Previous generations of cellular systems usually adopted low-frequency bands ranging from several hundred megahertz to several gigahertz due to the favorable propagation characteristics over these bands. To be specific, most of the first-generation systems operated in the frequency bands around 800 MHz while the second-generation systems utilized both 900 and 1800 MHz to support a higher system capacity. The third- and fourth-generation systems further extended the frequency bands for International Mobile Telecommunications (IMT) services to 2.1 and 2.6 GHz, respectively. Transmitting radio signals at low-frequency bands receives an advantage of small free-space path losses (FSPLs), facilitating wide-area coverage and long-range transmission with moderate radiation power. Furthermore, these signals are easy to penetrate buildings and water. Consequently, the spectrum up to 6 GHz, especially below 3 GHz, is the desired frequency bands for IMT services since mobile operators can offer good outdoor and indoor coverage with cost-efficient cellular networks.

With the proliferation of the mobile Internet, the demand for more spectral resources to satisfy the explosive traffic growth became urgent, whereas it is very challenging to find enough spectrum, especially with large contiguous bandwidths, within sub-6 GHz frequency band. On the one hand, spectral resources over these bands are pretty limited compared with the demand for mobile broadband. On the other hand, there are a wide variety of applications operating in these bands, such as radar, radio navigation, radiolocation, radio astronomy, broadcasting (AM, FM, and Television), maritime mobile, aeronautical mobile, satellite communications, and amateur radio, as illustrated in Figure 4.1, leading to a global bandwidth shortage. Therefore, it inspires mobile network operators to explore the underutilized millimeter-wave spectrum to offer mobile broadband services. Millimeter wave, also known as millimeter band, refers to electromagnetic spectrum with wavelengths between 1 mm (300 GHz) and 10 mm (equivalent to 30 GHz). It was envisaged that mobile communications could benefit from mmWave bands by a set of use cases, e.g. low-cost fiber replacement for mobile backhaul, dense mmWave small cells, wireless broadband access, and low-latency uncompressed high-definition media transfers [Rappaport et al., 2014].

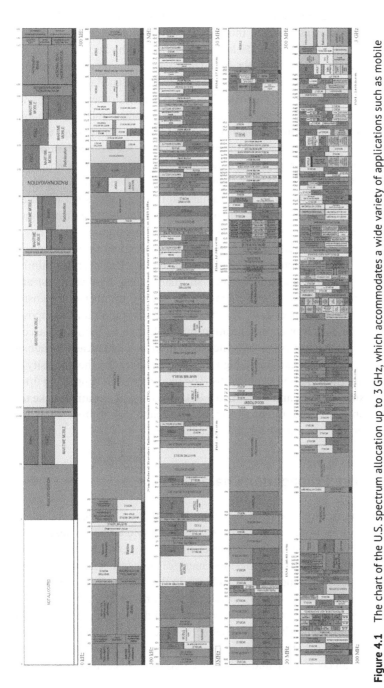

Figure 4.1 The chart of the U.S. spectrum allocation up to 3 GHz, which accommodates a wide variety of applications such as mobile cellular communications, wireless communications, radar, radio navigation, radio astronomy, broadcasting (AM, FM, and Television), maritime mobile, aeronautical mobile, satellite, and amateur radio. Source: The public website of the U.S. Department of Commerce, National Telecommunications and Information Administration, Office of Spectrum Management.

3GPP specified the relevant spectrum for 5G New Radio (NR), which was divided into two frequency ranges: the First Frequency Range (FR1), including sub-6 GHz frequency bands from 450 MHz to 6 GHz, and the Second Frequency Range (FR2) covering 24.25–52.6 GHz. Initial mmWave deployments are expected in 28 GHz (3GPP NR band n257 and n261) and 39 GHz (3GPP n260), followed by 26 GHz (3GPP n258). More mmWave bands are needed to be exploited with the rising demand for NR services. However, its practical rollout in mobile networks imposes significant technical challenges for the system design, wireless transmission techniques, and communication protocols. For example, mmWave signals suffer from high propagation losses and are vulnerable to atmospheric attenuation due to water vapor and oxygen absorption, which is negligible at low frequencies. In addition, mmWave signals cannot penetrate solid materials such as reinforced concrete walls, while human bodies can block a direct signal path. In order to compensate for such significant propagation and penetration losses, antenna arrays are required at both the base station and terminal sides to concentrate the transmission energy into a small region. However, beamforming over large-scale antenna arrays brings an unaffordable hardware cost and high power consumption due to the demand for a large number of expensive Radio Frequency (RF) components. The beam-based transmission also imposes challenges on the design of communication protocol, such as the initial access where omnidirectional coverage for synchronization and broadcasting signals are mandatory. Moreover, extremely high data throughput puts the necessity of applying broader transmission bandwidths. The requirement of supporting up to 400 MHz in a single carrier and more than 1 GHz in carrier aggregation make the implementation of RF and antenna elements, baseband signal processing, and associated communication protocol much challenging.

4.2 mmWave Propagation Characteristics

Wireless communication happens through the radiation of electromagnetic waves from the transmitter to the receiver. A wireless channel is susceptible to noise, interference, and various channel impairments, while such impairments dynamically change with the movement of mobile users. In mmWave wireless communications, most physical objects in the surrounding environment become very large relative to the tiny wavelengths at mmWave frequencies. Some propagation effects such as reflection, scattering, and diffraction exhibit distinct characteristics compared with Ultra High Frequency (UHF) and Super High Frequency (SHF, also referred to as microwave) bands. In addition, some negligible propagation effects at UHF and microwave frequencies become substantially severe at mmWave frequencies, e.g. atmospheric absorption caused by oxygen and water vapor, imposing the necessity of rethinking the wireless medium. This section will characterize the mmWave channel, its key physical features, and the modeling issues, which can be roughly divided into two types:

- *Large-scale fading* occurs over relatively large distances, consisting of free-space propagation loss caused by the dissipation of a transmitted signal, atmospheric absorption, weather effects, and shadowing due to the obstruction between the transmitter and receiver that attenuates signal power through penetration, reflection, diffraction, and scattering.
- *Small-scale fading* occurs over very short distances on the order of the carrier wavelength. It is caused by the constructive and destructive addition of multiple signal copies traveled through different paths between the transmitter and receiver.

4.2.1 Large-Scale Propagation Effects

Large-scale fading models the macroscopic properties of electromagnetic propagation, where the strength of a received signal depends on the distance between the transmitter and receiver. FSPL due to the dissipation of a transmitted signal become more significant since the increase of carrier frequency, while the characteristics of reflection, diffraction, and scattering at mmWave frequencies are also different compared with that of UHF and microwave bands. A mmWave signal suffers from more severe penetration loss than a low-frequency radio signal when going through obstacles such as foliage, buildings, and tinted glass walls. Moreover, mmWave wireless communications have to consider atmospheric loss due to the energy absorption of oxygen and water molecules and the impact of weather since the physical dimensions of raindrops, hailstones, and snowflakes are on the same order of the carrier wavelength.

4.2.1.1 Free-Space Propagation Loss

To begin with, we study the simplest of all possible transmission scenarios: electromagnetic waves transmitted through free space to a receiver at a distance d from the transmitter. There is no obstacle between the transmitter and receiver, and the signal propagates along a straight line called Line-of-Sight (LOS) without reflection or scattering. Assuming that an isotropic antenna generates a spherical wave, which follows the law of energy conservation where the power contained on the surface of a sphere of any radius d is equal to the Effective Isotropic Radiated Power (EIRP) of the transmitter. The power flux density of a transmitted signal, measured in units of Watts per square meter, is given by the EIRP divided by the surface area of a sphere with radius d. Since the received signal power captured by a receive antenna is proportional to its effective area denoted by A_r, we have

$$P_r(d) = \left(\frac{P_{EIRP}}{4\pi d^2}\right) A_r. \tag{4.1}$$

The gain of a typical receive antenna can be expressed in terms of its effective area and the operating frequency, i.e.

$$G_r = \eta A_r \left(\frac{4\pi}{\lambda^2}\right), \tag{4.2}$$

where λ stands for the wavelength of the operating frequency and η denotes the maximal efficiency of the antenna, which is no greater than 1 and might be dramatically less than 1 for inefficient antennas. Meanwhile, the EIRP of a transmitter is given by the product of its transmitted power P_t and its transmitter antenna gain G_t, i.e. $P_{EIRP} = P_t G_t$. Using the relationship between effective area and antenna gain given in Eq. (4.2), the received power in Eq. (4.1) can be rewritten as

$$P_r(d) = \frac{P_t G_t G_r}{\eta} \left(\frac{\lambda}{4\pi d}\right)^2, \tag{4.3}$$

where P_t and P_r are the transmitted and received power in absolute linear unites (usually Watts or milliwatts), respectively, while G_t and G_r denote the linear gains of the transmit and receive antennas relative to a unity-gain (0 dB) isotropic antenna. Equation (4.3), which is known as Friis' FSPL formula, is valid in the far electromagnetic field of the antenna. In wireless communications, it is typical to express propagation attenuation using decibel values since the range of signal power dynamically varies across several orders of magnitude over relatively small distances. Rewrite Eq. (4.3) on a logarithm scale:

$$P_r|_{dBm} = P_t|_{dBm} + \underbrace{10 \lg\left(G_t G_r\right) + 20 \lg\left(\frac{\lambda}{4\pi d_0}\right) - 20 \lg\left(\frac{d}{d_0}\right)}_{\text{Free-space path gain}}, \tag{4.4}$$

where $|_{dBm}$ stands for in units of decibel-milliwatt. The reference distance d_0 should be in the far field of the antenna so that the near-field effect does not affect the reference path loss. In traditional cellular systems with large coverage, 1 km or 100 m reference distances are commonly used, whereas 1 m is enough for mmWave frequencies.

FSPL is defined as the logarithmic ratio between the transmit and receive power without considering the antenna gains:

$$PL = 10 \lg\frac{P_t}{P_r} = -\left[10 \lg\left(G_t G_r\right) + 20 \lg\left(\frac{\lambda}{4\pi d_0}\right) - 20 \lg\left(\frac{d}{d_0}\right)\right]. \tag{4.5}$$

The opposite number of FSPL is called free-space path gain, i.e. $P_G = -PL$, as indicated in Eq. (4.4). It can be seen that the received power decays inversely proportional to the square of the propagation distance and is also proportional to the square of the carrier wavelength. As we know, the received power in free space decays 20 dB per decade in the distance due to the distance squared term in the denominator. Note that the dependence of received power on the wavelength λ is due to the effective area of the receive antenna (see Eq. (4.2)]. It implies that wireless transmission at mmWave frequencies will suffer from severe path losses due to shorter wavelengths compared with that of UHF and microwave bands. The wavelength λ in m is inversely proportional to its frequency f, namely $\lambda = c/f$ with the speed of light in free space $c \approx 3 \times 10^8$ m/s. To demonstrate the effect of higher frequency, Table 4.1 provides a comparison of FSPLs at early cellular bands of 460 MHz, the unlicensed Industrial, Scientific and Medical (ISM) band of 2.4 GHz, the Unlicensed National Information Infrastructure (U-NII) band close to 5 GHz, and a typical mmWave frequency of 60 GHz. These simple calculations reveal that there is an excess propagation loss of around 20–40 dB compared with sub-6 GHz communication environment.

4.2.1.2 NLOS Propagation and Shadowing

FSPL cannot properly reflect all characteristics of mmWave channels since the physical environment of terrestrial mobile systems does not follow LOS electromagnetic propagation. A transmitted signal will encounter many objects that reflect, diffract, or scatter this signal in a typical urban or indoor scenario, forming Non-Line-of-Sight (NLOS) propagation between the transmitter and receiver, as shown in Figure 4.2. Due to the differences in propagation distances and paths, these additional copies of the transmitted signal referred to as multi-path signals, have different attenuation in power, different delays in time, and different shifts in frequency and phase. The combination of these multi-path signals at the receiver further raises variation of the received signal power.

The reflection of electromagnetic radiation means the returning back of electromagnetic waves by a smooth surface. In general, a reflecting surface is a boundary

Table 4.1 Comparison of free-space path loss for various frequencies.

	$f_c = 460$ MHz	$f_c = 2.4$ GHz	$f_c = 5$ GHz	$f_c = 60$ GHz
$d = 1$ m	25.7 dB	40 dB	46.4 dB	68 dB
$d = 10$ m	45.7 dB	60 dB	66.4 dB	88 dB
$d = 100$ m	65.7 dB	80 dB	86.4 dB	108 dB
$d = 1$ km	85.7 dB	100 dB	106.4 dB	128 dB

Source: [Rappaport et al. [2014], chap. 3]/Pearson Education.

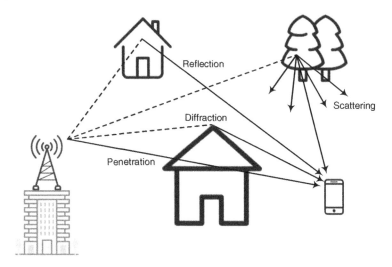

Figure 4.2 Illustration of NLOS radio propagation through reflection, diffraction, scattering, and penetration.

between two materials with different electromagnetic properties, such as the boundary between air and glass, air and metal, or air and water. Reflection is a critical propagation mechanism at traditional UHF and microwave frequencies to bypass the blockage between the transmitter and receiver. Results of channel measurement demonstrated that mmWave frequencies exhibit better reflective properties in indoor and outdoor environments than UHF and microwave bands. General physical objects such as building walls, lamp posts, trees, metal garbage cans, and even the heads of human beings can be very reflective to mmWave signals, generating strong multi-path signals by bouncing off surrounding objects. In electromagnetics, a metric called reflection coefficient is applied to describe how much of an electromagnetic wave is reflected by an impedance discontinuity in the transmission medium. It is equal to the ratio of the amplitude of the reflected wave to the incident wave. Reflection coefficients for outdoor objects such as tinted glass and reinforced concrete walls exceed 0.8. Table 4.2 provides an insight on reflection coefficients for common building materials measured at the mmWave frequency of 28 GHz.

The reflection of electromagnetic waves occurs when reflective surfaces are smooth enough, where the reflection angle is identical to the incident angle. When a plane wave impinges onto rough surfaces with a plane boundary that is not infinitely large, the incident electromagnetic waves reflect in many random directions rather than the specular one. Scattering is a negligible propagation mechanism in low-frequency bands due to its significant attenuation, e.g. decaying in terms of the distance at a rate of d^4. At mmWave frequencies, the

Table 4.2 Reflection coefficients of typical building materials measured at 28 GHz.

Environment	Material	Incident angle (°)	Reflection coefficient
Outdoor	Tinted glass	10	0.896
	Concrete	10	0.815
		45	0.623
Indoor	Clear glass	10	0.740
	Dry wall	10	0.704
		45	0.628

Source: Zhao et al. [2013]/With permission of IEEE.

dimensions of all physical objects such as human beings, buildings, and lamp posts become relatively large compared with tiny wavelengths. It implies that a scattered signal might be as substantial as a reflected signal. In addition to reflection and scattering, a transmitted signal can also bend around an object in its path to the receiver, resulting from the Earth's curved surface, hilly or irregular terrain, and building edges. Although diffraction is also an important propagation mechanism of low-frequency bands to form NLOS paths, it becomes the weakest due to shorter wavelengths at mmWave frequencies.

If a physical object blocks the LOS propagation path of a transmitted signal, the electromagnetic wave can penetrate the obstacles with a price of power attenuation to reach the receiver. Measurement results at mmWave frequencies shown that high-frequency bands suffer from more severe attenuation than UHF and microwave bands. In outdoor environments, tinted glass and thick walls have penetration losses of approximately 40 and 28 dB at 28 GHz, indicating that a base station deployed outdoors is hard to offer services for indoor users. However, most interior partitions and furniture in indoor environments do not substantially attenuate the signals, where penetration losses comparable to UHF and microwave bands are expected, e.g. 2–6 dB. It implies that indoor networks can operate well at mmWave frequencies while avoiding interference from co-channel outdoor networks. Table 4.3 shows the penetration losses for common building materials measured at the mmWave frequency of 28 GHz.

4.2.1.3 Atmospheric Attenuation

In traditional low-frequency bands used for cellular communications, atmospheric effect is not taken into account during the calculation of propagation loss. However, all electromagnetic waves suffer from atmospheric attenuation due to the absorption of gaseous molecules such as oxygen and water vapor. This effect is substantially magnified at certain mmWave frequencies. For instance, many oxygen absorption lines merged together at near 60 GHz to form a single, broad absorption band. Atmospheric attenuation can be accurately evaluated in terms of

Table 4.3 Penetration losses of typical building materials measured at 28 GHz.

Environment	Material	Thickness (cm)	Penetration loss (dB)
Outdoor	Tinted glass	3.8	40.1
	Brick wall	185.4	28.3
Indoor	Clear glass	<1.3	3.9
	Tinted glass	<1.3	24.5
	Partition	38.1	6.8

Source: Adapted from Zhao et al. [2013].

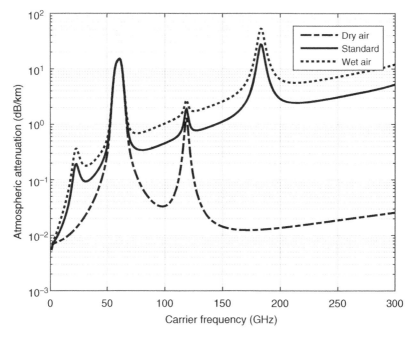

Figure 4.3 Atmospheric attenuation due to the absorption of oxygen and water vapor, calculated at an interval of 1 GHz, with standard atmosphere condition where air pressure 1013.25 hPa, temperature 15 °C, and water vapor density 7.5 g/m^3, according to ITU-R P.676 [2019]. In addition, the effect of humidity using water vapor density of 0 g/m^3 (dry air) and water vapor density 15 g/m^3 (wet air) is shown. Source: Adapted from ITU-R P.676 [2019].

air pressure, temperature, and humidity as a summation of the individual spectral lines from oxygen and water vapor [ITU-R P.676, 2019]. Figure 4.3 illustrates the effect of oxygen absorption, as indicated by *Dry Air*, calculated at an air pressure of 1013.25 hPa and a temperature of 15 °C while assuming the air is perfectly dry with a water vapor density of 0 g/m^3. Oxygen absorption dominates the atmospheric attenuation at near 60 GHz, leading to a peak loss of approximately 15 dB/km. We also show the effect of humidity for the *standard* atmosphere condition at the seal level (air pressure of 1013.25 hPa, temperature of 15 °C, and water vapor density of 7.5 g/m^3). Except for a few bands (i.e. 60 and 120 GHz), water absorption plays a dominant role to generate atmospheric attenuation. If the humidity increases, the atmospheric attenuation becomes substantially larger, accounting for a peak of approximately 50 dB/km at frequencies around 180 GHz. In contrast, such atmospheric attenuation at sub-6 GHz bands is on the order of 0.01 dB/km, which is negligible.

In addition to molecular absorption, the weather is also an important factor of atmospheric attenuation since the physical sizes of raindrops, hailstones, and snowflakes are on the order of carrier wavelengths at mmWave frequencies. The outcomes of the research in the 1970s and 1980s, which focused on weather characteristics on satellite communication links, provided the knowledge of mmWave propagation in various weather conditions. Rain, fog, hail, and snow impose excess unwanted signal losses on millimeter-wave propagation paths through the lower atmosphere. Using rainfall as an example, its attenuation is a function of distance, rainfall rate, and the mean dimension of raindrops, which can be estimated by rain attenuation models such as [Crane, 1980]. Such attenuation can be treated as an additional path loss that is simply added to the path loss raised by free-space propagation loss and atmospheric absorption. Measurement at 28 GHz demonstrated that a heavy rainfall with a rain rate of more than 25 mm/h bring attenuation of about 7 dB/km. Extreme attenuation of up to 50 dB/km occurs at a particular frequency of 120 GHz and an extreme rain rate of 100–150 mm/h. From the perspective of satellite communications, the weather attenuation makes mmWave wireless transmission unreliable, if not unusable, due to the long distance of a satellite link. However, terrestrial mmWave mobile communications with small cell sizes are viable (considering a maximal loss of several decibels at a distance of 100 m), especially when large-antenna arrays are applied to compensate for such a propagation loss.

4.2.2 Small-Scale Propagation Effects

When a transmitter sends out a sinusoidal signal $x(t) = \cos 2\pi f t$, the received signal at the receiver is a summation of multiple copies of this transmitted signal arrived from different propagation paths due to the reflection, scattering, and diffraction effects of physical objects in the surrounding environment. The received signal can be written as

$$y(t) = \sum_{l=1}^{L(t)} a_l(f, t) \cos 2\pi f \left(t - \tau_l(f, t) \right), \tag{4.6}$$

where $a_l(f, t)$ and $\tau_l(f, t)$ denote the attenuation and propagation delay on path l at time t, respectively, and $L(t)$ is the number of resolvable multipath components. Attenuation and delays are mainly determined by the distance from the transmitter to the reflector and from the reflector to the receiver. These two parameters usually change slowly with respective to frequency, and the transmission bandwidth is far smaller than the carrier frequency. Therefore, it is reasonable to assume that attenuation and delays on each propagation path are frequency-independent. Thus, Eq. (4.6) can be rewritten as

$$y(t) = \sum_{l=1}^{L(t)} a_l(t) \cos 2\pi f \left(t - \tau_l(t) \right). \tag{4.7}$$

The multipath fading channel can be modeled as a linear time-varying system, described by the response $h(\tau, t)$ at time t to an input impulse at time $t - \tau$. From Eq. (4.7), it is straightforward to obtain the impulse response

$$h(\tau, t) = \sum_{l=1}^{L(t)} a_l(t) \delta \left(t - \tau_l(t) \right). \tag{4.8}$$

The idea condition for wireless transmission is that there is only one propagation path with no attenuation and zero delay, namely $h(\tau, t) = \delta(t)$. In a special case when the transmitter, receiver, and scatters are all stationary, the channel has a linear time-invariant impulse response:

$$h(\tau) = \sum_{l=1}^{L} a_l \delta \left(\tau - \tau_l \right). \tag{4.9}$$

Additionally, we can get the frequency response for a time-varying impulse response through the Fourier transform, i.e.

$$H(f, t) = \int_{-\infty}^{+\infty} h(\tau, t) e^{-2\pi j f \tau} d\tau = \sum_{l=1}^{L(t)} a_l(t) e^{-2\pi j f \tau_l(t)}. \tag{4.10}$$

Practical wireless communications are passband transmission that is carried out in a bandwidth at carrier frequency f_c. However, most of the signal processing in wireless communications, such as channel coding, modulation, detection, synchronization, and estimation, are usually implemented at the baseband. From the perspective of system design, it makes sense to obtain a complex baseband equivalent model as follows [Tse and Viswanath, 2005]:

$$h_b(\tau, t) = \sum_{l=1}^{L(t)} a_l(t) \delta(\tau - \tau_l(t)) e^{-2\pi j f_c \tau_l(t)}. \tag{4.11}$$

In contrast to traditional UHF and microwave bands, mmWave transmission links suffer from higher free-space propagation loss, higher penetration loss, and excess losses due to atmospheric attenuation and weather effects. Such large-scale propagation effects at mmWave frequencies can be reflected on the path attenuation $a_l(t)$ without modifying the channel models given in Eqs. (4.8) or (4.11). Furthermore, due to the shrink of wavelength, mmWave signals experience stronger reflection that occurs on objects much larger than the wavelength and richer scattering that occurs on objects with similar dimensions of wavelength. Consequently, the number of resolvable multipath components L become significant, but the formulae mentioned earlier can still correctly model it.

4.2.3 Delay Spread and Coherence Bandwidth

When a signal pulse goes through a multipath channel, the received signal will appear as a pulse train, with each pulse corresponding to the direct path or an NLOS path. An important feature of radio propagation is *multipath delay spread* or *time dispersion* raised from the distinct arrival time of various propagation paths. Assuming $\tau_1(t)$ in the multipath channel model given by Eq. (4.11) denotes the propagation time of the first arriving multipath component, *minimum excess delay* is equal to $\tau_1(t)$, as a reference delay of zero. Meanwhile, the propagation time of the last arriving multipath component is $\tau_L(t)$. Multipath delay spread can be simply measured by the difference of the arrival time between the shortest and longest resolvable path, also referred to as *maximum excess delay*, in terms of

$$T_d := \tau_L(t) - \tau_1(t). \tag{4.12}$$

Power Delay Profile (PDP) gives the intensity of a received signal through a multipath channel as a function of the multipath delay τ, we have

$$S(\tau) = \mathbb{E}\left[|h(t, \tau)|^2\right]. \tag{4.13}$$

The average and Root Mean Square (RMS) delay spread can be calculated by

$$\mu_\tau = \frac{\int_0^{+\infty} \tau S(\tau) d\tau}{\int_0^{+\infty} S(\tau) d\tau} \tag{4.14}$$

and

$$T_\tau = \sqrt{\frac{\int_0^{+\infty} (\tau - \mu_\tau)^2 S(\tau) d\tau}{\int_0^{+\infty} S(\tau) d\tau}}. \tag{4.15}$$

The impulse response of a wireless channel varies both in time and frequency, and the delay spread determines how quickly it changes in frequency. Recalling that the frequency response given in Eq. (4.10), there is a phase difference $2\pi f(\tau_i - \tau_j)$ between multipath components i and j. Given the maximal phase difference among all paths as $2\pi f T_d$, the magnitude of the overall frequency response changes significantly when the phase difference increases or decreases with a value of π. Thus, the *coherence bandwidth*, which indicates the fading rate of wireless channels in the frequency domain, is defined as

$$B_c := \frac{1}{2T_d}. \tag{4.16}$$

In narrowband transmission, the bandwidth of a transmitted signal is usually far less than the coherence bandwidth, i.e. $B \ll B_c$. Thus, the fading across the entire bandwidth is highly correlated, called *frequency-flat fading*. In this case, the delay spread is considerably less than the symbol period $T_s = 1/B$, and therefore a single tap is sufficient to represent the channel filter. On the contrary, if the signal bandwidth $B \gg B_c$, two frequency points separated by more than the coherence bandwidth exhibit roughly independent response. Thus, wideband communications suffer from *frequency-selective fading* and Inter-Symbol Interference (ISI). The multipath delay spreads across multiple symbols, and the channel filter can be represented with a number of taps rather than a single tap. In wireless communications, the mechanisms of mitigating ISI play a vital role in designing wideband signal formatting and receiver structure.

Traditional macro base stations with a cell size of a few kilometers are very likely to have path distances that differ by more than 300–600 m, accounting for delays of 1 or 2 μs. This corresponds to a coherence bandwidth far less than 1 MHz. Due to the constraint of significant large-scale propagation loss, base stations operating at mmWave frequencies are suitable for providing small-cell coverage in indoor and outdoor environments. As cells become smaller, delay spread shrinks. Numerous measurement campaigns showed that the multipath delay spread of mmWave signals at typical small coverage falls into the range from 10 to 100 ns. When massive antenna arrays are applied to generate very narrow beams to concentrate the radiation energy to a small area, smaller multipath delay spreads are expected. It contributes a larger coherence bandwidth, which in turn facilitates the design of mmWave wireless transmission.

4.2.4 Doppler Spread and Coherence Time

Another fundamental characteristic of the multipath channel is its time-scale variation, arising from the movement of the transmitter, receiver, or surrounding objects. If signals are continuously transmitted from a moving transmitter or the receiver is moving, we will observe the changes in the number of resolvable paths, as well as the attenuation and propagation delay on each path, leading to time-varying channel response.

Consider a mobile station moves at a constant velocity of v, along a route that has a spatial angle of θ between the direction of signal propagation and the direction of movement. Within a time interval of Δt, the mobile station moves from one point to another with a distance of $v \Delta t$. It causes a difference in path lengths, i.e. $\Delta d = v \Delta t \cos \theta$, traveled by electromagnetic waves from the transmitter to receiver. Due to the difference in propagation distances, the phase variation of the received signal is

$$\Delta \phi = \frac{2\pi \Delta d}{\lambda} = \frac{2\pi v \Delta t \cos \theta}{\lambda}, \tag{4.17}$$

and therefore the change of frequency, referred to as *Doppler shift*, is given by

$$f_d = \frac{\Delta \phi}{2\pi \Delta t} = \frac{v \cos \theta}{\lambda}. \tag{4.18}$$

As we know, the signals arrive in distinct directions in a multipath propagation environment. If the mobile station moves toward the direction of incident electromagnetic waves, namely $-90° < \theta < 90°$, the Doppler shift is positive $f_d > 0$. On the contrary, the Doppler shift is negative $f_d < 0$ when the mobile station moves away from the direction of incident electromagnetic waves, i.e. $90° < \theta < 270°$. Suppose we transmit a sinusoidal signal $\cos 2\pi f_c t$, the received signal will be broadened in the frequency domain ranged from $f_c + v/\lambda$ when the mobile station moves exactly toward the transmitter when $\theta = 0°$ to $f_c - v/\lambda$ when the mobile station moves exactly away from the transmitter when $\theta = 180°$. Then, *Doppler Spread*

$$D_s = 2f_m \tag{4.19}$$

is defined as a measure of the spectral broadening caused by the relative movement, where $f_m = v/\lambda$ denotes the maximum Doppler shift.

Considering the effect of the Doppler shift, the frequency response of a wireless channel given in Eq. (4.10) can be rewritten as

$$H(f, t) = \int_{-\infty}^{+\infty} h(\tau, t)e^{-2\pi j f \tau} d\tau = \sum_{l=1}^{L(t)} a_l(t)e^{-2\pi j(f+f_d)\tau_l(t)}. \tag{4.20}$$

We can see that the Doppler shift raises a phase change on each path, e.g. $2\pi f_d \tau_l(t)$ on path l, and therefore the phase change significantly at a time interval of $\Delta \tau = 1/2f_d$. When multipath components combine at the receiver, such phase changes affect their constructive and destructive interference. This happens at a time interval of

$$T_c = \frac{1}{4D_s} = \frac{1}{8f_m}, \tag{4.21}$$

which is called *coherence time* to characterize the time duration over which the channel response can be regarded as invariant. This is a somewhat rough relation and there are different definitions for this parameter. The coherence time is approximately

$$T_c \approx \frac{9}{16\pi f_m},$$ (4.22)

if it is defined as the time over which the time correlation function is above 0.5. In Rayleigh fading, Eq. (4.22) is too restrictive. A popular rule of thumb for digital communications is to define the coherence time as the geometric mean of $1/f_m$ and Eq. (4.22), also known as Clarke's model, i.e.

$$T_c \approx \sqrt{\frac{9}{16\pi f_m^2}} = \frac{0.423}{f_m}.$$ (4.23)

Regardless of the diversity in definition, the critical knowledge to be recognized is that the time coherence is determined mainly by the Doppler spread in a reciprocal relation, i.e. the larger the Doppler spread, the smaller the time coherence. Depending on how quickly a transmitted baseband signal changes relative to the fading rate of the channel, wireless channels can be categorized into *slow fading* and *fast fading*. In slow fading, the symbol period is much smaller than the coherence time

$$T_s \ll T_c.$$ (4.24)

We can assume that the channel is constant across a number of symbol periods in the time domain, and the Doppler spread can be negligible compared with the signal bandwidth. In fast fading, where

$$T_s > T_c,$$ (4.25)

where the channel response varies within a single symbol period while frequency dispersion due to Doppler spreading is considerable.

From Eq. (4.18), we know that the Doppler effect magnifies with the shrink of wavelength. Consequently, a mmWave channel exhibits more significant time variation compared with UHF or microwave wireless channels. This effect was demonstrated in Tharek and McGeehan [1988], where a transmitter moving away from a receiver at a constant velocity of 1 m/s generated a Doppler shift of around 200 Hz in the millimetric waveband of 60 GHz. Theoretically, for vehicles with a velocity of 120 km/h and a carrier frequency of 60 GHz, the maximum Doppler shift equals to around 6667 Hz. For higher velocity, such as high-speed train or airplane speeds, and higher frequency bands above 60 GHz, the time-varying nature will be extreme.

In traditional wireless communications at low-frequency bands, the transmitter is able to adaptively choose its parameters such as the transmit power, constellation size, coding rate, transmit antenna, and precoding pattern to

Table 4.4 Exemplary comparison of key characteristics between microwave and millimeter-wave channels.

Parameter	Symbol	Representative values	
		Microwave	mmWave
Carrier frequency	f_c	1 GHz	60 GHz
Signal bandwidth	B	1 MHz	1 GHz
Cell size	d	1 km	50 m
Delay spread	T_d	1 μs	50 ns
Coherence bandwidth	$B_c = 1/2T_d$	500 kHz	10 MHz
Velocity	v	20 m/s	1 m/s
Doppler spread	D_s	133 Hz	400 Hz
Coherence time	$T_c = 1/4D_s$	1.9 ms	0.625 ms

achieve excellent performance with the knowledge of Channel State Information (CSI). The CSI is usually obtained by estimating received reference signals at the receiver and then fed back to the transmitter in a Frequency-Division Duplexing (FDD) system. Due to the feedback delay, it tends to become outdated under the rapid channel variation. It has been extensively recognized that the outdated CSI imposes overwhelming performance deterioration on a wide variety of wireless systems, such as massive MIMO, multi-user scheduling, interference alignment, beamforming, transmit antenna selection, closed-loop transmit antenna diversity, opportunistic relaying, coordinated multi-point, orthogonal frequency-division multiplexing, resource management, and physical layer security. Hence, the Doppler effect imposes a big challenge for the design of physical layer and radio resource control at mmWave frequencies since the channel changes more dramatically and the acquisition of accurate CSI is much difficult.

To provide the readers an insight into the characteristics of mmWave channels, Table 4.4 shows a comparison of key parameters between microwave and millimeter-wave channels.

4.2.5 Angular Spread

In addition to the time and frequency domain, the application of multiple-antenna techniques imposes the necessity of studying the degree of freedom empowered by the angular domain. For typical terrestrial propagation, electromagnetic waves arrive at the receiver from different angles, as shown in Figure 4.4. Without loss of generality, this figure only focuses on the azimuthal plane but neglecting the elevation direction. The Angle of Departure (AoD) for the LOS path with respect

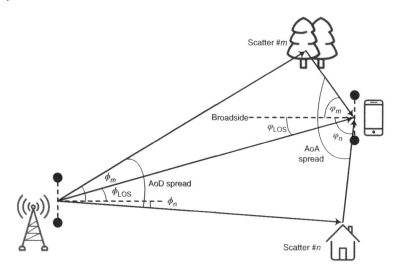

Figure 4.4 The angular-domain representation of multipath propagation.

to the broadside of the base station's antenna array is denoted by ϕ_{LOS}, while the Angle of Arrival (AoA) for the LOS path with respect to the broadside of the mobile station's antenna array is denoted by φ_{LOS}. The absolute AoD and AoA for the mth reflective path are denoted by ϕ_m and φ_m, respectively. Similarly, the absolute AoD and AoA for the nth reflective path are denoted by ϕ_n and φ_n, respectively. Assuming that the mth and nth paths stand for the *maximum* and *minimum* angles of signal propagation from the transmitter to the receiver, the instantaneous angular spread for AoD can be defined as

$$\phi_{\text{AS-AoD}} = \phi_m + \phi_n, \tag{4.26}$$

and the instantaneous angular spread for AoA is

$$\varphi_{\text{AS-AoA}} = \varphi_m + \varphi_n. \tag{4.27}$$

In generally, the angular spread at a macro base station that has a clear propagation space is smaller than that of a mobile station surrounded by many physical objects.

The instantaneous angular spread is deterministic but cannot properly characterize the time-varying multipath environment. Therefore, it makes sense to study the statistical model averaging instantaneous parameters temporally or spatially. The distribution of multipath power with respect to the angular domain is conveniently described by Power Angular Spectrum (PAS) denoted by $p(\theta)$, where $\theta \in [0, 2\pi]$ is the azimuth angle. The PAS is obtained by spatially averaging the instantaneous azimuth power profiles over several tens of wavelengths within the range where the same multipath components are maintained in order to suppress the variations due to rapid fading. It can be modeled either by zero-mean Gaussian or zero-mean Laplacian distributions, given by

$$p(\theta) = \frac{1}{\sqrt{2\pi\sigma_\theta^2}} \exp\left(-\frac{\theta^2}{2\sigma_\theta^2}\right), \tag{4.28}$$

and

$$p(\theta) = \frac{1}{\sqrt{2\sigma_\theta^2}} \exp\left(-\left|\frac{\sqrt{2}\theta}{\sigma_\theta}\right|\right), \tag{4.29}$$

respectively, with the standard deviation σ_θ.

The average AoD at the transmitter or the average AoA at the receiver (in the horizontal direction) can be calculated by

$$\mu_\theta = \frac{\int_0^{2\pi} \theta p(\theta) d\theta}{\int_0^{2\pi} p(\theta) d\theta}. \tag{4.30}$$

In order to compare different multipath channels and to develop some general design guidelines for wireless systems, a statistical channel parameter called RMS angular spread can be defined by the square root of the second central moment, i.e.

$$\theta_{\text{RMS}} = \sqrt{\frac{\int_0^{2\pi} (\theta - \mu_\theta)^2 p(\theta) d\theta}{\int_0^{2\pi} p(\theta) d\theta}}. \tag{4.31}$$

Angular spread represents the richness of a multipath environment that determines the effectiveness of spatial diversity and spatial multiplexing. Various channel measurement campaigns revealed that the PAS distribution at mmWave frequencies is typically modeled as the superposition of clusters, commonly used in microwave frequency channels.

The definition of angular spread is straightforwardly extended to the elevation angle, as shown in Figure 4.5. Suppose the PAS of elevation angle is $p(\psi)$, the average AoD or AoA in the vertical direction is given by

$$\mu_\psi = \frac{\int_0^{2\pi} \psi p(\psi) d\psi}{\int_0^{2\pi} p(\psi) d\psi}, \tag{4.32}$$

and the RMS angular spread is

$$\psi_{\text{RMS}} = \sqrt{\frac{\int_0^{2\pi} (\psi - \mu_\psi)^2 p(\psi) d\psi}{\int_0^{2\pi} p(\psi) d\psi}}. \tag{4.33}$$

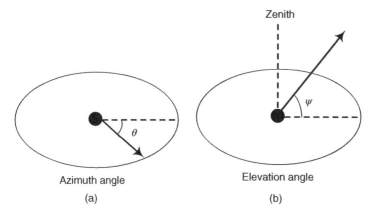

Figure 4.5 The azimuth (a) and elevation (b) angles of a propagation path.

4.3 Millimeter-Wave Channel Models

The prerequisite of designing a mobile communication system is a reliable and accurate channel model for the operating frequency band, allowing researchers and engineers to evaluate the performance of competing radio-transmission and medium-control techniques without having to make expensive and time-consuming field measurements on their own. Over the past decades, the wireless community developed a large number of statistical and empirical channel models by fitting curves or analytical expressions to measured data obtained from actual field measurements. This method implicitly takes into account all propagation effects, both known and unknown, and therefore works well. However, most of these models focus on sub-6 GHz frequency bands for traditional wireless systems. Since the wavelength of mmWave signals is far shorter than that of microwave signals below 6 GHz, the parameters for mmWave channel models are expected to be quite distinct. Therefore, it is essential to develop mmWave channel models for research, development, performance evaluation, and standardization. Extensive channel measurement campaigns covering potential mmWave bands up to 100 GHz in various deployment scenarios have been carried out by industry and research institutions over the past decade. Table 4.5 summarizes the major efforts on measurements and modeling of mmWave channels.

4.3.1 Large-Scale Fading

Theoretical and empirical propagation models indicate that the average power of any received signal decreased with the distance between the transmitter and the

Table 4.5 Summary of major mmWave measurement campaigns.

Model	Frequency	Description
METIS	2–60 GHz	• Identified 5G requirements (e.g. wide frequency range, high bandwidth, massive MIMO, three-dimensional (3D) and accurate polarization modeling) • Provided different channel model methodologies (map-based model, stochastic model, or hybrid model)
MiWEBA	60 GHz	• Addressed various challenges: shadowing, spatial consistency, environment dynamics, spherical wave modeling, dual mobility Doppler model, ratio between diffuse and specular reflections, and polarization • Proposed quasi-deterministic channel model
NYU	28/38/60/73 GHz	• Extensive urban measurements on both indoor and outdoor • LOS, NLOS, and blockage modeling • Wideband PDPs • Physics-based path loss modeling
802.11ad	60 GHz	• Conducted ray-tracing methodology on indoor channels • Intra-cluster parameters in terms of ray excess delay and ray power distribution • Human blockage models in terms of blockage probability and blockage attenuation
mmMAGIC	6–100 GHz	• Extensive channel measurements bringing together major vendors, European operators, research institutions, and universities • Advanced channel models for rigorous validation and feasibility analysis of the proposed concepts and system, as well as for usage in regulatory and standardization
QuaDRiGa	10/28/60/82 GHz	• Fully fledged 3D geometry-based stochastic channel model • Quasi-deterministic multi-link tracking of user movement in changing environments • Massive MIMO modeling enabled by a new multi-bounce scattering approach and spherical wave propagation
3GPP	6–100 GHz	• An extension of well-accepted 3GPP Spatial Channel Model (SCM) and 3D SCM models • Unified modeling for a wide range of frequency bands from 6 to 100 GHz • Support various scenarios: Rural Marco (RMa), Urban Macro (UMa), Urban Micro (UMi), Indoor Hotspot (InH), and Indoor Factory (InF)

receiver in a logarithmic scale. By generalizing the free-space propagation model in Eq. (4.3), we have

$$P_r(d) = P_t G_t G_r \left(\frac{\lambda}{4\pi d} \right)^n, \tag{4.34}$$

where the path loss exponent n indicates the rate at which path loss increases with distance. The value of n depends on the specific propagation environment, e.g. $n = 2$ in free space and $n = 4$ for a two-ray ground bounce. The particular propagation losses at mmWave frequencies such as atmospheric attenuation and weather effects can be reflected in such a model by getting a higher value of n. Therefore, the path loss

$$PL(d) \propto \left(\frac{d}{d_0} \right)^n \tag{4.35}$$

or

$$PL(d) = PL_{d_0} + 10 n \lg \left(\frac{d}{d_0} \right), \tag{4.36}$$

where the reference path loss is calculated by (see Eq. (4.5))

$$PL_{d_0} = - \left[10 \lg \left(G_t G_r \right) + 20 \lg \left(\frac{\lambda}{4\pi d_0} \right) \right]. \tag{4.37}$$

In particular, the difference between the microwave propagation loss and the mmWave propagation loss is primarily contained in the difference of the first meter propagation.

The above equation does not consider the shadow fading based on the fact that the surrounding environment may be distinct at two different locations with the same distance between the transmitter and receiver. It is raised from the random variation due to the blockage from physical objects along the signal path, reflecting surfaces, and scattering objects. This randomness is well described by a log-normal distribution. Then, we get the close-in free-space reference distance model to characterize the large-scale propagation effect, i.e.

$$PL(d) = PL_{d_0} + 10 n \lg \left(\frac{d}{d_0} \right) + \chi_\sigma, \tag{4.38}$$

where χ_σ (in dB) is a zero-mean Gaussian distributed random variable with standard deviation of σ (also in dB), i.e. $\chi_\sigma \sim \mathcal{N}(0, \sigma^2)$.

There exists a modification of the free-space reference model used mainly in 3GPP channel modeling. It has a similar mathematical expression but applies a floating intercept, where measured data are fitted to a least-squared curve using an arbitrary leverage point. This approach provides a bit smaller standard derivation to measured data but yields a model that has no physical basis. The floating intercept model, also called an (α, β) model, is given by

$$PL(d) = \alpha + 10\beta \lg(d) + \chi_\sigma. \tag{4.39}$$

4.3.2 3GPP Channel Models

Based on the outcomes of extensive measurement campaigns and channel modeling from academia and industry, 3GPP developed channel models for frequencies from 6 to 100 GHz. Considering the extensive acceptance of SCM and SCM-Extended models developed by 3GPP and its dominant role in the standardization of 5G and beyond 5G systems, it makes sense to introduce the mmWave channel models from 3GPP.

At the 3GPP Technology Specification Group (TSG) Radio Access Network (RAN) 69th meeting in September 2015, the Study Item *Study on Channel Model for Frequency Spectrum Above 6 GHz* was approved. This study item identified the status and expectation of information on high frequencies (e.g. spectrum allocation, scenarios of interest, and measurements) and the channel modeling for frequencies up to 100 GHz. Aligned with earlier sub-6 GHz models, such as the 3D SCM model (3GPP TR 36.873) or IMT-Advanced (ITU-R M.2135), the new model supports comparisons across different frequency bands. Subsequently, at the 3GPP TSG RAN 81th meeting, the Study Item *Study on Channel Modeling for Indoor Industrial Scenarios* was established. The findings from these two study items were released in the technical report [3GPP TR38.900, 2018].

This channel model is applicable for the link and system-level performance evaluation in the following conditions:

- For system-level simulations, supported scenarios are urban microcell street canyon, urban macrocell, indoor office, rural macrocell, and Indoor Factory (InF).
- Bandwidth is supported up to 10% of the center frequency but no larger than 2 GHz.
- Mobility of either one end of the link or both ends of the link is supported.
- For the stochastic model, spatial consistency is supported by the correlation of large-scale parameters and small-scale parameters, as well as LOS and NLOS.
- The support of large antenna arrays is based on far-field assumption and stationary channel over the size of the array.

4.3.2.1 Urban Micro Scenario

In the case of urban micro, Urban Micro (UMi) model describes the scenario, where the base stations are mounted below rooftop levels of surrounding buildings. It is intended to capture real-life scenarios such as a city or station square. The model depends heavily on whether there is a direct path, where the NLOS model is given by

$$PL_{\text{UMi-NLOS}} = 32.4 + 20\lg(f_c) + 31.9\lg\left(d_{\text{3D}}\right) + \chi_\sigma, \tag{4.40}$$

where f_c is the carrier frequency, the shadow fading has a standard derivation of $\sigma = 8.2$ dB, d_{3D} denotes the 3D distance between the transmitter and receiver, which is calculated by

$$d_{3D} = \sqrt{d^2 + \left(h_{BS} - h_{UE}\right)^2} \tag{4.41}$$

with the traditional distance d between the transmitter and receiver used for large-scale coverage, the antenna height of base station h_{BS}, and the antenna height of user equipment h_{UE}. The LOS channel of urban micro scenario is described by a two-slope model:

$$PL_{UMi-LOS} = \begin{cases} 32.4 + 20\lg(f_c) + 21\lg\left(d_{3D}\right) + \chi_\sigma, & 10\,\text{m} \le d \le d_{BP} \\ 32.4 + 20\lg(f_c) + 40\lg\left(d_{3D}\right) - \\ 9.5\lg\left[d_{BP}^2 + \left(h_{BS} - h_{UE}\right)^2\right] + \chi_\sigma, & d_{BP} \le d \le 5\,\text{km} \end{cases}, \tag{4.42}$$

where the standard derivation of shadow fading is $\sigma = 4$ dB, and d_{BP} denotes the break-point distance. This distance can be calculated by

$$d_{BP} = \frac{4(h_{BS} - h_E)(h_{UE} - h_E)f_c}{c}, \tag{4.43}$$

where h_E is the effective environment height, equaling to $h_E = 1$ m for the urban micro scenario, and $c = 3 \times 10^8$ m/s is the velocity of electromagnetic wave in free space.

4.3.2.2 Urban Macro Scenario

Urban Macro (UMa) also provides the modeling of dense populated scenarios such as city canyon. In contrast to the UMi model, the main difference is that its coverage area is larger, i.e. the Inter-Site Distance (ISD) is up to 500 m rather than 200 m in UMi. Besides, base stations are assumed to be amounted above the rooftop level of surrounding buildings.

The NLOS channel can be modeled by

$$PL_{UMa-NLOS} = 32.4 + 20\lg(f_c) + 30\lg\left(d_{3D}\right) + \chi_\sigma, \tag{4.44}$$

with $\sigma = 7.8$ dB.

The LOS channel is described by a two-slope model as

$$PL_{UMa-LOS} = \begin{cases} 28 + 20\lg(f_c) + 22\lg\left(d_{3D}\right) + \chi_\sigma, & 10\,\text{m} \le d \le d_{BP} \\ 28 + 20\lg(f_c) + 40\lg\left(d_{3D}\right) - \\ 9\lg\left[d_{BP}^2 + \left(h_{BS} - h_{UE}\right)^2\right] + \chi_\sigma, & d_{BP} \le d \le 5\,\text{km} \end{cases}, \tag{4.45}$$

with $\sigma = 4$ dB.

4.3.2.3 Indoor Scenario

This scenario is intended to capture various typical indoor deployment cases, which are categorized into two types: *Indoor Hotspot (InH)* or *InF*. InH selects two typical cases, including office environments and shopping malls. The typical office environment is comprised of open cubicle areas, walled offices, open areas, corridors, etc. Base stations are mounted at the height of 1–3 m either on the ceilings or walls. The shopping malls are often several stories high and may include an open area shared by several floors. Base stations are usually installed at the height of approximately 3 m on the walls or ceilings of the corridors and shops. The InF scenario can be further divided into four types in terms of the surrounding environment and the antenna height of base stations, i.e. InF-DL (Dense cluster and Low base station height), InF-SH (Sparse cluster and High base station height), InF-DH (Dense cluster and High base station height), and InF-HH (High transmitter and High receiver height). The models for indoor scenarios are summarized in Table 4.6.

To provide the readers a concrete view, the path losses of UMa and UMi scenarios as a function of distance and carrier frequency are illustrated in Figures 4.6 and 4.7, respectively. It is noted that 3GPP aims to provide universal modeling to cover a wide range of mmWave bands from 6 to 100 GHz. This generalization has to ignore some particular frequency-dependent channel characteristics such as strong oxygen attenuation at 60 GHz. Therefore, it is better to provide some modifications to the 3GPP models or adapt some dedicated models, e.g. IEEE

Table 4.6 3GPP indoor path loss models.

Scenario	Path	Path loss	Shadow fading
InH-Office	LOS	$32.4 + 20\lg(f_c) + 17.3\lg\left(d_{3D}\right)$	$\sigma = 3$
	NLOS	$32.4 + 20\lg(f_c) + 31.9\lg\left(d_{3D}\right)$	$\sigma = 8.29$
InH-Mall	LOS	$32.4 + 20\lg(f_c) + 17.3\lg\left(d_{3D}\right)$	$\sigma = 2$
InF	LOS	$PL_{LOS} = 31.84 + 19\lg(f_c) + 21.5\lg\left(d_{3D}\right)$	$\sigma = 4$
	NLOS	$PL_1 = 33 + 20\lg(f_c) + 25.5\lg\left(d_{3D}\right),$	$\sigma = 5.7$
		$PL_{InF\text{-}SL} = \max\left(PL_1, PL_{LOS}\right)$	
		$PL_2 = 18.6 + 20\lg(f_c) + 35.7\lg\left(d_{3D}\right),$	$\sigma = 7.2$
		$PL_{InF\text{-}DL} = \max\left(PL_2, PL_{LOS}, PL_{InF\text{-}SL}\right)$	
		$PL_3 = 32.4 + 20\lg(f_c) + 23\lg\left(d_{3D}\right),$	$\sigma = 5.9$
		$PL_{InF\text{-}SH} = \max\left(PL_3, PL_{LOS}\right)$	
		$PL_4 = 33.63 + 20\lg(f_c) + 21.9\lg\left(d_{3D}\right),$	$\sigma = 4$
		$PL_{InF\text{-}DH} = \max\left(PL_4, PL_{LOS}\right)$	

Source: Adapted from 3GPP TR38.900 [2018].

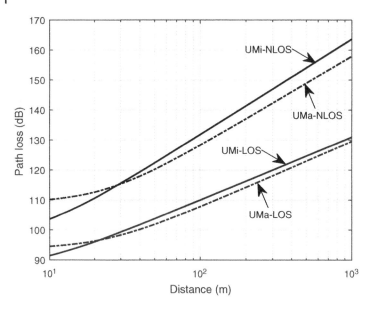

Figure 4.6 Path losses of 3GPP UMa and UMi scenarios, without shadow fading, at the carrier frequency of 60 GHz, where the antenna height of macro base station is 25 m, the antenna height of micro base station is 10 m, and the antenna height of mobile station is 1.5 m.

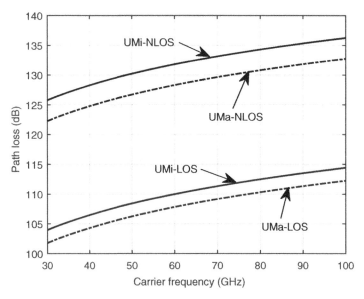

Figure 4.7 Path losses of 3GPP UMa and UMi scenarios as a function of carrier frequency ranged from 30 to 100 GHz, where the antenna height of macro base station is 25 m, the antenna height of micro base station is 10 m, the antenna height of mobile station is 1.5 m, and the observing distance is fixed to 100 m.

802.11ad channel models focusing on 60 GHz indoor scenarios for the operation of Wi-Fi systems, when the system design, research, or standardization is conducted on a particular frequency band.

4.3.3 Small-Scale Fading

Due to the constructive and destructive interference of multiple signal components, the strength of the received signal is time-varying in the scale of wavelength. According to *the central limit theorem*, the channel impulse response will be well-modeled as a Gaussian process irrespective of the distribution of individual components, if there is sufficiently much scattering. When there is no dominant LOS component, the envelope of the received signal is Rayleigh distributed. It is described by Probability Density Function (PDF)

$$p(r) = \frac{r}{\Omega} \exp\left(-\frac{r^2}{2\Omega}\right), \quad r \geq 0, \tag{4.46}$$

where Ω is the average power of the received signal. The mean value of the envelope of the Rayleigh distribution is

$$\mu_r = \mathbb{E}[r] = \int_0^{+\infty} rp(r)dr = \sqrt{\frac{\pi\Omega}{2}} = 1.2533\sqrt{\Omega}, \tag{4.47}$$

and its variance is

$$\sigma_r^2 = \mathbb{E}[r^2] - \mathbb{E}^2[r] = \int_0^{+\infty} r^2 p(r)dr - \frac{\pi\Omega}{2} = 0.4292\Omega. \tag{4.48}$$

With a dominant signal component such as a LOS path, the small-scale fading envelope exhibits a Rician distribution:

$$p(r) = \frac{r}{\Omega} \exp\left(-\frac{r^2 + A^2}{2\Omega}\right) I_0\left(\frac{Ar}{\Omega}\right), \quad \forall (r \geq 0, A \geq 0), \tag{4.49}$$

where A denotes the peak amplitude of the dominant signal and $I_0(\cdot)$ is the zeroth-order modified Bessel function of the first kind. The Rician distribution is often described by a parameter called K-factor, which is defined as the ratio of the deterministic signal power and the variance of the multipath components:

$$K = \frac{A^2}{2\Omega}. \tag{4.50}$$

If $K \to 0$, the Rician distribution degenerates to a Rayleigh distribution since the dominant path is vanished.

Statistical models such as 3GPP SCM are used for sub-6 GHz frequency bands with bandwidths up to 100 MHz to produce complex coefficients for simulating channel impulse response. These models provide essential statistical information such as multipath delays, cluster powers, AoA, and AoD, along with large-scale propagation loss based on real-world measurement. Although these models have been successfully applied in describing the stochastic nature of low-frequency wideband channels, they are not accurate enough in temporal resolution for mmWave channels and make a simplifying assumption that all clusters of multipath energy travel closely together in both time and space. Consequently, these models are difficult to accurately describe the mmWave channels, which have much wider signal bandwidths, e.g. 800 and 1.5 GHz used for the measurement campaigns of 28 and 60 GHz, respectively, where multiple multipath clusters can arrive within a particular spatial direction.

To properly model the mmWave channels, impulse response may be represented simultaneously in the temporal and spatial domains. The temporal characteristics are described by the PDP, containing statistical information such as the arrival time, delay spread, and power levels, while the power angular profile provides the spatial characteristics including AoA, AoD, and angular spread. For example, the small-scale channel models developed during the standardization of IEEE 802.15.3c and IEEE 802.11ad, which focused on the unlicensed frequency band of 60 GHz, are based on clustering in both the time and space domains as observed in measurement. These models were derived by extending the standard Saleh–Valenzuela (S-V) propagation model. The difference with the original S-V model is that they treat a LOS path between the transmitter and receiver separately. The separateness of the LOS path and its strength relative to the other multipath components suggest that the mmWave channel may consider as Rician when the LOS path is present. When the LOS path is blocked, the mmWave channel is well described by a Rayleigh distribution.

The channel impulse response in complex baseband can be given by Yong [2007]:

$$h(t, \phi) = \sum_{l=1}^{L} \sum_{k=1}^{K_l} \alpha_{l,k} \delta(t - T_l - \tau_{l,k}) \delta(\phi - \Phi_l - \phi_{l,k}), \tag{4.51}$$

where L denotes the total number of clusters, K_l is the number of rays in the lth cluster, inter-cluster parameters T_l and Φ_l represent the delay and mean AoA of the lth cluster, and intra-cluster parameters $\alpha_{l,k}$, $\tau_{l,k}$, and $\phi_{l,k}$ stand for the complex amplitude, excess delay, and relative azimuth angle of the kth ray in the lth cluster.

Utilizing directive antennas, there will be a distinct strong LOS path on top of the clustering multipath components. This LOS path can be included by adding a LOS component to Eq. (4.51) as given next

$$h(t, \phi) = \beta \delta(t, \phi_{\text{LOS}}) + \sum_{l=1}^{L} \sum_{k=1}^{K_l} \alpha_{l,k} \delta(t - T_l - \tau_{l,k}) \delta(\phi - \Phi_l - \phi_{l,k}), \tag{4.52}$$

where β is the gain of the LOS path arriving at zero excess delay from azimuth angle ϕ_{LOS}, which are deterministic by using ray tracing or simple geometry-based method. Equation (4.52) only contains azimuth angle-of-arrival, but it can be further extended to indicate azimuth angle-of-departure at the transmitter and elevation angles:

$$h(t, \phi, \varphi, \psi, \omega) = \beta\delta(t, \phi_{\text{LOS}}) + \sum_{l=1}^{L}\sum_{k=1}^{K_l} \alpha_{l,k}\delta(t - T_l - \tau_{l,k})\delta(\phi - \Phi_l^{\text{AoA}} - \phi_{l,k})$$
$$\cdot \delta(\varphi - \Phi_l^{\text{AoD}} - \varphi_{l,k}) \tag{4.53}$$
$$\cdot \delta(\psi - \Psi_l^{\text{AoA}} - \psi_{l,k})$$
$$\cdot \delta(\omega - \Psi_l^{\text{AoD}} - \omega_{l,k}),$$

where φ, ψ, and ω represent azimuth angle-of-departure, elevation angle-of-arrival, and elevation angle-of-departure, respectively.

Table 4.7 The parameters of the 3GPP CDL-A model.

Cluster	Delay (ms)	Power (dB)	AoD (°)	AoA (°)	ZoD (°)	ZoA (°)
1	0.0000	−13.4	−178.1	51.3	50.2	125.4
2	0.3819	0	−4.2	−152.7	93.2	91.3
3	0.4025	−2.2	−4.2	−152.7	93.2	91.3
4	0.5868	−4	−4.2	−152.7	93.2	91.3
5	0.4610	−6	90.2	76.6	122	94
6	0.5375	−8.2	90.2	76.6	122	94
7	0.6708	−9.9	90.2	76.6	122	94
8	0.5750	−10.5	121.5	−1.8	150.2	47.1
9	0.7618	−7.5	−81.7	−41.9	55.2	56
10	1.5375	−15.9	158.4	94.2	26.4	30.1
11	1.8978	−6.6	−83	51.9	126.4	58.8
12	2.2242	−16.7	134.8	−115.9	171.6	26
13	2.1718	−12.4	−153	26.6	151.4	49.2
14	2.4942	−15.2	−172	76.6	157.2	143.1
15	2.5119	−10.8	−129.9	−7	47.2	117.4
16	3.0582	−11.3	−136	−23	40.4	122.7
17	4.0810	−12.7	165.4	−47.2	43.3	123.2
18	4.4579	−16.2	148.4	110.4	161.8	32.6
19	4.5695	−18.3	132.7	144.5	10.8	27.2
20	4.7966	−18.9	−118.6	155.3	16.7	15.2
21	5.0066	−16.6	−154.1	102	171.7	146
22	5.3043	−19.9	126.5	−151.8	22.7	150.7
23	9.6586	−29.7	−56.2	55.2	144.9	156.1

Source: 3GPP TR38.900 [2018]/ETSI.

3GPP study item on mmWave channels provided the Clustered Delay Line (CDL) models for the entire frequency range from 6 to 100 GHz with a maximum bandwidth of 2 GHz. Three CDL models, namely CDL-A, CDL-B, and CDL-C, are constructed to depict three different channel profiles for NLOS, while CDL-D and CDL-E are for LOS. To provide a concrete view, the parameters of CDL-A are listed in Table 4.7. The acronyms of ZoA and ZoD are Zenith AoA and Zenith angel of Departure, respectively. In addition, 3GPP also provided the Tapped Delay Line (TDL) models for simplified evaluations, e.g. non-MIMO scenarios, at mmWave frequencies up to 100 GHz with a maximum bandwidth of 2 GHz. Three

Table 4.8 The parameters of the 3GPP TDL-A model.

Tap	Normalized delay (ms)	Power (dB)	Fading distribution
1	0.0000	−13.4	Rayleigh
2	0.3819	0	Rayleigh
3	0.4025	−2.2	Rayleigh
4	0.5868	−4	Rayleigh
5	0.4610	−6	Rayleigh
6	0.5375	−8.2	Rayleigh
7	0.6708	−9.9	Rayleigh
8	0.5750	−10.5	Rayleigh
9	0.7618	−7.5	Rayleigh
10	1.5375	−15.9	Rayleigh
11	1.8978	−6.6	Rayleigh
12	2.2242	−16.7	Rayleigh
13	2.1718	−12.4	Rayleigh
14	2.4942	−15.2	Rayleigh
15	2.5119	−10.8	Rayleigh
16	3.0582	−11.3	Rayleigh
17	4.0810	−12.7	Rayleigh
18	4.4579	−16.2	Rayleigh
19	4.5695	−18.3	Rayleigh
20	4.7966	−18.9	Rayleigh
21	5.0066	−16.6	Rayleigh
22	5.3043	−19.9	Rayleigh
23	9.6586	−29.7	Rayleigh

Source: 3GPP TR38.900 [2018]/ETSI.

TDL models, namely TDL-A, TDL-B, and TDL-C, are constructed to describe three different channel profiles for NLOS, while TDL-D and TDL-E are for LOS. The parameters of the TDL-A model can be found in Table 4.8.

4.4 mmWave Transmission Technologies

mmWave wireless communications are promising due to abundant spectral resources available in the spectrum band from 30 to 300 GHz and the large blocks of contiguous spectrum to overcome the bandwidth crunch problem at sub-6 GHz frequencies. However, the propagation characteristics at high frequencies are more severe than low frequencies in terms of path loss, atmospheric absorption, rain attenuation, diffraction, and blockage. As a result, mmWave signals suffer from more significant propagation loss and penetration loss than microwave signals. It is mainly applied to provide local-area coverage and therefore serve fewer users in each cell, unlike the conventional microwave cellular systems that seamlessly cover a large number of users in a wide area. A large-scale antenna array with tens or hundreds of elements is required to provide high power gains that compensate for the isotropic propagation loss. The antenna size of a mmWave transceiver is tiny due to the short wavelength of high-frequency signals being in the range of 1–10 mm, facilitating the implementation of a large-scale array in a compact device. These differences impose various challenges and constraints on the design of physical-layer transmission algorithms and medium-access protocols of mmWave systems. This section will present critical enabling techniques to realize the anticipated benefits of mmWave communications, focusing on beamforming, initial access, multi-beam synchronization and broadcasting, and heterogeneous deployment.

4.4.1 Beamforming

Beamforming, also known as spatial filtering or smart antenna, has a wide range of application fields such as sonar, radar, wireless communications, seismology, acoustic, radio astronomy, and biomedicine. In wireless communications, beamforming is mainly applied to increase power gains, suppress inter-cell or multi-user interference, and provide a spatial-domain degree of freedom called Space-Division Multiple Access (SDMA). It is implemented by adjusting the phase shifts of different elements in an antenna array so that signals at particular angles experience constructive interference while others undergo destructive interference.

The conventional beamforming is fully digital, where the desired beam is formed by simply multiplying the baseband signal with a weighting vector.

However, digital beamforming requires a RF chain for each antenna element, leading to unaffordable energy consumption and hardware cost for a mmWave transceiver equipped with a large-scale array. Therefore, another technical form that can lower implementation complexity, called analog beamforming, has been adopted in indoor mmWave communications, such as Wireless Local Area Network (WLAN) operated in 60 GHz. By employing analog phase-shifters to adjust the phases of signals, analog beamforming needs only a single RF chain to steer the beam, leading to low hardware cost and energy consumption. However, since an analog circuit can only partly adjust the phases of signals, it is difficult to adapt a beam to a particular channel condition appropriately, and this leads to a considerable performance loss. In addition, fully analog architecture can only support single-stream transmission, which cannot achieve the multiplexing gain to improve spectral efficiency.

Hybrid analog-digital beamforming to balance the benefits of fully digital and fully analog beamforming was recognized as fit for mmWave transmission. At the 3GPP RAN1 meeting in June 2016, the hybrid architecture was agreed to be adopted in the 5G system. The key idea is to divide the conventional baseband processing into two parts: a large-size analog signal processing (realized by an analog circuit) and a dimension-reduced digital signal processing (requiring only a few RF chains). Since the number of effective scatterers at mmWave frequencies is often small, the number of data streams is generally much smaller than the number of antennas. Consequently, hybrid beamforming can significantly reduce the number of RF chains, resulting in lower hardware cost and energy consumption.

4.4.1.1 Digital Beamforming

Digital beamforming, also known as smart antenna, adaptive antenna array, or digital antenna array, is a multi-antenna technique with advanced signal processing algorithms to identify spatial parameters such as the AoA of a signal and use such parameters to determine a weighting vector, which forms a beam concentrating on the direction of a target mobile user. The smart antenna technology has been employed as a critical feature of TD-SCDMA, earlier than the application of other multi-antenna techniques such as MIMO in cellular systems.

Consider an array of N omnidirectional elements, indexed by $n = 1, \ldots, N$, radiating signals into a homogeneous media in the far field of uncorrelated sinusoidal point sources of frequency f_0. As illustrated in Figure 4.8, the time taken by a plane wave propagating from the nth transmit element to a receive antenna located in the direction indicated by the AoD θ is

$$\tau_n(\theta) = \frac{\mathbf{r}_n \cdot \mathbf{u}(\theta)}{c}, \tag{4.54}$$

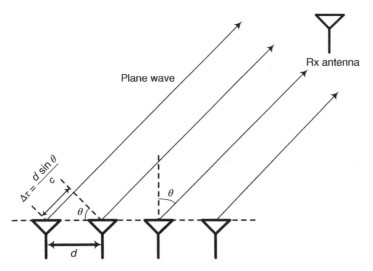

Figure 4.8 Far-field geometry of the radiation of a plane wave from a uniform linear array in the angle-of-departure of θ.

where $\mathbf{r}_n$ denotes the position vector of the nth element relative to the reference point, $\mathbf{u}(\theta)$ is the unit vector in angle θ, c is the speed of propagation of the plane wave front, and $\cdot$ denotes the inner product.

The receiver observes the signal transmitted by the reference element expressed in complex notation as

$$s(t)e^{j2\pi f_0 t} \tag{4.55}$$

with $s(t)$ stands for the complex bandband signal. The wavefront on the nth element arrives with a time difference $\tau_n(\theta)$ compared with that of the reference element. Thus, the signal induced at the receive antenna due to the nth element can be expressed as

$$s(t)e^{j2\pi f_0 [t-\tau_n(\theta)]}. \tag{4.56}$$

This expression is based on *the narrow-band assumption* for array signal processing, assuming that the signal bandwidth is narrow enough and the array dimension is small enough, so that the baseband signal keeps almost constant over the interval of $\tau_n(\theta)$, i.e. the approximation $s(t) \approx s(t - \tau_n(\theta))$ holds.

The overall received signal induced due to all N elements is

$$y(t) = \sum_{n=1}^{N} s(t)e^{j2\pi f_0 [t-\tau_n(\theta)]} + n(t), \tag{4.57}$$

where $n(t)$ indicates white Gaussian noise at the receive antenna. The principle of narrow-band beam-former is to impose a phase shift of the signal on each element by multiplexing a complex weight $w_n(t)$ with the baseband signal. Thus, the received signal with beamforming becomes

$$y(t) = \sum_{n=1}^{N} w_n^*(t)s(t)e^{j2\pi f_0[t-\tau_n(\theta)]} + n(t), \tag{4.58}$$

where the superscript $*$ denotes the complex conjugate.

For a Uniform Linear Array (ULA) with equally element spacing d aligned with a direction such that the first element is situated at the origin, Eq. (4.54) can be rewritten as

$$\tau_n(\theta) = \frac{d}{c}(n-1)\sin\theta. \tag{4.59}$$

In an Additive White Gaussian Noise (AWGN) channel, the received signal is obtained by substituting (4.59) into (4.58), yielding

$$
\begin{aligned}
y(t) &= \sum_{n=1}^{N} w_n^*(t)s(t)e^{j2\pi f_0 t}e^{-j\frac{2\pi}{\lambda}(n-1)d\sin\theta} + n(t) \\
&= \left(\sum_{n=1}^{N} w_n^*(t)e^{-j\frac{2\pi}{\lambda}(n-1)d\sin\theta} \right) s(t)e^{j2\pi f_0 t} + n(t) \\
&= g(\theta,t)s(t)e^{j2\pi f_0 t} + n(t),
\end{aligned}
\tag{4.60}
$$

where $g(\theta,t)$ is the effect of beamforming, referred to as the beam pattern.

Defining the weighting vector

$$\mathbf{w}(t) = \left[w_1(t), w_2(t), \dots, w_N(t) \right]^T, \tag{4.61}$$

and the steering vector of ULA

$$\mathbf{a}(\theta) = \left[1, e^{-j\frac{2\pi}{\lambda}d\sin\theta}, e^{-j\frac{2\pi}{\lambda}2d\sin\theta}, \dots, e^{-j\frac{2\pi}{\lambda}(N-1)d\sin\theta} \right]^T, \tag{4.62}$$

the beam pattern of ULA can be calculated by

$$g(\theta,t) = \mathbf{w}^H(t)\mathbf{a}(\theta), \tag{4.63}$$

where superscripts T and H denote the transpose and complex conjugate transpose of a vector or matrix, respectively.

Figure 4.9 illustrates two formed beam patterns over an eight-element ULA. The radiated energy can be concentrated in a particular direction with a

Figure 4.9 Beamforming over an eight-element ULA with inter-antenna spacing of $d = \lambda/2$. The weighting vector of beam pattern (a) is $\mathbf{w}_1 = [1,1,1,1,1,1,1,1]^T$ while the weighting vector of beam pattern (b) is $\mathbf{w}_2 = [1,-1,1,-1,1,-1,1,-1]^T$.

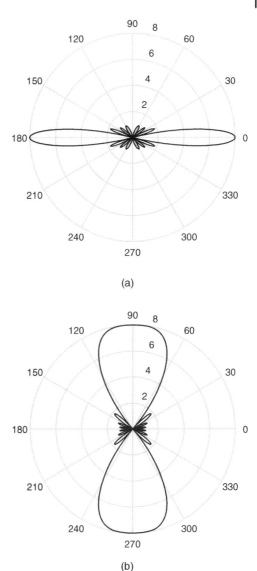

(a)

(b)

power gain equaling to the number of antenna elements. In other words, an eight-element ULA in this figure brings a power gain of 8 (in 0° and 180° of beam a while 90° and 270° of beam b) over an omnidirectional antenna with the same power. Through adjusting the weighting vector, the beam can be steered toward any particular direction in terms of the angular information of a mobile user, which can be estimated by using some classical algorithms such as MUltiple

SIgnal Classification (MUSIC) and Estimation of Signal Parameters via Rational Invariance Techniques (ESPRIT) [Gardner, 1988]. Suppose the angle of a mobile user is θ_0, letting

$$\mathbf{w} = \mathbf{a}^*(\theta_0) = \left[1, e^{j\frac{2\pi}{\lambda}d\sin\theta_0}, \dots, e^{j\frac{2\pi}{\lambda}(N-1)d\sin\theta_0}\right]^T \tag{4.64}$$

would form a beam pointing to the desired angle.

4.4.1.2 Analog Beamforming

In digital beamforming, each antenna element has its own RF chain. Such an architecture is believed to be too power-consuming for mobile devices supporting mmWave transmission. Therefore, analog beamforming, where the beam is formed in the analog domain and hence only requiring a single RF chain from the mixer stage down to digital baseband, is considered for the initial implementation of the mmWave transceiver.

As shown in Figure 4.10, a baseband signal is first transformed into an analog signal $s(t)$ and then up-converted to an RF signal $s(t)e^{j2\pi f_0 t}$. Thereafter, the signal is split in a power splitter and fed into a bank of analog phase shifters. The transmitted signal at the nth element is

$$s(t)e^{j2\pi f_0 t}e^{j\phi_n(t)} \tag{4.65}$$

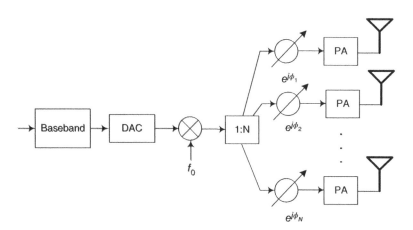

Figure 4.10 Block diagram of an analog beamforming transmitter. Going through the Digital-Analog Converter (DAC), a baseband signal is converted to an RF signal in a single up-conversion mixer and after that split in a power splitter and fed into a bank of phase shifters before being processed by Power Amplifier (PA).

with the added phase shift $\phi_n(t)$. Considering the difference of propagation time $\tau_n(\theta)$, as shown in Figure 4.8, the received signal induced at the receiver is then

$$
\begin{aligned}
y(t) &= \sum_{n=1}^{N} s(t) e^{j2\pi f_0[t-\tau_n(\theta)]} e^{j\phi_n(t)} + n(t) \\
&= \left(\sum_{n=1}^{N} e^{-j2\pi f_0 \tau_n(\theta)} e^{j\phi_n(t)} \right) s(t) e^{j2\pi f_0 t} + n(t) \\
&= g_a(\theta, t) s(t) e^{j2\pi f_0 t} + n(t),
\end{aligned}
\tag{4.66}
$$

with the generated beam pattern

$$
g_a(\theta, t) = \sum_{n=1}^{N} e^{-j2\pi f_0 \tau_n(\theta)} e^{j\phi_n(t)}.
\tag{4.67}
$$

Compared Eq. (4.66) with Eq. (4.60), it is concluded that analog beamforming can generate the same effect as digital beamforming. However, analog beamforming can only focus on a single direction, which limits its multiplexing capability. Due to the limited dynamic and resolution of analog phase shifters, there is also implementation loss compared with digital beamforming with optimal phase control in digital signals.

4.4.1.3 Hybrid Beamforming

Implementing digital beamforming in a mmWave system equipped with large-scale antenna arrays requires a massive large of hardware components, including mixers, analog-to-digital converters, digital-to-analog converters, and Power Amplifiers (PAs). It raises prohibitive hardware cost and power consumption, especially for mobile terminals, and thus is not viable. Moreover, the significantly increased dimension of beamforming brings a high computational burden on the baseband signal process. These constraints have driven mmWave transmission to pick analog beamforming, which uses a single RF chain. Hence, analog beamforming is implemented as the de-facto approach for indoor mmWave systems. However, it only supports single-stream transmission and suffers from performance loss due to the hardware impairment of analog phase shifters. As a trade-off, hybrid beamforming has been proposed as an efficient approach to support multi-stream transmission with only a few RF chains and a phase shifter network. Compared with analog beamforming, hybrid beamforming supports spatial multiplexing and spatial division multiple access. In addition, it achieves spectral efficiency comparable to digital beamforming with much lower hardware complexity. Therefore, it is promising to be adopted as the mmWave transceiver structure in beyond 5G systems [Zhang et al., 2019].

Hybrid beamforming can be implemented with different structures of circuit networks, leading to various signal processing designs, different hardware

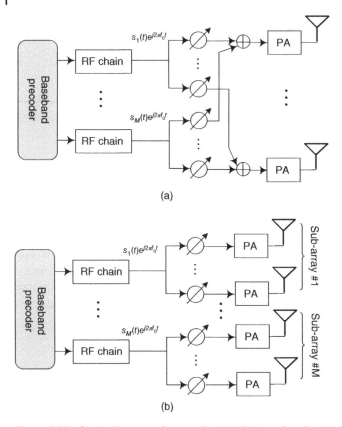

Figure 4.11 Block diagrams of two typical hybrid beamforming: (a) fully connected and (b) partially connected.

constraints, and different performance of mmWave systems. Hybrid beamforming has two basic structures:

- *Fully Connected Hybrid Beamforming*
  The transmit data is first precoded in the baseband into M data streams, and each stream is processed by an independent RF chain. Each RF chain is connected to all N antennas via an analog network, as shown in Figure 4.11. Hence, a total of MN analog phase shifters are required. A highly directive beam can be formed by adjusting the phases of transmitted signals on all antennas.

- *Partially Connected Hybrid Beamforming*
  In this structure, each RF chain is only connected to a subset of all antenna elements called sub-array. Due to this limitation, each data stream can only achieve a power gain equaling the number of elements in the sub-array, and beam directivity is reduced proportionally to the number of sub-arrays. However, this structure might be preferred from the perspective of practice

since it substantially reduces hardware complexity by dramatically reducing the number of required analog phase shifters, i.e. N.

Mathematically, the output of the baseband precoder is denoted by $s_m(t)$, $1 \leq m \leq M$, as shown in Figure 4.11. After the Digital-Analog Converter (DAC) and up-conversion, the mth RF chain feeds

$$s_m(t)e^{j2\pi f_0 t}, \quad 1 \leq m \leq M, \tag{4.68}$$

into the analog network. Suppose the phase shift between the mth RF chain and the nth antenna in a fully connected network is ϕ_{nm}, for all $1 \leq n \leq N$ and $1 \leq m \leq M$, the output of this phase shifter is $s_m(t)e^{j2\pi f_0 t}e^{j\phi_{nm}}$. Thus, the transmitted signal at antenna n is obtained by

$$x_n(t) = \sum_{m=1}^{M} s_m(t)e^{j2\pi f_0 t}e^{j\phi_{nm}}. \tag{4.69}$$

Build a weighting matrix as

$$\mathbf{W} = \begin{bmatrix} e^{j\phi_{11}} & e^{j\phi_{12}} & \cdots & e^{j\phi_{1M}} \\ e^{j\phi_{21}} & e^{j\phi_{22}} & \cdots & e^{j\phi_{2M}} \\ \vdots & \vdots & \ddots & \vdots \\ e^{j\phi_{N1}} & e^{j\phi_{N2}} & \cdots & e^{j\phi_{NM}} \end{bmatrix}, \tag{4.70}$$

and define

$$\mathbf{x}(t) = \left[x_1(t), x_2(t), \ldots, x_N(t) \right]^T, \tag{4.71}$$

and

$$\mathbf{s}(t) = \left[s_1(t), s_2(t), \ldots, s_M(t) \right]^T, \tag{4.72}$$

yielding

$$\mathbf{x}(t) = \mathbf{P}\mathbf{s}(t)e^{j2\pi f_0 t}. \tag{4.73}$$

Within an AWGN channel, the receiver observes an induced signal

$$\begin{aligned} y(t) &= \sum_{n=1}^{N} x_n(t - \tau_n(\theta)) + n(t) \\ &= \sum_{n=1}^{N}\sum_{m=1}^{M} s_m(t)e^{j2\pi f_0[t-\tau_n(\theta)]}e^{j\phi_{nm}} + n(t) \\ &= \left(\sum_{n=1}^{N}\sum_{m=1}^{M} s_m(t)e^{-j2\pi f_0 \tau_n(\theta)}e^{j\phi_{nm}} \right) e^{j2\pi f_0 t} + n(t) \\ &= \left(\sum_{m=1}^{M} s_m(t)\mathbf{a}^T(\theta)\mathbf{w}_m \right) e^{j2\pi f_0 t} + n(t), \end{aligned} \tag{4.74}$$

where $\mathbf{a}^T(\theta)$ is the steering vector of the array and $\mathbf{w}_m = [e^{j\phi_{1m}}, e^{j\phi_{2m}}, \ldots, e^{j\phi_{Nm}}]^T$ is the mth column of the weighting matrix.

Suppose there are M users located in different angles θ_m, $1 \leq m \leq M$, a typical user m waits for its desired signal $s_m(t)$. Theoretically, if we find a set of weighing vectors to satisfy

$$\begin{cases} \mathbf{a}^T(\theta_{m'})\mathbf{w}_m = N, & \text{if } m = m', \\ \mathbf{a}^T(\theta_{m'})\mathbf{w}_m = 0, & \text{if } m \neq m'. \end{cases}$$

Substituting Eq. (4.75) into Eq. (4.66), we get the received signal of the mth user as

$$y_m(t) = N s_m(t) e^{j2\pi f_0 t} + n(t).$$

Through the down-conversion and analog-to-digital conversion, the desired signal $s_m(t)$ is successfully delivered to user m, achieving the spatial multiplexing gain of M and a power gain of N for each data stream.

In the partially connected analog beamforming, the antenna array is divided into several sub-arrays and each antenna is allocated to only one RF chain. Each sub-array contains $N_s = N/M$ antennas (suppose N_s is an integer). In contrast to the transmit signal of the fully connected analog beamforming given in Eq. (4.69), the transmit signals for the mth sub-array is given by

$$\mathbf{x}_m(t) = \left[s_m(t) e^{j2\pi f_0 t} e^{j\phi_{1m}}, \ldots, s_m(t) e^{j2\pi f_0 t} e^{j\phi_{N_s m}} \right]^T. \tag{4.75}$$

The received signal is then

$$y(t) = \sum_{m=1}^{M} \sum_{n_s=1}^{N_s} s_m(t) e^{j2\pi f_0 [t - \tau_{[(m-1)N_s + n_s]}(\theta)]} e^{j\phi_{n_s m}} + n(t)$$

$$= \left(\sum_{m=1}^{M} s_m(t) \sum_{n_s=1}^{N_s} e^{-j2\pi f_0 \tau_{[(m-1)N_s + n_s]}(\theta)} e^{j\phi_{n_s m}} \right) e^{j2\pi f_0 t} + n(t). \tag{4.76}$$

Denoting the steering vector of sub-array m as

$$\mathbf{a}_m(\theta) = \left[e^{-j2\pi f_0 \tau_{[(m-1)N_s + 1]}(\theta)}, \ldots, e^{-j2\pi f_0 \tau_{[mN_s]}(\theta)} \right]^T, \tag{4.77}$$

and its corresponding weighting vector

$$\mathbf{w}_m = \left[e^{j\phi_{1m}}, \ldots, e^{j\phi_{N_s m}} \right]^T, \tag{4.78}$$

Eq. (4.76) is rewritten as

$$y(t) = \left(\sum_{m=1}^{M} \mathbf{a}_m^T(\theta) \mathbf{w}_m s_m(t) \right) e^{j2\pi f_0 t} + n(t)$$

$$= \sum_{m=1}^{M} g_m(\theta, t) s_m(t) e^{j2\pi f_0 t} + n(t), \tag{4.79}$$

where $g_m(\theta, t)$ indicates the beam pattern generated by the mth sub-array. As we can see from Eq. (4.79), the partially connected analog beamforming generates M beams but each beam has only a power gain of N_s, that is M times lower than that of the fully connected analog beamforming.

4.4.1.4   3D Beamforming

The aforementioned beamforming schemes focus on generating and steering the beam pattern in the horizontal plane, also known as two-dimensional (2D) beamforming, whereas the vertical domain is not exploited. In contrast to such 2D beamforming, 3D beamforming adapts the beam pattern in both elevation and azimuth planes to provide more degrees of freedom. Higher user capacity, less inter-cell and multi-user interference, higher energy efficiency, improved coverage, and increased spectral efficiency are the advantages of 3D beamforming.

An antenna array used for mmWave transmission usually has a massive number of elements to generate high power gain to compensate for severe propagation loss. For instance, the 5G mmWave system supports 64–256 antennas at the base station side and 4–16 antennas at the mobile station side. One of the major challenges is how to pack such a large number of elements within a restricted volume, especially for the traditional one-dimensional antenna arrays, such as the linear and circular array. For example, when 100 elements are uniformly spaced with a half wavelength separation at a carrier frequency of 20 GHz, the length of a ULA will be approximately 75 cm, and the diameter of a Uniform Circle Array (UCA) is about 24 cm [Tan et al., 2017]. The solution is to use planar antenna arrays such as Uniform Rectangular Planar Array (URPA), Uniform Hexagonal Planar Array (UHPA), and Uniform Circular Planar Array (UCPA), which can deploy higher numbers of antenna elements in a compact device, achieve more directive beam, and provide higher antenna gains to overcome path loss.

As illustrated in Figure 4.12, a uniform planar array having N_x elements in the x-direction with inter-element spacing of d_x, and N_y elements in the y-direction

Figure 4.12   Far-field geometry of the radiation of a plane wave from a uniform planar array in the direction with an elevation angle φ and an azimuthal angle θ.

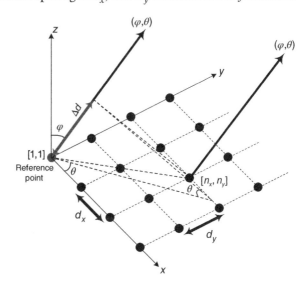

separated by d_y. In a 3D coordinate system, the direction of arrival or departure of a plane wave is described by an elevation angle φ and an azimuthal angle θ, denoted by (φ, θ). The difference of propagation distance between the reference point [1,1] and a typical element located in the n_x row and the n_y column, i.e. $[n_x, n_y]$, is given by

$$\Delta d_{n_x n_y}(\varphi, \theta) = (n_x - 1)d_x \sin \varphi \cos \theta + (n_y - 1)d_y \sin \varphi \sin \theta, \tag{4.80}$$

which raises a time difference of

$$\tau_{n_x n_y}(\varphi, \theta) = \frac{\Delta d_{n_x n_y}(\varphi, \theta)}{c} \tag{4.81}$$

$$= \frac{\sin \varphi \left[(n_x - 1)d_x \cos \theta + (n_y - 1)d_y \sin \theta\right]}{c}.$$

In the far-filed regime, the steering vector for the URPA is expressed by Chen [2013]

$$\mathbf{a}(\varphi, \theta) = \mathbf{v}_x(\varphi, \theta) \otimes \mathbf{v}_y(\varphi, \theta), \tag{4.82}$$

where $\mathbf{v}_x(\varphi, \theta)$ and $\mathbf{v}_y(\varphi, \theta)$ can be viewed as the steering vectors on the x- and y-direction, respectively, with

$$\mathbf{v}_x(\varphi, \theta) = \left[1, e^{j\frac{2\pi}{\lambda}d_x \sin \varphi \cos \theta}, \ldots, e^{j\frac{2\pi}{\lambda}(N_x-1)d_x \sin \varphi \cos \theta}\right]^T \tag{4.83}$$

and

$$\mathbf{v}_y(\varphi, \theta) = \left[1, e^{j\frac{2\pi}{\lambda}d_y \sin \varphi \sin \theta}, \ldots, e^{j\frac{2\pi}{\lambda}(N_y-1)d_y \sin \varphi \sin \theta}\right]^T. \tag{4.84}$$

Denote the weight for the antenna element $[n_x, n_y]$ as $w_{n_x n_y}(t)$ and assume the planar array transmits a single-stream signal $s(t)$ toward a mobile user. Similar to Eq. (4.66), the receiver observes an induced signal

$$y(t) = \sum_{n_x=1}^{N_x} \sum_{n_y=1}^{N_y} w_{n_x n_y}(t)s(t)e^{j2\pi f_0\left[t - \tau_{n_x n_y}(\varphi, \theta)\right]} + n(t)$$

$$= \left(\sum_{n_x=1}^{N_x} \sum_{n_y=1}^{N_y} w_{n_x n_y}(t)e^{-j\frac{2\pi \sin \varphi}{\lambda}\left[(n_x-1)d_x \cos \theta + (n_y-1)d_y \sin \theta\right]}\right)s(t)e^{j2\pi f_0 t} + n(t)$$

$$= g(\varphi, \theta, t)s(t)e^{j2\pi f_0 t} + n(t). \tag{4.85}$$

If the direction of a target mobile user is (φ_0, θ_0), the weights should be the reciprocal of the corresponding phase differences to maximize the signal strength in this direction, i.e.

$$w_{n_x n_y}(t) = e^{j\frac{2\pi \sin \varphi_0}{\lambda}\left[(n_x-1)d_x \cos \theta_0 + (n_y-1)d_y \sin \theta_0\right]}. \tag{4.86}$$

Substituting Eq. (4.86) into Eq. (4.85), we can get the pattern of a 3D beam as

$$g(\varphi, \theta, t) = \sum_{n_x=1}^{N_x} \sum_{n_y=1}^{N_y} w_{n_x n_y}(t) e^{-j\frac{2\pi \sin \varphi}{\lambda}[(n_x-1)d_x \cos \theta + (n_y-1)d_y \sin \theta]}$$

$$= \sum_{n_x=1}^{N_x} \sum_{n_y=1}^{N_y} e^{j\frac{2\pi \sin \varphi_0}{\lambda}[(n_x-1)d_x \cos \theta_0 + (n_y-1)d_y \sin \theta_0]}$$

$$\times e^{-j\frac{2\pi \sin \varphi}{\lambda}[(n_x-1)d_x \cos \theta + (n_y-1)d_y \sin \theta]}. \tag{4.87}$$

Setting $\varphi = \varphi_0$ and $\theta = \theta_0$ in Eq. (4.87), yields

$$g(\varphi_0, \theta_0, t) = N_x N_y, \tag{4.88}$$

which implies that a maximal power gain of $N_x N_y$, equivalent to the total number of antenna elements, is achieved in the direction of (φ_0, θ_0).

4.4.2 Initial Access

In all cellular communication systems, when a terminal powers on, performs the transition from IDEL to CONNECTED mode, or initially enters the coverage area of a system, it needs to search for a suitable cell to launch initial access and random access. Before a terminal can communicate with a base station, the following access procedure is performed:

- *Cell Search*: When initially accessing the system, a mobile device performs a cell search at power-up. Afterward, this device must continuously search for neighboring cells to determine whether a handover needs to be triggered. Its main functionalities include:
  1. Acquisition of frequency and symbol synchronization to a cell,
  2. Acquisition of frame timing of the cell – that is, determine the start of the downlink frame, and
  3. Determination of the physical-layer cell identity of the cell.
  To perform a cell search, two types of synchronization signals (SSs), i.e. Primary Synchronization Signal (PSS) and Secondary Synchronization Signal (SSS), are periodically transmitted on the downlink from each cell.
- *Extraction of System Information*: Once the frequency and time synchronization is achieved, the subsequent procedure for a terminal is to extract the system information of the serving cell. The network periodically broadcasts this information, by which a terminal can access and operate appropriately within the network and a specific cell. In both Long-Term Evolution (LTE) and NR, the system information is categorized into two different types: Master Information Block (MIB) and System Information Block (SIB). MIB contains a limited amount of information, which is necessary for acquiring the remaining system

information. More specifically, the MIB includes information such as the downlink cell bandwidth, the control channel configuration such as Physical Hybrid ARQ Indicator Channel (PHICH), and the System Frame Number (SFN). SIB contains the main part of the system information, which is broadcasted repeatedly using the downlink shared channel. The typical message in the SIB might include the uplink cell bandwidth, random-access parameters, and parameters related to uplink power control, neighboring-cell-related information, the allocation of subframes, etc. MIB is carried by the Physical Broadcast Channel (PBCH), whereas SIB is in the shared downlink control channel.

- *Random Access*: The device needs to request a connection setup by which a dedicated resource can be assigned for the initial transmission. The system allocates a specific resource block for a terminal to transmit its request, known as Physical Random Access Channel (PRACH) in either LTE or NR. In addition to the initial access, random access is also used for re-establishing a radio link after radio-link failure, handover when uplink synchronization needs to be established to a new cell, uplink scheduling, and positioning. Random access operates either in a contention-free or contention-based manner. Contention-free random access can only be used for re-establishing uplink synchronization upon downlink data arrival, handover, and positioning. Contention-based random access usually adopts a four-step procedure as follows:

  1. *Preamble Transmission*: The transmission of a random-access preamble, allowing the base station to estimate the uplink timing. Uplink synchronization is necessary as the terminal otherwise cannot transmit any uplink data.
  2. *Random Access Response*: The network transmits a timing-advance command to adjust the terminal transmit timing based on the timing estimate obtained in the first step. In addition to establishing uplink synchronization, the second step also assigns uplink resources to the terminal to be used in the third step in the random-access procedure.
  3. *Uplink Request*: The terminal transmits its identification information to the network. The exact content of this signaling depends on the state of the terminal, in particular whether it is previously known to the network or not.
  4. *Contention Resolution*: The final step consists of the transmission of a contention-resolution message from the network to the terminal on the downlink control channel. This step also resolves any contention due to multiple terminals trying to access simultaneously using the same random-access resource.

4.4.2.1 Multi-Beam Synchronization and Broadcasting

To combat severe propagation loss suffered from mmWave frequencies, highly directional antennas are required at both the base station and the mobile terminal to achieve sufficient power gains for wide-area coverage. This reliance on direc-tionality has a significant impact on the design of control layer procedures. In LTE systems, SSs and master system information are broadcasted in the down-link channel with omnidirectional coverage. Beamforming or other directional transmission are applied merely after a physical link has been successfully estab-lished. However, 5G mmWave cellular systems need to apply the directional beams to transmit both control and user data. Therefore, the initial access must provide a mechanism by which the base station and the mobile terminal can determine the initial broadcast directions.

Basically, mmWave cellular networks can adopt multi-beam approaches for syn-chronization and broadcasting, including exhaustive search, iterative search, and context information-based search [Giordani et al., 2016], which are described as follows:

- *Exhaustive Search* is a brute-force method to sequentially scan the 360° angular space by multiple narrow beams, potentially in time-varying random directions. In other words, base stations periodically transmit SSs and system information by continuously changing the weight vectors. Both terminals and base stations have a predefined codebook consisting of a set of weighting vectors, where each weighting vector can form a narrow beam to cover a particular direction, and all beams together seamlessly cover the whole angular space, as demonstrated in Figure 4.13. This method provides good coverage without any loss of power gain compared with the transmission of user data, especially for the users at the cell edge, whereas it suffers from high overhead and long discovery delay.

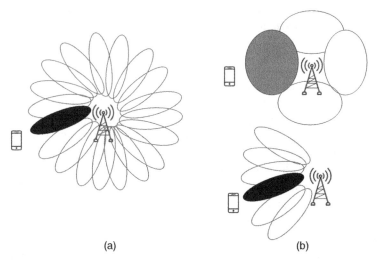

(a) (b)

Figure 4.13 Illustration of two basic multi-beam synchronization approaches: exhaustive search (a) and iterative search (b).

- *Iterative Search* is a two-phase hierarchical procedure based on a faster user discovery technique in order to alleviate the exhaustive search delay. In the first phase, the base station transmits SSs sequentially over a few wide beams, which can be formed by selecting particular weighting vectors. Relying on the feedback of the terminal, the base station knows the rough direction of a terminal. In the second phase, the base station only needs to scan the angular space covered by the best wide beam, rather than 360°. Although this method can determine the direction of a terminal faster than the exhaustive search, the power gain of the wide beam is smaller than that of the narrow beam, which in turn brings smaller coverage and high misdetection probability for edge users.

- *Context Information-Based Search*: This method exploits the geographical knowledge of the terminal and base station, which are enabled by a separate control plane, aiming at improving the cell edge discovery and minimize the delay. Small cells operating at mmWave frequencies are deployed under the coverage of an anchor cell operating at traditional low frequencies. A macro base station gets control over the initial access of several small cells, informing the mmWave base stations about the terminals' locations to enable mmWave base stations to steer their beams toward the target terminals directly. This procedure is performed mainly in three steps:

  1. The macro base station distributes the location information such as Global Navigation Satellite System (GNSS) coordinates of all mmWave base stations within its coverage area.

  2. Each terminal gets its GNSS coordinates and orientation.

  3. According to the location information obtained in previous steps, each terminal geometrically selects the closest mmWave base station and steers a beam toward it. Similarly, each mmWave base station can get the location information of terminals and can steer a particular beam to serve a terminal.

4.4.2.2 Conventional Initial Access in LTE

Predefined sequences with particular characteristics are periodically transmitted on radio frames in the downlink to facilitate mobile terminals to achieve frequency and timing synchronization during the cell search. In LTE, there are three different PSSs, each of which is a 62-symbol frequency-domain Zadoff–Chu sequence extended with five null subcarriers at either side and mapped to 72 subcarriers (six resource blocks) centered around the Direct Current (DC) subcarrier. Zadoff–Chu sequences are polyphase codes that have the particular property of having zero periodic autocorrelation at all nonzero lags. When used as a SS, the correlation between the ideal and received sequences is the largest if the lag is zero. When there is any lag between the two sequences, the correlation is zero. There is no restriction on the lengths of Zadoff–Chu sequences [Chu, 1972].

Table 4.9 The mapping between the cell identity within the group and root sequence indices.

Cell identity within the group $N_{ID}^{(2)}$	Root index u
0	25
1	29
2	34

Source: 3GPP TS36.211 [2009]/ETSI.

The sequence used for the PSS of LTE is generated according to

$$d_u(n) = \begin{cases} e^{-j\frac{\pi u n(n+1)}{63}}, & n = 0,1,\dots,30 \\ e^{-j\frac{\pi u(n+1)(n+2)}{63}}, & n = 31,32\dots,61, \end{cases} \tag{4.89}$$

where u stands for the Zadoff–Chu root sequence index, which has three values depending on the cell identity within the group, as given in Table 4.9.

The radio frame of LTE lasts 10 ms that is divided into ten *subframes*, each of which is composed of two *time slots* with a duration of 0.5 ms. Given a sampling rate of 30.72 MHz, each time slot contains 15 360 samples that are grouped into seven Orthogonal Frequency-Division Multiplexing (OFDM) symbols. According to its cell identity within the group, a cell selects a PSS among three different Zadoff–Chu sequences and transmits it periodically with a period of 5 ms. To be specific, two identical PSSs are inserted into the last OFDM symbol of the first slot (the indexing starts from 0) and the 10th slot in the FDD mode, while the TDD mode places two identical PSSs in the third OFDM symbol of the second slot and the 12th slot, respectively, as shown in Figure 4.14. Through searching the PSS, a mobile terminal can achieve:

- The frequency and symbol synchronization to a cell
- Five-millisecond timing of radio frames in the downlink
- Partial information of cell identity
- The position of the SSS

Similar to the PSS, the SSS occupies the center 72 subcarriers (not including the DC subcarrier) and locates in subframes 0 and 5 (for both FDD and TDD). The main functions of the SSS are

- Determine the frame timing in the downlink
- Determine the physical-layer cell identity

There are 504 unique physical-layer cell identities for LTE. The physical-layer cell identities are grouped into 168 cell-identity groups, and each group contains three

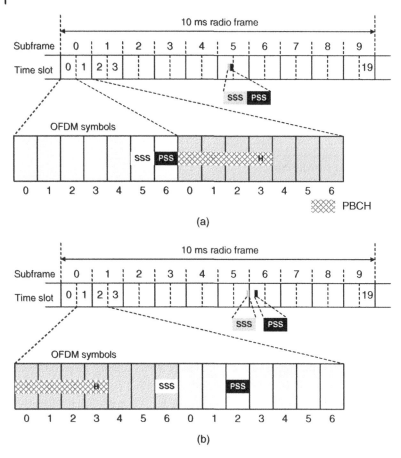

Figure 4.14 Timing of synchronization and broadcasting signals in the FDD (a, synchronization and broadcasting signals in FDD LTE) and TDD (b, synchronization and broadcasting signals in TDD LTE) mode of LTE.

unique identities. Thus, the physical-layer cell identity is uniquely determined by a value in the range of 0–167, representing the cell-identity group $N_{\mathrm{ID}}^{(1)}$, and a value in the range of 0–2, representing the cell identity within a cell-identity group $N_{\mathrm{ID}}^{(2)}$. The sequence used for the SSS is an interleaved concatenation of two length-31 m-sequences. The concatenated sequence is scrambled with a scrambling sequence given by the PSS. It has 168 different values by cycling through every possible state of a shift register of length m. The values of the SSS are applied to differentiate 168 cell-identity groups, i.e. $N_{ID}^{(1)} = 0,1, \dots, 167$. Thus, the physical-layer cell identity, N_{ID}^{cell}, is determined by

$$N_{\mathrm{ID}}^{\mathrm{cell}} = 3N_{\mathrm{ID}}^{(1)} + N_{\mathrm{ID}}^{(2)}. \tag{4.90}$$

The scrambled sequences are interleaved to alternate the sequences transmitted in the first and second SSS transmission in each radio frame. It allows the receiver to determine the frame timing from observing only one of the two sequences. Depending on which frame type (TDD or FDD) is applied, the SSS is transmitted in the same subframe as the PSS but one OFDM symbol earlier in FDD while it is three OFDM symbols before the PSS in TDD, as shown in Figure 4.14.

After frequency and timing synchronization is achieved in the downlink, a mobile terminal needs to extract the master system information carried by the PBCH. PBCH transmission is spread over four contiguous radio frames to span a period of 40 ms. In each radio frame, PBCH occupies the center 72 subcarriers of the first four OFDM symbols in the second slot. Excluding reference signal resource elements, the whole PBCH occupies about 960 resource elements. As PBCH uses Quadrature Phase Shift Keying (QPSK) modulation, it amounts to 1920 information bits. The block of bits transmitted on the PBCH, equaling 1920 for normal Cyclic Prefix (CP) and 1728 for extended CP, shall be scrambled with a cell-specific sequence before modulation. The timing of radio frame, SSs, and PBCH of both the FDD and TDD mode of LTE are illustrated in Figure 4.14.

4.4.2.3 Beam-Sweeping Initial Access in NR

Even though the terms PSS, SSS, and PBCH were defined in LTE already, NR first defined the term SS block, which consists of the PSS, SSS, and PBCH. In the NR system, as shown in Figure 4.15, an SS block spans four OFDM symbols in the time domain and 240 subcarriers in the frequency domain.

- PSS is transmitted using 127 subcarriers of the first OFDM symbol of the SS block, whereas the remaining subcarriers at both sides are empty.
- SSS occupies the same set of subcarriers as the PSS in the third OFDM symbol. There are eight and nine empty subcarriers on each side of the PSS.
- PBCH occupies the whole subcarriers of the second and fourth OFDM symbols of the SS block and also uses 48 subcarriers on each side of the SSS. Hence, the total number of resource units used for PBCH within a single SS block equals 576 (Table 4.10).

The length of an NR radio frame is also 10 ms, which is divided into ten subframes with each having 1 ms duration as LTE. NR supports flexible OFDM with variable subcarrier spacing given by

$$\Delta f = 2^v \cdot 15 \text{ kHz}, \tag{4.91}$$

where v is an integer, satisfying

$$0 \leq v \leq 4, \tag{4.92}$$

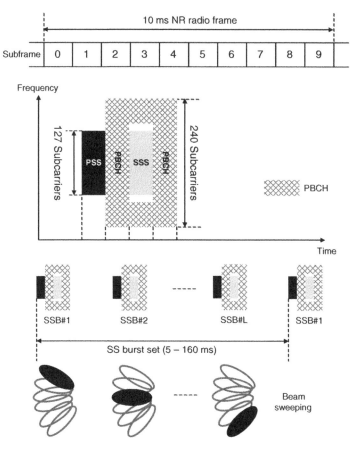

Figure 4.15 Multiple time-multiplexed SS blocks within an SS-burst-set period ranging from 5 ms to 160 ms to enable beam sweeping.

Table 4.10 OFDM numerology in NR radio frames and SS block parameters.

v	$\triangle f$ (kHz)	N_{OFDM}^{Slot}	$N_{Slot}^{Subframe}$	SSB bandwidth (MHz)	Duration (μs)
0	15	14	1	3.6	285
1	30	14	2	7.2	143
2	60	14	4	N/A	N/A
3	120	14	8	28.8	36
4	240	14	16	57.6	18

for the possible subcarrier spacing of 15, 30, 60, 120, and 240 kHz. Each slot contains a fixed number of $N_{\text{OFDM}}^{\text{Slot}} = 14$ normal CP OFDM symbols, so that each subframe consists of $N_{\text{Slot}}^{\text{Subframe}} = 2^{\nu}$ slots in terms of the subcarrier spacing. One key difference between the SS block in NR and the synchronization and broadcasting signals in LTE is the possibility of employing beam sweeping for SS-block transmission. The set of SS blocks within a beam-sweeping procedure is referred to as an *SS burst set*. A burst is transmitted periodically with a period that can vary from 5 to 160 ms. The maximum number (L) of SS blocks in a single SS burst set is frequency dependent, i.e.

- At low-frequency bands below 3 GHz, there can be up to $L = 4$ SS blocks within an SS burst set to support beam sweeping over four beams.
- At frequency bands between 3 and 6 GHz, there can be up to $L = 8$ SS blocks within an SS burst set to support beam sweeping over eight beams.
- For high-frequency bands (FR2), there can be up to $L = 64$ SS blocks within an SS burst set to support beam sweeping over 64 beams. That is because a large antenna array has to be applied to compensate for high propagation and penetration losses at mmWave frequencies and therefore there are a large number of beams with more narrow beam-width. An SS block is transmitted over each beam to guarantee that all directions can receive the synchronization and broadcasting signals, as illustrated in Figure 4.15.

4.4.3 Omnidirectional Beamforming

As mentioned in the previous section, some particular signals such as physical SSs and system broadcasting information must be sent to all mobile users within the whole coverage area of a cell or sector. Therefore, besides the dedicated channels with high power or multiplexing gains, broadcasting channels that can provide omnidirectional coverage ensure that every user in a cell or sector receives signals with acceptable quality. The traditional approach to transmitting synchronization and broadcast signals is to use a single antenna with an omnidirectional radiation pattern. Consequently, signal broadcasting was never a concern in earlier mobile systems that employed single-antenna base stations. However, an antenna array has become an essential part of advanced wireless communication systems to improve spectral efficiency. Intuitively, one can employ a specific antenna in the antenna array to transmit the broadcast signals. Nevertheless, the selected antenna needs a much higher PA, which is more expensive and power-consuming than the remaining antennas to achieve similar coverage as the unicast signal leveraging a beamforming gain. Therefore, it is meaningful to re-use multiple low-powered antennas to transmit broadcast signals to guarantee a cost-efficient and power-efficient system.

There have been several such multi-antenna schemes for omnidirectional coverage. A space-time block code, particularly the Alamouti code, has been successfully applied in the Universal Mobile Telecommunications System (UMTS) in the case of two transmit antennas. Cyclic delay diversity (CDD) is a simple multi-antenna transmission scheme recommended for Digital Video Broadcasting (DVB) and 3GPP LTE. Some research studies revealed that CDD is essentially a beamforming technique in the frequency domain. However, the performance of CDD has not yet been theoretically proved and generally verified by practices, particularly in cases of four antennas and above. In the 3GPP LTE, the space-frequency block code and frequency-switched transmit diversity were employed for broadcast channels. This scheme for broadcasting is efficient but closely dependent on the number of antennas. In the TD-SCDMA system, where the smart-antenna technique was used, a particular beam, known as *the broadcast beam*, with a flat amplitude within a certain angular range, such as 120°, was designed for broadcast channels. However, this scheme has two deficiencies. The first is the low power efficiency caused by small weight coefficients. For example, a broadcast beam with 120° is generated by the weighting vector [0.55, 1, 1, 0.55, 0.85, 1, 0.85], where two RF channels have utilized approximately only 30% (due to $0.552^2 = 0.304704$) of the full capability of the PA. The second is the adverse deformation of the broadcast beam stemmed from the character deviation and failure of RF channels. Then, regular calibration of RF channels and maintenance of the weight vectors would become mandatory, imposing a heavy burden on mobile operators.

The number of antenna elements used for mmWave transmission is generally large in order to provide high power gains to compensate for high propagation and penetration at mmWave frequencies. The beamforming is applied not only to the data plane but also to the control plane. As mentioned in the previous section, NR adopted beam sweeping to provide omnidirectional coverage for the synchronization and broadcasting signals. However, the sweeping procedure with a large number of beams such as 64 beams supported by NR is time-consuming, and the overhead due to the time-frequency resources consumed by the SS blocks is significant. This section introduces a technique called random beamforming (RBF) that aims to achieve omnidirectional coverage for broadcast channels in multiple-antenna systems [Yang et al., 2013]. A random weighting vector is applied on each communication resource element in the time-frequency domain, with an average pattern being omnidirectionally equal. It is an open-loop scheme, without the need for user feedback, in contrast to the opportunistic beamforming technique. Moreover, the maximum PA utilization efficiency is achieved by using equal weight coefficients.

4.4.3.1 Random Beamforming

The basic principle of RBF is to generate a random beam pattern over each time-frequency resource unit by applying a particular random weighting vector. For a large enough number of resource units, the average power of these random patterns in each direction is nearly equal since no direction is preferable to others. Thus, omnidirectional coverage can be realized, reusing the existing low-power antennas, without an extra high-power antenna or RF chain.

Design Criteria Let us use a ULA as an example. Assuming that there are N elements with inter-element spacing of d, the steering vector of the ULA is expressed as

$$\mathbf{a}(\theta) = \left[1, e^{-j\frac{2\pi}{\lambda} d \sin \theta}, e^{-j\frac{2\pi}{\lambda} 2d \sin \theta}, \dots, e^{-j\frac{2\pi}{\lambda} (N-1)d \sin \theta} \right]^T. \tag{4.93}$$

Applying a weighting vector on time-frequency resource unit t

$$\mathbf{w}(t) = \left[w_1(t), w_2(t), \dots, w_N(t) \right]^T, \tag{4.94}$$

which can be implemented by multiplexing a weighting coefficient over each baseband branch in digital beamforming or adjusting the signal phase on each antenna directly by analog shifters in analog beamforming. Then, the beam pattern of ULA can be given by

$$g(\theta, t) = \mathbf{w}^H(t)\mathbf{a}(\theta). \tag{4.95}$$

In a flat-fading channel, the received signal at resource unit t for a single antenna system is expressed by

$$y_{\text{SISO}}(t) = h(t)s(t) + n(t), \tag{4.96}$$

where $s(t)$ is a narrow-band transmitted signal, $h(t)$ is the channel response, and $n(t)$ stands for the additive Gaussian noise. From the perspective of the angular domain, the single antenna system has optimal omnidirectional coverage that is the benchmark of the RBF technique over multi-antenna systems. Consider a macro-cell scenario with a far-field assumption, where there are no reflectors around the base station and the mobile station locates in a scattering area with L reflectors. The diameter of the scattering area is small enough compared with the separation distance between the transmitter and receiver. Then, the AoA for various multi-path components, denoted as θ_l, is approximately the same, namely $\theta_l = \theta$. Denote the channel response of the lth path of the nth transmit antenna as $h_{n,l}$, we have [Yang et al., 2011]

$$h_{n,l} = h_{1,l} e^{-j\frac{2\pi}{\lambda} (n-1)d \sin \theta_l} \approx h_{1,l} e^{-j\frac{2\pi}{\lambda} (n-1)d \sin \theta}. \tag{4.97}$$

For a typical mobile user, the model for an RBF system with multiple antennas is identical to that of a single-antenna system superposed with extra time- and frequency-selective fading caused by random beam pattern, which can be expressed by

$$
\begin{aligned}
y_{\text{RBF}}(t) &= \sum_{l=1}^{L}\sum_{n=1}^{N} h_{n,l}(t)w_n^*(t)s(t) + n(t) \\
&\approx \sum_{l=1}^{L} h_{1,l}(t)\sum_{n=1}^{N} w_n^*(t)s(t)e^{-j\frac{2\pi}{\lambda}(n-1)d\sin\theta} + n(t) \\
&= h_1(t)g(\theta,t)s(t) + n(t),
\end{aligned}
\tag{4.98}
$$

where $h_1(t)$ is the channel response between the reference antenna and the receiver at time-frequency resource unit t. Compared Eq. (4.96) with Eq. (4.98), it is revealed that the only factor to affect the omnidirectional coverage is the beam pattern $g(\theta,t)$. Since generating beam pattern with equal instantaneous transmit energy in all direction is not possible, the design of RBF focuses on the average transmit energy. To ensure the average power is unchanged after adopting beamforming for a particular angle, it requires

$$
\int |g(\theta,t)|^2 p_g(\theta,t)dt = 1,
\tag{4.99}
$$

where $p_g(\theta,t)$ is the power density function of random pattern in terms of the angle and the time-frequency resource unit. From the discussion, the criteria in designing the random pattern sequence can be summarized as follows [Yang et al., 2013]

- Keep equal average power in each direction for omnidirectional coverage
- Set equal power in each antenna to maximize PA efficiency
- Use random patterns with the minimum variance

To achieve the objective of equal average power in each direction, *the pattern variance* in the angular dimension in defined as a metric to measure the pattern's degree of deviation from a circle:

$$
\sigma_g^2 = \sqrt{\frac{1}{2\pi}\int_0^{2\pi}\left[|g(\theta,t)|^2 - \mathbb{E}\left(|g(\theta,t)|^2\right)\right]^2 d\theta}.
\tag{4.100}
$$

Random Beam Pattern   It is convenient to get a basis weighting vector with low variance by a computer search. To satisfy the second and third criteria, the basis weighting vector is the vector with the minimum variance and unit module for each entry, i.e.

- $|w_1| = |w_2| = \cdots = |w_N| = 1,$
- $\mathbf{w}_0 = \arg\min(\sigma_g^2).$

Using an eight-element ULA with $d = \lambda/2$ as an example, the optimal weighting vector with the minimal variance is

$$\mathbf{w}_0 = \frac{\sqrt{2}}{2}\left[-\sqrt{2}, -1 + i, \sqrt{2}i, 1 - i, -1 + i, 1 - i, \sqrt{2}, 1 + i\right]^T. \tag{4.101}$$

To get enough beam patterns randomly from the basis pattern, an equip-amplitude transform can be applied, i.e.

$$\mathbf{w}_r = \mathbf{D}\mathbf{w}_0, \tag{4.102}$$

where $\mathbf{D}$ is a diagonal transform matrix given by

$$\mathbf{D} = \begin{bmatrix} 1 & 0 & 0 & \cdots & 0 \\ 0 & e^{j\phi(t)} & 0 & \cdots & 0 \\ 0 & 0 & e^{j2\phi(t)} & \cdots & 0 \\ & & & \ddots & \\ 0 & 0 & 0 & \cdots & e^{j(N-1)\phi(t)} \end{bmatrix}. \tag{4.103}$$

The formed beam pattern of $\mathbf{w}_r$ is

$$g_r(\theta, t) = \sum_{n=1}^{N} w_n^* e^{-j\frac{2\pi}{\lambda}(n-1)d\sin\theta} e^{j(n-1)\phi(t)}$$

$$= \sum_{n=1}^{N} w_n^* e^{-j(n-1)\left[\frac{2\pi}{\lambda}d\sin\theta - \phi(t)\right]}, \tag{4.104}$$

which shows that the transformed pattern only change the form of the basis pattern while keeping its value set unchanged. The basis beam pattern and three transformed patterns are illustrated in Figure 4.16.

4.4.3.2 Enhanced Random Beamforming

For a typical user in a particular direction, the beamforming gain fluctuates when the pattern changes over different time-frequency resource units, with a similar effect as time- and frequency-selective fading, even though the user is not moving. Such extra imperfection degrades the performance, leading to a gap between the single-antenna broadcasting and RBF. In addition to the traditional channel coding and diversity techniques that can be utilized to overcome these effects, a technique called Enhanced Random Beamforming (ERBF) based on the Alamouti coding has been proposed in Yang et al. [2013].

Alamouti coding is a simple but effective transmit diversity technique, which is usually applied in a system with two uncorrelated transmit antennas to achieve maximum rate and maximum diversity. The transmitted symbols are coded in the spatial and temporal domain instead of directly feeding into the beamformer. To be specific, two consecutive symbols $\mathbf{s} = [s_1, s_2]^T$ are encoded to a matrix $\mathbf{S}$ as follows

$$\mathbf{s} = \begin{bmatrix} s_1 \\ s_2 \end{bmatrix} \longrightarrow \mathbf{S} = \begin{bmatrix} s_1 & -s_2^* \\ s_2 & s_1^* \end{bmatrix}. \tag{4.105}$$

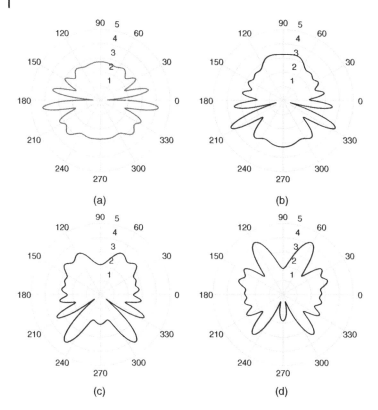

Figure 4.16 Random beams over an eight-element ULA with inter-antenna spacing of $d = \lambda/2$. The basis pattern formed by the weighting vector of Eq. (4.101) is shown in (a) while the transformed patterns with a phase shift of (b) 60°, (c) 120°, and (d) 180°.

As illustrated in Figure 4.17, the two output streams of the Alamouti encoder are multiplied by two independent weighting vectors, obtaining their corresponding signals in each antenna element, which are then added together to generate the transmitted signal for each element. Alamouti coding requires two uncorrelated antennas to achieve diversity gain. In the ERBF scheme, patterns 1 and 2 have uncorrelated beamforming gains and can be treated as two virtual antennas.

Denote the beamforming gains of patterns 1 and 2 are g_1 and g_2, respectively, with unit variances, i.e. $\mathbb{E}[|g_1|^2] = \mathbb{E}[|g_2|^2] = 1$. Therefore, the received signals for a typical user at two consecutive time-frequency resource units are expressed as

$$\begin{cases} r_1 = hg_1s_1 + hg_2s_2 + n_1 \\ r_2 = -hg_1s_2^* + hg_2s_1^* + n_2, \end{cases}$$

where $\mathbb{E}[|s_1|^2] = \mathbb{E}[|s_2|^2] = P/2$ with P being the sum transmit power in all antennas. Defining a received signal vector $\mathbf{r} = [r_1, r_2^*]^T$ and a noise vector $\mathbf{n} = [n_1, n_2^*]^T$,

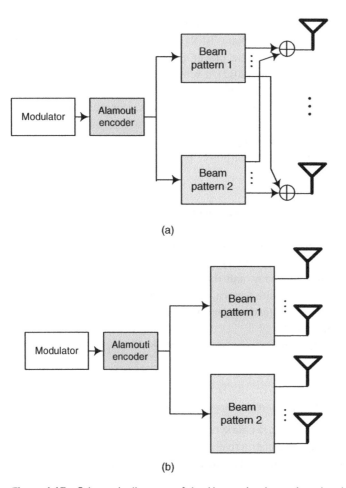

Figure 4.17 Schematic diagrams of the Alamouti-enhanced random beamforming (a) and complementary random beamforming (b).

and then building a channel matrix

$$\mathbf{H} = \begin{bmatrix} hg_1 & hg_2 \\ h^*g_2^* & -h^*g_1^* \end{bmatrix}, \tag{4.106}$$

Eq. (4.106) can be rewritten as

$$\mathbf{r} = \mathbf{Hs} + \mathbf{n}, \tag{4.107}$$

which can be detected, for example, through the Minimum Mean-Square Error (MMSE) estimation as

$$\hat{\mathbf{s}} = \left(\mathbf{H}^H\mathbf{H} + \sigma^2\mathbf{I}\right)^{-1}\mathbf{H}^H\mathbf{r}. \tag{4.108}$$

Although the enhanced RBF can improve the performance of omnidirectional coverage, its capacity and Bit Error Rate (BER) performance are still worse than that of the single antenna since the power fluctuates across time-frequency resource units. In Jiang and Yang [2012], a scheme called *Complementary Random Beamforming* that achieves the upper-bound performance (as same as the signal-antenna broadcasting) over antenna arrays was proposed. Determining the basis weighting vector with a pattern amplitude as flat as possible in the angular domain and forming a pair of complementary patterns on each time-frequency resource block, the resulting transmit power becomes isotropic instantaneously in the whole cell.

4.4.3.3 Complementary Random Beamforming

Firstly, a physical antenna array is divided into two sub-arrays, which are utilized to carry either branch of orthogonal signals generated by the Alamouti encoder. As RBF, a basis weighting vector with a pattern achieving the minimum variance is determined for either sub-array, and this pair of beam patterns complement each other so that their composite pattern is isotropic instantaneously at any particular angle rather than statistically equal by averaging over a large amount of time-frequency resource units.

Divide an N-element antenna array denoted by $\mathbf{e} = \{e_1, e_2, \ldots, e_N\}$ into two sub-arrays as

$$\begin{cases} \mathbf{e}_1 = \{e_1, e_2, \ldots, e_{N/2}\} \\ \mathbf{e}_2 = \{e_{N/2+1}, e_{N/2+2}, \ldots, e_N\} \end{cases} \tag{4.109}$$

In conventional beamforming, a data symbol is weighted correctly and transmitted at all elements to achieve constructively superposition in desired directions. In the complementary beamforming, the transmitted symbols are first encoded following the Alamouti scheme, and then two orthogonal streams are independently beam formed over $\mathbf{e}_1$ and $\mathbf{e}_2$, respectively. The electromagnetic interference phenomenon occurs only among elements transmitting correlated signals. Thereby, the pattern of different sub-arrays can be regarded as independent.

A pair of basis weighting vectors $\mathbf{w}_1$ and $\mathbf{w}_2$ are selected to satisfy the following criteria:

I. Minimize the variance of individual beam pattern over either sub-array, i.e.

$$\hat{\mathbf{w}}_k = \arg\min (\sigma_{g_k}^2),$$

where g_k, $k = 1,2$ denote the beam patterns over sub-arrays 1 and 2, respectively.

II. Minimize the variance of the composite pattern, i.e.

$$[\hat{\mathbf{w}}_1, \hat{\mathbf{w}}_2] = \arg\min (\sigma_g^2).$$

The composite pattern points to the resultant pattern over the whole antenna array, and its amplitude is defined as

$$|g| = \sqrt{\frac{|g_1|^2 + |g_2|^2}{2}}.$$ (4.110)

III. Set equal transmit power in each antenna to maximize the PA efficiency, i.e.

$$|w_1| = |w_2| = \cdots = |w_N| = 1.$$

Using an eight-element with inter-element spacing of $d = \lambda/2$ as an example, the steering vectors for two sub-arrays are expressed as (Figure 4.18)

$$\begin{cases} \mathbf{a}_1(\theta) = [1, e^{-j\pi\sin\theta}, e^{-j2\pi\sin\theta}, e^{-j3\pi\sin\theta}]^T \\ \mathbf{a}_2(\theta) = [e^{-j4\pi\sin\theta}, e^{-j5\pi\sin\theta}, e^{-j6\pi\sin\theta}, e^{-j7\pi\sin\theta}]^T. \end{cases}$$ (4.111)

Figure 4.18 Complementary random beams over an eight-element ULA with inter-antenna spacing of $d = \lambda/2$.
(a) Solid and dash-dot lines denote a pair of complementary patterns corresponding to weighting vectors in Eqs. (4.112) and (4.113) over two four-element sub-arrays, and their composite power pattern $(|g_1|^2 + |g_2|^2)/2 = 1$, achieving isotropic instantaneous radiation power as same as that of the single antenna. (b) Solid and dash-dot lines denote the transformed patterns through the equi-amplitude transform using $\phi_n = 45°$.

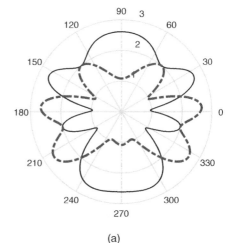

(a)

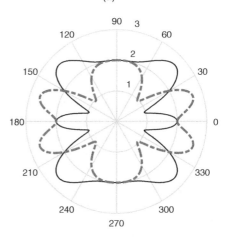

(b)

Through a simple numerical search by computer, a pair of basic weight vector:

$$\mathbf{w}_1 = [1, e^{j\frac{7\pi}{4}}, e^{j\frac{\pi}{2}}, e^{j\frac{5\pi}{4}}]^T \tag{4.112}$$

and

$$\mathbf{w}_2 = [1, e^{j\frac{3\pi}{4}}, e^{j\frac{\pi}{2}}, e^{j\frac{\pi}{4}}]^T \tag{4.113}$$

can be found. As illustrated in Figure 4.17, although pattern of either sub-arrays fluctuates in the angular domain as well as random patterns in the RBF scheme, their composite power pattern is isotropic in all angles, namely

$$|g|^2 = \frac{|g_1|^2 + |g_2|^2}{2} = 1. \tag{4.114}$$

In addition to the basis beam patterns, transformed patterns can be obtained using Eq. (4.102).

Thus, the received signal-to-noise ratio (SNR) of the complementary RBF equals to that of the single antenna, which can be mathematically depicted as

$$\gamma_{\text{CRBF}} = \frac{(|g_1|^2 + |g_2|^2)|h|^2 P}{2\sigma^2} = \frac{|h|^2 P}{\sigma^2} = \gamma_{\text{SISO}}. \tag{4.115}$$

4.5 Summary

The focus of this chapter was on the characteristics of mmWave propagation, which are fundamental to the design of mmWave transmission algorithms and communication protocols. mmWave signals exhibit similar behaviors as microwave and UHF radio signals in free-space propagation. However, the receiving power of mmWave signals is tinier due to the limited size of its antenna, imposing an illusion that the mmWave signals have higher free-space propagation losses. Also, mmWave signals suffer from high penetration loss, and even the human body can affect the direct transmission. In particular, the atmospheric effect and weather play a considerable role due to the absorption of oxygen and water molecules, which is usually negligible at low-frequency bands. This chapter introduced the channel models of large-scale and small-scale fading for the mmWave channel up to 100 GHz, which are essential for the system design, performance evaluation, validation of mmWave communications systems. To compensate for high propagation and penetration losses at mmWave frequencies, large-scale antenna arrays are applied to obtain high beamforming gain. The principles of three types of implementation, i.e. digital beamforming, analog beamforming, and hybrid beamforming, were introduced. Beamforming raises the problem of multi-beam synchronization and broadcasting in the initial access of mmWave communication systems. In final, this chapter provided the technological solutions to achieve omnidirectional coverage for broadcast channels, i.e. RBF and its enhanced schemes.

References

3GPP TR38.900 [2018], Study on channel model for frequency spectrum above 6 GHz, Technical Report TR38.900 v15.0.0, 3GPP.

3GPP TS36.211 [2009], Evolved universal terrestrial radio access (E-UTRA); physical channels and modulation (Release 8), Technical Specification TS36.211 v8.6.0, 3GPP.

Chen, J. [2013], When does asymptotic orthogonality exist for very large arrays?, *in* 'Proceedings of IEEE 2013 IEEE Global Communications Conference (GLOBECOM)', Atlanta, USA, pp. 4146–4150.

Chu, D. C. [1972], 'Polyphase codes with good periodic correlation properties', *IEEE Transactions on Information Theory* **18**(4), 531–532.

Crane, R. K. [1980], 'Prediction of attenuation by rain', *IEEE Transactions on Communications* **28**(9), 1717–1733.

Gardner, W. [1988], 'Simplification of MUSIC and ESPRIT by exploitation of cyclostationarity', *Proceedings of the IEEE* **76**(7), 845–847.

Giordani, M., Mezzavilla, M. and Zorzi, M. [2016], 'Initial access in 5G mmWave cellular networks', *IEEE Communications Magazine* **54**, 40–47.

ITU-R P.676 [2019], Attenuation by atmospheric gases and related effects, Recommendation P676-12, ITU-R.

Jiang, W. and Yang, X. [2012], An enhanced random beamforming scheme for signal broadcasting in multi-antenna systems, *in* 'Proceedings of IEEE 23rd International Symposium on Personal, Indoor and Mobile Radio Communications (PIMRC)', Sydney, Australia, pp. 2055–2060.

Rappaport, T. S., Heath, R. W., Daniels, R. C. and Murdock, J. N. [2014], *Millimeter wave wireless communications*, Pearson Education, Englewood Cliffs, NJ, USA.

Tan, W., Assimonis, S. D., Matthaiou, M., Han, Y., Li, X. and Jin, S. [2017], Analysis of different planar antenna arrays for mmWave Massive MIMO systems, *in* 'Proceedings of IEEE 85th Vehicular Technology Conference (VTC Spring)', Sydney, Australia, pp. 1–5.

Tharek, A. R. and McGeehan, J. P. [1988], Propagation and bit error rate measurements within buildings in the millimeter wave band about 60 GHz, *in* 'Proceedings of the eighth European Conference on Electrotechnics (EUROCON 1988)', Stockholm, Sweden, pp. 318–321.

Tse, D. and Viswanath, P. [2005], *Fundamentals of Wireless Communication*, Cambridge University Press, Cambridge, United Kingdom.

Yang, X., Jiang, W. and Vucetic, B. [2011], A random beamforming technique for broadcast channels in multiple antenna systems, *in* 'Proceedings of 2011 IEEE Vehicular Technology Conference (VTC Fall)', San Francisco, CA, USA, pp. 1–6.

Yang, X., Jiang, W. and Vucetic, B. [2013], 'A random beamforming technique for omnidirectional coverage in multiple-antenna systems', *IEEE Transactions on Vehicular Technology* **62**(3), 1420–1425.

Yong, S.-K. [2007], Channel modeling sub-committee final report, Technical Report IEEE 15-07-0584-01-003c, IEEE P802.15 Working Group for Wireless Personal Area Networks (WPANs) TG3c.

Zhang, J., Yu, X. and Letaief, K. B. [2019], 'Hybrid beamforming for 5G and beyond millimeter-wave systems: A holistic view', *IEEE Open Journal of the Communications Society* **1**, 77–91.

Zhao, H., Mayzus, R., Sun, S., Samimi, M., Schulz, J. K., Azar, Y., Wang, K., Wong, G. N., Jr., F. G. and Rappaport, T. S. [2013], 28 GHz millimeter wave cellular communication measurements for reflection and penetration loss in and around buildings in New York city, *in* 'Proceedings of the 2013 IEEE International Conference on Communications (ICC)', Budapest, Hungary, pp. 5163–5167.

5

Terahertz Technologies and Systems for 6G

To satisfy the demand of extreme transmission rate on the magnitude order of tera-bits-per-second envisioned in the 6G system, the available spectral resources over the millimeter-wave bands below 100 GHz are still limited, and wireless communications have to exploit the abundant spectrum in the Terahertz (THz) band. It refers to the spectrum range from 0.1 to 10 THz, or from 300 GHz to 3 THz, or sometimes from 100 GHz to 3 THz, as used in this book. The ITU-R has already identified the open of the spectrum between 275 and 450 GHz to land mobile and fixed services, paving the way of deploying THz commutations. In addition to THz communications, the THz band is also applied for other particular applications, such as imaging, sensing, and positioning, which are expected to achieve synergy with THz communications in the upcoming 6G system. Despite its high potential, wireless transmission over the THz band suffers from extremely worse channels raised from high free-space path loss, atmospheric absorption, rainfall attenuation, blockage, and high Doppler fluctuation. To overcome such high propagation losses, THz communications heavily rely on high directive transmission enabled by advanced technologies such as array-of-subarrays beamforming in an ultra-massive MIMO system or lens antenna arrays.

This chapter will focus on the key aspects of THz communications, consisting of

- The necessity of exploiting the terahertz band in the 6G era.
- Radio regulation of the terahertz band and state-of-the-art spectrum identification over the frequency range from 275 to 450 GHz.
- Potential use cases of THz communications in the future mobile and wireless networks.
- The applications of THz technologies in imaging, sensing, and positioning.
- Challenges of THz transmission, including high free-space spreading loss, atmospheric gaseous absorption, rainfall attenuation, blockage, and high Doppler fluctuation.

6G Key Technologies: A Comprehensive Guide, First Edition. Wei Jiang and Fa-Long Luo.
© 2023 The Institute of Electrical and Electronics Engineers, Inc. Published 2023 by John Wiley & Sons, Inc.

- The principle of array-of-subarrays beamforming in an ultra-massive MIMO system.
- The principle of lens antenna and the THz MIMO transmission based on lens antenna arrays.
- Introducing the world's first THz communications standard, i.e. IEEE 802.15.3d, operating in 252–322 GHz.

5.1 Potential of Terahertz Band

5.1.1 Spectrum Limit

For the past decades, we have witnessed one crucial trend in mobile communications, i.e. a new-generation cellular system employed wider signal bandwidth to realize a higher data transmission rate than its predecessor. Initially, the signal bandwidth of 1G analog systems was only around 30 kHz that is already sufficient to carry the voice signal of a mobile user. At that time, system designers naturally selected low-frequency bands with favorable propagation and penetration characteristics for mobile communications. Consequently, cellular networks until the 4G system operated in the conventional frequency bands below 6 GHz, which are now referred to as the sub-6 GHz band. To support the peak rate of 20 Gbps, the maximal bandwidth of the 5G system has been broadened to 1 GHz, imposed the necessity of exploiting higher frequencies. On the one hand, the total quantity of spectrum assigned to the IMT services in many regions is usually less than 1 GHz, while determining a large-volume contiguous spectrum below 6 GHz is impossible. On the other hand, there are enormous spectral resources at higher frequencies that were already used for a wide variety of non-cellular applications (APPs) such as satellite communications, remote sensing, radio astronomy, radar, to name a few. With the advancement in antenna technology and radio-frequency components, the millimeter-wave band previously considered unsuitable for mobile communications due to their unfavorable propagation characteristics becomes technologically usable. As a consequence, 5G became the first commercial cellular system to exploit high-frequency bands.

At the International Telecommunication Union (ITU) World Radio Conference (WRC)-15, an agenda item was appointed to identify high-frequency bands above 24 GHz for the IMT-2020 mobile services. Based on the studies conducted by the ITU-R after WRC-15, WRC-19 noted that ultra-low latency and very-high data-rate APPs require larger contiguous spectrum blocks. As a result, a total of 13.5 GHz spectra consisting of a set of high-frequency bands as follows were assigned for the deployment of 5G mmWave communications:

- 24.25–27.5 GHz
- 37–43.5 GHz

- 45.5–47 GHz
- 47.2–48.2 GHz
- 66–71 GHz

Meanwhile, 3GPP specified the relevant spectrum for 5G NR, which was divided into two frequency ranges:

- *FR1*: the First Frequency Range, including sub-6 GHz frequency bands from 450 MHz to 6 GHz
- *FR2*: the Second Frequency Range, covering 24.25 GHz to 52.6 GHz.

Initial mmWave deployments are expected in 28 GHz (3GPP NR band n257 and n261) and 39 GHz (3GPP n260), followed by 26 GHz (3GPP n258), as specified in Table 5.1. From this perspective, perhaps the most significant feature of 5G from the previous generations of mobile systems is the APP of millimeter-wave bands, which temporarily removed the spectrum shortage in 4G mobile communication systems.

Despite its current abundance in spectral redundancy, the mmWave bands are not necessarily adequate to tackle the increasing cravenness on bandwidth for another decade. The proliferation of new APPs such as virtual reality, augmented reality, ultra-high-definition video delivery, Internet of Things, Industry 4.0, connected and automatic vehicles, and wireless backhaul, and the emergence of disruptive use cases that have not been conceived yet, e.g. holographic-type communications, fully immersive gaming, Tactile Internet, and Internet of Intelligence, will impose the demand for extreme data rates and far more stringent quality-of-service requirements than what 5G networks can offer. It is envisioned that the 6G system will need to support wireless Terabit-per-second (Tbps) wireless links to satisfy the demand for information and communication technology in 2030 and beyond [Jiang et al., 2021].

Table 5.1 Operating frequency bands specified by 3GPP for NR in FR2.

NR band	Frequency range [GHz]	Duplex mode	Regions
n257	26.5–29.5	TDD	Asia, Americas
n258	24.25–27.5	TDD	Asia, Europe
n259	39.5–43.5	TDD	Global
n260	37.0–40.0	TDD	Americas
n261	27.5–28.35	TDD	Americas

Source: [Dahlman et al., 2021, ch. 3]/With permission of Elsevier.

5.1.2 The Need of Exploiting Terahertz Band

The whole electromagnetic spectrum is illustrated in Figure 5.1 while the main characteristics of different bands are explained in Table 5.2. In comparison with other bands, the THz band attracted more focus from academia and industry for the past years. The reasons can be summarized as follows:

- *The radio and microwave frequency bands are not able to support Terabits communications.* Thanks to the excellent propagation characteristics at sub-6 GHz frequencies, advanced transmission technologies, e.g. Orthogonal Frequency Division Multiplexing (OFDM) and Non-Orthogonal Multiple-Access (NOMA), high-order modulation such as 1024 Quadrature Amplitude Modulation (QAM), and radical spatial-multiplexing schemes over massive Multiple-Input Multiple-Output (MIMO) systems, have been invented to achieve a very high spectral efficiency. However, the scarcity of the available bandwidth limits the achievable transmission rate. For example, the LTE-A system realized the peak data rate of 1 Gbps when using an eight-by-four MIMO scheme over a 100 MHz aggregated bandwidth. Assuming that the sub-6 GHz band ultimately determines 1 GHz bandwidth for the IMT service, the target Terabits links can only be reached by means of wireless transmission technologies with an extreme spectral efficiency of 100 bps/Hz, which is infeasible in the foreseeable future.

- *The available mmWave spectrum is still limited to support Terabits transmission.* As of today, within the global mmWave cellular bands up to 100 GHz, there is a total of 13.5 GHz spectral resources available. In such a bandwidth, data rates on the order of 1 Tbps can only be achieved with transmission schemes having a spectral efficiency of approaching 100 bps/Hz, which requires symbol fidelity that is not feasible using currently known digital modulation techniques or transceiver components. The available bandwidth at mmWave frequency is still limited relative to such an extreme target, effectively imposing an upper bound on the data rates. Therefore, Terabits communications will flourish at frequencies above 100 GHz, where the available spectrum is massively abundant.

- *Today's hardware technologies constrain the optical bands to support Terabits links.* Despite the enormous available spectrum in the optical bands at infrared, visible-light, and ultraviolet frequencies, several issues limit the practicality of Optical Wireless Communications (OWC). The restriction of low transmission power due to eye-safety purposes, the effects of several types of atmospheric attenuation on the signal propagation (e.g. fog, rain, dust, or pollution), high diffuse reflection losses, and the impact of misalignment between transmitter and receiver restrict both the achievable data rates and transmission range of OWC systems [Akyildiza et al., 2014]. For example, an Infra-Red (IR) communication system that supports 10 Gbps wireless links in Line-of-Sight (LOS) propagation

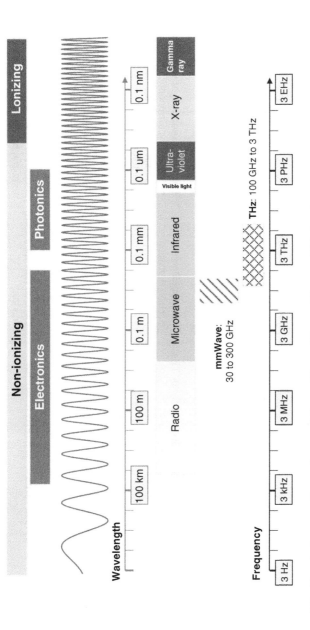

Figure 5.1 The electromagnetic spectrum and the positions of the mmWave and THz bands.

Table 5.2 The main features of different frequency bands over the whole electromagnetic spectrum.

Band	Wavelength	Frequency	Origin	Applications
Electric power	$>10^5$ m	<100 Hz	Vibrating atoms or molecules over macroscopic distances	Electric energy transmission
Radio	>1 m	3 Hz to 300 MHz	Vibrating atoms or molecules over macroscopic distances	Radio/TV broadcasting, communications, satellite, navigation, radiolocation, radio astronomy, radar, remote sensing
Microwave	1 mm to 1 m	300 MHz to 300 GHz	Vibrating atoms or molecules	Communications, navigation, satellite, radar, radio astronomy, heating, remote sensing, and spectroscopy
Infrared	0.75 µm to 1 mm	300 GHz to 400 THz	Vibrating atoms or electron transition	Communications, night vision, thermography, spectroscopy, astronomy, heating, tracking, hyperspectral imaging, meteorology, and climatology
Visible light	0.38 – 0.76 µm	400 – 790 THz	Vibrating atoms or electron transition	Lighting, heating, power generation, biological systems, and spectroscopy
Ultraviolet	10 nm to 0.38 µm	790 THz to 30 PHz	Vibrating atoms or electron transition	Photography, electrical and electronics industry, semiconductor manufacturing, material science, biology-related uses, analytic uses, and fluorescent dye uses
X-ray	10 pm to 10 nm	30 PHz to 30 EHz	Electron transition and braking	Medicine, projectional radiograph, computed tomography, fluoroscopy, and radiotherapy
Gamma ray	<10 pm	>30 EHz	Nuclear transition	Medicine (radiotherapy), industry (sterilization and disinfection), and the nuclear industry

for Wireless Local Area Network (WLAN) and an indoor free-space optical communication system that supports a 1 Gbps link at visible-light frequencies have become a reality already. Furthermore, a long-distance free-space optical communication system achieving an impressive transmission rate of 1.28 Tbps was successfully demonstrated ten years ago. However, it happens only with LOS paths, whereas much lower data rates occur in diffused Non-Line-of-Sight (NLOS) environments. High-capacity optical signals generated by typical optic fiber communication equipment were injected into an optical front-end with a large volume of 12 cm × 12 cm × 20 cm and a weight of almost 1 kg. All these constraints limit the feasibility of this large-scale optical approach for personal and mobile wireless communications.

- *The extreme high bands are not suitable for wireless communications.* Ionizing radiation, including ultraviolet, X-rays, and Gamma rays, is dangerous since it is known to have sufficiently high particle energy to dislodge electrons and create free radicals that can lead to cancer. Ionizing radiation has been applied for many fields, including radiotherapy, photography, semiconductor manufacturing, material science, gauging the thickness of metals, astronomy, nuclear medicine, sterilizing medical equipment, and pasteurizing certain foods and spices. The adverse health effects of ionizing radiation may be negligible if used with care, but it is still not suitable for personal communications [Rappaport et al., 2019]. Unlike ionizing radiation, mmWave and THz radiation are non-ionizing because the photon energy is not nearly sufficient (0.1–12.4 meV, which is more than three orders of magnitude weaker than ionizing photon energy levels) to release an electron from an atom or a molecule, where typically 12 eV is required for ionization. Since ionizing radiation is not determined to be a concern at mmWave and THz bands, heating might be the only primary cancer risk. The International Commission on Non-Ionizing Radiation Protection (ICNIRP) standards have been set up to protect against thermal hazards, particularly for the eyes and skin, which are most sensitive to heat from radiation due to lack of blood flow.

The term *terahertz* first occurred in the 1970s, where the term was used to describe the spectral line frequency coverage of a Michelson interferometer or to describe the frequency coverage of point contact diode detectors [Siegel, 2002]. Spectroscopists had much earlier coined the term for emission frequencies that fell below the far Infra-Red (IR). The far IR points to the lowest frequency part of infrared radiation with a wavelength of 15 micrometers (μm) to 1 millimeter (mm), corresponding to a frequency range of about 300 GHz to 20 THz. Millimeter wave refers to the frequency band from 30 to 300 GHz, corresponding to the range of signal wavelengths from 1 to 10 mm. The border between the far IR and THz, and the border between mmWave and THz, are still rather blurry. A usual definition of the THz band refers to the electromagnetic waves that fill the wavelength range from 100 to 1000 µm or the frequency band from 300 GHz to 3 THz. There is another definition with a much wider range, covering the whole bands from 0.1 to 10 THz. The difference of using frequency (THz) and wavelength (mmWave) for naming leaves an ambiguity of the range from 100 to 300 GHz, which is also referred to as *sub-THz* or *sub-mmWave* by some researchers. It is envisaged that 5G mainly focuses on the frequency bands below 100 GHz while 6G will cross over this frequency point. Therefore, we like to define the THz band as the frequency range from 100 GHz to 3 THz for the discussion in this book.

The THz band offers much larger spectral resources, ranging from tens of GHz to several THz depending on the transmission distance. As a result, the available bandwidth is more than one order of magnitude above the mmWave bands, while the frequency of operation is at least one order of magnitude below the optical bands. It opens the possibility of Terabits communications from the perspective of the spectrum. In addition, the technology required to make THz communication a reality is rapidly advancing, and the development of new transceiver architectures and antennas built upon novel materials with remarkable properties are finally overcoming one of the significant challenges. The high frequency allows a tiny antenna size for a highly compact level. Thus, it is envisaged that more than one thousand antennas can be embedded into a single base station to provide multiple super-narrow beams simultaneously and overcome the high propagation loss. It makes THz communications an exemplary supplementary to conventional networks operating at low frequencies for specific use cases, such as indoor communications and wireless backhaul, and a competitive option for future cyber-physical (PHY) APPs with extreme performance requirements.

There are many technical barriers ahead for achieving the potential of THz wireless communications, but the research community is already focusing on addressing these challenges with innovative solutions. Using the THz bands for access and backhaul connectivity brings the necessity of rethinking some of conventional transmission and networking mechanisms. Similar to mmWave, THz signals suffer from high path loss and considerable frequency-specific atmospheric attenuation, imposing a significant constraint on the transmission distance. Large-scale antenna arrays will be an essential ingredient of a THz communication system to achieve high power gains, compensating for the high propagation loss. The ultra-wideband and highly directive nature of THz wireless links impose challenges in terms of ultra-broadband antennas, radio frequency front-end, channel modelling, waveform design, single processing, beamforming, modulation, coding, and hardware constrains. The fundamental difference of interference due to pencil beams calls for a thorough characterization and detailed modeling of interference. For propagation and channel modeling, LOS and NLOS reflected and scattered signal components should be considered, as well as the inherent molecular noise, misalignment impairments, and blockage probability. Medium access control and radio resource management protocols need to operate with pencil beams and must therefore be based on radically new principles. Fast handover procedures need to incorporate the time required for discovery, synchronization, localization, and tracking functionalities given beam-based transmission.

5.1.3 Spectrum Regulation on Terahertz Band

In addition to all these technical challenges, the realization of THz communications still needs to tackle the issues of spectrum regulation. Regulatory agencies such as the ITU and The Federal Communications Commission (FCC) solicit comments to regulate frequencies above 100 GHz for point-to-point use, broadcasting services, and other wireless transmission APPs. In order to avoid harmful interference to Earth Exploration Satellite Service (EESS) and radio astronomy operating in the spectrum between 275 and 1000 GHz, the ITU-R World Radiocommunication Conference 2015 has initiated the activity called *Studies toward an identification for use by administrations for land-mobile and fixed services APPs operating in the frequency range* 275–450 GHz. The sharing study identified EESS as the more critical service due to its operational characteristics. In contrast, radio astronomy, with its large antennas typically located in a remote area and pointing to the sky, can be protected by simply applying a minimum distance to THz communication devices. At the WRC-19 conference, a new footnote was added to the radio regulations, allowing for the open of the spectrum between 275 and 450 GHz to land mobile and fixed services. Together with the already assigned spectrum below 275 GHz, a total of 160 GHz spectrum, containing two big contiguous spectrum bands with 44 GHz (i.e. from 252 to 296 GHz) and 94 GHz bandwidth, respectively, as depicted in Table 5.3, is available for THz communications without specific conditions necessary to protect EESS. In addition, there are three frequency bands with a total bandwidth of 38 GHz that can only be applied for land mobile and fixed services when specific conditions to ensure the protection of passive APPs have been determined in accordance with Resolution 731 of WRC-19, which deals with the consideration of sharing and adjacent-band compatibility between passive and active services.

The mmWave Coalition, a group of innovative companies and universities united in the objective of removing regulatory barriers to technologies using frequencies ranging from 95 to 275 GHz in the USA, submitted comments in January 2019 to the FCC and the National Telecommunications and Information Administration (NTIA) for developing a sustainable spectrum strategy and urged

Table 5.3 Available THz communications frequency bands recommended by the ITU.

Frequency (GHz)	Bandwidth (GHz)	Radio regulations
252 ~ 275	23	Land mobile and fixed services on a co-primary basis.
275 ~ 296	21	The use for land mobile and fixed services, coexisting with EESS without specific conditions to protect EESS.
306 ~ 313	7	
318 ~ 333	15	
356 ~ 450	94	
296 ~ 306	10	The spectrum of 38 GHz only for the use of land mobile and fixed services under specific conditions.
313 ~ 318	5	
333 ~ 356	23	

Source: Adapted from Kuerner and Hirata [2020].

Table 5.4 Unlicensed THz communications spectrum bands in the United States assigned by the FCC.

Frequency band (GHz)	Contiguous bandwidth (GHz)
116–123	7
174.8–182	7.2
185–190	5
244–246	2
Total	21.2

Source: Rappaport et al. [2019]/With permission of IEEE.

NTIA to facilitate the access to spectrum above 95 GHz. In March 2019, the FCC announced that it opens up the use of frequencies between 95 GHz and 3 THz in the United States, provided 21.2 GHz of spectrum for unlicensed use and permitted experimental licensing for 6G and beyond, as shown in Table 5.4. In addition, the IEEE formed the IEEE 802.15.3d task force in 2017 for global Wi-Fi use at frequencies from 252 to 325 GHz, creating the first THz wireless standard, with a nominal data rate of 100 Gbps and channel bandwidths from 2.16 to 69.12 GHz.

5.2 Terahertz Applications

The vast amount of spectral resources and ultra-wide bandwidth provided by the THz frequency band allow for a variety of wireless APPs that demand ultra-high transmission speed, e.g. holographic-type communications, high-definition virtual reality, fully immersive experience, ultra-high-definition video delivery, over-the-air computing, extreme-speed mobile Internet, autonomous driving, remote control, information shower, Tactile Internet, and high-speed wireless connectivity in data centers. It also provides new degree of freedom for the design of mobile systems. For instance, using THz links for wireless backhaul among base stations, which can enable flexible and ultra-dense architecture, speed up network deployment, and reduce the costs of site acquisition, installation, and maintenance. Due to very small wavelengths of THz signals, the antenna dimension is tiny. It opens the possibility of developing a plethora of novel APPs such as nanoscale communications for nanoscale devices or nanomachines, on-chip communications, Internet of Nano-Things (IoNTs), and intra-body network. It can also be combined with biocompatible and energy-efficient nanodevices to realize molecular communication using chemical signals. In addition to wireless transmission operating in the THz band, there are non-communication THz APPs that are likely to be integrated into 6G networks and beyond. For example, it is potential to make use of the particular PHY characteristics of THz signals to provide high-definition sensing of the surrounding PHY environment, which is promising for efficiently implementing harmonized communication and sensing. Some of these APPs can already be foreseen, such as wireless cognition, sensing, imaging, and positioning, whereas others will undoubtedly emerge as technology progresses.

5.2.1 Terahertz Wireless Communications

5.2.1.1 Terabit Cellular Hotspot

The increasing number of mobile or fixed users with extreme-throughput demand in dense urban environments or particular places such as industrial sites requires the deployment of ultra-dense networks. The THz band can provide abundant spectral resources and ultra-wide bandwidth for small cells, with a relatively short coverage distance and high occurrence of LOS paths, to offer Terabits communication links. These small cells cover static and mobile users, both in indoor and outdoor scenarios. Specific APPs are ultra-high-definition video delivery, information shower at the entrance, high-quality virtual reality, or holographic-type communications. Combined with conventional cellular networks operated in low-frequency bands, a heterogeneous network consisting of a macro-base-station tier and a small-cell tier can enable seamless connectivity and fully transparency over a wide-coverage area and global roaming, satisfying the extreme performance requirements for next-generation mobile networks. In addition, highly directive THz links can be used to provide an ultra-high-speed wireless backhaul to small cells to substantially reduce the time and expenditure of site acquisition, installation, and maintenance.

5.2.1.2 Terabit Wireless Local-Area Network

The THz band enables the implementation of terabit communications within a wireless local-area network. It can deliver excellent quality of experience like that of optical networks, allowing for the seamless interconnection between ultra-high-speed wired networks and personal wireless devices such as smartphones, tablet computers, laptops, and wearable electronics, with no speed/delay difference between wireless and wired links. This will facilitate the use of bandwidth-intensive APPs across static and portable users, mainly in indoor scenarios. Some specific APPs are high-definition virtual reality, holographic-type services, fully immersive gaming, or ultra-high-speed wireless data distribution in data centers. It also facilitates the deployment of industrial networks to interconnect a massive number of sensors and actuators in a factory, or campus networks to offer high data-throughput, low-latency, and high-reliability for equipment and machines such as Automated Guided Vehicles (AGVs) in a logistic center.

5.2.1.3 Terabit Device-To-Device Link

THz communications are a good candidate for providing Terabit direct links among devices in proximity. The usage scenario of Device-To-Device (D2D) links can be indoor, such as an office or home, where a set of personal or commercial devices are interconnected to form an ad hoc network, called Wireless Personal Area Networks. Specific APPs include multimedia kiosks and ultra-high-speed data transfer between personal devices. For example, transferring the equivalent content of a blue-ray disk to a high-definition large-size display could take less than one second with a Tbps link, boosting the data rates of existing technologies such as Wi-Fi Direct, Apple Airplay, or Miracast. A promising APP would be the Brain–Computer Interface (BCI), where THz communications can be applied to transfer a vast volume of collected brain-wave data to the computer that processes the data. In computer vision, the THz communications can also play an essential role in transferring the collected high-definition video to a platform running machine learning-based analytical software. Terabit D2D links can also be applied in an outdoor environment for Vehicle-to-X communications, providing high-throughput, low-latency connectivity among vehicles or between vehicles and surrounding infrastructure.

5.2.1.4   Secure Wireless Communication

The signal transmission at THz frequencies suffers from a significant path loss because the collected radiation power is proportional to the aperture of the antenna, which is on a tiny scale for such high frequencies. Meanwhile, THz communications must undergo more severe atmospheric attenuation than microwaves and millimeter waves. Consequently, the use of very large-scale antenna arrays with hundreds or even thousands of elements at both the transmitter and receiver becomes necessary to compensate for such propagation and penetration losses and achieve a reasonable communication distance. A highly directive beam concentrating the radiation energy in a very narrow, almost razor-sharp direction can drastically increase the difficulty of eavesdropping. An eavesdropper has to place its receiver at the direct link of two communicating parties. It might block the communication and cause two communicating parties to set up a new connection through another direction. Furthermore, THz communications adopt ultra-wide signal bandwidth, over which spread spectrum techniques can substantially alleviate narrow-band interference and common jamming attacks. Hence, THz communications can be regarded as secure transmission, enabling ultra-broadband secure communication links in critical scenarios such as the military and defense fields.

5.2.1.5   Terabit Wireless Backhaul

Fiber optical connections can provide high data throughput and reliability, but installation is usually time-consuming and costly since one should wait for scheduled road reconstructions. Sometimes, it is hard to deploy the public optical network of a mobile operator within some buildings or particular areas due to the objection of the property owners. However, a next-generation mobile network is envisioned to be highly heterogeneous, consisting of macro base stations, small cells, relays, cell-free nodes, distributed antennas, distributed baseband units, remote radio-frequency units (RFUs), roadside units, and intelligent reflecting surfaces, all of which require high-throughput backhaul or fronthaul connectivity. Highly directive THz links can be used to provide ultra-high-speed wireless backhauls or fronthauls to small cells to interconnect such network elements. It will substantially reduce the time and expenditure of site acquisition, installation, and maintenance. Also, it provides a new degree of freedom for the design of network architecture and communication mechanisms. In addition, mobile or fixed users in rural or remote areas nowadays suffer from worse coverage and low quality of service. If a cost-efficient and flexible solution cannot be guaranteed, the digital divide between rural areas and major cities will increase. As a wireless backhaul extension of the optical fiber, THz wireless links can work well as an essential building block to guarantee a universal telecommunication service with high-quality, ubiquitous connections everywhere. In such scenarios, apart from the extreme data rates, the critical parameter is transmission range, which should be several hundred meters or even on the order of kilometers.

5.2.1.6 Terahertz Nano-Communications

As we know, the minimal size of an antenna used for the transmission of terahertz signals can be on the magnitude order of micrometers. Intuitively, it will enable wireless connection among nanoscale machines or nanomachines using nanoscale antennas for very tiny specific equipment that performs particular tasks at the nanoscale, such as a biosensor injected into the human blood vessel. Each component of a nanomachine is up to a few hundred cubic nanometers in size, and the size of the entire device is in the order of a few cubic micrometers at most. It is not that nanomachines are developed to communicate in the THz band, but the tiny size and compact properties of nanoscale transceivers and antennas facilitate nanomachines to transfer volume collected data. Bz getting rid of the restriction of wirelines or optical fibers, these nanomachines are flexible to move and can enter more extreme areas such as a human body or a biochemical site with high temperatures.

Several specific use cases of THz nano-communications are provided by Aky-ildiza et al. [2014], i.e.

- *Health Monitoring*: Sodium, glucose, and other ions in the blood, cholesterol, cancer biomarkers, or the presence of different infectious agents can be detected utilizing nanoscale biosensors injected into the human body or embedded under the skin. A set of biosensors distributed within or around the body, comprising a body sensor network, could collect relevant PHY or biochemical data related to a human's health. Through a wireless interface, these sensory data can be delivered to a healthcare APP running on a personal smartphone or specialized medical equipment for real-time monitoring or sent to the cloud of a healthcare provider for professional analysis with specific intelligent tools and the long-term historical record.
- *Nuclear, Biological, and Chemical Defense*: Chemical and biological nanosensors are able to detect harmful chemicals and biological threats in a distributed manner. One of the main benefits of using nanosensors rather than classical macroscale or microscale sensors is that a chemical composite can be detected in a concentration as low as one molecule and much more timely than classical sensors. Considering these sensors need direct contact with molecules, a wireless sensor network interconnecting a vast number of nanosensors becomes necessary. Wireless THz nano-communication will be able to converge the information of the molecular composition of the air, water, or soil in a specific location to monitoring equipment with high throughput and low latency.
- *IoNTs*: Using THz nano-communications to interconnect nanoscale machines, devices, and sensors with existing wireless networks and the Internet makes

a truly cyber-PHY system that can be named as the IoNTs. The IoNT enables disruptive APPs that will reshape the way the human work or live. For example, in a smart office or smart home, a nano-transceiver and nano-antenna can be embedded in every object to allow them to be permanently connected to the Internet. As a result, a user can keep track of all its professional and personal items effortlessly.

- *On-Chip Communication*: THz communication can provide an efficient and scalable approach to inter-core connections in on-chip wireless networks using planar nano-antenna arrays to create ultra-high-speed links. This novel approach will expectedly fulfill the stringent requirements of the area-constraint and communication-intensive on-chip scenario by its high bandwidth, low latency, and low overhead. More importantly, the use of graphene-based THz communication would deliver inherent multicast and broadcast communication capabilities at the core level. THz communications provide a new degree of freedom for the design of chips, which might bring a new way to improve hardware performance.

5.2.2 Non-Communication Terahertz Applications

5.2.2.1 Terahertz Sensing

With high frequencies, the spatial resolution of a propagated signal becomes much finer due to tiny wavelengths, thereby enabling high-definition spatial differentiation at THz frequencies. THz sensing techniques exploit the tiny wavelength on the order of micrometers and the frequency-selective resonances of various materials over the measured environment to gain unique information based on the observed signal signature. THz signals can penetrate various non-conducting materials such as plastics, fabrics, paper, wood, and ceramics. However, it is much difficult for THz signals to penetrate metal materials while water heavily attenuates its radiation power. Different thicknesses, densities, or chemical compositions of various materials raise the particular strength and phase variations on THz signals, allowing accurate identification of PHY objects. In addition, THz sensing can make use of vast channel bandwidths over 100 GHz and the ability to implement very high gain antennas in a small PHY size. Furthermore, THz photon energies are several orders of magnitude weaker than those of X-rays, thereby allowing for safe interactions with the human body in medical sensing and security screening APPs.

It will become feasible to generate images of PHY spaces by systematically monitoring received signal signatures at a wide array of different angles. Since beam steering can be real-timely implemented and radio propagation distances are short (e.g. several meters in a room), leading to propagation times that are less than 10 ns, it shall be feasible to measure the properties of a room, an office, or a complex environment in a matter of seconds or less. This ability opens up a new degree of freedom of wireless APP that enables future wireless devices to carry out wireless reality sensing and gather a map or view of any place, leading to detailed three-dimensional maps created on the fly and shared on the cloud. Also, since certain materials and gasses have vibrational absorption at particular frequencies throughout the THz band, it becomes possible to detect the presence of certain items based on frequency scanning spectroscopy. For example, the presence of certain chemicals or allergens in food, water, and air, or other defects in the surrounding environment, can be detected based on spectroscopy. THz sensing will enable new APPs such as miniaturized radars for gesture detection and touchless smartphones, spectrometers for explosive detection and gas sensing, THz security body scanning, air quality detection, personal health monitoring, and wireless synchronization. By building real-time maps of any environment, it may be possible to predict channel characteristics at a mobile device, aid in the alignment of directional antennas, provide on-the-fly localization, and adapt wireless transmission parameters. This capability could also be fed to the cloud to enable a real-time collection capability for mapping and sensing the world, which could be used in commercial APPs for transportation, shopping, and other retail uses [Rappaport et al., 2019].

5.2.2.2 Terahertz Imaging

Using THz radiation to form images, referring to as Terahertz imaging, has many particular technical advantages over microwaves and visible light. THz imaging exhibits high spatial resolution due to smaller wavelengths and ultra-wide bandwidths with moderately sized hardware than imaging using low frequencies. Compared with infrared and visible light, THz waves have better penetration performance, making common materials relatively transparent before THz imaging equipment. There are many security screening APPs, such as checking postal packages for concealed objects, allowing THz imaging through envelopes, packages, parcels, and small bags to identify potential hazardous items. Based on the property that THz radiation is non-ionizing and therefore no known health risk to biological cells except for heating has motivated its APP in the human body, where ionizing radiation, i.e. Ultraviolet, X-Ray, and Gamma Ray, will raise high health risks. Therefore, THz imaging is suitable for the stand-off detection of items such as firearms, bombs, and explosive belts hidden beneath clothing in airports, train stations, and border crossings. THz waves can be applied to build a thickness measurement system specifically designed for determining the thickness of individual layers in multiple layer systems on metallic and non-metallic substrates with an accuracy on the order of magnitude of micrometers.

The APPs such as security screening and medical imaging take advantage of the penetration capabilities of THz waves while reflection or scattering have been treated as an unwanted component. THz waves can bounce off most building materials to achieve indirect imaging of scenes hidden behind an occlusion that is opaque in the visible spectrum, similar to using the flat surface of an open door as a mirror to observe the inside of a room. Visible light and infrared waves have very short wavelengths, and unless the surface is carefully polished, the light is scattered almost in all directions, leading to image ambiguity. Meanwhile, lower frequencies feature longer wavelengths limiting the capability to single object identification or motion detection rather than detailed images of the occluded scene. Unlike microwaves that only have strong specular reflection and optical waves that exhibit just diffuse scattering, THz radiation has both specular reflection and diffuse scattering from most building surfaces. The strong specular component turns the surfaces into something close to "electrical mirrors", thus allowing to image objects around obstacles, while maintaining spatial coherency (narrow beams) and high spatial resolution. Hence, THz waves are situated in a unique position in the electromagnetic spectrum, featuring smaller wavelengths, while the surface roughness of most building materials will provide sufficient specular reflection and diffusion scattering. Therefore, THz waves can augment human and computer vision to see around corners and to view NLOS objects, enabling unique capabilities in rescue and surveillance, autonomous navigation, and localization. A building surface (e.g. wall, floor, and door) typically behaves to first order as mirrors (e.g. perfect reflectors of THz energy), thus allowing THz imaging to see around corners and behind walls if there are sufficient reflection or scattering paths. NLOS THz imaging uses a backpropagation synthetic aperture radar (SAR) to generate three-dimensional scene by computing the time of flight of multipath backscattered signals. In contrast, the optical wavelength is smaller than the surface roughness of most surfaces. Thus, optical NLOS imaging requires complex hardware and computationally expensive reconstruction algorithms while exhibiting short imaging distances.

Moreover, THz imaging is more effective than visible light or infrared-based imaging such as Light Detection and Ranging (LIDAR), due to its robustness against weather and ambient light. It is worth noting that although LIDAR can provide higher resolution, LIDAR cannot work in bad weather when it is foggy, raining, snowy, or cloudy. However, THz imaging can be used for assisting driving or flying in foul weather, as well as in military and national security. High-definition video resolution radars that operate at several hundred gigahertz will be sufficient to provide a TV-like picture quality and will complement radars at lower frequencies that provide long-range detection but with poor resolution. Dual-frequency imaging systems will enable driving or flying in heavy fog or rain [Rappaport et al., 2019].

5.2.2.3 Terahertz Positioning

It is envisioned that the next-generation cellular network will offer high-accurate positioning and localization in both indoor and outdoor environments, in addition to communication services, which Global Navigation Satellite System (GNSS) and conventional multi-cell-based localization techniques using low-frequency bands fail to provide. Leveraging THz imaging for localization has unique benefits compared with other methods. The THz imaging can localize users in the NLOS areas, even if their travel paths to the base station experience more than one reflection (e.g. multiple bounces). Devices incorporating THz imaging and THz communications will likely also provide centimeter-level localization anywhere. High-frequency localization techniques are based on the concept of Simultaneous Localization And Mapping (SLAM), in which the accuracy improves by collecting high-resolution images of the environment, where the THz imaging mentioned earlier can provide such high-resolution images. SLAM-based techniques consist of three main steps: imaging the surrounding environment, estimation of ranges to the user, and fusion of images with the estimated ranges. For instance, a sub-centimeter level of accuracy can be achieved by constructing three-dimensional images of the environment using signals between 200 and 300 GHz and projecting the angle- and time-of-arrival information from the user to estimate locations. Since SLAM deals with relatively slow-moving objects, there is sufficient time to process high-resolution THz measurements. Such measurements can hold sensing information, resulting in complex state models comprising the fine-grained location, size, and orientation of target objects, as well as their electromagnetic properties and material types [Sarieddeen et al., 2020].

5.3 Challenges of Terahertz Communications

Despite its high potential, there are some barriers ahead for achieving the benefits of THz wireless communications. Similar to mmWave, terahertz signals suffer from high propagation loss, imposing a significant constraint on the transmission distance. The antenna aperture of THz communications is relatively small due to its tiny wavelength, leading to a weak capability to capture the radiation power. It leads to the fact that the higher frequency has a more considerable free-space path loss. Unlike low frequencies, where wireless communications never consider the atmospheric absorption, water vapor and oxygen molecules impose significant attenuation ranging from a few hundred dB to at most around 20 000 dB per kilometer. In addition to the gaseous absorption from water molecules, liquid water droplets, either in the form of suspended particles into clouds or rain falling hydrometeors, can attenuate the signal strength since its dimension is comparable to the wavelength of high-frequency signals. Due to the shrink of the wavelength in the THz band, the size of surrounding PHY objects becomes relatively large enough for signal scattering, and ordinary surfaces become too rough to make specular reflection. As a consequence, THz transmission relies heavily on the availability of the LOS link. Nevertheless, a direct path between the transmitter and receiver can be easily blocked by buildings, furniture, vehicles, foliage, and even humans, leading to the drop of signal power or even outage. Last but not least, the wireless channel at the THz band fluctuates more quickly than that of microwave and mmWave bands because the same velocity causes a higher Doppler shift.

5.3.1 High Free-Space Path Loss

When an electromagnetic wave propagates in the free space, the energy continuously spreads over an ever-increasing surface area as it propagates away from an ideal radiator that is a single point. The radiation from this isotropic antenna generates a spherical wave, as shown in Figure 5.2. The law of energy conservation tells us that the power contained on the surface of a sphere of any radius d

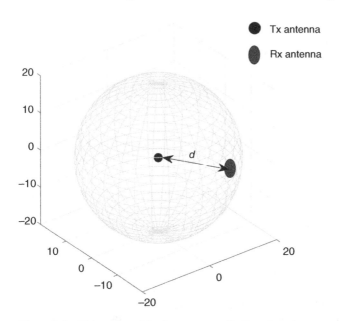

Figure 5.2 Illustration of the free-space radiation of an electromagnetic wave at the radio, millimeter-wave, or terahertz band, where the energy of a signal radiated from a transmit antenna spreads isotropically. At a propagation distance of d, therefore, the energy evenly distributes over the surface of a sphere with a radius of d.

keeps constant and equals to the Effective Isotropic Radiated Power (EIRP) of the transmitter P_{EIRP}. The power flux density, measured in units of Watts per square meter, is given by the EIRP divided by the surface area of a sphere with radius d, i.e. $\frac{P_{\text{EIRP}}}{4\pi d^2}$. Since the energy captured by a receiver is proportional to its antenna area denoted by A_r, the received power is

$$P_r = \left(\frac{P_{\text{EIRP}}}{4\pi d^2} \right) A_r. \tag{5.1}$$

The EIRP of a transmitter represents the maximal power in a particular direction compared with an isotropic antenna, which radiates power with unit gain in all directions. Hence, the EIRP is the product of the transmitted power P_t and the transmit antenna gain G_t, i.e.

$$P_{\text{EIRP}} = P_t G_t. \tag{5.2}$$

Meanwhile, the gain of a receive antenna can be determined by its effective aperture and the operating frequency, i.e.

$$G_r = \eta A_r \left(\frac{4\pi}{\lambda^2} \right), \tag{5.3}$$

where λ stands for the wavelength of the operating frequency and η denotes the maximal efficiency of the antenna. From Eq. (5.3), we have

$$A_r = G_r \left(\frac{\lambda^2}{4\pi\eta} \right). \tag{5.4}$$

Substituting Eqs. (5.2) and (5.4) in Eq. (5.1), yields the Friis' Free-Space equation

$$P_r = P_t G_t G_r \frac{\lambda^2}{(4\pi d)^2 \eta}, \tag{5.5}$$

where P_t and P_r are the transmitted and received power in absolute linear unites (usually Watts or milliwatts), respectively, while G_t and G_r denote the linear gains of the transmit and receive antennas relative to a unity-gain (0 dB) isotropic antenna. In wireless communications, it is typical to express propagation attenuation using decibel values since the range of signal power dynamically varies across several orders of magnitude over relatively small distances.

Rewrite Eq. (5.5) on a logarithm scale:

$$P_r|_{\text{dBm}} = 10 \lg \left(\frac{P_t G_t G_r}{\eta} \right) + 20 \lg \left(\frac{\lambda_0}{4\pi d} \right) + 20 \lg \left(\frac{\lambda}{\lambda_0} \right), \tag{5.6}$$

where $|_{\text{dBm}}$ denotes that the value is in unit of decibel-milliwatt, and we write λ_0 to stand for the reference wavelength. It can be seen from the aforementioned equation that the received power is proportional to the square of wavelength, implying that the received power in free space decays 20 dB per decade with the shrink of wavelength. Note that the phenomena that the received power

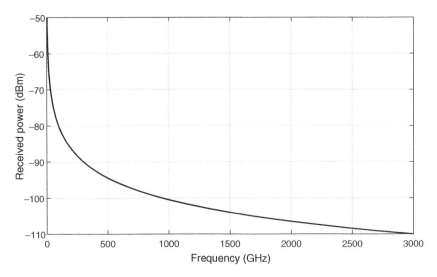

Figure 5.3 An example to illustrate the decay rate of the received power as a function of frequency ranging from 3 to 3000 GHz.

decreases with the increase of frequency is due to the dependence of the effective area (or aperture) of the receive antenna on wavelength, see Eq. (5.3). As a consequence, wireless transmission at terahertz frequencies suffers from much higher free-space path losses due to tiny wavelengths compared with that of microwave and millimeter-wave bands. To provide a concrete view, Figure 5.3 gives an example to illustrate the decay rate of the received power as a function of frequency within the range from 3 to 3000 GHz, where the received power at 3 GHz is assumed to be −50 dBm. The power drops 20 dB per decade with the rise of frequency, i.e. −70 dBm at 30 GHz, −90 dBm at 300 GHz, and −110 dBm at 3 THz.

5.3.2 Atmospheric Attenuation

Traditional cellular communications systems operating in low-frequency bands did not take into account the atmospheric effect during the calculation of the link budget. However, all electromagnetic waves suffer from more or less atmospheric attenuation due to the absorption of gaseous molecules such as oxygen and water vapor. This effect substantially magnifies at certain high frequencies across millimeter-wave and terahertz bands.

Under clear air conditions without condensed water in cloud or rain, the absorption of natural gases is the dominant atmospheric effect of decaying the strength of electromagnetic wave propagation. The gaseous composition of the atmosphere, as a mix of different species, is determined mainly by molecular oxygen and nitrogen.

Therefore, the absorption of oxygen molecules plays a major role in atmospheric attenuation. Water vapor suspending in the air is a minor gaseous component, but it deserves special attention because its presence strongly affects the propagation of electromagnetic radiation. It is notably the main contributor to gaseous attenuation, dominating the majority of the millimeter-wave and terahertz bands, except for only a few specific spectral regions where the absorption of oxygen is more evident. The gaseous attenuation arises from the interaction between oxygen or water molecules and electromagnetic waves. At millimeter-wave and terahertz frequencies, incident radiation causes rotational and vibrational transitions in polar molecules. These processes have a quantum nature, i.e. molecular motions take place at specific frequencies. Therefore, depending on its internal molecular structure, a gaseous component has its particular spectral absorption lines, consisting of the central frequency, the line intensity defining the depth of the absorption level, and a variety of spectral parameters. The spectral absorption lines for oxygen and water vapor are given in Tables 5.5 and 5.6, respectively.

A more detailed study of atmospheric attenuation, such as the effects of minor gaseous components, is often carried out in radio astronomy and remote sensing due to the characteristics of the signals under observation. However, from the perspective of wireless communications, the attenuation caused by some additional molecular species, e.g. oxygen isotopic species, oxygen vibrationally excited species, stratospheric ozone, ozone isotopic species, ozone vibrationally excited species, a variety of nitrogen, carbon, and sulfur oxides, is usually negligible since it represents small values compared with that of water vapor and oxygen [Siles et al., 2015].

In response to the need of the communications community to model the attenuation characteristics of atmospheric gases, the ITU-R carried out a study item and provided a mathematical procedure to figure out this specific attenuation. As demonstrated in ITU-R P.676 [2019], the atmospheric attenuation in the frequency range from 1 to 1000 GHz can be accurately estimated at any air pressure, temperature, and humidity as a summation of the individual spectral lines from oxygen and water vapor, along with small additional factors for the non-resonant Debye spectrum of oxygen below 10 GHz, pressure-induced nitrogen absorption over 100 GHz, and a wet continuum to account for the excess absorption from water vapor.

The total gaseous attenuation γ, in decibels per kilometer, is expressed as the combined effect

$$\gamma = \gamma_o + \gamma_w, \tag{5.7}$$

where γ_w is the attenuation due to water vapor, and γ_o is the attenuation due to oxygen. The latter results from the effect of oxygen spectral lines plus a very small

Table 5.5 Spectroscopic absorption lines for oxygen attenuation.

f_i	a_1	a_2	a_3	a_4	a_5	a_6
50.474214	0.975	9.651	6.690	0.0	2.566	6.850
50.987745	2.529	8.653	7.170	0.0	2.246	6.800
51.503360	6.193	7.709	7.640	0.0	1.947	6.729
52.021429	14.320	6.819	8.110	0.0	1.667	6.640
52.542418	31.240	5.983	8.580	0.0	1.388	6.526
53.066934	64.290	5.201	9.060	0.0	1.349	6.206
53.595775	124.600	4.474	9.550	0.0	2.227	5.085
54.130025	227.300	3.800	9.960	0.0	3.170	3.750
54.671180	389.700	3.182	10.370	0.0	3.558	2.654
55.221384	627.100	2.618	10.890	0.0	2.560	2.952
55.783815	945.300	2.109	11.340	0.0	−1.172	6.135
56.264774	543.400	0.014	17.030	0.0	3.525	−0.978
56.363399	1331.800	1.654	11.890	0.0	−2.378	6.547
56.968211	1746.600	1.255	12.230	0.0	−3.545	6.451
57.612486	2120.100	0.910	12.620	0.0	−5.416	6.056
58.323877	2363.700	0.621	12.950	0.0	−1.932	0.436
58.446588	1442.100	0.083	14.910	0.0	6.768	−1.273
59.164204	2379.900	0.387	13.530	0.0	−6.561	2.309
59.590983	2090.700	0.207	14.080	0.0	6.957	−0.776
60.306056	2103.400	0.207	14.150	0.0	−6.395	0.699
60.434778	2438.000	0.386	13.390	0.0	6.342	−2.825
61.150562	2479.500	0.621	12.920	0.0	1.014	−0.584
61.800158	2275.900	0.910	12.630	0.0	5.014	−6.619
62.411220	1915.400	1.255	12.170	0.0	3.029	−6.759
62.486253	1503.000	0.083	15.130	0.0	−4.499	0.844
62.997984	1490.200	1.654	11.740	0.0	1.856	−6.675
63.568526	1078.000	2.108	11.340	0.0	0.658	−6.139
64.127775	728.700	2.617	10.880	0.0	−3.036	−2.895
64.678910	461.300	3.181	10.380	0.0	−3.968	−2.590
65.224078	274.000	3.800	9.960	0.0	−3.528	−3.680
65.764779	153.000	4.473	9.550	0.0	−2.548	−5.002
66.302096	80.400	5.200	9.060	0.0	−1.660	−6.091
66.836834	39.800	5.982	8.580	0.0	−1.680	−6.393
67.369601	18.560	6.818	8.110	0.0	−1.956	−6.475
67.900868	8.172	7.708	7.640	0.0	−2.216	−6.545
68.431006	3.397	8.652	7.170	0.0	−2.492	−6.600
68.960312	1.334	9.650	6.690	0.0	−2.773	−6.650
118.750334	940.300	0.010	16.640	0.0	−0.439	0.079

Source: ITU-R P.676 [2019]/International Telecommunication Union.

Table 5.6 Spectroscopic absorption lines for water-vapor attenuation.

f_j	b_1	b_2	b_3	b_4	b_5	b_6
22.23508	0.1079	2.144	26.38	0.76	5.087	1
67.80396	0.0011	8.732	28.58	0.69	4.93	0.82
119.99594	0.0007	8.353	29.48	0.7	4.78	0.79
183.310087	2.273	0.668	29.06	0.77	5.022	0.85
321.22563	0.047	6.179	24.04	0.67	4.398	0.54
325.152888	1.514	1.541	28.23	0.64	4.893	0.74
336.227764	0.001	9.825	26.93	0.69	4.74	0.61
380.197353	11.67	1.048	28.11	0.54	5.063	0.89
390.134508	0.0045	7.347	21.52	0.63	4.81	0.55
437.346667	0.0632	5.048	18.45	0.6	4.23	0.48
439.150807	0.9098	3.595	20.07	0.63	4.483	0.52
443.018343	0.192	5.048	15.55	0.6	5.083	0.5
448.001085	10.41	1.405	25.64	0.66	5.028	0.67
470.888999	0.3254	3.597	21.34	0.66	4.506	0.65
474.689092	1.26	2.379	23.2	0.65	4.804	0.64
488.490108	0.2529	2.852	25.86	0.69	5.201	0.72
503.568532	0.0372	6.731	16.12	0.61	3.98	0.43
504.482692	0.0124	6.731	16.12	0.61	4.01	0.45
547.67644	0.9785	0.158	26	0.7	4.5	1
552.02096	0.184	0.158	26	0.7	4.5	1
556.935985	497	0.159	30.86	0.69	4.552	1
620.700807	5.015	2.391	24.38	0.71	4.856	0.68
645.766085	0.0067	8.633	18	0.6	4	0.5
658.00528	0.2732	7.816	32.1	0.69	4.14	1
752.033113	243.4	0.396	30.86	0.68	4.352	0.84
841.051732	0.0134	8.177	15.9	0.33	5.76	0.45
859.965698	0.1325	8.055	30.6	0.68	4.09	0.84
899.303175	0.0547	7.914	29.85	0.68	4.53	0.9
902.611085	0.0386	8.429	28.65	0.7	5.1	0.95
906.205957	0.1836	5.11	24.08	0.7	4.7	0.53
916.171582	8.4	1.441	26.73	0.7	5.15	0.78
923.112692	0.0079	10.293	29	0.7	5	0.8
970.315022	9.009	1.919	25.5	0.64	4.94	0.67
987.926764	134.6	0.257	29.85	0.68	4.55	0.9
1780	17506	0.952	196.3	2	24.15	5

Source: ITU-R P.676 [2019]/International Telecommunication Union.

continuum arising from pressure-induced nitrogen and non-resonant Debye spectrum of oxygen, and it can be determined by

$$\gamma_o = 0.182f \sum_i S_i F_i(f) + \Upsilon(f), \tag{5.8}$$

where f presents frequency, S_i is the strength of the ith oxygen line, and F_i denotes the shape factor of this oxygen line. The line strength is given by

$$S_i = a_1 \times 10^{-7} P\theta^3 e^{a_2(1-\theta)}, \tag{5.9}$$

where P stands for dry air pressure in the unit of hectopascal (hPa), θ is a constant $\theta = 300/T$ with temperature T on the Kelvin scale.

The second item of Eq. (5.8) stands for the dry air continuum stemming from the pressure-induced nitrogen attenuation above 100 GHz and the non-resonant Debye spectrum of oxygen below 10 GHz. The dry continuum is frequency-dependent, as given by

$$\Upsilon(f) = \theta^2 Pf \left\{ \frac{6.14 \times 10^{-5}}{\omega \left[1 + \left(\frac{f}{\omega}\right)^2\right]} + \frac{1.4 \times 10^{-12} P\theta^{\frac{3}{2}}}{1 + 1.9 \times 10^{-5} f^{\frac{3}{2}}} \right\}, \tag{5.10}$$

where ω denotes the width parameter for the Debye spectrum, i.e.

$$\omega = 5.6 \times 10^{-4}(P + E)\theta^{0.8}, \tag{5.11}$$

and E denotes the water-vapor partial pressure that is obtained from the water-vapor density ρ and temperature T in terms of

$$E = \frac{\rho T}{216.7}. \tag{5.12}$$

Meanwhile, the shape factor in Eq. (5.8) is a function of f, which is given by

$$F_i(f) = \frac{f}{f_i} \left[\frac{\Delta f_o - \delta(f_i - f)}{(f_i - f)^2 + \Delta f_o^2} + \frac{\Delta f_o - \delta(f_i + f)}{(f_i + f)^2 + \Delta f_o^2} \right], \tag{5.13}$$

where f_i represents the frequency of oxygen spectral lines, as listed in Table 5.5, δ is a correlation factor due to interference effect in oxygen lines

$$\delta = 10^{-4} \times (a_5 + a_6\theta)(P + E)\theta^{0.8}, \tag{5.14}$$

and Δf_o stands for the width of the spectral line, i.e.

$$\Delta f_o = a_3 \times 10^{-4} \left[P\theta^{(0.8-a_4)} + 1.1E\theta \right], \tag{5.15}$$

which is needed to be modified to account for the Zeeman splitting of oxygen lines, following

$$\Delta f_o = \sqrt{\Delta f_o^2 + 2.25 \times 10^{-6}}. \tag{5.16}$$

The specific attenuation due to water vapor in Eq. (5.7) is calculated by

$$\gamma_w = 0.182 f \sum_j T_j K_j(f), \tag{5.17}$$

where T_j is the strength of the jth water-vapor line, and K_j is the shape factor for this water-vapor line. The line strength is given by

$$T_j = b_1 \times 10^{-1} E \theta^{3.5} e^{b_2(1-\theta)}, \tag{5.18}$$

and the shape factor equals

$$K_j(f) = \frac{f}{f_j} \left[\frac{\Delta f_w}{(f_j - f)^2 + \Delta f_w^2} + \frac{\Delta f_w}{(f_j + f)^2 + \Delta f_w^2} \right], \tag{5.19}$$

where f_j denotes the water-vapor line frequencies, as given in Table 5.6, and Δf_w stands for the width of the spectral line, i.e.

$$\Delta f_w = b_3 \times 10^{-4} (P \theta^{b_4} + b_5 E \theta^{b_6}), \tag{5.20}$$

which needs to be modified to account for the Doppler broadening of the water-vapor lines, following

$$\Delta f_w = 0.535 \, \Delta f_w + \sqrt{0.217 \, \Delta f_w^2 + \frac{2.1316 \times 10^{-12} f_j^2}{\theta}}. \tag{5.21}$$

The resulting atmospheric attenuation as a function of frequency is illustrated in Figure 5.4, covering the spectrum range from 1 to 1000 GHz. The absorption of oxygen molecules, as indicated by *Oxygen* in the figure, is calculated at an air pressure of 1013.25 hPa and the temperature of 15 °C while assuming the air is perfectly dry with a water vapor density of 0 g/m³. Oxygen absorption forms two peaks at the frequencies centered on 60 and 118.7 GHz since many oxygen absorption lines merged there. We also use *Standard Air* to show the standard atmospheric condition at the seal level (with an air pressure of 1013.25 hPa, temperature of 15 °C, and a water-vapor density of 7.5 g/m³). Except for these two frequency windows (i.e. 60 and 118.7 GHz), the absorption of water vapor plays a dominant role in atmospheric attenuation over the whole mmWave and THz bands. At high frequencies, atmospheric attenuation becomes substantially large, accounting for the peak of approximately 20 000 dB/km at frequencies around 560 GHz. In other words, only a short distance of 10 m transmission brings a power loss of approximately 200 dB, which is prohibitively too high for the implementation of wireless communications. In contrast, the atmospheric attenuation at the sub-6 GHz band is on the order of magnitude around 0.01 dB/km, which is negligible. By far, we are

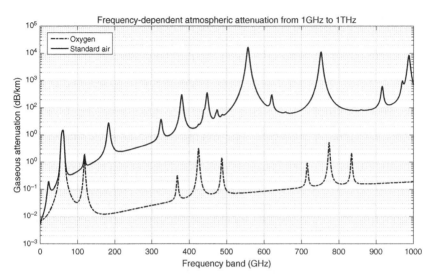

Figure 5.4 Atmospheric attenuation at millimeter-wave and terahertz frequencies due to the absorption of oxygen and water vapor, where *Standard Air* stands for the standard atmosphere condition with air pressure 1013.25 hPa, temperature 15 °C, and water-vapor density 7.5 g/m³, according to ITU-R P.676 [2019], while *Oxygen* highlights the effect of oxygen absorption by setting water-vapor density to 0 g/m³. Except for frequencies centered at 60 GHz and 118.7 GHz, the effect of water vapor dominates most of millimeter-wave bands and all of terahertz bands. Source: ITU-R P.676 (2019)/With permission of International Telecommunication Union.

able to identify a variety of low-absorption windows for wireless communications, between the following frequency ranges:

- 30–50 GHz
- 70–110 GHz
- 130–170 GHz
- 200–310 GHz
- 330–370 GHz
- 390–440 GHz
- 625–725 GHz
- 780–910 GHz

For instance, the frequency band centered at 150 GHz has already been applied in experimental point-to-point fixed links with a transmission distance of about 1 km, where atmospheric attenuation is on the order of magnitude about 1 dB in standard atmospheric conditions. In addition, a breakthrough work in Germany has reached 40 Gbps wireless data transmission along with a distance of 1 dB at 240 GHz with atmospheric attenuation of around 3 dB [Hirata et al., 2009]. Within the THz band, the atmospheric attenuation spans from 10 dB at 350 GHz to about 80 dB at 840 GHz. For short-range wireless communications, small cells with limited coverage, and nano-communications, such attenuation is acceptable and can be compensated by technologies such as beamforming.

5.3.3 Weather Effects

In addition to gaseous absorption, the weather is also an important factor of atmospheric attenuation since the PHY dimensions of raindrops, hailstones, and snowflakes are on the same order of carrier wavelengths at high frequencies. The research in the 1970s and 1980s focusing on the weather characteristics of satellite communication links provided much insight into mmWave and THz propagation under various weather conditions. The outcomes revealed that liquid water droplets, in contrast to water molecules imposing gaseous attenuation, either in the form of suspended particles into clouds or rain falling hydrometeors, deserve special attention as frequency increases [Crane, 1980].

On the one hand, a cloud that is an aggregate of tiny water droplets (over 0 °C) or ice crystals (−40 °C and −20 °C) causes an energy loss upon the electromagnetic waves going through it in the troposphere at various altitudes. On the other hand, water droplets in the form of rain, fog, hail, and snow impose excess unwanted signal losses on millimeter-wave and terahertz propagation paths through the lower atmosphere. The dimension of water particles within clouds ranges from a minimum of approximately 1–30 μm while the size of ice crystals varies from 0.1 to 1 mm. Rain droplets are oblate spheroids with radii up to a few tens of millimeters or generally perfect spheres with radii below 1 mm. The size of water droplets is comparable to wavelengths at mmWave frequencies (1–10 mm) and wavelengths at the THz band (0.1–1 mm). Consequently, water droplets attenuate the energy of electromagnetic waves at these bands through absorption and scattering. From the perspective of terrestrial networks, cloud attenuation is not an issue since the possibility of a signal going through a cloud is relatively very low. The analysis of the weather effects in this book will focus on the attenuation due to rainfall.

The rain attenuation is a function of distance, rainfall rate, and the mean dimension of raindrops, which can be estimated by attenuation models such as [ITU-R P.838, 2005]. Such attenuation can be treated as an additional loss that is simply added on top of the free-space path loss and gaseous absorption. The measurement at 28 GHz demonstrated that a heavy rainfall with a rain rate of more than 25 mm/h brings attenuation of about 7 dB/km. Extreme attenuation of up to 50 dB/km occurs at a particular frequency of 120 GHz and an extreme rain rate of 100–150 mm/h. As a rule of thumb, rain provides an excess attenuation of approximately 10–20 dB over a distance of 1 km at the THz band.

The ITR-R provided a procedure to approximate this specific attenuation in terms of the rain rate R in millimeter per hour (mm/h) by using a power-low equation [ITU-R P.838, 2005]:

$$\gamma_R = k(f)R^{\alpha(f)}. \tag{5.22}$$

Two frequency-dependent coefficients k and α, which have been developed from curve-fitting to power-low coefficients derived from scattering calculations, are determined as a function of frequency f in the unit of GHz, i.e.

$$k = 10^{\left\{\sum_{j=1}^{4}\left(a_j \exp\left[-\left(\frac{\lg(f)-b_j}{c_j}\right)^2\right]\right)+m_k \lg(f)+n_k\right\}}, \tag{5.23}$$

and

$$\alpha = \sum_{i=1}^{5}\left(a_i \exp\left[-\left(\frac{\lg(f)-b_i}{c_i}\right)^2\right]\right) + m_\alpha \lg(f) + n_\alpha. \tag{5.24}$$

In Eq. (5.23), k can be either a constant for horizontal polarization k_H or a constant for vertical polarization k_V, and in Eq. (5.24), α can be the values for either horizontal polarization α_H or vertical polarization α_V, which are given in Table 5.7. For linear and circular polarization, and for all path geometries, k and α can be calculated using

$$k = \frac{\left[k_H + k_V + (k_H - k_V)\cos^2(\theta)\cos(2\tau)\right]}{2}, \tag{5.25}$$

and

$$\alpha = \frac{\left[k_H\alpha_H + k_V\alpha_V + (k_H\alpha_H - k_V\alpha_V)\cos^2(\theta)\cos(2\tau)\right]}{2k}, \tag{5.26}$$

where θ denotes the path elevation angle, and τ is the polarization tilt angle.

The resulting rain attenuation derived from this ITU-R model is illustrated in Figure 5.5, in terms of frequency ranging from 1 to 1000 GHz and rain rate from a light rain (1 mm/h) to a heavy rain (200 mm/h). Note that we assume $\theta = 90°$ in this figure for the wireless links in terrestrial networks.

Besides the ITU-R model, there are other models such as a simplified one given in Smulders and Correia [1997], i.e.

$$\lambda_{[dB/km]}(f_{[GHz]}, R) = k(f)R^{\alpha(f)}, \tag{5.27}$$

where

$$k(f) = 10^{1.203\lg(f)-2.290}, \tag{5.28}$$

and

$$\alpha(f) = 1.703 - 0.493\lg(f). \tag{5.29}$$

Table 5.7 The values for two coefficients in the calculation of rain attenuation.

	j	a_j	b_j	c_j	m_k	n_k
k_H	1	−5.3398	−0.1001	1.1310	−0.1896	0.7115
	2	−0.3535	1.2697	0.4540		
	3	−0.2379	0.8604	0.1535		
	4	−0.9416	0.6455	0.1682		
k_V	1	−3.8060	0.5693	0.8106	−0.1640	0.6330
	2	−3.4497	−0.2291	0.5106		
	3	−0.3990	0.7304	0.1190		
	4	0.5017	1.0732	0.2720		
	i	a_i	b_i	c_i	m_α	n_α
α_H	1	−0.1432	1.8244	−0.5519	0.6785	−1.9554
	2	0.2959	0.7756	0.1982		
	3	0.3218	0.6377	0.1316		
	4	−5.3761	−0.9623	1.4783		
	5	16.1721	−3.2998	3.4399		
α_V	1	−0.0777	2.3384	−0.7628	−0.0537	0.8343
	2	0.5673	0.9555	0.5404		
	3	−0.2024	1.1452	0.2681		
	4	−48.2991	0.7917	0.1162		
	5	48.5833	0.7915	0.1165		

Source: Adapted from ITU-R P.838 [2005].

From the perspective of satellite communications, weather attenuation makes wireless transmission unreliable, if not unusable, due to the long distance of a satellite link. However, terrestrial THz mobile communications with small cell sizes are viable (considering a maximal loss of several decibels at a distance of 100 m), especially when large-antenna arrays are applied to compensate for such an excess loss.

5.3.4 Blockage

Due to the shrink of the wavelength in the THz band, the size of surrounding PHY objects becomes relatively large enough for signal scattering, and ordinary surfaces become too rough to make specular reflection. As a consequence, THz transmission relies heavily on the availability of the LOS link. Nevertheless, a direct path

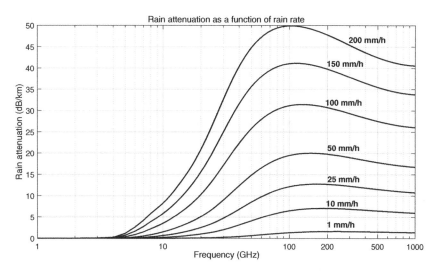

Figure 5.5 Rain attenuation measured in dB/km for terrestrial communications links as a function of rain rate and frequency, covering the range from 1 to 1000 GHz. The rain rate is measured in millimeter per hour (mm/h), averaging over a period of time such as one hour. The values from light rain (1 mm/h) to heavy rain (200 mm/h) are illustrated. The peak of attenuation occurs on the frequency band from 100 to 300 GHz since the wavelength in this band matches the size of raindrops.

between the transmitter and receiver can be easily blocked by objects or humans in between. The terahertz signals and their mmWave counterparts are highly susceptible to blockages raised from buildings, furniture, vehicles, foliage, and even humans, compared with the signals at lower frequencies. A single blockage can lead to a power loss with a few tens of dB. For example, the presence of vegetation can cause a foliage loss in terms of the depth of vegetation. The attenuation of 17, 22, and 25 dB are observed at 28, 60, and 90 GHz, respectively. Moreover, the presence of humans has a more profound influence due to the dynamic movement of humans. The fading of THz signals attributed to the self-body blockage is continuously changing with a dynamic range of up to 35 dB. These blockages can dramatically decay the signal strength and may even lead to a total outage. Therefore, it is necessary to clarify the traits of blockage and find effective solutions to avoid blockage or quickly recover the connection when a link gets blocked.

Statistical modeling can be applied to quantify the effect of blockage, e.g. modeling self-body blockage by means of a Boolean model in which a human is processed as a three-dimensional cylinder with centers forming a two-dimensional (2D) Poisson Point Process (PPP). In indoor environments, self-body blockage has also been modeled as 2D circles of fixed radius r with centers forming a PPP. A LOS probability model assumes that a link of distance d will be LOS with probability $p_L(d)$ and NLOS otherwise. The expressions of $p_L(d)$ are usually obtained empirically for different settings. The LOS probability for a LOS link with self-body blockage can be estimated by Tripathi et al. [2021]

$$p_L = 1 - e^{-\mu(rd+\pi r^2)}, \tag{5.30}$$

where μ denotes the density of blockages, and d is the 2D distance in meters.

For the urban macro-cell scenario defined in 3GPP, the LOS probability is

$$p_L(d) = \min\left(\frac{d_1}{d}, 1\right)\left(1 - e^{-\frac{d}{d_2}}\right) + e^{-\frac{d}{d_2}}, \tag{5.31}$$

where d_1 and d_2 are the fitting parameters, equaling to 18 and 63 m, respectively. The same model is also applicable for the urban micro-cell scenario, with $d_2 = 36$ m. There are some variations in the LOS probability expressions across different channel measurement campaigns and environments. For example, the LOS probability model developed by NYU is

$$p_L(d) = \left(\min\left(\frac{d_1}{d}, 1\right)\left(1 - e^{-\frac{d}{d_2}}\right) + e^{-\frac{d}{d_2}}\right)^2, \tag{5.32}$$

where the fitting parameters $d_1 = 20$ m and $d_2 = 160$ m, respectively.

In a cellular network with random rectangular blockages where blockages are modelled as the Boolean process, the LOS probability is given by

$$p_L(d) = e^{-\beta d}, \tag{5.33}$$

with

$$\beta = \frac{2\mu(\mathbb{E}[W] + \mathbb{E}[L])}{\pi}, \tag{5.34}$$

where L and W are the length and width of a typical rectangular blockage.

5.3.5 High Channel Fluctuation

When a mobile station moves at a velocity of v along a path that has a spatial angle of θ with the signal propagation, it causes a difference in path lengths, i.e. $\Delta d = v \Delta t \cos\theta$. Due to the difference in propagation distances, there is a phase variation of

$$\Delta \phi = \frac{2\pi \Delta d}{\lambda} = \frac{2\pi v \Delta t \cos\theta}{\lambda}, \tag{5.35}$$

and therefore the *Doppler shift*

$$f_d = \frac{\Delta \phi}{2\pi \Delta t} = \frac{v \cos\theta}{\lambda}. \tag{5.36}$$

The received signal will be broadened in the frequency domain ranged from $f_c + v/\lambda$ when the mobile station moves exactly toward the transmitter to $f_c - v/\lambda$

when the mobile station moves exactly away from the transmitter. Then, *Doppler Spread*

$$D_s = 2f_m \tag{5.37}$$

is defined as a measure of the spectral broadening caused by the relative movement, where $f_m = v/\lambda$ denotes the maximum Doppler shift.

Transmitting a sinusoidal signal $\cos 2\pi f_c t$, the Doppler shift raises a phase change on each path, e.g. $2\pi f_d \tau_l(t)$ on path l, and therefore the phase change significantly at a time interval of $\triangle \tau = 1/2f_d$. When multipath components combine at the receiver, such phase changes affect their constructive and destructive interference. This happens at a time interval of

$$T_c = \frac{1}{4D_s} = \frac{1}{8f_m}, \tag{5.38}$$

which is called *coherence time* to characterize the time duration over which the channel response can be regarded as invariant. This is a somewhat rough relation and there are other definitions such as

$$T_c \approx \frac{9}{16\pi f_m}, \tag{5.39}$$

or

$$T_c \approx \sqrt{\frac{9}{16\pi f_m^2}} = \frac{0.423}{f_m}. \tag{5.40}$$

Regardless of the diversity in definition, the critical knowledge to be recognized is that the time coherence is determined mainly by the Doppler spread in a reciprocal relation, i.e. the larger the Doppler spread, the smaller the time coherence. Depending on how quickly a transmitted baseband signal changes relative to the fading rate of the channel, wireless channels can be categorized into *slow fading* and *fast fading*. In slow fading, the symbol period is much smaller than the coherence time

$$T_s \ll T_c. \tag{5.41}$$

We can assume that the channel is constant across a number of symbol periods in the time domain, and the Doppler spread can be negligible compared with the signal bandwidth. In fast fading, where

$$T_s > T_c, \tag{5.42}$$

where the channel response varies within a single symbol period while frequency dispersion due to Doppler spreading is considerable.

From Eq. (5.36), we know that the Doppler effect magnifies with the shrink of wavelength. Consequently, a THz channel exhibits more significant fluctuation compared with microwave wireless channels, and even mmWave channels.

When a transmitter moving away from a receiver at a constant velocity of 1 m/s, it generates a Doppler shift of around 200 Hz in the millimetric waveband of 60 GHz. This value will increase to 2000 Hz in the terahertz band of 600 GHz. In the design of a THz transceiver, the effect of such high Doppler shifts should be taken into account.

5.4 Array-of-Subarrays Beamforming

Compared with their mmWave counterparts, terahertz signals suffer from more severe free-space spreading loss since the acquired power is proportional to the receive-antenna aperture. Additionally, the atmospheric attenuation becomes far more severe in the terahertz band since the absorption of water vapor and oxygen molecules is significant. Specifically, the attenuation can reach hundreds of dB per kilometer, whereas that of mmWave is only a few tens of dB per kilometer. Moreover, due to tiny wavelengths in the terahertz band, reflection and scattering losses along the propagation path are more significant than those of mmWave. Thus, terahertz signal transmission has fewer paths and is sparser than mmWave in the angular domain. Accordingly, terahertz communications are LOS-dominant, where the transmit power concentrates on the LOS path, and the power gap between the LOS and NLOS paths becomes larger. With higher reflection loss and fewer rays, the overall angular spread of terahertz signals is small. For instance, a maximal angular spread of 40° has been observed for indoor environments in the terahertz band, while up to 120° for indoor scenarios at 60 GHz mmWave frequencies. To overcome this large propagation loss, the beamforming technique is essential to be applied in terahertz communications systems. Thanks to the tiny wavelengths, a large number of antenna elements can be tightly packed in a small device to generate high power gains.

The fully digital beamforming forms the desired beam by simply multiplying the transmitted signal with a weighting vector at the baseband. However, it leads to unaffordable energy consumption and hardware cost for a transceiver equipped with a large-scale antenna array. Therefore, another technical form that can lower implementation complexity, called analog beamforming, has been extensively studied in mmWave communications. By employing analog phase-shifters to adjust the phases of signals, analog beamforming needs only a single Radio Frequency (RF) chain to steer the beam, leading to low hardware cost and energy consumption. However, since an analog circuit can only partly adjust the phases of signals, it is challenging to appropriately adapt a beam to a particular channel condition, leading to considerable performance loss. In addition, fully analog architecture can only support single-stream transmission, which cannot achieve the multiplexing gain to improve spectral efficiency.

Given the hardware constraints, hybrid digital-analog architecture is the best choice for terahertz large-scale antenna systems rather than fully digital or fully analog architecture from the perspective of performance and complexity trade-off. The key idea is to divide the conventional baseband processing into two parts: a large-size analog signal processing (realized by an analog circuit) and a dimension-reduced digital signal processing (requiring only a few RF chains). Since the number of effective scatterers at mmWave and THz frequencies is often minor, the number of data streams is generally much smaller than the number of antennas. Consequently, hybrid beamforming can significantly reduce the required number of RF chains, resulting in simple hardware and low energy consumption.

In a hybrid architecture, antenna elements can be connected to RF chains in two typical ways, i.e. Fully Connected (FC) and Array-of-Subarrays (AoSA) [Lin and Li, 2016]. In the FC structure, the signal of an RF chain radiates over all antenna elements via an individual group of phase shifters and corresponding signal combiners. In contrast, for the AoSA structure, an RF chain uniquely drives a disjoint subset of antennas, each of which is attached to an exclusive phase shifter. Consequently, antenna elements are shared among different RF chains for the fully connected structure, whereas an Antenna Subarray (AS) is only accessible to one specific RF chain for the array-of-subarrays structure. In particular, for the FC structure, a typical RF chain should have the capability to drive the entire large-scale antenna array, which is power-aggressive considering the limited power generated by terahertz sources and could also result in high loss. Besides, the use of a large number of phase shifters and combiners will exacerbate the power consumption. However, signal processing is carried out at a subarray level using an adequate number of antennas for the AoSA structure. Therefore, with fewer phase shifters, the hardware complexity, power consumption, and signal power loss can be dramatically reduced. In addition, power gain, multiplexing gain, and spatial-diversity gain can be jointly reaped by cooperating with precoding in the baseband, as illustrated in Figure 5.6.

Consider a three-dimensional ultra-massive MIMO system that integrates a very large number of plasmonic nano-antennas in a very compact footprint using nanomaterials such as graphene. An antenna array comprises active graphene elements over a common metallic ground layer, with a dielectric layer in between. Assume the array at the transmitter and receiver side consist of $M_t \times N_t$ and $M_r \times N_r$ subarrays, respectively, and each subarray is composed of $Q \times Q$ antenna elements. Consequently, the resultant configuration can be represented as an $M_t N_t Q^2 \times M_r N_r Q^2$ MIMO system by vectorizing the 2D antenna indices on each side. Such a doubly massive MIMO system differs from conventional massive MIMO, which applies a large-scale array with a massive number of elements only at the base-station side to serve multiple users with a single or a few antennas.

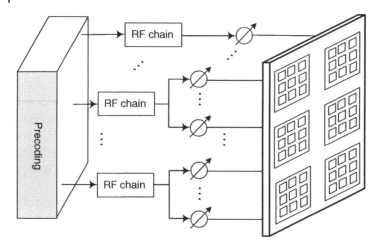

Figure 5.6 The array-of-subarrays structure of hybrid beamforming in a terahertz communication system.

The THz signal transmission is LOS-dominant due to limited reflection, and negligible scatters and diffracted components. Consequently, a single-carrier Los transmission over a frequency-flat channel can be modeled as [Sarieddeen et al., 2019]

$$\mathbf{y} = \mathbf{W}_r^H \mathbf{H} \mathbf{W}_t^H \mathbf{x} + \mathbf{W}_r^H \mathbf{n}, \tag{5.43}$$

where $\mathbf{x} = [x_1, x_2, \ldots, x_{N_s}]^T$ is an information-bearing symbol vector, $\mathbf{y} \in \mathbb{C}^{N_s \times 1}$ denotes the received symbol vector, $\mathbf{H} \in \mathbb{C}^{M_r N_r \times M_t N_t}$ expresses the channel matrix, $\mathbf{W}_t \in \mathbb{C}^{N_s \times M_t N_t}$ and $\mathbf{W}_r \in C^{M_r N_r \times N_s}$ stand for the precoding and combining matrices, respectively, and $\mathbf{n} \in \mathbb{C}^{M_r N_r \times 1}$ is the vector of additive noise. Each Standards Association (SA) is assumed to form a single beam pattern, and therefore the channel response between the (m_t, n_t) and (m_r, n_r) SAs is defined as

$$h_{(m_r, n_r, m_t n_t)} = \mathbf{a}_r^H(\phi_r, \theta_r) G_r \alpha_{(m_r, n_r, m_t n_t)} G_t \mathbf{a}_t^H(\phi_t, \theta_t), \tag{5.44}$$

for $1 \leq m_t \leq M_t$, $1 \leq m_r \leq M_r$, $1 \leq n_t \leq N_t$, and $1 \leq n_r \leq N_r$, where α is the LOS path gain, G_t and G_r are the transmit and receive antenna gains of the Friis Free-Space formula as given in Eq. (5.5), $\mathbf{a}_t$ and $\mathbf{a}_r$ denote the steering vectors of the transmit and receive SAs, respectively, ϕ is the azimuth angle, and θ is the elevation angle of the propagation path.

Neglecting the effect of mutual coupling among antenna elements, the ideal steering vector of the transmit SA can be expressed by

$$\mathbf{a}_0(\phi_t, \theta_t) = \frac{1}{Q} \left[e^{j\Phi_{1,1}}, \ldots, e^{j\Phi_{1,Q}}, e^{j\Phi_{2,1}}, \ldots e^{j\Phi_{Q,1}}, \ldots, e^{j\Phi_{Q,Q}} \right]^T \tag{5.45}$$

with $\Phi_{p,q}$ corresponds to the phase difference of antenna element (p, q), as defined by

$$
\Phi_{(p,q)}(\phi_t, \theta_t) = \psi_x^{(p,q)} \frac{2\pi}{\lambda} \cos \phi_t \sin \theta_t
$$

$$
+ \psi_y^{(p,q)} \frac{2\pi}{\lambda} \sin \phi_t \sin \theta_t + \psi_z^{(p,q)} \frac{2\pi}{\lambda} \cos \theta_t, \tag{5.46}
$$

where $\psi_x^{(p,q)}$, $\psi_y^{(p,q)}$, and $\psi_z^{(p,q)}$ are the coordinate positions of elements in the three-dimensional space. At the receiver side, $\mathbf{a}_0(\phi_r, \theta_r)$ can be similarly defined. Suppose the desired beam for a transmit SA has the angle of departure denoted by $(\hat{\phi}_t, \hat{\theta}_t)$, the optimal weighting vector is determined by

$$
\mathbf{w}_t = \mathbf{a}_t(\hat{\phi}_t, \hat{\theta}_t), \tag{5.47}
$$

which is obtained by substituting $(\hat{\phi}_t, \hat{\theta}_t)$ into Eq. (5.46), i.e.

$$
\hat{\Phi}_{(p,q)}(\hat{\phi}_t, \hat{\theta}_t) = \psi_x^{(p,q)} \frac{2\pi}{\lambda} \cos \hat{\phi}_t \sin \hat{\theta}_t
$$

$$
+ \psi_y^{(p,q)} \frac{2\pi}{\lambda} \sin \hat{\phi}_t \sin \hat{\theta}_t + \psi_z^{(p,q)} \frac{2\pi}{\lambda} \cos \hat{\theta}_t. \tag{5.48}
$$

The resulting beam pattern can then be expressed by

$$
g_t(\phi, \theta) = \mathbf{w}_t^H \mathbf{a}_t(\phi, \theta)
$$

$$
= \frac{1}{\sqrt{M_t N_t}} \sum_{m_t=1}^{M_t} \sum_{n_t=1}^{N_t} e^{j \frac{2\pi}{\lambda} \left[\Phi_{(p,q)}(\phi_t, \theta_t) - \hat{\Phi}_{(p,q)}(\hat{\phi}_t, \hat{\theta}_t) \right]}, \tag{5.49}
$$

implying that there is a maximal power gain of $M_t N_t$ at the desired direction of $(\hat{\phi}_t, \hat{\theta}_t)$.

Similarly, the weighting vector for a receive SA is given by $\mathbf{w}_r = \mathbf{a}_r(\hat{\phi}_r, \hat{\theta}_r)$ with the angles of arrival of $(\hat{\phi}_r, \hat{\theta}_r)$, resulting in a maximal power gain of $M_r N_r$ at the angle of arrival of the LOS path $(\hat{\phi}_r, \hat{\theta}_r)$.

5.5 Lens Antenna

The antenna of THz communications must be highly directive to mitigate the severe propagation attenuation due to free-space path loss, atmospheric absorption, and rainfall attenuation. The straightforward way is to apply a large-scale antenna array to generate high beamforming gain to compensate for such a loss. Although the hybrid digital-analog architecture can lower the required number of RF chains, the hardware complexity and power consumption of the hybrid beamforming continue to be challenging due to the need for a large number of analog RF components (e.g. analog phase shifters). Some studies demonstrated that the power consumed by phase shifters (and their variable gain amplifiers) becomes critical. This problem becomes exacerbated in the hybrid architecture since the number of analog phase shifters might be multiple of the number of RF chains and antennas. High-frequency communications starve for a disruptive antenna technology. Therefore, a promising low-cost solution for high frequencies and highly directive antennas called *lens antenna* draws the attention of researchers.

5.5.1 Refraction of Radio Waves

A lens antenna is a particular antenna that employs a shaped piece of radio-transparent material to bend and concentrate electromagnetic waves by refraction, as an optical lens does for visible light. It usually comprises an emitter radiating radio waves and a piece of dielectric or composite material in front of the emitter as a converging lens to force the radio waves into a narrow beam. Conversely, the lens directs the incoming radio waves into the feeder in a receive antenna, converting the induced electromagnetic waves into electric currents. To generate narrow beams, the lens needs to be much larger than the wavelength of the electromagnetic waves. Hence, a lens antenna is more suitable for mmWave and THz communications, with tiny wavelengths. Like an optical lens, radio waves have a different speed within the lens material than in free space so that the varying lens thickness delays the waves passing through it by different amounts, changing the shape of the wavefront and the direction of the waves, as demonstrated in Figure 5.7.

The first experiments using lenses to refract and focus radio waves occurred during the earliest research on radio waves in the 1890s (referring to Wikipedia). James Clerk Maxwell predicted the existence of electromagnetic waves in 1873 and proposed the inference that visible light consists of electromagnetic waves with tiny wavelengths. In 1887, Heinrich Hertz successfully proved the existence of electromagnetic waves by discovering radio waves. Early scientists thought

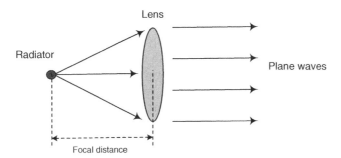

Figure 5.7 Schematic diagram of a lens antenna consisting of a radiator and a radio lens.

of radio waves as a form of invisible light. To verify Maxwell's theory that light is electromagnetic waves, these researchers concentrated on duplicating classic optics experiments with short-wavelength radio waves, diffracting them with various lenses. Hertz first demonstrated the refraction phenomena of radio waves at 450 MHz using a prism. These experiments confirmed that both visible light and radio waves are composed of the electromagnetic waves predicted by Maxwell, differing only in frequency.

The feasibility of concentrating radio waves by forcing them into a narrow beam as light waves in an optical lens attracted the interest of many researchers at that time. In 1889, Oliver Lodge and James L. Howard attempted to refract 300 MHz waves with cylindrical lenses but failed to find a focusing effect because the equipment was smaller than the wavelength. In 1894, Lodge successfully focused radio waves at 4 GHz with a glass lens. In the same year, Indian physicist Jagadish Chandra Bose has constructed lens antennas in his microwave experiments over 6–60 GHz, using a cylindrical sulfur lens to collimate the microwave beam and patenting a receiving antenna consisting of a glass lens focusing microwaves on a crystal detector. In 1894, Augusto Righi focused radio waves at 12 GHz with 32 cm lenses in his microwave experiments at the University of Bologna. The development of modern lens antennas occurred during a significant expansion of research into microwave technology around the World War II to develop military radar. In 1946, R. K. Luneberg invented the famous Luneberg lens, which is applied as a radar reflector that is sometimes attached to stealth fighters to make it detectable during training operations, or to conceal their true electromagnetic signature.

5.5.2 Lens Antenna Array

On top of lens antennas, an advanced antenna structure referred to as a lens antenna array has been developed. A lens antenna array usually consists of two major components: an Electro-Magnetic (EM) lens and an array with antenna elements positioned in the focal region of the lens. EM lenses can be implemented in different ways, e.g. dielectric materials, transmission lines with variable lengths, and periodic inductive and capacitive structures. Despite its various implementations, the function of EM lenses is to provide variable phase shifting for electromagnetic waves at different angles. In other words, a lens antenna array can direct the signals emitted from different transmitting antennas to different beams with sufficiently separated angles of departure. Conversely, a lens antenna array at the receiver can concentrate the incident signals from sufficiently separated directions to different receiving antennas. This array can be applied to massive MIMO systems to achieve significant performance gains and lower cost and complexity compared with a conventional array.

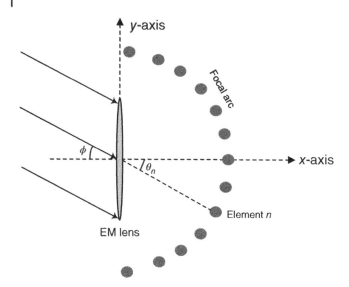

Figure 5.8 Schematic diagram of a lens antenna array with an incident signal in the azimuth angle of ϕ.

Without any loss of generality, see Zeng and Zhang [2016], only the azimuth plane is considered by placing a planar EM lens with size $D_y \times D_z$ and negligible thickness centered at the origin point on the y–z plane. The antenna elements locate on the focal arc of the lens in the azimuth plane, as shown in Figure 5.8. The coordinate of a typical element n, where $n \in \{0, \pm1, \dots, \pm(N-1)/2\}$ and N stands for the total number of antenna elements (assumed to be odd), can be expressed by $x_n = F \cos \theta_n$, $y_n = -F \sin \theta_n$, and $z_n = 0$ with the focal length of the lens F and the angle of the nth element relative to the x-axis $\theta_n \in [\pi/2, \pi/2]$.

The inter-element spacing is critical, i.e. the elements are placed on the focal arc so that $\tilde{\theta}_n = \sin \theta_n$ are equally spaced in the interval $[-1,1]$ as

$$\tilde{\theta}_n = \frac{n}{\tilde{D}}, \tag{5.50}$$

where $\tilde{D} = D_y/\lambda$ is the lens dimension in the azimuth plane normalized by the carrier wavelength λ. If a plane wave arrives with an incident angle of ϕ, the impinging signal at the lens center is $r_0(\phi)$ and the resulting received signal of the nth element is $r_n(\phi)$.

The array response vector can be given by

$$\mathbf{a}(\phi) = \left[a_1(\phi), a_2(\phi), \dots, a_N(\phi)\right]^T, \tag{5.51}$$

with the element response

$$a_n(\phi) = \frac{r_n(\phi)}{r_0(\phi)}. \tag{5.52}$$

If the lens antenna array is critical as specified in Eq. (5.50), the array response vector can be given by

$$a_n(\phi) \approx e^{-j\Phi_0}\sqrt{A}\mathrm{sinc}(n - \tilde{D}\tilde{\phi}), \tag{5.53}$$

where $A = D_y D_z/\lambda^2$ is the normalized aperture, Φ_0 is a common phase shift from the lens's aperture to the array, and $\tilde{\phi} = \sin\phi \in [-1,1]$ is referred to as the spatial frequency corresponding to ϕ.

Different from the conventional arrays, whose responses are determined by the phase difference across different elements, the lens antenna array exhibits a *sinc*-function response with angle-dependent focusing capability. To be specific, the power of the signal with the incident angle of ϕ can be magnified by around A times by the element located in the close proximity of the focal point $\tilde{D}\tilde{\phi}$, whereas it is almost negligible for those elements located far from the focal point. Consequently, two signals arrive at two sufficiently different angles can be effectively separated by selecting different elements.

Figure 5.9 illustrates a MIMO system equipped with a lens antenna array with Q and N elements at the transmitter and receiver side, respectively. In a multi-path environment, the channel impulse response can be modeled as

$$\mathbf{H} = \sum_{l=1}^{L}\alpha_l\mathbf{a}_r(\phi_{r,l})\mathbf{a}_T^H(\phi_{t,l})\delta(t - \tau_l), \tag{5.54}$$

where $\mathbf{H}$ is an $N \times Q$ matrix with a typical entry $h_{nq}(t)$ denoting the channel impulse response from the transmitting element q to the receiving element n, L is the total number of resolvable signal paths, α_l and τ_l are the complex-valued gain and delay for the lth path, $\phi_{t,l}$ and $\phi_{r,l}$ express the angle of departure and

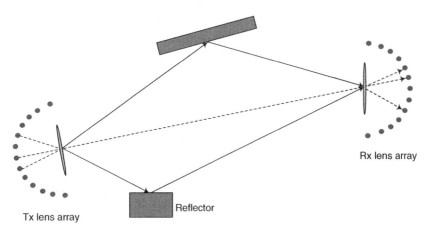

Figure 5.9 Schematic diagram of a MIMO system with lens antenna arrays at both the transmitter and receiver in a multi-path channel.

the angle of arrival for the lth path, respectively, and $\mathbf{a}_t$ and $\mathbf{a}_r$ stand for the response vectors for the lens antenna arrays at the transmitter and receiver respectively.

5.6 Case Study – IEEE 802.15.3d

One of the most significant novelties brought by 5G NR is applying the mmWave band in mobile communications. 3GPP identified several mmWave bands over the frequency range from 24.25 to 52.6 GHz during the definition of Frequency Range 2 (FR2) for the initial deployment of NR networks. Meanwhile, the IEEE ratified the utilization of directional mmWave communications at 60 GHz, initially in the standard IEEE 802.15.3c for high-definition video delivery and then in IEEE 802.11ad for multi-Gigabit-per-second WLANs. The successor of IEEE 802.11ad, named IEEE 802.11ay, aims to achieve a transmission rate over 100 Gbps by using multi-stream MIMO, higher channel bandwidth, and high-order modulation over this license-exempt mmWave band. Future 6G systems are expected to achieve much higher data rates, where wireless communications over frequencies above 100 GHz, particularly the THz band communications and Visible Light Communications (VLC), both providing contiguous spectrum of tens, hundreds, or even thousands of gigahertz, are promising [Petrov et al., 2020].

The major players from academia, industry, standardization bodies, and regulatory agencies have been actively exploring the feasibility of such communications over the last decades. The efforts on VLC have resulted in the establishment of IEEE 802.15.7 [Rajagopal et al., 2012], aiming to specify the PHY and medium access control layers for prospective OWCs. The first attempt in building a wireless communication system at THz frequencies started in 2008 with the foundation of a Terahertz Interest Group (IGthz) under the IEEE 802.15 umbrella. In May 2014, Task Group 3d was formed to standardize a switched point-to-point communication system operating in the frequencies from 60 GHz to the lower THz bands. The initial outcomes of this group included four supporting documents, i.e. the APP Requirements Document, the Technical Requirements Document, the Channel Modeling Document, and the Evaluation Criteria Document. In September 2014, it was split into two separate teams: Task Group 3d focusing on the activities to the lower THz frequency bands, and Task Group 3e worked on developing an amendment for close-proximity communication links at 60 Gbps. During the meeting in March 2016, the supporting documents for IEEE 802.15.3d were approved, and the call for proposals was issued. Based on the proposal reviews and two sponsor recirculation ballots, IEEE 802.15.3d-2017 was ratified by the IEEE SA Standards Board in September 2017.

IEEE 802.15.3d-2017 is an amendment to IEEE 802.15.3-2016 that specifies an alternative PHY layer at the lower THz frequency band, between 252 and 325 GHz, for switched point-to-point connections, along with the necessary modifications on the Medium Access Control (MAC) layer to support this PHY [IEEE 802.15.3d-2017, 2017]. This amendment builds on a new paradigm called *pairnet*, introduced in IEEE 802.15.3e-2017, and inherits the corresponding MAC changes defined there. Some of the key features are as follows:

- Designed for a peak rate over 100 Gbps.
- Effective coverage from tens of centimeters to a few hundred meters.
- Operation in the THz frequency band from 252 to 352 GHz.
- Use of eight bandwidths between 2.16 and 69.12 GHz.
- A pairnet structure supporting intra-device communications, close proximity communications, wireless data centers, and wireless backhaul.
- Two PHY options (i.e. single carrier and on-off keying (OOK)) to achieve either ultra-high-speed or low-complexity modes.

5.6.1 IEEE 802.15.3d Usage Scenarios

The deployment scenarios of IEEE 802.15.3d are limited to point-to-point wireless connections between stationary or quasi-static devices, allowing for low complexity on system design and implementation. Compared with mobile communications and WLANs (particularly IEEE 802.11ad and IEEE 802.11ay), some procedures such as multiple access, initial access, resource allocation, and interference mitigation can be relaxed or avoided to simplify the first standardization effort for THz wireless communications. Despite offering only point-to-point links, IEEE 802.15.3d is capable of playing a particular role in a variety of APPs [Kürner, 2015], such as

- *Close-Proximity Communications*: IEEE 802.15.3d aims to offer high-rate data exchange for close-proximity communications with two use cases, i.e. device-to-device communications and kiosk download. As of today, transferring large-volume files using WLANs takes a long time, degrading the quality of the user experience. THz communications offer supreme data rates with an order of magnitude far higher than existing and emerging close-proximity wireless transmission over the microwave or mmWave bands. Remarkably, the download time of a high-definition movie with a volume of 1 GB gets reduced from around 10 seconds with IEEE 802.11ac to less than 0.1 seconds using IEEE 802.15.3d. Kiosk downloading service enables mobile terminals to instantly synchronize with content providers or storage clouds. Wireless connection between a mobile terminal and a kiosk terminal is provided by neither a conventional cellular system nor a local wireless network with coverage at least tens of meters but by non-contact wireless communications whose transmission range is only a few centimeters or less. Kiosk terminals are usually located in public areas such as train stations, airports, shopping malls, convenience stores, subways, libraries, and public parks. When a user touches a kiosk terminal with a mobile terminal, data files are uploaded to the network or downloaded to the terminal in a very short time. Additionally, close-proximity D2D APPs such as file exchange enable the high-speed transfer of large-volume multimedia data (photo, video, etc.) between two electronic devices such as smartphones, tablets, laptops, digital cameras, and printers. A user can push any data file from one terminal to another terminal with just a touching action. For example, students can share music with friends merely by touching the smartphone to the music player. A tourist can get digital video used for introducing the sightseeing places simply by placing the smartphone close to the tourist information desk or at the entrance of a museum. The devices used in the close-proximity D2D APPs can also be wireless storage devices such as wireless flash memory, wireless solid-state drive, game cards, and smart posters, which are applied to store large-volume data.

- *Wireless Backhaul and Fronthaul Links*: One of the trends in wireless networking is to utilize the ultra-dense deployment of small cells to increase system capacity in populated areas. It demands a high-speed backhaul connection between these small-cell base stations and the core network. When a small cell is equipped with several Remote Radio Heads (RRHs), a communication link referred to as fronthaul is mandatory to connect each RRH with the Base-Band Unit (BBU). In a Centralized Radio Access Network (C-RAN), where a BBU pool controls a number of distributed RFUs, a reliable and high-speed fronthaul network is also needed. Recent research outcomes for beyond 5G or 6G further move forward toward more complicated and heterogeneous networking, imposing a higher demand of backhauling and fronthauling. For instance, the functional split discussed within 3GPP and Open Radio Access Network (O-RAN) association further divides a radio access network into different functional nodes called Centralized Unit (CU), Distributed Unit (DU), or Radio Unit (RU). The deployment of CU, DU, and RU is highly flexible and scalable, requiring more advanced interconnection approaches than ever. Recently, there is a hotly discussed architecture for 6G called cell-free massive MIMO [Jiang and Schotten, 2021], which is potential but brings a challenge to interconnect a central processing unit with a massive number of distributed antennas. In cell-free architecture, huge-volume data consisting of transmitting data and instantaneous channel state information needs to exchange in real-time. Here, the massive utilization of optical fiber connections for backhaul and fronthaul links is time-consuming and expensive in site acquisition, installation, and maintenance. Therefore, a wireless alternative such as Integrated Access and Backhaul (IAB) has been recently proposed to complement the optical links in such challenging setups. In these scenarios, IEEE 802.15.3d can offer point-to-points communications with high data rates and interference exemption. It supports the data rate of over 100 Gbps, an order of magnitude much higher than the state-of-the-art mmWave technologies, to notably simplify the cabling.

- *Cable Replacement in Data Centers*: With the proliferation of information technology and artificial intelligence, many fundamental service provisioning and critical data storage have become more dependent on data centers, leading to a booming number of data centers for the past decade. The computing and storage nodes in data centers are usually connected with high-speed Ethernet via fiber optic. However, wired cabling is time-consuming and expensive for installation and maintenance, wastes much space, and affects the cooling system as some air streams become blocked. In addition, it is weak from the perspective of flexibility and scalability since it is not easy to reconfigure. Today, an alternative solution is actively studied, suggesting the utilization of high-rate point-to-point wireless links when connecting the racks to each other. This approach enables a more flexible design of a data center and also reduces the amount of cabling, resulting in cost-efficient deployment and operation. IEEE 802.15.3d aims to provide this "inter-rack" connectivity, thus complementing the fiber optics in specific deployments [Petrov et al., 2020].

- *Intra-Device Communications*: The first idea of using wireless links for intra-device communication was proposed already as early as 2001 in Chang et al. [2001], which identified two major challenges: the lack of efficient digital to analog converters allowing many levels of quantization at high speeds of sampling, and the absence of allocated large bandwidths allowing simple modulation with maybe two levels of quantization. The THz frequency band can offer abundant spectrum, solving these two issues by allowing the use of simple modulation schemes such as OOK. To be specific, consumer devices and the electronic components inside a single device may be connected with IEEE 802.15.3d. Today, modern computers are already equipped with several high-rate wired links, such as the bus between the Central Processing Unit (CPU) and the random access memory and between the chipset and the network interface. In addition, highly complex solutions are utilized in emerging massive multi-core CPUs to connect the computing cores and the shared cache memory. The rapid growth in the data exchange inside the computers challenges their design. For instance, modern motherboards already have 12 layers, while the emerging "networks-on-chip" for future CPUs occupy over 30% of the processor space and power consumption. Here, IEEE 802.15.3d provides a possible alternative, enabling high-rate data links between the critical components while simultaneously simplifying the layout design [Petrov et al., 2020].

5.6.2 Physical Layer

The THz PHY layer is designed for switched point-to-point wireless transmission operating in the frequency band from 252 to 322 GHz. It supports spectrum flexibility with eight different channel bandwidths ranging from 2.16 to 69.12 GHz to realize a peak data rate over 100 Gbps with a fallback to lower data rates on demand. The following two PHY modes are specified:

- *THz Single Carrier (THz-SC) mode* for extremely high payload data rates. Depending on the combination of modulation, bandwidth, and channel coding, a wide variety of rates are supported, where the theoretical maximal data rate can reach approximately 300 Gbps. The THz-SC PHY supports six modulation schemes: Binary phase-shift keying (BPSK), quadrature phase-shift keying (QPSK), 8-phase-shift keying (PSK), 8-amplitude and phase-shift keying (APSK), 16-QAM, and 64-QAM. It has two Forward Error Correction (FEC) coding modes, i.e. Low-Density Parity-Check (LDPC) codes with rates of 14/15 and 11/15.
- *THz On-Off Keying (THz-OOK) mode* for cost-effective devices that require low complexity and simple design. The THz-OOK PHY supports a single modulation scheme, i.e. OOK, and three FEC schemes. The Reed Solomon (RS) code is mandatory and allows simple decoding without soft-decision information. The LDPC codes with rates of 14/15 and 11/15 are optional and need soft-decision information.

5.6.2.1 Channelization

IEEE 802.15.3d operates over the THz frequency range between 252.72 GHz and 321.84 GHz, amounting to a total of 69.12 GHz available spectrum. Depending on its application and hardware capability, the entire frequency range can be either assigned to a single ultra-wide channel or a magnitude of channels with smaller bandwidths. The standard specified eight different channel bandwidths where the minimal one is 2.16 GHz while others are integer multiples of 2.16 GHz, up to 69.12 GHz. As shown in Figure 5.10, the channelization can be grouped into eight sets in terms of its bandwidths. For example, the first set divides the total spectrum into 32 orthogonal channels with a bandwidth of 2.16 GHz. If the bandwidth doubles to 4.32 GHz, this spectrum can accommodate 16 orthogonal channels. The bandwidth continues to enlarge until all 69.12 GHz spectrum is allocated to a single channel. As a result, there are 69 overlapping channels differentiated by Channel Identity (*CHNL_ID*) from 1 to 69. A channel (*CHNL_ID* = 41) staring from 287.28 to 291.6 GHz with a bandwidth of 4.32 GHz is defined as the default for the operation of an IEEE 802.15.3d system. The control word *phyCurrentChannel* in the PHY header contains the *CHNL_ID* of the current

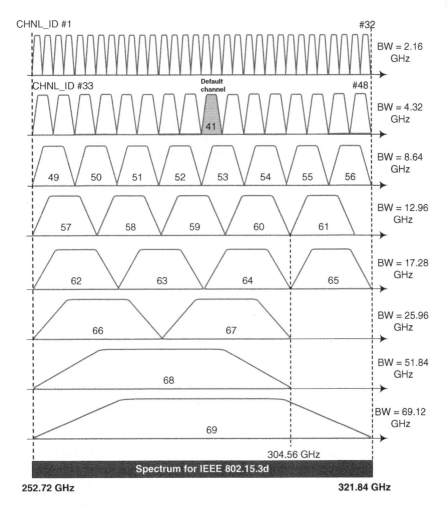

Figure 5.10 Channelization of IEEE 802.15.3d, where a total of 69.12 GHz spectrum from 252.72 to 321.84 GHz are divided into transmission channels with eight different bandwidths from BW = 2.16 to 69.12 GHz. All possible channels are indexed with Channel Identity (CHNL_ID) from 1 to 69, while the 41th channel with a bandwidth of 4.32 GHz is the default channel.

channel. Following the decision at World Radio Conference 2019 (WRC-2019), all the aforementioned channels are globally available for THz communications if specific conditions to protect radio astronomy and Earth exploration satellite service are required. These conditions do not explicitly specify any transmit power limits and are applicable, in practice, primarily to the narrow areas surrounding the ground radio astronomy stations.

5.6.2.2 Modulation

After channel encoding and spreading, the information and control bits shall be fed into the modulator for constellation mapping. The THz-SC PHY mode employs six different modulation schemes to support the adaptive Modulation and Coding Scheme (MCS) in terms of the quality of wireless channels. To be specific, the coded and interleaved binary serial data b_i, where $i = 0,1,2,...$, is modulated using $\pi/2$-BPSK, $\pi/2$-QPSK, $\pi/2$-8PSK, $\pi/2$-8APSK, 16QAM, or 64QAM in the single-carrier mode, depending on the selected MCS. The binary serial stream is divided into groups with 1, 2, 3, 4, or 6 bits and converted into complex numbers representing $\pi/2$-BPSK, $\pi/2$-QPSK, $\pi/2$-8PSK, $\pi/2$-8APSK, 16QAM or 64QAM constellation points, following the Gray-coded mapping. A transmitted symbol c_k is formed by multiplying the value of the mapped constellation point $(I_k + jQ_k)$ by a normalization factor K, i.e.

$$c_k = (I_k + jQ_k) \times K, \tag{5.55}$$

where I_k and Q_k correspond to the in-phase and quadrature components, respectively, and j is the imaginary unit. The purpose of the normalization factor is to guarantee the same average power for different modulation schemes. In practical implementations, an approximate value of the normalization can be used as long as the device conforms to the modulation accuracy requirements. The normalization factors for $\pi/2$-BPSK, $\pi/2$-QPSK, $\pi/2$-8PSK, $\pi/2$-8APSK, 16QAM, and 64QAM are 1, 1, 1, $\sqrt{2}/\sqrt{11}$, $1/\sqrt{10}$, and $1/\sqrt{42}$, respectively.

In contrast, the THz-OOK mode adopts the simplest modulation scheme called OOK for the purpose of low complexity and low power consumption. This PHY is designed for payload bit rates between 1.3 Gbps using a single channel with a bandwidth of 2.16 GHz, and the maximum 52.6 Gbps using a bandwidth of 69.12 GHz, covering a transmission range of a few tens of centimeters. The OOK modulation is a kind of two-point constellation mapping by representing a binary "1" with the presence of the signal and a binary "0" with its absence. The difference between BPSK and OOK lies in that the former can be represented by two arbitrary constellation points, whereas the constellation points of the latter are fixed to 0 and 1. As a result, the normalization factor of OOK is $\sqrt{2}$, rather than 1.

5.6.2.3 Forward Error Correction

The IEEE 802.15.3d standard specifies two types of FEC codes for the THz-SC PHY, i.e. a rate-14/15 LDPC(1440,1344) code and a rate-11/15 LDPC(1440,1056) code. The THz-OOK PHY supports three FEC schemes, where a (240,224)-RS code is mandatory while the LDPC (1440,1344) and the LDPC (1440,1056) are both optional. The error rate requirement is a Frame Error Rate (FER) of less than 10^{-7} with a frame payload length of 214 octets, corresponding to a Bit Error Rate (BER) of 10^{-12}. Measuring error rate is conducted at the PHY Service Access Point (SAP) interface over the Additive White Gaussian Noise (AWGN) channel after any error correction methods (excluding retransmission) have been applied.

Reed-Solomon Block Code   The RS(240,224) is used for encoding the frame payload of High Rate Close Proximity (HRCP)-OOK. RS(n+16,n), a shortened version of RS(240,224), where n is the number of octets in the frame header, is applied to encode the frame header of HRCP-OOK. The generator polynomial is given by

$$g(x) = \prod_{k=1}^{16} (x + a^k), \tag{5.56}$$

where $a = 0x02$ is a root of the binary primitive polynomial $p(x) = 1 + x^2 + x^3 + x^4 + x^8$. A block of information octets $\mathbf{m} = (m_{223}, m_{222}, \ldots, m_0)$ is encoded into a codeword $\mathbf{c} = (m_{223}, m_{222}, \ldots, m_0, r_{15}, r_{14}, \ldots, r_0)$ by computing the remainder polynomial

$$r(x) = \sum_{k=0}^{15} r_k x^k = x^{16} m(x) \mod g(x), \tag{5.57}$$

where $m(x)$ is the information polynomial

$$m(x) = \sum_{k=0}^{223} m_k x^k, \tag{5.58}$$

and $r_k (k = 0,1, \ldots, 15)$ and $m_k (k = 0,1, \ldots, 223)$ are elements of GF(2^8). More details of the RS encoder and its configuration can refer to IEEE 802.15.3e-2017 [2017].

Low-Density Parity-Check Code   The LDPC codes are systematic. In other words, the LDPC encoder builds a codeword

$$\mathbf{c} = (i_0, i_1, \ldots, i_{k-1}, p_0, p_1, \ldots, p_{1440-k-1}) \tag{5.59}$$

by adding $1440 - k$ parity bits $(p_0, p_1, \ldots, p_{1440-k-1})$ at the tail of an information block $\mathbf{i} = (i_0, i_1, \ldots, i_{k-1})$. In IEEE 802.15.3d, the length of information blocks can be either $k = 1056$ for rate-11/15 LDPC or $k = 1344$ for rate-14/15 LDPC codes. The codeword needs to satisfy

$$\mathbf{Hc}^T = \mathbf{0}, \tag{5.60}$$

where $\mathbf{H}$ represents a parity-check matrix with the dimension of $(1440 - k) \times 1440$. Each entry of this matrix denoted by $h_{i,j}$, where $0 \le i < 1440 - k$ and $0 \le j < 1440$, is either 1 or 0. Table 5.8 lists the parameters of two LDPC codes specified in IEEE 802.15.3d, including code rates, information-block lengths, parity lengths, and the matrix entries whose values are 1 in the first 15 columns

Table 5.8 Parameters of LDPC codes in IEEE 802.15.3d.

Parameter	Value	
Code rate	11/15	14/15
Codeword length (bits)	1440	1440
Information-block length, k (bits)	1056	1344
Parity length (bits)	384	96
Matrix entries with the value of 1	$h_{4,0}, h_{96,0}, h_{193,0}$	$h_{0,0}, h_{1,0}, h_{4,0}$
	$h_{34,1}, h_{135,1}, h_{320,1}$	$h_{32,1}, h_{34,1}, h_{39,1}$
	$h_{70,2}, h_{270,2}, h_{352,2}$	$h_{64,2}, h_{70,2}, h_{78,2}$
	$h_{104,3}, h_{287,3}, h_{306,3}$	$h_{8,3}, h_{18,3}, h_{95,3}$
	$h_{31,4}, h_{150,4}, h_{234,4}$	$h_{31,4}, h_{42,4}, h_{54,4}$
	$h_{91,5}, h_{159,5}, h_{364,5}$	$h_{63,5}, h_{76,5}, h_{91,5}$
	$h_{45,6}, h_{286,6}, h_{302,6}$	$h_{14,6}, h_{45,6}, h_{94,6}$
	$h_{126,7}, h_{239,7}, h_{371,7}$	$h_{30,7}, h_{47,7}, h_{83,7}$
	$h_{17,8}, h_{158,8}, h_{272,8}$	$h_{17,8}, h_{62,8}, h_{80,8}$
	$h_{28,9}, h_{178,9}, h_{336,9}$	$h_{28,9}, h_{48,9}, h_{82,9}$
	$h_{60,10}, h_{214,10}, h_{369,10}$	$h_{22,10}, h_{60,10}, h_{81,10}$
	$h_{145,11}, h_{219,11}, h_{372,11}$	$h_{27,11}, h_{49,11}, h_{84,11}$
	$h_{7,12}, h_{173,12}, h_{245,12}$	$h_{7,12}, h_{53,12}, h_{77,12}$
	$h_{19,13}, h_{140,13}, h_{373,13}$	$h_{19,13}, h_{44,13}, h_{85,13}$
	$h_{6,14}, h_{238,14}, h_{363,14}$	$h_{6,14}, h_{46,14}, h_{75,14}$

Source: Adapted from IEEE 802.15.3-2016 [2016] and IEEE 802.15.3e-2017 [2017].

of the parity-check matrix. For other columns $j \geq 15$, the matrix entries can be determined by the following equation

$$h_{i,j} = h_{\left[\mod\left(i+\left\lfloor \frac{j}{15}\right\rfloor,96\right),\ \mod\ (j,15)\right]}, \tag{5.61}$$

where $\lfloor x \rfloor$ stands for the floor function that gives the greatest integer less than or equal to x, and mod (x, y) is the modulo function that is defined as $x - n \times y$, where n is the nearest integer less than or equal to x/y. Each LDPC code is a quasi-cyclic code such that every cyclic shift of a codeword by 15 symbols yields another codeword.

5.6.3 Medium Access Control

In contrast to cellular communication standards and IEEE 802.11-family WLAN standards, IEEE 802.15.3d supports only point-to-point communications.

Nevertheless, its wireless connection can be switched compared with fixed point-to-point communications. The standard follows the MAC layer specified in IEEE 802.15.3e that employs a simple networking paradigm connecting no more than two devices, referred to as *pairnet*. Although the point-to-point transmission restricts the range of possible use cases, it simplifies the MAC procedures such as multiple access, initial access, resource allocation, and interference mitigation.

A pairnet is comprised of at most two devices, i.e. Pairnet Coordinator (PRC) and Pairnet Device (PRDEV), with a typical transmission range of 10 cm or less for close-proximity applications and the coverage is up to several hundred meters for other cases. A device that is able to serve as the PRC establishes a pairnet by initializing the Sequence Number field and Last Received Sequence Number field and then sending the beacon frame in the default channel (i.e. the channel with $CHNL\_ID = 41$). Once a pairnet connection is set up, the beacon transmission is turned off. After the communication process is completed, the PRC or PRDEV terminates the pairnet by sending a Disassociation Request command. Then, the PRC reactivates the beacon transmission in order to prepare for creating a new pairnet.

The communication process is divided into two periods:

1. Pairnet Setup Period (PSP)
2. Pairnet Associated Period (PAP)

During the PSP, the PRC forms a pairnet and starts periodically sending beacons with the necessary network parameters. For instance, the beacon frame includes the information on the number and duration of the access slots that can be used by the device willing to join the pairnet. Also, the beacon sent at the beginning of every superframe contains timing signals necessary to synchronize the PRDEV with the clock of the PRC. When a PRC-capable device forms a pairnet, the type of pairnet depends on the supported PHY modes. For example, if the PRC supports only the THz-SC mode, it will start a THz-SC pairnet in which the beacon frame is sent with the THz-SC mode. The same process is used for the THz-OOK mode. If a PRC supports both THz PHY modes, it can select the type of pairnet. It allows connection from each type of device by transmitting both the THz-SC mode beacon frame and the THz-OOK mode beacon frame. Another device ready to join the pairnet responds to the beacon by sending the Association Request at the beginning of one of the defined access slots. After a successful reception of the Association Request, the PRC stops the beacon transmission and sends an Association Response to the PRDEV, ending the PSP. In addition to the pairnet setup, higher-layer protocol setups, such as Internet Protocol (IP) or Object Exchange (OBEX) file transfer, may be performed during the PSP. These additions are supported as both the command frames of Association Request and Association Response have the fields to carry higher-layer protocol information.

The switched point-to-point nature of the connection means that the pairnet connection can be terminated, and a new pairnet can be established with a different device. With the successful reception of the Association Response, the actual data exchange starts in the second period. Both the PRC and the PRDEV transmit data frames and optional acknowledgments during this period. The frames are separated by the defined Short Inter-Frame Space (SIFS). When either of the nodes wants to terminate the communications, it sends a Disassociation Request. The PRC may also end PAP if no message was received from the PRDEV during the defined timeout. If the PRDEV does not want to terminate the communications but has no actual data, it may transmit a Probe Request to restart the PRC timeout timer. Whenever the PAP is over (either via a Disassociation Request or by a timeout), the PRC switches back to the PSP, continues transmitting beacons, and waits for the new connections [Petrov et al., 2020].

5.6.4 Frame Structure

The structure of IEEE 802.15.3d THz frame consists of three major parts: the PHY preamble, frame header, and frame payload, as shown in Figure 5.11.

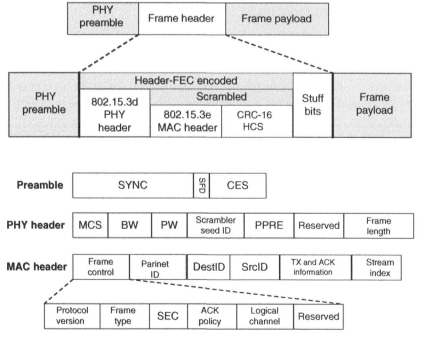

Figure 5.11 The structure of IEEE 802.15.3d frame.

5.6.4.1 Preamble

A PHY preamble is added at the beginning of a frame to assist the receiver in an auto-gain control setting, frame detection, timing acquisition, frequency offset estimation, frame synchronization, and channel estimation. There are two types of PHY preambles, i.e. the PHY long preamble and short preamble. A PHY long preamble is used during the PSP, and a PHY short preamble is applied during the PAP. The structure of the preambles is shown in x, consisting of three fields: frame Synchronization (SYNC), Start Frame Delimiter (SFD), and Channel Estimation Sequence. The SYNC field is used for frame detection and utilizes the repetition of a 128-bit-length Golay sequence. The SYNC field for the long preamble is a 28-times repetition of the Golay sequence, and the SYNC field for the short preamble is a 14-times repetition. The function of the SFC field is to establish frame timing, implemented by the sign inversion of the original Golay sequence. The CES field is a 1408-bit sequence built from the Golay sequence and a Golay complementary sequence. The preamble is transmitted with a chip rate of 1.76 giga-samples-per-second. Hence, the duration of SFD is 0.07 µs while CES is 0.8 µs. The duration of SYNC for the long and short preambles is 2.01 and 1.02 µs, respectively, amounting to two lengths of the whole preamble, i.e. 2.91 and 1.89 µs.

5.6.4.2 PHY Header

A frame header, consisting of a PHY header and a MAC header, is added at the tail of the PHY preamble to convey the system information necessary for decoding the frame at the receiver. As shown in the figure, the PHY header is comprised of the following fields: MCS, Bandwidth (BW), Pilot Word (PW), Scrambler Seed ID, Pilot Preamble (PPRE), Reserved, and Frame Length, which are briefly explained as follows:

- *The MCS field* utilizes four bits to identify twelve MCS combinations from six modulation schemes (i.e. BPSK, QPSK, 8PSK, 8APSK, 16QAM, and 64QAM) and two coding types (i.e. 11/15-LDPC and 14/15-LDPC).
- *The BW field* uses three bits to carry the index of eight types of bandwidths of IEEE 802.15.3d, ranging from 2.16 to 69.12 GHz.
- *The PW field* shall be set to one if the pilot word is used in the frame and shall be set to zero otherwise.
- Except for the PHY preamble and PHY header, other parts, including the MAC header, Header Check Sequence (HCS), and the frame body, need to be scrambled by module-2 addition of the data with the output of a pseudo-random bit sequence generator. The initialization vector is determined from *the Scrambler Seed ID field* contained in the PHY header of the received frame.
- *PPRE* is an optional feature that allows a device to adjust the receiver algorithms periodically. It is inserted into the scrambled, encoded, spread, and modulated

MAC frame body with an interval of 1024, 2048, or 4096 blocks. Like the PHY preamble, the PPRE shall be the concatenation of SYNC, SFD, and CES with the modulation of $\pi/2$-BPSK.

- *The Frame Length Field* shall be an unsigned integer equal to the number of octets in the MAC frame body of a regular frame, excluding the frame check sequence. The maximum frame length allowed is 2 099 200 octets, including the MAC frame body, but not the PHY preamble, the base header (PHY header, MAC header, and HCS), or the stuff bits.

5.6.4.3   MAC Header

The MAC header comprises the following fields: Frame Control, Pairnet Identity, Destination Identity (DestID), Source Identity (SrcID), Transmission (TX) and Acknowledgement (ACK) Information, and Stream Index, as shown in Figure 5.11, which are briefly introduced as follows:

- *Frame Control* is 2-octet long, which is further divided into six parts, i.e. Protocol Version, Frame Type, Security (SEC), ACK Policy, Logical Channel, and Reserved. The *Protocol Version* field is invariant in size and placement across all revisions of IEEE 802.15.3 standards, where its value is a binary number 0 for piconet and 1 for parinet. The protocol version will be incremented only when a fundamental incompatibility exists between a new revision and the prior revision of the standard. A device that receives a frame with a higher revision level than it supports may discard the frame without any indication. The *Frame Type* field uses three bits to classify the beacon frame, data frame, command frame, and multi-protocol data frame, while some reserved for the further revisions. The *SEC field* shall be set to one when the frame payload is protected using the key specified by the security ID and be set to zero otherwise. The *ACK Policy* field is applied to indicate the type of acknowledgement procedure that the addressed recipient is required to perform. The *Logical Channel* field is available for use by the higher layer protocol users.

- *Pairnet ID* is the unique identifier for the pairnet, which normally remains constant during an instantiation of a pairnet and may be persistent for multiple sequential instantiations of the pairnets by the same PRC.

- There are two fields in the MAC frame used to indicate the source device and destination device. The device identifier is assigned by the PRC in the beacon frame before the association of a device and is unique within a pairnet.

5.6.4.4   Construction Process of Frame Header

A frame header shall be inserted at the tail of the PHY preamble, conveying information in the PHY and MAC headers necessary for successfully decoding the frame. The detailed process of constructing the frame header is shown in Figure 5.12, i.e.

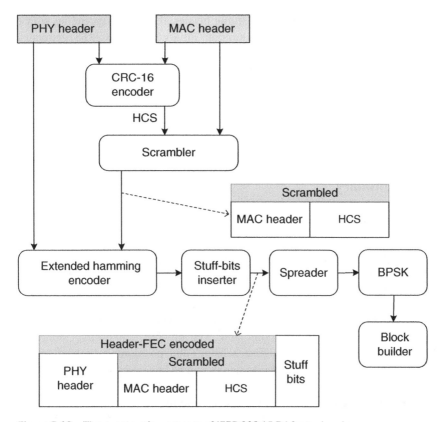

Figure 5.12 The construction process of IEEE 802.15.3d frame header.

- Form the base frame header as follows:
  1. Construct the PHY header based on information provided by the MAC.
  2. Compute the HCS over the combination of the PHY header and MAC header. An error-detecting code based on a Cyclic Redundancy Check sequence with a length of sixteen bits (CRC-16) is specified.
  3. Append the HCS to the MAC header.
  4. Scramble the combination of the MAC header and HCS.
  5. Encode the concatenation of the PHY header, scrambled MAC header, and scrambled HCS, into concatenated codewords of an Extended Hamming (EH) code to increase the robustness of the frame header. For each four-bit input sequence, the EH encoder generates four parity bits and forms an eight-bit codeword.
  6. Form the base frame header by concatenating the coded PHY header, coded and scrambled MAC header, coded and scrambled HCS, and header stuff bits.

- To increase the robustness of the frame header, the coding spreading with a spreading factor of 4 using Pseudo-Random Binary Sequence (PRBS) generated by Linear Feedback Shift Register (LFSR) is applied. For IEEE 802.15.3, the binary input 0 is spread to 1010 while 1 is spread to 0101.
- Modulate the frame header using $\pi/2$-BPSK.
- Build blocks from the resulting frame header by inserting pilot words. For the pilot word with length 0, the length of data is 64 symbols, while there are 56 data symbols for the pilot word with length 8.

5.7 Summary

The current spectral resources available over the microwave and millimeter-wave bands are still challenging to support the implementation of the tera-bits-per-second transmission rate envisioned by the upcoming 6G mobile networks. Consequently, the wireless communication community has to exploit even higher frequencies at the terahertz band. The THz-communications technology is expected to play a critical role in the 6G system by providing Tbps small cells, wireless backhauling and fronthauling, device-to-device close-proximity connectivity, intra-device communications, and even on-chip networking. Moreover, the functionalities such as imaging, sensing, and positioning enabled by the particular characteristics of terahertz signals can be integrated with THz communications to achieve a great synergy in the diverse 6G use cases and deployment scenarios. Although the research on THz communications is still in its infancy, the ITU-R has already identified the open of the spectrum between 275 and 450 GHz to land mobile and fixed services, removing the barrier of deploying THz commutations from the perspective of spectrum regulation. However, the most significant challenge for achieving the benefits of THz wireless communications is the constraint of transmission range due to the significant propagation loss raised from high free-space path loss, atmospheric gaseous absorption, rainfall attenuation, and blockage. On the other side of the coin, the tiny wavelength of the terahertz waves facilitates the utilization of an ultra-massive MIMO antenna array, where novel schemes such as array-of-subarrays beamforming are promising. Moreover, lens antennas that provide a low-cost solution to generate high directive beams also exhibit high potential in THz communications. As the world's first wireless standard operated in the terahertz band, IEEE 802.15.3 can provide many hints for future THz mobile communication systems. However, it is only limited to point-to-point links, and thus there is a big gap to be filled by exploring new schemes such as multiple access, initial access, radio resource allocation, interference mitigation to build a THz-based mobile system.

References

Akyildiza, I. F., Jornet, J. M. and Han, C. [2014], 'Terahertz band: Next frontier for wireless communications', *Physical Communication* **12**, 16–32.

Chang, M., Roychowdhury, V., Zhang, L., Shin, H. and Qian, Y. [2001], 'RF/wireless interconnect for inter- and intra-chip communications', *Proceedings of the IEEE* **89**(4), 456–466.

Crane, R. K. [1980], 'Prediction of attenuation by rain', *IEEE Transactions on Communications* **28**(9), 1717–1733.

Dahlman, E., Parkvall, S. and Sköld, J. [2021], *5G NR - The Next Generation Wireless Access Technology*, Academic Press, Elsevier, London, the United Kindom.

Hirata, A., Yamaguchi, R., Kosugi, T., Takahashi, H., Murata, K., Nagatsuma, T., Kukutsu, N., Kado, Y., Lai, N., Okabe, S., Kimura, S., Ikegawa, H., Nishikawa, H., Nakayama, T. and Inada, T. [2009], '10-Gbit/s wireless link using InP HEMT MMICs for generating 120-GHz-band millimeter-wave signal', *IEEE Transactions on Microwave Theory and Techniques* **57**(5), 1102–1109.

IEEE 802.15.3-2016 [2016], 802.15.3-2016 - IEEE standard for high data rate wireless multi-media networks, Standard 802.15.3-2016, IEEE Computer Society, New York, USA.

IEEE 802.15.3d-2017 [2017], 802.15.3d-2017 - IEEE standard for high data rate wireless multi-media networks–amendment 2: 100 Gb/s wireless switched point-to-point physical layer, Standard 802.15.3d-2017, IEEE Computer Society, New York, USA.

IEEE 802.15.3e-2017 [2017], 802.15.3e-2017 - IEEE standard for high data rate wireless multi-media networks–amendment 1: High-rate close proximity point-to-point communications, Standard 802.15.3e-2017, IEEE Computer Society, New York, USA.

ITU-R P.676 [2019], Attenuation by atmospheric gases and related effects, Recommendation P676-12, ITU-R.

ITU-R P.838 [2005], Specific attenuation model for rain for use in prediction methods, Recommendation P838-3, ITU-R.

Jiang, W. and Schotten, H. D. [2021], 'Cell-free massive MIMO-OFDM transmission over frequency-selective fading channels', *IEEE Communications Letters* **25**(8), 2718–2722.

Jiang, W., Han, B., Habibi, M. A. and Schotten, H. D. [2021], 'The road towards 6G: A comprehensive survey', *IEEE Open Journal of the Communications Society* **2**, 334–366.

Kürner, T. [2015], IEEE P802.15 working group for wireless personal area networks (WPANs): TG3d applications requirements document (ARD), Document Nr. 14/0304r16, IEEE P802.15.

Kuerner, T. and Hirata, A. [2020], On the impact of the results of WRC 2019 on THz communications, *in* 'Proceedings of 2020 Third International Workshop on Mobile Terahertz Systems (IWMTS)', Essen, Germany, pp. 1–3.

Lin, C. and Li, G. Y. [2016], 'Terahertz communications: An array-of-subarrays solution', *IEEE Communications Magazine* **54**(12), 124–131.

Petrov, V., Kurner, T. and Hosako, I. [2020], 'IEEE 802.15.3d: First standardization efforts for sub-terahertz band communications toward 6G', *IEEE Communications Magazine* **58**(11), 28–33.

Rajagopal, S., Roberts, R. D. and Lim, S.-K. [2012], 'IEEE 802.15.7 visible light communication: Modulation schemes and dimming support', *IEEE Communications Magazine* **50**(3), 72–82.

Rappaport, T. S., Xing, Y., Kanhere, O., Ju, S., Madanayak, A., Mandal, S., Alkhateeb, A. and Trichopoulos, G. C. [2019], 'Wireless communications and applications above 100 GHz: Opportunities and challenges for 6G and beyond', *IEEE Access* **7**, 78729–78757.

Sarieddeen, H., Alouini, M.-S. and Al-Naffouri, T. Y. [2019], 'Terahertz-band ultra-massive spatial modulation MIMO', *IEEE Journal on Selected Areas in Communications* **37**(9), 2040–2052.

Sarieddeen, H., Saeed, N., Al-Naffouri, T. Y. and Alouini, M.-S. [2020], 'Next generation Terahertz communications: A rendezvous of sensing, imaging, and localization', *IEEE Communications Magazine* **58**(5), 69–75.

Siegel, P. H. [2002], 'Terahertz technology', *IEEE Transactions on Microwave Theory and Techniques* **50**(3), 910–928.

Siles, G. A., Riera, J. M. and del Pino, P. G. [2015], 'Atmospheric attenuation in wireless communication systems at millimeter and THz frequencies', *IEEE Antennas and Propagation Magazine* **57**(1), 48–61.

Smulders, P. F. M. and Correia, L. M. [1997], 'Characterisation of propagation in 60 GHz radio channels', *Electronics and Communication Engineering Journal* **9**(2), 73–80.

Tripathi, S., Sabu, N. V., Gupta, A. K. and Dhillon, H. S. [2021], 'Millimeter-wave and terahertz spectrum for 6G wireless'. arXiv:2102.10267.

Zeng, Y. and Zhang, R. [2016], 'Millimeter wave MIMO with lens antenna array: A new path division multiplexing paradigm', *IEEE Transactions on Communications* **64**(4), 1557–1571.

6

Optical and Visible Light Wireless Communications in 6G

Optical Wireless Communications (OWC) points to wireless communications that use the optical spectrum, including infrared (IR), visible light, and ultraviolet (UV), as the transmission medium. It is a promising complementary technology for traditional wireless communications operating over Radio-Frequency (RF) bands. The OWC systems use visible light to convey information, commonly referred to as Visible Light Communications (VLC), which has attracted much attention over the last years. The optical band can provide almost unlimited bandwidth without permission from regulators worldwide. It can be applied to realize high-speed access with low costs thanks to the availability of off-the-shelf optical emitters and detectors (i.e. light-emitting diodes, laser diodes, and photodiodes). Since the IR and UV waves have similar behavior as the visible light, the security risks and RF interference can be significantly confined, and the concern about the radio radiation to human health can be eliminated. It is expected to have obvious advantages in deployment scenarios such as home networking, vehicular communications in intelligent transportation systems, airplane passenger lighting, and electronic medical equipment that is sensitive to RF interference. Outdoor point-to-point OWC, also known as Free-Space Optical (FSO) communications, use high-power, high-concentrated laser beam at the transmitter to achieve high data rate over long distance up to several thousand kilometers. It offers a cost-effective solution for the backhaul bottleneck in terrestrial networks, enables crosslinks among space, air, and ground platforms, and facilitates high-capacity inter-satellite links for the emerging LEO satellite constellation. In addition, underwater OWC provides higher transmission rates than the traditional acoustic communications systems with significantly low power consumption and low computational complexity.

This chapter will focus on the key aspects of OWC, consisting of

- Definition and characteristics of the optical spectrum consisting of IR, visible light, and UV radiation.
- Main technical advantages and challenges of optical wireless communications.

6G Key Technologies: A Comprehensive Guide, First Edition. Wei Jiang and Fa-Long Luo.

- Potential application scenarios of optical wireless communications.
- The evolution of wireless IR communications, visible light communications, wireless UV communications, and FSO communications.
- The setup of an optical transceiver.
- Main characteristics of optical sources and detectors, including light-emitting diodes, laser diodes, and photodiodes.
- Different optical link configurations, and their pros and cons.
- Optical Multi-Input Multi-Output (MIMO) techniques, including optical spatial multiplexing and optical spatial modulation.

6.1 The Optical Spectrum

Optical Wireless Communication refers to wireless communications operating in the optical spectrum, using optical signals to convey information. It is a promising complementary technology for traditional wireless communications based on radio-frequency bands and electronics technology. Visible Light Communications, i.e. the OWC systems using visible light as the transmission medium, has attracted much attention over the last years. In addition, the OWC systems using infrared and ultraviolet signals, also known as IR and UV communications, respectively, have been applied over the last decades.

The optical spectrum consists of three parts: IR, visible light, and UV bands, as shown in Figure 6.1 and Table 6.1.

6.1.1 Infrared

Infrared radiation, sometimes also referred to as IR light, is a type of electromagnetic radiation, where a continuum of frequencies is produced when atoms absorb and then release energy. From the lowest to highest frequencies, electromagnetic radiation includes radio waves, microwaves, IR, visible light, UV, X-rays, and gamma-rays. IR radiation has wavelengths longer than those of visible light. Therefore, it is invisible to human eyes, but we can feel it as heat. All objects in the universe emit some levels of IR radiation, where the most prominent sources are the sun and fire. IR radiation is one of the three ways heat is transferred from one place to another, the other two being convection and conduction. Everything with a temperature above around $5\,°K$ ($-450\,°F$ or $-268\,°C$) emits IR radiation. The sun gives off half of its total energy as IR, and much of the star's visible light is absorbed and re-emitted as IR,

Infrared was discovered in the year 1800 by British astronomer William Herschel, who noticed a type of invisible radiation in the spectrum lower than red light through its effect on a thermometer. In an experiment to measure the

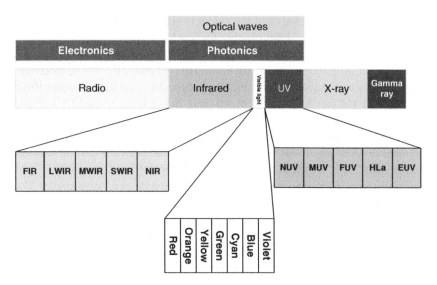

Figure 6.1 The optical spectrum consisting of infrared, visible light, and ultraviolet bands.

Table 6.1 The optical spectrum.

Band		Wavelength	Frequency
Infrared	Far infrared (FIR)	15–1000 μm	300 GHz to 20 THz
	Long-wavelength infrared (LWIR)	8–15 μm	20–37 THz
	Mid-wavelength infrared (MWIR)	3–8 μm	37–100 THz
	Short-wavelength infrared (SWIR)	1.4–3 μm	100–214 THz
	Near infrared (NIR)	0.75–1.4 μm	214–400 THz
Visible	Red	625–750 nm	400–480 THz
	Orange	590–625 nm	480–510 THz
	Yellow	565–590 nm	510–530 THz
	Green	500–565 nm	530–600 THz
	Cyan	485–500 nm	600–620 THz
	Blue	450–485 nm	620–670 THz
	Violet	380–450 nm	670–790 THz
Ultraviolet	Near-UV (NUV)	300–400 nm	750 THz to 1 PHz
	Middle-UV (MUV)	200–300 nm	1–1.5 PHz
	Far-UV (FUV)	122–200 nm	1.5–2.46 PHz
	Hydrogen Lyman-α	121–122 nm	2.46–2.48 PHz
	Extreme-UV (EUV)	10–121 nm	2.48–30 PHz

difference in temperature between the colors in the visible spectrum, he placed thermometers in the path of light within each color of the visible spectrum. He observed an increase in temperature from blue to red, and he found an even warmer temperature measurement just beyond the red end of the visible spectrum.

Within the electromagnetic spectrum, IR waves occur at frequencies above those of microwaves and just below those of red light, hence being named *IR*. Infrared extends from the nominal red edge of the visible light spectrum at 750 nanometers (nm) to one micrometer (µm), corresponding to a frequency range of from approximately 300 GHz to 400 THz, although these values are not definitive. Similar to the visible light spectrum, which ranges from violet (the shortest visible-light wavelength) to red (longest wavelength), IR has its own range of wavelengths. The shorter *NIR* waves, which are closer to visible light on the electromagnetic spectrum, do not emit any detectable heat. It has been applied for TV remote control, such as changing the channels. The longer *far-infrared (FIR)* waves, which are closer to the microwave section on the electromagnetic spectrum, can be felt as intense heat, such as the heat from sunlight or fire. The FIR are partly overlapped with the terahertz band.

6.1.2 Visible Light

Light or visible light refers to the portion of the electromagnetic spectrum, which can be detected by human eyes to perceive brightness and color. Light penetrates as far as the retina in the eye and the dermis in the skin. Most people can perceive wavelengths between about 400 and 700 nm visually. Through observing rainbows, humans recognized that the white light consists of different colored lights: red, orange, yellow, green, cyan, blue, and violet. The visible light band starts from the longest wavelength (violet light) at around 400 nm to the shortest wavelength (red light) of approximately 700 nm. Rather than being clear cut, the boundaries of the visible spectrum region for humans exhibit fluid transitions. The low end of the optical spectrum below the red side of the visible spectrum where light has a wavelength longer than any light in the visible spectrum is called IR, while the high end beyond the violet side of the visible spectrum where light has a wavelength shorter than any light in the visible spectrum is called UV. Moreover, a person's eyesight and sensitivity to light vary over their lifetime due to aging processes in the eye. Especially for the short-wavelength section of the visible spectrum (blue light), the transparency of the lens decreases with age. The principal natural radiation source for light is the sun and other sources such as lighting. However, our everyday lives also feature a multitude of artificial light sources. Light is not only responsible for allowing us to see our surroundings but also has other biological effects and influences the sleep cycle. However, there is a risk of damage, especially to the eyes and the skin, if the intensity of light exceeds certain effect thresholds.

6.1.3 Ultraviolet

UV light refers to the region of the electromagnetic spectrum between visible light and X-rays, with wavelengths falling between 10 nm (corresponding to frequency 30 PHz) and 400 nm (750 THz), shorter than that of visible light but longer than that of X-rays. Around 10% of the total electromagnetic radiation from the Sun is UV light. It is not visible to human eyes because it has a shorter wavelength than the light our brain perceives as images, whereas many birds and insects can visually perceive UV light. UV radiation was discovered in 1801 when the German physicist Johann Wilhelm Ritter observed that invisible lights just beyond the violet end of the visible spectrum darkened silver chloride-soaked paper more quickly than violet light. He named it *de-oxidizing* rays to emphasize its chemical reactivity to distinguish from *heat rays* (IR) discovered in the previous year at the other side of the visible spectrum. In 1893, German physicist Victor Schumann discovered UV radiation with wavelengths below 200 nm, named "vacuum ultraviolet" because the oxygen in air strongly absorbs it.

The natural source of UV radiation is sunlight, which is categorized into three different types:

- **UV-A radiation** (320 nm to 400 nm) is more commonly known as *black light*, which is applied to cause objects to emit fluorescence for artistic and celebratory purposes. Many insects and birds can perceive UV-A radiation visually.
- **UV-B radiation** (290 nm to 320 nm) raises sunburns in the skin under long-time exposure to sunlight, increasing the risk of skin cancer and other cellular damage. However, around 95% of all UV-B radiation from the Sun can be absorbed by the ozone layer in Earth's atmosphere.
- **UV-C radiation** (100 nm to 290 nm) is the highest energy portion of the UV radiation. It is highly harmful to human health, but the Earth's ozone layer almost wholly absorbs it. It is commonly applied as a disinfectant in food, air, and water to kill microorganisms by destroying their cells' nucleic acids.

Exposure to UV-B light helps the skip produce a type of vitamin D, which promotes calcium absorption and plays a vital role in bone and muscle health. However, prolonged exposure to UV-A and UV-B waves without adequate protection can cause severe skin burns and eye injuries. Sunscreen is a necessary precaution against UV radiation since it provides a protective layer to absorb UV-A and UV-B waves before affecting the skin. With much exposure to sunlight without protection, a person's risk of skin cancer and other dangerous cellular afflictions dramatically increases. The eyes should also be protected from UV radiation outdoors by wearing sunglasses designed to block UV-A and UV-B rays. Otherwise, it causes short-term effects like Photokeratitis (known in some cases as arc-eye or snow blindness) or severe long-term conditions, including cataracts which lead to blindness.

When studying light passing through outer space, scientists often use a different set of UV subtypes dealing with astronomical objects, i.e.

- **Near Ultraviolet (NUV) Radiation**
- **Middle Ultraviolet (MUV) Radiation**
- **Far Ultraviolet (FUV) Radiation**
- **Hydrogen Lyman-alpha Radiation**
- **Extreme Ultraviolet (EUV) Radiation** can only travel through a vacuum and is wholly absorbed in Earth's atmosphere. EUV radiation ionizes the upper atmosphere, creating the ionosphere. In addition, Earth's thermosphere is heated mainly by EUV waves from the Sun. Since solar EUV waves cannot penetrate the atmosphere, scientists must measure them using space satellites.

6.2 Advantages and Challenges

Compared with the traditional wireless communications operated in the RF bands, OWCs exhibit some particular advantages:

- *Massive Spectral Resources* The optical spectrum spans approximately from 300 GHz to 30 PHz, amounting to a highly massive available spectral resources of around 30 000 THz, which is hundreds of thousands of times that of the bandwidth provided by radio waves. It implies that spectral resources are no longer the bottleneck of wireless communications and we need to rethink the traditional approaches aiming to high spectral efficiency to achieve a high data rate over a limited bandwidth.
- *License-Free Operation* Radio frequencies, especially the sub-6 GHz band, carry a lot of critical services, such as communications, navigation, radar, radio and TV broadcasting, satellite, and radio astronomy, most of which desire as many bandwidth as possible. Hence, regional and worldwide regulation bodies have to impose stringent rules on radio spectrum to allocation the resources and avoid mutual interference. Mobile operators suffers from high expenditure to get spectrum licenses. In contrast, the optical spectrum do not have such constraints until now due to its massive spectrum, high-directive transmission property, and immunity to interference.
- *Immune to Radio Interference* RF interference causes disturbance in an electrical circuit due to electromagnetic radiation or electromagnetic induction. Meanwhile, inter- or intra-system interference of wireless communication systems, such as multi-user interference, inter-cell interference, adjacent-channel interference, and out-of-band radiation, degrade the system performance

substantially. Hence, potential sources of radio interference, including consumer electronic equipment, are asked to turn off in airplanes or hospitals. Furthermore, radio radiation has potentially dangerous in hazardous operations, such as power plants, nuclear generators, and underground mines. In addition, the transmission power of radio equipment should be below a certain regulated level since there are serious health risks for human beings exposed to high radio radiation for a long time. In contrast, light does not create radio interference. Thus, OWC systems can provide unique advantages when deploying in particular places where traditional wireless communication is forbidden.

- *Low-Cost Transceiver* OWC employs Light-Emitting Diode (LED) or Laser Diode (LD) as the optical source to convert an electrical signal to an optical signal at the transmitter while the receiver uses a Photo-Diode (PD) to convert the optical signal into electrical current. The information is conveyed by modulating the intensity of optical pulse simply through on-off keying or pulse-position modulation. Unlike expensive RF components and complex transceivers used in RF communications, these optical components (i.e. LEDs, LDs, and PDs) have been off-the-shelf for a long time and are readily available in the market. Furthermore, these optical components are cheap, lightweight, high power-efficient, low heat radiation, and long lifetime.

- *Intrinsic Security* Wireless transmission at RF frequencies has good broadcasting properties with wide-area coverage, and wireless signals can easily penetrate walls due to the long wavelengths of radio waves. Therefore, an eavesdropper can easily intercept the signals anywhere without any possibility of being detected. In contrast, optical waves cannot penetrate objects in their path, and they are more likely to be reflected. Meanwhile, optical signals have an intrinsic feature of propagating highly directive rather than broadcasting. Hence, wireless communications over optical bands provide high-level security.

- *Adaptive Transmission Range* OWC provides flexible transmission distances depending on its deployment scenarios and applied optical sources. For example, with low-power, cheap LEDs, optical signals can be transmitted over several millimeters or several meters for short-range communications, whereas the transmission distance reaches up to several thousand kilometers using high power laser beams for some long-range scenarios, such as inter-satellite FSO links and ground-satellite laser communications.

Despite its advantage, OWC suffers from some particular challengers that are not severe in radio communications, such as weather attenuation, atmospheric absorption, scintillation, alignments errors, and hazard protection for humans [Wang et al., 2020].

- **Weather Attenuation** Adverse weather conditions can affect the operation of outdoor OWC (or FSO communications), degrading the performance or leading to a system outage. Scattering means the deflection of incident light from its initial direction, which causes spatial, angular, and temporal spread. It attenuates the energy of optical signals on the order of magnitude tens or hundreds of decibels per kilometer and distorts the transmitted signal. The precise scattering mechanism is dependent on the ratio of particle size to radiation wavelength. When this ratio is near to unity, optical signals severely scatter. At the millimeter-wave and terahertz frequencies, rain and snow can hamper signal propagation. However, raindrops and snowflakes are much larger than optical wavelengths. The most harmful atmospheric conditions are fog and haze due to their comparable radii on the same order of magnitude as the wavelengths of optical signals [Al-Kinani et al., 2018].

- **Atmospheric Absorption** Molecules in the atmosphere such as oxygen and water vapor can absorb the energy of light, imposing attenuation of the optical power. This absorption depends on the wavelength so that a given atmospheric situation may be transparent to some types of lights while completely blocking others. For example, IR light is primarily absorbed by water vapor and Carbon Dioxide (CO_2) in the atmosphere, while absorption by oxygen (O_2) and ozone (O_3) strongly attenuate UV light. Regions of fair transmission are called free absorption windows and are found in the visible and near- IR bands. At middle IR wavelengths, there are more absorption windows in the region of $3\,\mu m$ to $5\,\mu m$ and $8\,\mu m$ to $12\,\mu m$, but above 22 μm, water vapor absorption again rises to prohibitive levels for transmission. Weather can often change quickly, and as the atmosphere's molecular content alters, the propagation channel's transmission characteristics change too. These changes can seriously influence the availability of an optical transmission link [Kedar and Arnon, 2004].

- **Turbulence** The varying atmospheric conditions cause turbulence due to different temperatures, atmospheric pressures, and humidity levels within the propagation media. It leads to scintillation of the light and beam wander. In other words, the optical power acquired by a receiver fluctuates over time and the position of the incident light shifts in space. This phenomenon steps from air conditioning vents in the vicinity of the transceiver, irradiative heat from the roof, or currents of pollutants in the atmosphere. Temperature and humidity gradients cause changes in the atmospheric refractive index, which is the source of optical distortions. Winds and cloud coverage also influence the level of turbulence, and even the time of day can alter temperature gradients. Much previous research in the related fields developed statistical tools to model the dependence of the refractive index on the ambient temperature, pressure, and humidity as well as on the radiation wavelength.

- **Beam Alignment** Outdoor OWCs heavily rely on line-of-sight transmission. Therefore, it is imperative to maintain the alignment between transmitter and receiver during communication continuously. However, accurate alignment is particularly challenging with narrow beam divergence angles at the transmitter and narrow Fields of View (FOV) at the receiver. Outdoor OWC equipment is generally installed on the top of high buildings. Building sway due to thermal expansion of building frame parts or weak earthquakes is possible to cause misalignment. The former may have a daily cycle and seasonal characteristics, while the latter is usually unpredictable. Another common cause of misalignment is strong winds, especially when OWC equipment is mounted on tall buildings. Under the influence of dynamic wind loads, many high-rise buildings sway in the along-wind and cross-wind directions or twist due to torsion. The horizontal movement of the building due to these effects can vary from 1/200 to 1/800 of the building height. Building sway leads to pointing errors at the beam steering that degrades system performance. Hence, outdoor OWC needs a tracking and pointing system to maintain accurate alignment. This alignment process comprises two steps, i.e. coarse tracking and fine pointing. It makes use of GPS location information or other *a priori* knowledge to achieve coarse tracking firstly. Then, it requires electro-optic mechanisms such as a quadrature or a matrix detector to steer beams finely [Kedar and Arnon, 2004].

- **Safety and Regulations** Exposure to light may cause injury to both the skin and the eye, but the damage to the eye is far more significant because the eye can concentrate the energy of light, covering the wavelengths from around 0.4 to 1.4 μm, on the retina. The light with other wavelengths is filtered out by the front part of the eye (i.e. the cornea) before the energy is concentrated. In addition, invisible light at the near-IR band also imposes a safety hazard to human eyes if operated incorrectly. Consequently, the design, deployment, and operation of optical communication systems must ensure that the optical radiation is safe to avoid any injury to the people who might contact it. Optical sources and detectors at 0.7–1.0 μm are cheap, but the eye safety regulations are particularly stringent. At longer wavelengths over 1.5 μm, the eye safety regulations are much relaxed, but optical components are relatively expensive. There are already several guidelines on laser equipment or laws for the safety of optical beams, issued by international standardization or regulatory bodies, such as American National Standards Institute (ANSI) Z136.1 standards for laser use, International Electrotechnical Commission (IEC) standards for laser and laser equipment (IEC60825-1), and the law (21 CFR 1040) from the U.S. Food and Drug Administration (FDA). Table 6.2 summarizes the main characteristics and requirements for the classification system as specified by the IEC 60825-1 standard.

Table 6.2 Laser classification.

Type	Description
Class 1	Low-power device emitting radiation at a wavelength in the band 302.5–4000 nm. Device intrinsically without danger from its technical design under all reasonably foreseeable usage conditions, including vision using optical instruments (binoculars, microscope, monocular).
Class 1M	Same as Class 1, but there is the possibility of danger when viewed with optical instruments (binoculars, telescope, etc.). Class 1M lasers produce large-diameter beams or beams that are divergent.
Class 2	Low-power device emitting visible radiation (400–700 nm). Eye protection is normally ensured by the defense reflexes, including the palpebral reflex (closing of the eyelid). The palpebral reflex provides effective protection under all reasonably foreseeable usage conditions, including vision using optical instruments (binoculars, microscope, monocular).
Class 2M	Low-power device emitting visible radiation (400–700 nm). Eye protection is normally ensured by the defense reflexes, including the palpebral reflex (closing of the eyelid). The palpebral reflex provides an effective protection under all reasonably foreseeable usage conditions, with the exception of vision using optical instruments (binoculars, microscope, monocular).
Class 3R	Average-power device emitting radiation in the 302.5–4000 nm band. Direct vision is potentially dangerous. Generally located on rooftops.
Class 3B	Average-power device emitting radiation in the 302.5–4000 nm band. Direct vision of the beam is always dangerous. Medical checks and specific training required before installation or maintenance is carried out. Generally located on rooftops.
Class 4	There is always danger to the eye and skin, fire risk exists. Must be equipped with a key switch and a safety interlock. Medical checks and specific training required before installation or maintenance is carried out.

Source: Ghassemlooy et al. [2018]/With permission of Taylor & Francis.

6.3 OWC Applications

The OWC system cannot wholly replace the role of radio communications due to some constraints such as low mobility and point-to-point features. However, it is promising as a complementary technology to radio communications, deploying in particular scenarios where other means of communications cannot perform well. Therefore, it could be adopted in a multitude of applications, including

- *Last-Mile Broadband Access*: FSO is applied to solve the problem of the last-mile bottleneck between users and the fiber optic backbone where installing an optical fiber network is difficult or too costly. It is attractive in particular deployment scenarios, such as inter-building connectivity in dense-populated areas, connectivity across a river, rail tracks and streets, inter-island communications, and campus networking. Point-to-point FSO equipment ranging from 50 m to a few kilometers are readily available in the market with data rates on the order of magnitude Gbps.

- *Intelligent Transportation System*: VLC can be applied in vehicle-to-vehicle and vehicle-to-infrastructure communications. For example, most current vehicles have equipped with LED lamps, which work as transmitters to send traffic, safety, and other information between two vehicles. In addition, traffic lights, street lamps, and some infrastructure such as gas stations can become communication nodes to carry out data communications and exchange instantaneous traffic information.

- *On-Board Communications*: Another promising application for OWC is the passenger communication service in airplanes, trains, or ships. In this scenario, passengers can access the network via reading lamps using white LEDs installed in the overhead, acquiring data with on-board servers (e.g. real-time travel messages, gastronomy booking, and entertainment), or connecting to the Internet through the on-board gateway. It can avoid the rule of turning off personal electronic equipment due to the possible radio interference to planes.

- *Interference-Sensitive Scenarios*: OWC is immune to radio interference, which is attractive for particular scenarios that are sensitive to such interferences. In hospitals, notably the respiratory and anesthesia areas, some medical equipment is prone to interference with various radio waves. Therefore, the implementation of the OWC system can benefit their operation. A similar situation happens in airplanes where passengers' electronic equipment must turn off during take-off and landing. In Industry 4.0 factories, a large number of autonomous devices such as robotic arms and Automatic Guided Vehicles (AGVs) need wireless connectivity, imposing a complex radio environment with the high possibility of mutual interference, where OWC can also play an important role.

- *Extreme Environment*: In underground mines, using radio communication may cause explosion because of the transmission power. On the other hand, no wireless techniques except OWC can establish high-speed communication link underwater. In this context, OWC is a safe and high adaptability technology that provides illumination and data transmission simultaneously [Hou et al., 2015].

- *Long-Range Connectivity*: Using high-power, high-concentrative laser beams, FSO can effectively connect two communication parties in the distance ranging from several kilometers to several thousand kilometers. It is beneficial to some deployment scenarios such as space-to-ground links, inter-satellite links, inter-aerial platform links, and inter-connectivity among non-terrestrial networking.

- *Redundant Link and Disaster Recovery*: FSO communications can be deployed quickly to provide a backup transmission link in damage or outage of an optical fiber connection. Natural disasters, terrorist attacks, and temporary events require flexible deployment and timely responses. Temporary FSO links can be readily established within hours in emergencies where local infrastructure could be damaged or saturated. A painful example of the FSO deployment efficiency as a redundant link was witnessed after the 911 terrorist attacks in New York City. FSO links were rapidly deployed in this area for financial corporations, which were left out with no landlines.

- *Indoor Illumination-Communication*: Through the proper layout of LED lamps, an OWC system can provide well illumination and data communications to indoor scenarios such as schools, universities, libraries, shopping malls, train stations, airports, and offices [Elgala et al., 2011]. It is an alternative way to provide high-throughput networking in addition to current wireless local area networks operating in radio frequencies. In this context, power-line communication and OWC can be naturally combined, where the power cable serve as both power supply and back-hauling to LEDs.

- *Military Communications*: The advantages of applying optical wireless military communications [Juarez et al., 2006] over conventional radio communications are three-fold: (i) In contrast to the broadcast nature of radio waves, optical signals are highly directive and venerable to blockage, raising the level of security since eavesdroppers are hard to intercept these signals. (ii) OWC can improve the stealthiness of military equipment since there is no radio signature for being detected by passive radars. Also, the high directivity of optical beams substantially lowers the possibility of exposure to any monitoring systems. (iii) Military equipment using OWC is immune to radio interference and jamming, while the interference technology over the optical spectrum is still challenging.

6.4 Evolution of Optical Wireless Communications

Broadly speaking, exchanging information-bear messages through smoke, beacon fires, torches, semaphore, and sunlight from ancient times can be recognized as the historical forms of OWCs. The earliest use of light for communication purposes is attributed to ancient Greeks and Romans who used their polished shields to flash sunlight for delivering simple messages in battles around 800 BC. In ancient China, soldiers along the Great Wall sent smoke signals on its beacon towers to warn of enemy invasion. By building the beacon towers at regular intervals, long-range communications can be realized along the Wall over several thousands of kilometers. In the late nineteenth century, heliographs were commonly utilized for military communication, using a pair of mirrors to direct a controlled beam of sunlight during the day and some other forms of bright light at night. Another historical milestone in the field of OWC is the photophone invented by Alexander Graham Bell and his assistant Charles Sumner Tainter in 1880. The photophone transmitted voice signals using optical signals at a distance of around 200 m. Voice-caused vibrations on a mirror at the transmitter were reflected by sunlight to the receiver, which converted such vibrations back to voice [Uysal and Nouri, 2014].

In the modern sense, LEDs and LDs are used as optical sources in OWC systems, divided into *wireless IR communications* [Kahn and Barry, 1997], *VLCs* [Pathak et al., 2015], and *wireless UV communications* [Xu and Sadler, 2008] in terms of the wavelengths of optical signals. In addition, outdoor OWC for long-range transmission using high-power laser is dedicated to a technique called *FSO communications* [Khalighi and Uysal, 2014].

6.4.1 Wireless Infrared Communications

Wireless transmission in the IR band was first proposed as a method for short-range communication several decades ago. Since IR transceivers are usually lightweight, low cost, low power-consumption, and easy to manufacture, they have been widely applied for remote control of televisions, air conditioners, DVD players, and electronic toys. In the 1970s, an indoor wireless broadcast system using IR at 950 nm wavelength to interconnect a cluster of data terminals with the range up to 50 m was demonstrated [Gfeller and Bapst, 1979].

Initially, commercial products from different manufacturers could not be interoperable, which necessitated the establishment of a universal IR communications standard. Consequently, the Infrared Data Association (IrDA), an industry-sponsored standardization organization, was established in 1993 to specify the IR physical interface specifications and communications protocols for short-range data communications in applications such as personal area networks. IrDA Commercial IrDA products provide data rates ranging from 9.6 kbps to 16 Mbps in short line-of-sight links varying from less than a meter to several meters. In the MAC layer, different protocols such as Infrared Mobile Communications (IrMC), Infrared Communications (IrCOMM), and Object Exchange (OBEX) can be applied, as well as Ultra-Fast Infrared (UFIR) protocol that supports data rates up to 100 Mbps. Since the optical receiver of an IR device could be blinded by its transmitted light, the half-duplex operation mode is applied. Diffuse IR physical layer and medium access control layer technologies for wireless local area networks were specified under IEEE 802.11 working group as early as 1999. Unlike IrDA, an IR WLAN system is non-line-of-sight by reflecting signals on walls and ceilings. It can operate in adjacent rooms independently without any mutual interference, with a very low possibility of eavesdropping. Last but not least, IR laser communication systems can also be used for long-range transmission. Typical deployment scenarios include inter-building links for metropolitan or campus-area networks. Such systems are strictly point-to-point and are sensitive to fog and other weather conditions.

6.4.2 Visible Light Communications

In contrast to IR and UV communications, VLC can provide illumination and wireless connectivity simultaneously. White LEDs have a high luminous efficacy (the ratio of luminous flux to consumed electricity power) of more than 100 lumens/watt compared with 15 lumens/watt and 60 lumens/watt for incandescent bulbs and fluorescent lamps, respectively. Furthermore, white LEDs have a long lifetime of around 50 000 hours, whereas incandescent bulbs and fluorescent lamps can usually use 1200 and 10 000 hours, respectively. As a consequence, the popularization of LED lighting has been boosted around the world.

The idea of simultaneous illumination and communications by using high-speed switching LEDs to modulate visible light was firstly proposed in 1999 [Pang et al., 1999]. In 2001, Twibright Labs' Reasonable Optical Near Joint Access (RONJA) project implemented a VLC system to build a 10 Mbps full-duplex point-to-point wireless connection with a range of 1.4 km. Using white colored LEDs for a home wireless communication link was proposed by a research group at Keio University in Japan in 2000 [Tanaka et al., 2000]. In order to promote and standardize safe, efficient communications using visible light, the Visible Light Communication Consortium (VLCC) was founded in 2003 by Nakagawa Laboratories in partnership with CASIO, NEC, and Toshiba in Japan. In 2006, researchers from the Pennsylvania State University proposed to use the power-line communication technology as the backhauling of VLC to provide broadband access for indoor applications. In 2010, a joint research team from Siemens and Heinrich-Hertz Institute, Fraunhofer demonstrated a transmission rate at 500 Mbps with a white LED over a distance of 5 m, and 100 Mbps over a longer distance using five LEDs in Berlin. In 2011, IEEE 802.15 working group approved the first short-range OWCs standard using visible light, known as IEEE 802.15.7. It aims to deliver data rates sufficient to support audio and video multimedia services and also considers mobility of the visible link. IEEE 802.15.7 specifications cover the physical layer design using light wavelengths from 380 to 780 nm, and a MAC sublayer that accommodates the unique needs of visible links. Based on IEEE VLC standard, the term Light Fidelity (Li-Fi) was coined by Harald Haas, providing a bi-directional VLC system by utilizing IR waves or radio frequencies for the uplink. Later, IEEE 802.15.7 task group continued the development of the IEEE 802.15.7 r1 standard, aiming to provide three main functions including Optical Camera Communication (OCC), Li-Fi, and Light-Emitting Diode-Identification (LED-ID), as a convergence of illumination, data communications, and positioning. A transmission rate of more than 3 Gbps from a single LED have been demonstrated in 2014 [Tsonev et al., 2014], and a rate of 56 Gbps was achieved using a vertical-cavity surface-emitting laser [Lu et al., 2017].

6.4.3 Wireless Ultraviolet Communications

Within the entire UV spectrum (10–400 nm), the UV-C band (100–290 nm) exhibits high potential as a medium for wireless communications due to its capability of non-line-of-sight transmission and the effect of negligible ambient noise. Primarily, UV radiation in the UV-C band is solar blind because most of the solar radiation is absorbed by the ozone layer in the Earth's upper atmosphere. Thus, it results in nearly negligible in-band ambient noise and an ideal terrestrial transmission channel. Second, the UV light within this band has a high degree of relatively angle-dependent scattering due to the presence of suspended particles, fog, and haze in the atmosphere. The high signal-to-noise ratios within the transmission channel itself and the strong scattering enable the establishment of non-line-of-sight communication links with large field-of-view photodiodes, which capture large amounts of scattered light.

The idea of applying wireless UV communications for outdoor navel applications was conceived before World War II at the Naval Research Laboratory in the United States. In 1968, an experimental work for a long distance link over a 26 km range was implemented in MIT Lincoln Laboratory. The researchers used a xenon flash tube as an UV source to radiate waves of the continuous spectrum at high power, with the shortest wavelength at 280 nm and a photomultiplier tube at the receiver [Xu and Sadler, 2008]. Since UV transceivers were bulky, power-hungry, and expensive, the research on wireless UV communications did not progress for the next several decades. In the 2020s, the commercialization of semiconductor optical sources, offering the implementation of transceivers with low cost, small size, low-power consumption, and large bandwidth, gave the rebirth of wireless UV communications. In the United States, a variety of outdoor experiments for a short-range UV-C link (up to 100 m) took place by the researchers at MIT Lincoln laboratory. The researchers built an appropriate hardware platform, used either LED arrays or laser diodes as an optical source and a photomultiplier tube as a detector, obtained channel measurements, and developed suitable scattering models. In China, a research team from the Beijing Institute of Technology designed and implemented a UV-C platform using a mercury lamp and a photomultiplier tube for transmission experiments over a distance of more than 1 km [Vavoulas et al., 2019].

6.4.4   Free-Space Optical Communications

Outdoor OWCs, known as FSO communications, are usually operated using the near-Infrared spectrum as the communication medium due to low attenuation compared with visible light and UV light. FSO systems often employ laser diodes rather than light-emitting diodes to generate high-concentrated, high-power optical signals for data transmission. Narrow beams with small divergence angles are used to establish communication links between a transmitter and a receiver, leading to pointing and tracking problems. Thus, FSO systems are used for high-data-rate communication between two fixed points over distances ranging from a few meters to several thousand kilometers. Compared with RF-based communication, FSO links have very high available bandwidth, thus providing considerably higher data rates. Transmission rates of 10 Gbps have already been implemented for long-distance communication, and a 40-Gbps FSO link was implemented.

6.5   Optical Transceiver

In an OWC system, as illustrated in Figure 6.2, the transmitter modulates information bits into optical waveforms. Then, the generated optical signals are radiated through the atmosphere towards a remote receiver. A light-emitting diode

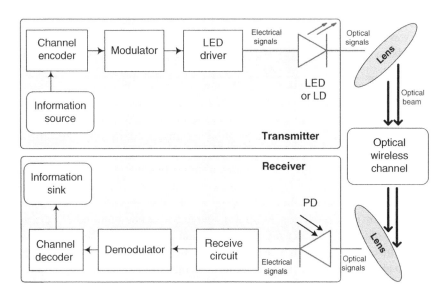

Figure 6.2   Block diagram of an end-to-end optical wireless communications system.

or laser diode is applied to convert an electrical signal into an optical signal at the transmitter, and the receiver uses a photodiode to convert the optical signal into an electrical current. Finally, the receiver processes this electrical current and then demodulates and decodes the baseband signal to recover the original information bits.

The transmitter comprises an optical source and its driving circuit, a channel encoder, a modulator, and a lens to focus or form an optical beam. Information bits from the information source are first encoded, then modulated to an electrical signal. The modulated signal is then passed through the optical sources to adjust the optical intensity. The optical beam is concentrated utilizing an optical lens or beam-forming optics before being transmitted. LEDs with beam collimators are usually applied for short-range wireless communications, while the typical optical source in FSO systems is a high-power laser diode for long-range transmission. An optical component should be small in footprint and have low power consumption. Meanwhile, it should deliver a relatively high optical power over a wide temperature range with a long Mean Time Between Failures (MTBF). Most commercial FSO systems operate in two particular wavelength windows at 850 and 1550 nm, where the atmospheric attenuation is less than 0.2 dB/km, and off-the-shelf components are available since these two windows coincide with the standard transmission windows of fiber optic communication systems. Vertical-Cavity Surface-Emitting Laser (VCSEL) are primarily utilized for transmission around 850 nm, and Fabry–Perot (FP) and Distributed Feedback (DFB) lasers are primarily used for operation at 1550 nm [Khalighi and Uysal, 2014].

The information can be conveyed by simply modulating the intensity of optical pulse through widely-used schemes such as on-off keying or pulse-position modulation, as well as coherent modulation or advanced multi-carrier schemes [Ohtsuki, 2003] such as Orthogonal Frequency-Division Multiplexing (OFDM) to get a higher transmission rate. To support multiple users in a single optical access point, OWC can apply not only typical electrical multiplexing technologies such as time-division, frequency-division, and code-division multi-access, but also optical multiplexing such as wavelength-division multi-access. Optical Multi-Input Multi-Output (MIMO) technology [Zeng et al., 2009] is also implemented in OWC, where multiple LEDs and multiple PDs are applied, in contrast to the multiple antennas in a typical MIMO system operating in the radio band.

OWC systems can be classified into two categories based on the detection method, i.e. non-coherent and coherent. Non-coherent systems generally employ Intensity Modulation with Direct Detection (IM/DD), where information bits are represented by the intensity of the emitted light at the transmitter. As shown in Figure 6.3, the optical intensity is adjusted in terms of the forward current through the LED. Ideally, a drive current I_d is comprised of a constant direct

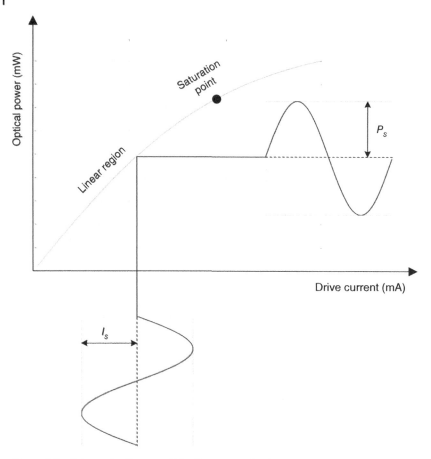

Figure 6.3 Schematic diagram of the linear mapping between output optical power and input drive current in a typical LED.

current (DC) I_{DC} and current swing I_s, i.e.

$$I_d = I_{DC} + I_s,\tag{6.1}$$

generating an output optical power

$$P_o = P_{DC} + P_s,\tag{6.2}$$

when operating within the linear region of the LED [Karunatilaka et al., 2015]. At the receiver, a photodiode directly detects changes in the light intensity to generate a proportional current. IM/DD systems are commonly used in terrestrial OWC links due to their simplicity and low cost. In contrast, coherent systems also utilize frequency or phase modulation in addition to amplitude. With the aid of a local oscillator, the received signal is optically mixed before photo-detection at the receiver. Although coherent systems offer superior performance in background noise rejection, mitigating turbulence-induced fading, and higher receiver sensitivity, it is too complex and power-consuming from the practical perspective. As illustrated in Figure 6.2, the front-end of an IM/DD OWC system has an optical lens to collect and focus the received beam on the photodiode. The photodiode converts the optical signal into an electrical current employing a trans-impedance circuit, usually a low-noise optical amplifier with a load resistor. The output of the trans-impedance circuit is low-pass filtered in order to limit the thermal noise. Finally, the receiver demodulates and decodes the received signal to recover the original information bits.

Solid-state photodiodes are usually utilized in commercial OWC systems since they have excellent quantum efficiency for the commonly used wavelengths. The junction material can be of Silicon (Si), Indium Gallium Arsenide (InGaAs), or Germanium (Ge), which are primarily sensitive to the commonly used wavelengths and have an extremely short transit time, resulting in high bandwidth and fast-response detectors. Si photodiodes have a high sensitivity around 850 nm, whereas InGaAs photodiodes are suitable for longer wavelengths around 1550 nm. Ge photodiodes are rarely used because of their relatively high level of dark current. Solid-state photodiodes can be a Positive-Intrinsic-Negative (PIN) diode or an Avalanche Photodiode (APD). PIN diodes are usually used for outdoor OWC (a.k.a FSO) systems working at distances up to a few kilometers. The main drawback of PIN diodes is that the receiver performance becomes very limited by the thermal noise. For long-range transmission, APDs are preferred due to the process of impact ionization. However, this advantage comes at the price of increased implementation complexity. In particular, a relatively high voltage is needed for APD reverse biasing that necessitates the application of special electronic circuits, leading to higher power consumption [Khalighi and Uysal, 2014]. In addition to photodiodes, the OWC system applying image sensors to detect the optical pulse also called optical camera system. The imagine sensor can convert the optical signal into the electrical signal, which has the advantage of easier implementation due to the wide spread of camera-embedded smart phones nowadays.

6.6 Optical Sources and Detectors

An atom is the smallest matter unit retaining all of the chemical properties of an element. According to classical physics, an atom consists of a nucleus in the center of the atom and one or more bound electrons surrounding the nucleus. Electrons form notional shells, where a shell is as an orbit followed by electrons around the nucleus. The further an electron orbiting an atomic nucleus, the larger

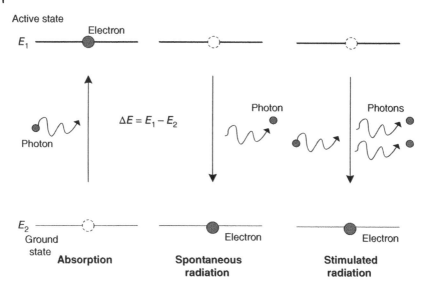

Figure 6.4 The principle of spontaneous and stimulated radiation.

of its energy. The closest shell is commonly referred to as a ground state. When an electron absorbs energy from light (photons), magnetic fields, or heat, it is excited from a lower energy level to an upper level by moving away from the nucleus, as illustrated in Figure 6.4. An excited state eventually decay to a lower state by emitting a photon since upper levels are generally less stable than lower levels. That is called spontaneous radiation, where the phase and direction associated with the photon are random. Suppose the energy of the upper and lower levels are E_1 and E_2, respectively, the emitted photon has the energy

$$\Delta E = E_1 - E_2 = hf_0. \tag{6.3}$$

where f_0 denotes frequency and the Planck constant

$$h = 6.62607015 \times 10^{-34} \text{ J/Hz}. \tag{6.4}$$

Alternatively, stimulated radiation takes place when the excited electron interact with an incoming photon, generating a new photon with copied direction and phase that are all identical to the incident photon. As a consequence, stimulated radiation creates highly directional and coherent light emission.

Henry Joseph Round was puzzled in 1907 when he observed electroluminescence from a solid-state diode, leading to the discovery of the light-emitting diode. This optical-electronic phenomenon is called recombination radiation, where light emission arises from the junction between the regions of p- and n-type semiconductors when electrons in the conduction band of a semiconductor drop into holes in the valence band. Recombination radiation is the basis of light emission from both LEDs and LDs.

In OWCs, the most commonly used light sources are the incoherent light-emitting diodes and the coherent laser diodes, which are used as optical transmitters. LEDs are used for short-range, low- to medium-data-rate indoor applications, whereas LDs that are monochromatic, coherent, high-power, and directional are mainly employed for high-speed outdoor applications. As for optical receivers in OWC, both PIN photodiodes and APDs can be used, though the latter is costly but offers higher sensitivity and bandwidth than the PIN PDs. This section discusses the types of light sources, their structures, and their optical characteristics.

6.6.1 Light-Emitting Diode

Light-emitting diodes are made of semiconductor materials that can modify the conductivity by introducing different impurities into the crystal lattice. For example, an intrinsic semiconductor (silicon or germanium) doped with Phosphorus (P), Arsenic (As), or Antimony (Sb) as an impurity is called an *n-type* semiconductor because it provides electrons to create negative current carriers. Reversely, adding trivalent impurities such as Boron (B), Gallium (G), Indium (In), and Aluminum (Al) into an intrinsic semiconductor generates a *p-type* semiconductor, containing an excess of holes as positive current carriers. When fusing these two semiconductor materials, their junction behaves very differently than either type of material alone. In such a PN junction, current flows easily from the p-side to the n-side, whereas the flow becomes very hard, if not inflexible, in the reverse direction. The p-side of a semiconductor is also called the *anode*, while the n-side is also referred to as the *cathode*.

Without a voltage difference (i.e. no bias) across the junction, charge carriers are distributed roughly the same as impurities, resulting in no net current flows. An equilibrium condition is reached near the PN junction, where the electrons drop from the conduction band to fill the holes in the valence band. Moreover, no current flows under reverse biasing by adding a positive voltage to the n side and a negative voltage to the p side. The negative electrode attracts holes from the p-type material, while the positive electrode attracts electrons from the n-type material [Figure 6.5]. It draws carriers away from the junction, and the current flow called the leakage current is only slightly above zero due to the semiconductor's high resistance. Increasing the reverse-biasing voltage across the junction causes no current increase until a large reverse current flow called the avalanche breakdown occurs [Bergh and Dean, 1972].

An LED radiates light waves through spontaneous emission when a forward bias voltage is applied across the PN junction, namely a positive voltage to the p-side and a negative voltage to the n-side. It attracts the holes to the n-side of the PN junction and vice versa, making them recombine and release their extra energy as the electrons in the conduction band fall into the holes in the valence band. The conversion process from electronic power to optical power is pretty

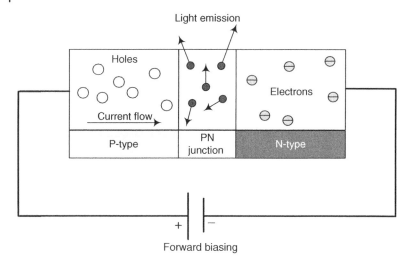

Figure 6.5 Schematic diagram of a light-emitting diode, where a semiconductor PN junction radiates light (infrared, visible light, or ultraviolet) with a forward-biasing voltage.

efficient, resulting in very little heat compared with conventional incandescent bulbs. Radiated photons can be IR, visible light, or UV lightwave, whose nominal wavelength depends on the bandgap energy of the semiconductor material. The mixture of materials determines the bandgap of a compound semiconductor. Table 6.3 provides some examples of different inorganic semiconductor materials used to generate a wide variety of light with various wavelengths. The emission characteristics of LEDs can be modeled into an emission spectrum with

Table 6.3 Some typical semiconductor materials and their emission ranges.

Semiconductor materials	LED emission	Wavelengths (nm)
Aluminum gallium arsenide (AlGaAs)	Red and infrared	230–350
Aluminum nitrate (AlN)	Ultraviolet	210
Indium gallium nitrate (InGaN)	Green, blue, near-UV	360–525
Zinc selenide (ZnSe)	Blue	459
Silicon carbide (SiC)	Blue	470
Gallium phosphide (GaP)	Red, yellow, green	550–590
Gallium arsenide (GaAs)	Infrared	910–1020
Gallium arsenide phosphide (GaAsP)	Red and orange	700

Source: Held [2008]/With permission of Taylor & Francis.

a peak wavelength, at which the maximum emission occurs, and a linewidth corresponding to the wavelength spread, which is determined by Schubert [2006]

$$\Delta\lambda = \frac{1.8\lambda^2 kT}{hc},\qquad(6.5)$$

where T is the temperature, h denotes the Plank constant, and c is the speed of light. For example, the theoretical peak wavelength and linewidth of a GaAs LED emission at room temperature are 625 and 28 nm, respectively. Figure 6.6 illustrates three typical emission spectra of LED materials (red, green, and blue).

Because LEDs emit light generated by spontaneous emission, radiated photons from the PN junction spreads in arbitrary directions. Hence, LEDs emit light across a wider range of angles than laser that is raised from stimulated emission, making it more desirable than lasers for some applications, such as illumination and display. To get the most efficient output, an optical lens can be mounted in the front of an LED to focus the emitted light, usually perpendicular to the surface. But its optical power is still substantially lower than that of laser. Consequently, LEDs are usually applied for low-data-rate, short-range OWCs where their output power is sufficient but the advantages of simple drive circuitry, low cost, wide temperature operation range, and much higher reliability can be exploited

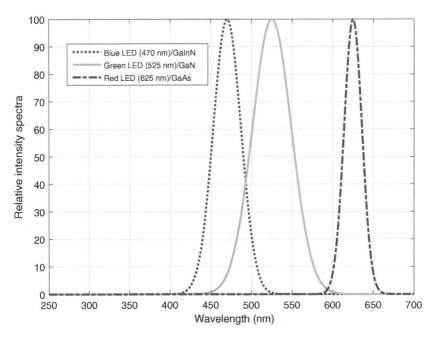

Figure 6.6 Typical radiation spectra of different LEDs, including blue (470 nm), green (525 nm), and red (625 nm).

[Bergh and Copeland, 1980]. In contrast, laser-based OWCs are more adaptive for high-data-rate, long-range applications with higher complexity, higher cost, and larger power consumption.

6.6.2 Laser Diode

Semiconductor materials such as gallium arsenide and indium phosphide can also be used to fabricate *laser diodes*, which are interchangeably called *diode lasers* or *semiconductor lasers*. Like other types such as gas lasers, solid-state lasers, and fiber lasers, semiconductor diode lasers form optical beams when spontaneous emission triggers a cascade of stimulated emission from a population inversion inside a resonant optical cavity when an electric current passes through a semi-conductor material. The laser exhibits drastically increased brightness and optical power, and significantly narrower spectra, than the light radiated from its LED counterpart.

The structure of a simple laser diode is similar to that of a light-emitting diode, consisting of an n-type semiconductor and a p-type semiconductor. Like an LED, an LD generates light from the recombination of electron-hole excitons with a forward-biased voltage across the PN junction. When the driving current is low, electron-hole excitons release their energy due to spontaneous emission, the same as an LED. However, LDs have reflective surfaces for optical feedback. The feedback has little impact at a low current below the point (known as the laser threshold) needed to produce a population inversion. As the driving current increases, more electron-hole pairs are generated spontaneously, increasing the possibility that a spontaneously emitted photon stimulates emission from an exciton that has yet to release its extra energy. Once the current reaches a high enough level beyond the threshold, it originates a population inversion between the exciton state and the atoms with the extra electron bound in the valence band, leading to stimulated emission.

As shown in Figure 6.7, a laser diode operates as a relatively inefficient LED if the driving current is below the threshold. Above the threshold, at which the output shifts from low-power spontaneous emission to higher-power stimulated emission, the laser diode converts a much higher portion of the input electrical power into light energy, as shown by the steeper slope. Therefore, the threshold current is a crucial factor to determine the performance of semiconductor laser. Electrical power needed to reach the threshold, namely the fraction of below-threshold current that is not converted into light energy, becomes thermal heating that must be dissipated in laser diodes. This heat not only wastes power but also degrades laser performance with shortened lifetime.

Figure 6.8 shows the basic structure of an edge-emitting laser diode, consisting of at least three layers: a p-type layer, an n-type layer, and a junction layer. The

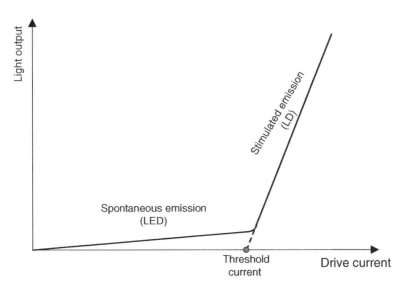

Figure 6.7 Light radiation of a diode rises sharply above the laser threshold.

reflective surfaces are perpendicular to the thin junction layer and the reflective cavity is aligned along the junction plane. Since semiconductors have a high refractive index, an uncoated solid air interface reflects much of the stimulated emission back into the semiconductor, providing feedback for the laser resonator. The large population inversion at high current makes gain high in semiconductor lasers, so a reflective cavity with only a few hundred micrometers long can make sustainable oscillation.

The emitting area on an edge-emitting laser diode is thin, leading to rapidly diverging beams. Laser diodes can also be fabricated to emit from their surface, referred to as surface-emitting laser diodes. A larger, circular emitting area can handle more power and produce high-quality beams with much lower divergence. Figure 6.8 illustrates the structure of a surface-emitting laser with a resonant cavity perpendicular to the active layer, known as VCSEL. Instead of oscillating along the long side of a thin active layer, VCSELs oscillate perpendicular to the surface of a thin disk of the active layer. To this end, two multiple-layer mirrors are integrated on the top and bottom of the junction layer, respectively, with the beam emerging through the surface. High-reflectivity mirrors required by a laser cavity can be directly fabricated in the semiconductor by depositing multiple alternating thin layers with slightly different reflectivity. This multi-layer structure forms a multi-layer interference coating designed to reflect a particular wavelength strongly. As a result, the reflector on the substrate side transmits a small fraction

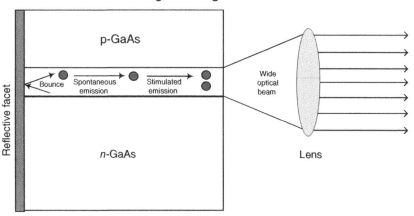

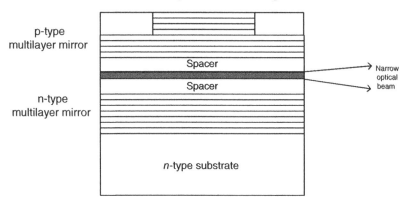

Figure 6.8 The structures of an edge-emitting laser diode (the upper) and a vertical-cavity surface-emitting laser diode.

of the cavity light, while the reflector above the active layer reflects all the light back to the cavity.

These structural differences make VCSELs behave much differently compared with edge-emitting diodes. Overall gain within a VCSEL cavity is low because light oscillating between the top and bottom mirrors passes through only a thin slice of the active layer. Although the gain per unit length is high, the active layer itself is so thin that the total gain in a round-trip of the VCSEL cavity is small. This structure is limited to generating powers the order of magnitude in milliwatt, well below the maximum available from edge emitters. Consequently, VCSELs have low threshold currents, making them significantly more efficient. Their high efficiency and low drive current also give them a long lifetime [Hecht, 2018].

Unlike edge-emitting diodes, surface emission comes from a region that usually is circular, generating a circular beam. The advantage is that the output beam can be coupled directly into optical fibers by putting the output face directly against the fiber core. Meanwhile, the emitting aperture in an edge-emitting diode is narrow. A beam emitted from a small aperture has a relatively large divergence angle. Its typical beam divergence are 10° in the direction parallel to the active layer and 40° perpendicular to the active layer. This spreading angle is larger than the beam from a high-quality flashlight, making the beam quite different from the tightly focused beam of a helium neon laser or a solid-state laser. Fortunately, external optics can be employed to correct this broad beam divergence. A cylindrical lens, which focuses light in one direction but not in the perpendicular direction, can make the beam circular in shape. Collimating lenses can focus the rapidly diverging beam from an edge emitter so it looks as narrow as the beam from a helium neon laser. In contrast, VCSELs are much better because they emit light from a wider aperture with a typical divergence of 10°.

Similar to light emitting diodes, the wavelengths emitted by laser diodes depend on the composition of semiconductor materials, as listed in Table 6.4. In addition, a comparison of the main characteristics of light-emitting diodes and laser diodes are provided in Table 6.5.

Table 6.4 Typical laser-diode materials and their emission wavelength ranges.

Semiconductor materials	Peak wavelength
Aluminum gallium nitrate (AlGaN)	350–400 nm
Gallium indium nitrate (GaInN)	375–440 nm
Zinc sulfoselenide (ZnSSe)	447–480 nm
Aluminum gallium indium phosphide (AlGaInP)/GaAs	620–680 nm
Gallium aluminum arsenide (GaAlAs)/GaAs	750–900 nm
GaAs/GaAs	904 nm
Indium gallium arsenide (InGaAs)/GaAs	915–1050 nm
Indium gallium arsenide phosphide (InGaAsP)	1100–1650 nm
Indium gallium arsenide phosphide antimony (InGaAsSb)	2–5 μm
Lead sulfoselenide (PbSSe)	4.2–8 μm
Lead tin selenide (PbSnSe)	8–30 μm
Quantum cascade	3–50 μm

Source: Hecht [2018]/With permission of John Wiley & Sons.

Table 6.5 A comparison of light-emitting diodes and laser diodes.

Characteristics	LED	LD
Optical output power	Low	High
Optical linewidth	25–100 nm	0.01–5 nm
Modulation bandwidth	kHz to hundreds of MHz	kHz to tens of GHz
Conversion efficiency	10–20%	30–70%
Divergence	Broad beam	Highly collimated
Reliability	High	Moderate
Coherence	No	Yes
Temperature dependence	Low	High
Drive circuitry	Simple	Complex
Cost	Low	Moderate to high

Source: Adapted from Ghassemlooy et al. [2018].

6.6.3 Photodiode

A photodiode is a semiconductor device with a PN junction that converts incident light photons into an electric current or voltage. Sometimes it is also called a photodetector, a light detector, or a photo-sensor. Figure 6.9 shows a rough

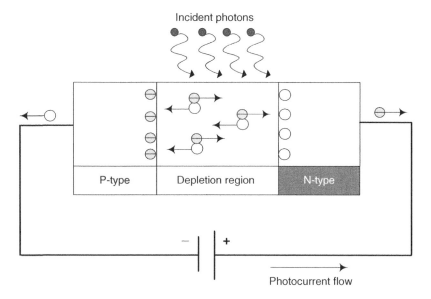

Figure 6.9 The basic construction of a photodiode, and the principle of generating a photoelectric current with incident light photons.

cross-section of a typical photodiode, which looks like a light-emitting diode, and its principle of conducting the photon-electron conversion. The p-side of the semiconductor creates excess holes, and the n-side has an abundance of electrons. A particular area between the n-side and p-side is formed when electrons diffuse from the n-side to the p-side while holes diffuse in the reverse direction. As a result, it produces a region with no free carriers and a built-in voltage, referred to as the depletion region, allowing electrical currents to flow only in one direction from the anode to the cathode.

Photodiodes usually operate under a reverse-bias voltage across the PN junction; namely, the p-side of the photodiode is connected to the negative terminal of a power supply, and the n-side is connected to the positive terminal. When photons strike the diode, valence electrons in the photodiode absorb light energy. If the energy is greater than the band-gap of the semiconductor material, the valence electrons break bonding with their parent atoms and become free electrons. Free electrons move freely from one place to another place by carrying the electric current. When a valence electron leaves the valence shell, a corresponding hole is created. The mechanism of forming an electron-hole exciton by using light energy is known as the inner photoelectric effect. If a photon is absorbed in either the p-side or n-side semiconductor materials, the electron-hole exciton will be recombined as heat if they are far enough away from the depletion region. Electron-hole pairs will move to opposite ends if the absorption arises in the depletion region due to the in-build electric field and the external electric field. In other words, electrons move toward the positive potential on the cathode, whereas the holes move toward the negative potential of the anode. These moving charge carriers generate a current flow known as photocurrent in the photodiode.

Although the basic operation of photodiodes is similar, different types of photodiodes are designed for specific applications [Lam and Hussain, 1992]. For example, PIN photodiodes are developed for those applications that require high response speed. In terms of their structures and functions, photodiodes can be classified into the following major types:

- *PN Photodiode*: The first developed and the most basic type is the PN photodiode. It was the most widely used photodiode before the development of PIN photodiodes. As it is relatively small, its sensitivity performance is not good compared with its successors (i.e. the PIN photodiode and APD). The depletion region contains few free charge carriers, and the region width can be manipulated by adding voltage bias. The dark current of a PN photodiode is low, resulting in low noise.

- *PIN Photodiode*: At present, the most commonly applied photodiode is a PIN type. The major difference compared with the PN type is that an intrinsic layer is fabricated between the P and N layers to create a larger depletion region. This diode can gather more light photons since the wider intrinsic area between the p-type and n-type semiconductors. This intrinsic layer is highly resistive, and therefore increases the electric field strength in the photodiode. Due to the increased depletion region, it offers a lower capacitance, and a higher response speed.

- *APD Photodiode*: With a high reverse bias (near the breakdown voltage), APDs applies impact ionization (avalanche effect) to create an internal current gain in the material, which in turn increases the effective responsivity (larger current generated per photon). It generally has a higher response speed and the ability to detect low light. The typical spectral response range is around 300–1000 nm. In an APD, noise due to dark current is higher than in a PIN photodiode, but the increased signal gain is much greater, making a high signal-to-noise ratio.

Photodiodes can be fabricated by a variety of semiconductor materials. Each material has particular features to generate different photodiodes with various sensitivity, wavelength ranges, noise levels, and response speeds. There are four major performance indicators applied to decide the right materials, i.e.

- *Response speed* of the photodiode is determined by the capacitance of the PN junction. It is measured by the time taken by charge carriers to cross the PN junction. The capacitance is directly determined by the width of the depletion region.
- *Responsivity* is the ratio of photocurrent to the power of incident light, which is generally expressed in units of A/W (current over power), as given by

$$R = \frac{\eta q \lambda}{ch} \approx \frac{\eta \lambda}{1.24}, \tag{6.6}$$

where η denotes quantum efficiency, which is defined as the carrier generation rate divided by the photon incident rate, q is the elementary charge in Coulomb, and h stands for the Planck constant. Since the quantum efficiency is less than 1, the maximum responsivity of PN and PIN photodiodes is around 0.6 A/W, imposing a severe limitation on the receiver sensitivity. That is why APDs have been considered, where responsivity can be improved on one or two orders of magnitude by operating with a very high electric field.

- *Dark current* is the current in the photodiode when there is no incident light. This can be one of the main sources of noise in a photodiode. Without biasing, the dark current can be very low.
- *Breakdown Voltage* is the largest reverse voltage that can be applied to the photodiode before there is an exponential increase in leakage current or dark current. Photodiodes should be operated below this maximum reverse bias. Breakdown voltage decreases with an increase in temperature.

To facilitate a reader to get a concrete view, Table 6.6 provides typical performance parameters of some photodiodes with different semiconductor materials.

Table 6.6 Typical performance parameters of photodiodes.

Parameter	Silicon		Germanium		InGaAs	
	PIN	APD	PIN	APD	PIN	APD
Wavelength range (nm)	400–1100		800–1800		900–1700	
Peak wavelength (nm)	900	830	1550	1300	1300/1550	1300/1550
Responsivity (A/W)	0.6	77–130	0.65–0.7	3–28	0.63–0.8	—
Quantum efficiency (%)	65–90	77	50–55	55–75	60–70	60–70
Gain	1	150–250	1	5–40	1	10–30
Bias voltage (V)	45–100	220	6–10	20–35	5	< 30
Dark current (nA)	1–10	0.1–1	50–500	10–500	1–20	1–5
Capacitance (pF)	1.2–3	1.3–2	2–5	2–5	0.5–2	0.5
Rise time (ns)	0.5–1	0.1–2	0.1–0.5	0.5–0.8	0.06–0.5	0.1–0.5

Source: Ghassemlooy et al. [2018]/With permission of Taylor & Francis.

6.7 Optical Link Configuration

Optical links have various configurations, which can be classified in terms of two criteria: the existence of a Line-Of-Sight (LOS) path and the degree of directivity, as illustrated in Figure 6.10. A LOS link between the transmitter and receiver provides the most significant path strength and the most minimal propagation delay. Without an available LOS link, signal transmission has to rely upon the reflection of ceilings, walls, or some other diffusely reflecting surface, forming a Non-Line-Of-Sight (NLOS) propagation environment. But an NLOS link can increase link robustness, allowing an effective connection even when barriers block the direct link between the transmitter and receiver. Unlike an antenna in the RF band with a broadcasting pattern, an optical transceiver has another angle-dependent feature, i.e. the beam angle of a light-emitting diode/laser diode and FOV for a detector. If both sides are directional, i.e. a narrow beam at the transmitter and a narrow FOV at the receiver, a directed link establishes. A directed link increases the signal power, minimizes the multi-path effect, and suppresses ambient light noise. However, accurate positioning and tracking must be employed to establish and maintain a link, imposing high complexity on the design of communication systems, especially in a mobile environment. The link is called nondirected when employing a wide-angle-beam transmitter and a wide-FOV receiver. A nondirected link alleviates the need for such pointing and is more convenient to use, particularly for mobile terminals. It is also possible to

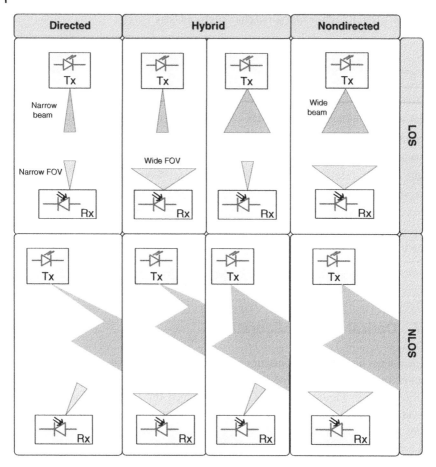

Figure 6.10 Classification of wireless optical communications links in terms of the existence of a LOS path and the degree of directivity. An optical transmitter can generate either a wide or narrow beam while an optical receiver is possible to have a wide or narrow field of view.

establish s hybrid link, which combines a transmitter and a receiver with different degrees of directivity [Kahn and Barry, 1997].

In a *directed LOS* link, the transmitter focuses the optical energy in a narrow beam, resulting in a high-power flux density at the receiver. Meanwhile, a receiver with narrow FOV alleviates multipath-induced signal distortion and ambient light noise. The transmission rate is restricted by free-space path loss rather than the effect of multipath dispersion. Consequently, the directed LOS link offers the highest data rate and the longest communication distance at the same condition. In addition, it can improve security since a highly directive optical beam is hard to interrupt and vulnerable to the blockage of an eavesdropper. Directed LOS has been applied for many years in low bit rate, simple remote control applications for domestic electrical equipment, such as televisions and audio equipment. Over the last decade, there was a rising interest in applying directed point-to-point communications for various outdoor applications such as campus networks, inter-building connectivity, last-miles access, cellular network backhauling and front-hauling, non-terrestrial connectivity, inter-satellite links, disaster recovery, fiber replacement in some terrains, and temporary links. However, a directed LOS link is problematic concerning coverage and roaming. It is hard to support the movement of a mobile user because of the requirement for adjusting the established alignment between the transmitter and receiver. It is also challenging to support multiple users simultaneously with a simple transceiver design.

A *non-directed LOS* link is considered to be the most flexible configuration for indoor communication applications. An optical communication system with a wide-beam transmitter and a wide-FOV receiver can achieve a wide coverage area and excellent support for mobility similar to a RF system. It overcomes blocking and shadowing by relying on reflections from surfaces of objects indoors such that a higher portion of the transmitted light energy is detected at the photodiode. In addition, the need for accurate beam alignment and tracking is alleviated. However, a non-directed LOS link suffers from high attenuation along the optical path, and they must also contend with multipath-induced dispersion. Although the multipath propagation does not result in multipath fading since detector sizes are large compared with the wavelength, it does raise inter-symbol interference, which is the main factor limiting the data rate. Furthermore, a wide-FOV receiver is susceptible to intense ambient light noise in indoor environments with a high probability of strong background lighting, degrading the link performance [Ghassemlooy et al., 2018].

A non-directed NLOS optical link is also referred to as *a diffuse link*, where a wide-beam transmitter is connected with a wide-FOV receiver through an NLOS link. Due to high robustness and high flexibility, it is the most convenient configuration for indoor wireless networks since it does not require any alignment and maintenance of optical beams and is immune to blockage and shadowing. These benefits come at a price of high path loss, typically 50–70 dB for a horizontal separation of 5 m. This loss increases further if a temporary obstruction, such as humans and furniture, blocks the main signal path. In addition, a photodiode with a wide FOV collects signals that have undergone one or more reflections from ceiling, walls, and room objects. Reflections attenuate the signal severely, with typical reflection coefficients between 0.4 and 0.9. Furthermore, multipath propagation may raise severe dispersion-induced ISI, limiting the maximal transmission rate.

6.8 Optical MIMO

In conventional RF communications, MIMO has already become the essential technique for excising mobile systems and wireless local area networks. Without consuming precious time or frequency resources, it can realize high data throughput or high reliability by simply adding extra antennas at the transmitter or receiver side. The MIMO technology can be divided into spatial multiplexing for a high capacity, spatial diversity for increased reliability, and beamforming for a high power gain. OWCs can easily deploy multiple optical transmitters and receivers since the optical devices, i.e. LEDs, LDs, and PDs, are all off-the-shelf, widely available, low cost, low power consumption, and easy to install. Consequently, there is potential to use an array of optical devices to achieve parallel data transmission or multiple redundant paths using optical MIMO techniques. In terms of their functionalities, the optical MIMO techniques can be implemented in the forms of spatial multiplexing or spatial modulation, which are introduced in the following sections:

6.8.1 Spatial Multiplexing

A non-imaging optical MIMO system, consisting of N_T transmitters and N_R receivers, is illustrated in Figure 6.11. Each transmitter is equipped with an LED array with K LEDs, and each receiver has an individual non-imaging

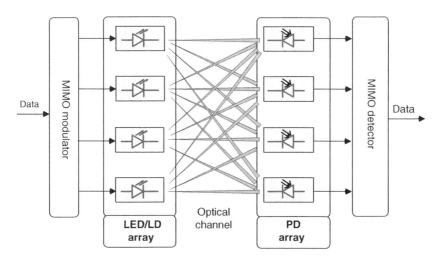

Figure 6.11 Schematic diagram of a non-imaging optical MIMO system where multiple data streams are transmitted and received by an LED/LD array at the transmitter and a PD array at the receiver, respectively.

concentrator. An MIMO modulator modulates a serial binary sequence in terms of optical modulation schemes such as On-Off Keying into a transmitted data stream and then converts it into N_T parallel data streams. At a typical time instant, the transmitted signal vector can be expressed by

$$\mathbf{s} = [s_1, s_2, \ldots, s_{N_T}]^T, \tag{6.7}$$

where $s_i, i = 1, 2, \ldots, N_T$ is the transmitted signal of the ith transmitter [Zeng et al., 2009].

Light from all LED arrays is received by separate receivers through LOS and diffused paths. DC gain between transmitter i and receiver j can be modeled by summing the optical power emitted from all K LEDs at transmitter i, i.e.

$$h_{ij} = \begin{cases} \sum_{k=1}^{K} \dfrac{A_{rx}^j}{d_{ijk}^2} R_0(\phi_{ijk}) \cos(\psi_{ijk}), & 0 \leq \psi_{ijk} \leq \psi_c \\ 0, & \psi_{ijk} > \psi_c \end{cases}, \tag{6.8}$$

where A_{rx}^j denotes the collection area of receiver j, d_{ijk} is the distance between the kth LED at the ith transmitter and the jth receiver, ϕ_{ijk} and ψ_{ijk} represent the angles of emission and incidence, respectively, ψ_c stands for the photodiode FOV, and $R_0(\cdot)$ expresses the *Lambertian* radiant intensity of LEDs, which is given by

$$R(\phi) = \frac{m+1}{2\pi} \cos^m(\phi), \tag{6.9}$$

where m denotes the order of Lambertian emission, and ϕ is the angle of emission relative to the optical axis of the emitter. We write P_{LED} to denote the average power emitted by an LED, and the instantaneous transmitted power equals to $P_t = P_{\text{LED}}R(\phi)$, ranging from 0 to $2P_{\text{LED}}$. Thus, a matrix modeling this optical MIMO channel is formed as

$$\mathbf{H} = \begin{pmatrix} h_{11} & h_{12} & \cdots & h_{1N_T} \\ h_{21} & h_{22} & \cdots & h_{2N_T} \\ \vdots & \vdots & \ddots & \vdots \\ h_{N_R 1} & h_{N_R 2} & \cdots & h_{N_R N_T} \end{pmatrix}. \tag{6.10}$$

A receiver is comprised of an optical concentrator, followed by a detector and a preamplifier. Assume that the concentrator gain is the theoretical maximum, and the collection area in Eq. (6.8) is determined by

$$A_{rx}^j = \frac{n^2}{\sin^2(\psi_c)} A_{\text{PD}} \tag{6.11}$$

with the photodiode area A_{PD}, and the concentrator refractive index n. Then, the received signal at the jth receiver can be written as

$$r_j = \gamma P_{\mathrm{LED}} \sum_{i=1}^{N_T} h_{ij} s_i + n_j, \tag{6.12}$$

where γ denotes the responsivity of a photodiode, and n_j is the noise current, which can be calculated by

$$n_j = 2e\gamma(P_j + P_a)B + i_a^2 B, \tag{6.13}$$

where P_j is the average received power

$$P_j = P_{\mathrm{LED}} \sum_{i=1}^{N_T} h_{ij} s_i, \tag{6.14}$$

P_a stands for the power of ambient light

$$P_a = 2\pi \chi_a A_{rx}^j (1 - \cos(\psi_c)), \tag{6.15}$$

B is the receiver bandwidth, and i_a is the noise current density of the preamplifier. Denoting the received vector as

$$\mathbf{r} = [r_1, r_2, \dots, r_{N_R}]^T, \tag{6.16}$$

and the noise vector

$$\mathbf{n} = [n_1, n_2, \dots, n_{N_R}]^T, \tag{6.17}$$

yields

$$\mathbf{r} = \gamma P_{\mathrm{LED}} \mathbf{H} \mathbf{s} + \mathbf{n}. \tag{6.18}$$

The transmitted symbols can be estimated in terms of

$$\hat{\mathbf{s}} = \frac{1}{\gamma P_{\mathrm{LED}}} (\mathbf{H}^H \mathbf{H})^{-1} \mathbf{H}^H \mathbf{r}, \tag{6.19}$$

when the knowledge of $\mathbf{H}$ is known at the receiver.

Unlike a radio system, an OWC system can apply an imaging detector at the receiver. Light from each LED array arrives the receiver and strikes any pixel on the detector array. The imaging optical MIMO system follows the same model of the non-imaging case, with a modification of the channel matrix $\mathbf{H}$, where each element is then determined by

$$h_{ij}^{\mathrm{IMG}} = a_{ij} h_i', \tag{6.20}$$

where h_i' is the normalized power in the image of the ith LED array at the aperture of imaging lens, and a_{ij} quantifies how much of this power falls on the jth receiver.

The first parameter is given by

$$
h'_i = \begin{cases} \sum_{k=1}^{K} \dfrac{A'_{rx}}{(d'_{ik})^2} R_0(\phi_{ik}) \cos(\psi_{ik}), & 0 \le \psi_{ik} \le \psi_c \\ 0, & \psi_{ik} > \psi_c \end{cases},
\tag{6.21}
$$

where A'_{rx} denotes the collection area of an imaging receiver, d'_{ik} means the distance between the kth LED in transmitter i and the center of the collective lens [Zeng et al., 2009]. The second parameter is determined by

$$
a_{ij} = \frac{A_{ij}}{\sum_{s=1}^{N_R} A_{is}},
\tag{6.22}
$$

where A_{is} is the image area of the ith transmitter on the sth pixel.

6.8.2 Spatial Modulation

Optical Spatial Modulation (OSM) [Mesleh et al., 2011] is a power-efficient, bandwidth-efficient single-carrier transmission technique for OWCs. The transmitter is equipped with an optical transmitter array consisting of multiple spatially separated LEDs or LDs. The layout of these emitters is similar to the diagram of constellation points in a digital modulation scheme. Each transmitter in the array is assigned an index, corresponding to a unique binary sequence referred to as the spatial symbol. The incoming data bits are grouped, and then each sequence is mapped into one of the spatial symbols. A single optical transmitter is activated to transmit an optical signal at a typical instance, while other transmitters keep silent when the active transmitter emits light. The active transmitter radiates a certain intensity level at a particular time instance. At the receiver side, the optimal OSM detector is used to estimate the index of the active transmitter. If there are N_T transmitters, the system can deliver $\log_2 N_T$ bits per symbol period.

Consider an OSM system consisting of N_T transmit units and N_R receive units. Without loss of generality, assume $N_T = 4$ for illustration purposes. The data bits to be transmitted in three consecutive symbol periods are $[1, 1, 1, 0, 0, 0]^T$, which are grouped into $[\{1, 1\}, \{1, 0\}, \{0, 0\}]^T$. Each symbol period selects an LED in terms of the data sequence. The selected LED transmits a non-return to zero pulse with an optical intensity $s_l = P_t$, which ignore the frequency and phase information.

The resultant transmitted matrix is given by

$$
\mathbf{s}(t) = \begin{pmatrix} 0 & 0 & s_l \\ 0 & 0 & 0 \\ s_l & 0 & 0 \\ 0 & s_l & 0 \end{pmatrix}.
\tag{6.23}
$$

Each column of this matrix corresponds to the transmitted signals at a typical time instance, and each row corresponds to a particular LED. For example, the third LED emits light with an optical intensity of P_t while other LEDs are turned off at the first symbol period. The fourth LED is activated in the next symbol period, and so on.

The received signal can be written as

$$\mathbf{y}(t) = \sqrt{\rho}\mathbf{H}(t) \otimes \mathbf{s}(t) + \mathbf{n}(t), \tag{6.24}$$

where $\otimes$ stands for the convolution operation, and the average electrical signal-to-noise ratio at each receiver unit is given by

$$\rho = \frac{r^2 \overline{P}_r^2}{\sigma^2}, \tag{6.25}$$

where r denotes the photodiode responsivity, the average received optical power at each receiver unit

$$\overline{P}_r = \frac{1}{N_r} \sum_{i=1}^{N_r} P_r^i \tag{6.26}$$

with

$$P_r^i = \sum_{k=0}^{K} \overline{h}_{il}^k P_t \tag{6.27}$$

being the average received optical power at receiver unit i when transmit unit l emits, and $\overline{h}_{il}^k$ is the channel path gain between transmit unit l and receive unit i for the kth path.

The noise in an optical communications system is the sum of thermal noise at the receiver and shot noise due to ambient light, i.e.

$$\sigma^2 = \sigma_{\text{shot}}^2 + \sigma_{\text{thermal}}^2, \tag{6.28}$$

which can be modeled as independent and identically distributed additive white Gaussian noise with double-sided power spectral density σ^2 [Kahn and Barry, 1997]. Meanwhile, the OSM channel is modeled into an $N_r \times N_t \times (K+1)$ matrix, where K denotes the number of propagation paths, defined as

$$\mathbf{H}(t) = \begin{pmatrix} \mathbf{h}_{11}(t) & \mathbf{h}_{12}(t) & \dots & \mathbf{h}_{1N_t}(t) \\ \mathbf{h}_{21}(t) & \mathbf{h}_{22}(t) & \dots & \mathbf{h}_{2N_t}(t) \\ \vdots & \vdots & \ddots & \vdots \\ \mathbf{h}_{N_r 1}(t) & \mathbf{h}_{N_r 2}(t) & \dots & \mathbf{h}_{N_r N_t}(t) \end{pmatrix}, \tag{6.29}$$

where $\mathbf{h}_{ij}(t) = [h_{ij}^0(t), h_{ij}^1(t), \dots, h_{ij}^K(t)]$ denotes a channel vector containing the channel response between transmit unit j and receive unit i.

The receiver applies modified maximal-likelihood detection to retrieve the index of the active transmitted LED,

$$\hat{l} = \arg \max_l p_\mathbf{y}(\mathbf{y}|\bar{\mathbf{s}}, \hat{\mathbf{H}}) \tag{6.30}$$

$$= \arg \min_l \sqrt{\rho}\| \mathbf{h}_l s_l \|^2 - 2(\mathbf{y}^T \mathbf{h}_l s_l), \tag{6.31}$$

where $\hat{\mathbf{H}}$ denotes the channel knowledge at the receiver, $\bar{\mathbf{s}}$ is the transmitted vector at the current symbol period, and

$$p_\mathbf{y}(\mathbf{y}|\bar{\mathbf{s}}, \hat{\mathbf{H}}) = \pi^{-N_t} \exp\left(-\left\|\mathbf{y} - \sqrt{\rho}\hat{\mathbf{H}}\bar{\mathbf{s}}\right\|^2\right) \tag{6.32}$$

is the probability density function of $\mathbf{y}$ conditioned on $\hat{\mathbf{H}}$ and $\bar{\mathbf{s}}$.

Furthermore, spatial modulation can simultaneously transmit data in both the spatial and signal domain. In other words, the selected transmit unit can emit light in multiple intensity levels denoted by $P_1, P_2, \ldots, P_M$, rather than only a single intensity, to enable another degree of freedom. The transmitted data bits are mapped to the respective signal and transmitter index, providing an enhanced spectral efficiency of $[\log_2(N_T) + \log_2(M)]$bps/Hz. As illustrated in Figure 6.12, the bit sequence $(1, 1, 1, 0)$ is represented by the transmitter index 4 and the intensity level P_3.

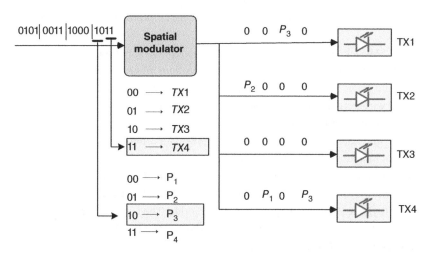

Figure 6.12 Schematic diagram of spatial modulation using $N_T = 4$ and $M = 4$ as an example. Source: Adapted from Fath and Haas [2013].

6.9 Summary

Using the terms of wireless IR communications, VLCs, wireless UV communications, or FSO communications, OWCs have been researched and developed for several decades. It is promising to be a complementary approach for current RF systems in the upcoming generation of mobile communication systems. OWC has some particular advantages, e.g. the provisioning of massively sufficient spectral resources over the IR, visible-light, and UV bands, and widely available off-the-shelf optical sources and detectors, which are cheap, lightweight, power-efficient, reliable, and long lifetime. Through the study of this chapter, the readers should have a complete view of OWCs, including its technical advantages and challenges, the evolution, various application scenarios, the basic setup of an OWC system, characteristics of optical sources and detectors, different optical link configuration, and optical MIMO techniques.

References

Al-Kinani, A., Wang, C.-X., Zhou, L. and Zhang, W. [2018], 'Optical wireless communication channel measurements and models', *IEEE Communication Surveys and Tutorials* **20**(3), 1939–1962. Third Quarter.

Bergh, A. and Copeland, J. [1980], 'Optical sources for fiber transmission systems', *Proceedings of the IEEE* **68**(10), 1240–1247.

Bergh, A. and Dean, P. [1972], 'Light-emitting diodes', *Proceedings of the IEEE* **60**(2), 156–223.

Elgala, H., Mesleh, R. and Haas, H. [2011], 'Indoor optical wireless communication: Potential and state-of-the-art', *IEEE Communications Magazine* **49**(9), 56–62.

Fath, T. and Haas, H. [2013], 'Performance comparison of MIMO techniques for optical wireless communications in indoor environments', *IEEE Transactions on Communications* **61**(2), 733–742.

Gfeller, F. and Bapst, U. [1979], 'Wireless in-house data communication via diffuse infrared radiation', *Proceedings of the IEEE* **67**(11), 1474–1486.

Ghassemlooy, Z., Popoola, W. and Rajbhandari, S. [2018], *Optical wireless communications: System and channel modelling with MATLAB*, 2nd edn, CRC Press, Taylor & Francis Group, Boca Raton, Florida, The Unites States.

Hecht, J. [2018], *Understanding Lasers: An Entry-Level Guide*, 4th edn, Wiley-IEEE Press, New York, The Unites States.

Held, G. [2008], *Introduction to Light Emitting Diode Technology and Applications*, 1st edn, CRC Press, Taylor & Francis Group, New York, The Unites States.

Hou, R., Chen, Y., Wu, J. and Zhang, H. [2015], A brief survey of optical wireless communication, *in* 'Proceedings of the 13th Australasian Symposium on Parallel and Distributed Computing (AUSPDC)', Sydney, Australia, pp. 41–50.

Juarez, J. C., Dwivedi, A., Hammons, A. R., Jones, S. D., Weerackody, V. and Nichols, R. A. [2006], 'Free-space optical communications for next-generation military networks', *IEEE Communications Magazine* **44**(11), 46–51.

Kahn, J. and Barry, J. [1997], 'Wireless infrared communications', *Proceedings of the IEEE* **85**(12), 265–298.

Karunatilaka, D., Zafar, F., Kalavally, V. and Parthiban, R. [2015], 'LED based indoor visible light communications: State of the art', *IEEE Communication Surveys and Tutorials* **17**(3), 1649–1678.Third Quarter.

Kedar, D. and Arnon, S. [2004], 'Urban optical wireless communication networks: The main challenges and possible solutions', *IEEE Communications Magazine* **42**(5), S2–S7.

Khalighi, M. A. and Uysal, M. [2014], 'Survey on free space optical communication: A communication theory perspective', *IEEE Communications Surveys & Tutorials* **16**(4), 2231–2258. Fourth Quarter.

Lam, A. and Hussain, A. [1992], 'Performance analysis of direct-detection optical CDMA communication systems with avalanche photodiodes', *IEEE Transactions on Communications* **40**(4), 810–820.

Lu, H.-H., Li, C.-Y., Chen, H.-W., Ho, C.-M., Cheng, M.-T., Yang, Z.-Y. and Lu, C.-K. [2017], 'A 56 Gb/s PAM4 VCSEL-based LiFi transmission with two-stage injection-locked technique', *IEEE Photonics Journal* **9**(1), 1–6.

Mesleh, R., Elgala, H. and Haas, H. [2011], 'Optical spatial modulation', *Journal of Optical Communications and Networking* **3**(3), 234–244.

Ohtsuki, T. [2003], 'Multi-subcarrier modulation in optical wireless communication', *IEEE Communications Magazine* **41**(3), 74–79.

Pang, G., Kwan, T., Chan, C.-H. and Liu, H. [1999], LED traffic light as a communications device, *in* 'Proceedings of the IEEE/IEEJ/JSAI International Conference on Intelligent Transportation Systems', Tokyo, Japan, pp. 788–793.

Pathak, P. H., Feng, X., Hu, P. and Mohapatra, P. [2015], 'Visible light communication, networking, and sensing: A survey, potential and challenges', *IEEE Communication Surveys and Tutorials* **17**(4), 2047–2077. Fourth Quarter.

Schubert, E. F. [2006], *Light-Emitting Diodes*, 2nd edn, Cambridge University Press, New York, The Unites States.

Tanaka, Y., Haruyama, S. and Nakagawa, M. [2000], Wireless optical transmissions with white colored LED for wireless home links, *in* 'Proceedings of the 11th IEEE International Symposium on Personal Indoor and Mobile Radio Communications (PIMRC)', London, UK, pp. 1325–1329.

Tsonev, D., Chun, H., Rajbhandari, S., McKendry, J. J. D., Videv, S., Gu, E., Haji, M., Watson, S., Kelly, A. E., Faulkner, G., Dawson, M. D., Haas, H. and O'Brien, D.

[2014], 'A 3-Gb/s single-LED OFDM-based wireless VLC link using a gallium nitride uLED', *IEEE Photonics Technology Letters* **26**(7), 637–640.

Uysal, M. and Nouri, H. [2014], Optical wireless communications – an emerging technology, *in* 'Proceedings of the 16th International Conference on Transparent Optical Networks (ICTON)', Graz, Austria.

Vavoulas, A., Sandalidis, H. G., Chatzidiamantis, N. D., Xu, Z. and Karagiannidis, G. K. [2019], 'A survey on ultraviolet C-band (UV-C) communications', *IEEE Communications Surveys & Tutorials* **21**(3), 2111–2133. Third Quarter.

Wang, C.-X., Huang, J., Wang, H., Gao, X., You, X. and Hao, Y. [2020], '6G wireless channel measurements and models: Trends and challenges', *IEEE Vehicular Technology Magazine* **15**(4), 22–32.

Xu, Z. and Sadler, B. M. [2008], 'Ultraviolet communications: Potential and state-of-the-art', *IEEE Communications Magazine* **46**(5), 67–73.

Zeng, L., O'Brien, D. C., Minh, H. L., Faulkner, G. E., Lee, K., Jung, D., Oh, Y. and Won, E. T. [2009], 'High data rate multiple input multiple output (MIMO) optical wireless communications using white LED lighting', *IEEE Journal on Selected Areas in Communications* **27**(9), 1654–1662.

Part III

Smart Radio Networks and Air Interface Technologies for 6G

7

Intelligent Reflecting Surface-Aided Communications for 6G

The traditional methods to realize high-capacity wireless systems, i.e. *deploying dense and heterogeneous network equipment, improving spectral efficiency,* and *acquiring more bandwidth,* incur high capital and operational expenditures, high energy consumption, and severe mutual interference. To meet the stringent performance requirements of the next-generation system, it is highly desirable to find a disruptive and revolutionary technology to achieve sustainable capacity and performance growth with affordable cost, low complexity, and low energy consumption. Recently, Intelligent Reflecting Surface (IRS), a.k.a., Reconfigurable Intelligent Surface (RIS) attract a lot of interests from both academia and industry. It is a promising new paradigm to achieve smart and reconfigurable wireless channel propagation environment. Generally speaking, IRS is a planar surface consisting of a large number of small, passive, and low-cost reflecting elements, each of which is able to independently induce a phase shift and amplitude attenuation to an incident electromagnetic wave. In contrast to current wireless transmission techniques that have to *passively* adapt to wireless propagation environment, IRS *proactively* modifies it through smartly controllable reflection, thereby collaboratively achieving fine-grained passive or reflect beamforming. By judiciously designing its reflection coefficients, the signals reflected by IRS can be added either constructively with those via other signal paths to increase the desired signal strength at the receiver, or destructively to mitigate the co-channel interference. Thus, it provides a new degree of freedom to the design of wireless systems with the aid of a smart and programmable wireless environment.

This chapter is comprised of the following major parts:

- The basic concept of the IRS, the technical advantages of IRS-aided wireless communications, and its potential applications.
- The fundamentals of IRS-aided single-antenna transmission, including the cascaded IRS channel, the signal model of passive beamforming, optimal reflecting coefficients, and a special feature called the product-distance path loss.

6G Key Technologies: A Comprehensive Guide, First Edition. Wei Jiang and Fa-Long Luo.
© 2023 The Institute of Electrical and Electronics Engineers, Inc. Published 2023 by John Wiley & Sons, Inc.

- The fundamentals of IRS-aided multi-antenna transmission, including joint active and passive beamforming, and joint precoding and reflecting optimization.
- The introduction of the dual-beam IRS technique, the forming of dual beams over hybrid analog-digital transceivers, and the optimization design.
- The basics of IRS-aided wideband transmission, including the cascaded frequency-selective fading channel, the system model of IRS-aided Orthogonal Frequency-Division Multiplexing (OFDM) transmission, and rate maximization.
- The impact of channel aging on the IRS, the modeling of outdated channel state information, and the analysis of performance loss. In addition, the principle of machine learning-based channel prediction, and the basics of recurrent neural networks, long-short term memory, and deep learning are introduced.

7.1 Basic Concept

Traditionally, stringent performance requirements of wireless communications, for example, ultra-high data rate, high energy efficiency, ubiquitous coverage, massive connectivity, ultra-high reliability, and low latency, as specified in 5G, have been achieved through three major approaches:

- *Dense and Heterogeneous Network Deployment*: Deploying increasingly more network equipment such as base stations, access points, remote radio heads, relays, and distributed antennas can increase the reuse of spectral resources in a given geographical area and shorten the propagation distance between a serving point and a user. Although this approach can substantially extend network coverage and boost system capacity, it incurs high capital and operational expenditures, high energy consumption, and severe mutual interference.
- *High Spectral Efficiency*: Integrating a massive number of antennas at the base station harnesses the enormous spatial multiplexing gain through massive Multi-Input Multi-Output (MIMO) technologies. This approach needs sophisticated signal processing techniques, imposing high hardware costs and energy consumption. Due to the fundamental limits of propagation environment, e.g. low-rank channels and high inter-antenna correlation, as well as practical constraints such as large array size, it is hard, if not infeasible, to further substantially improve spectral efficiency simply by scaling up in terms of the number of antennas.
- *More Bandwidth*: One of the technological trends in the wireless transmission is that the signal bandwidth becomes increasingly wider, from tens of kHz in the first-generation systems to hundreds of MHz in the fifth-generation

systems, aiming to support a higher transmission rate. Correspondingly, a large number of spectral resources are required, leading to spectrum shortage. In the next generation, migrating to higher frequency bands such as Millimeter Wave (mmWave), Terahertz (THz), and even light waves to exploit their abundant bandwidth becomes a necessity for achieving stringent performance, for instance, an extreme transmission speed of Terabits-per-Second (Tbps). Severe propagation loss and blockage susceptibility of high-frequency transmission inevitably call for denser network deployment and mounting more antennas (i.e. large-scale antenna array for high beamforming gain). This paradigm further exaggerates the problems of high capital and operational expenditures, high energy consumption, and severe mutual interference.

In addition to further evolving the aforementioned technologies toward the demand of the next-generation systems, it is highly desirable to find a disruptive and revolutionary technology to achieve sustainable capacity and performance growth with affordable cost, low complexity, and low energy consumption. On the other hand, the fundamental challenge to severely restrict the performance of wireless communications is attributed to elusive wireless channels due to their significant pass loss, shadowing, time variation, frequency selectivity, and multi-path propagation. Traditional approaches for tackling this fundamental limit either compensate for the channel loss and randomness by exploiting various robust modulation, coding, and diversity techniques, or adapt to it via adaptive controlling of transmission parameters. Nevertheless, these techniques not only need a large amount of overhead but also have limited adaptability over the largely random wireless channels, thus leaving a solid barrier to achieving high-reliable wireless communications.

In this regard, Intelligent Reflecting Surface (IRS) [Wu and Zhang, 2020], a.k.a., Reconfigurable Intelligent Surface (RIS) [Yuan et al., 2021], Large Intelligent Surface [Hu et al., 2018], Large Intelligent Metasurface (LIM), Programmable Metasurface [Tang et al., 2020], Reconfigurable Metasurface, Intelligent Walls, and Reconfigurable Reflect-array, has been proposed as a promising new paradigm to achieve smart and reconfigurable wireless channel propagation environment. Generally speaking, IRS is a planar surface consisting of a large number of small, passive, and low-cost reflecting elements, each of which is able to independently induce a phase shift and/or amplitude attenuation (collectively termed as reflection coefficient) to an incident electromagnetic wave. In contrast to current wireless transmission techniques that have to *passively* adapt to wireless propagation environment, IRS *proactively* modifies it through smartly controllable reflection, thereby collaboratively achieving fine-grained passive or reflect beamforming. By judiciously designing its reflection coefficients, the signals reflected by IRS can be added either constructively with those via other signal paths to increase the desired signal strength at the receiver, or destructively to mitigate the co-channel interference. Thus, it provides a new degree of freedom to the design of wireless systems with the aid of a smart and programmable wireless environment. Since its reflecting elements (e.g. low-cost printed dipoles) only passively reflect the impinging electromagnetic wave, Radio-Frequency (RF) chains for signal transmission and reception become unnecessary. Thus, it can be implemented with orders-of-magnitude lower hardware cost and power consumption than traditional active antenna arrays. Moreover, reflecting elements are generally low profile, lightweight, and conformal geometry. Therefore, IRS can be practically fabricated to conform to mount on arbitrarily shaped surfaces to cater to a wide variety of deployment scenarios and be integrated into existing wireless networks transparently as auxiliary equipment, thus providing great flexibility and compatibility. In a nutshell, IRS is recognized as a disruptive technology with the prominent features of *low complexity, low cost, low power consumption*, along with the potential of high performance.

IRS exhibits several particular advantages compared with other related technologies, i.e. wireless relays, backscatter communications, and active surface-based massive MIMO [Hu et al., 2018]. Wireless relays usually operate in a half-duplex mode and thus suffer from low spectrum efficiency compared with IRS operating in a full-duplex mode. Although a full-duplex relay is also achievable, it requires sophisticated self-interference cancelation techniques that are costly to implement. Furthermore, IRS is free of noise amplification since it only reflects the impinging electromagnetic waves as a passive array without any active transmit module (e.g. power amplifier). Unlike traditional backscatter such as Radio Frequency IDentification (RFID) tags that communicate with an RFID reader by modulating its reflected signal emitting from the reader, IRS is mainly applied to assist the existing communication link without its own data. In contrast, the reader in backscatter communication needs to implement self-interference cancelation at its receiver to decode the tag's message. Both the direct link and the reflected link in an IRS-aided transmission carry an identical signal and thus can be coherently superimposed at the receiver to boost the signal strength for better detection. Third, IRS also differs from the active surface-based massive MIMO due to their different array architectures (passive versus active) and operating mechanisms (reflecting versus transmitting) [Wu and Zhang, 2020].

Due to the advantages mentioned earlier, IRS is suitable for massively deployed in wireless networks to significantly enhance their spectral efficiency and energy efficiency in a cost-effective manner. It is envisioned that IRS will lead to a fundamental paradigm shift of designing a wireless system, namely, from scaling up the order of massive MIMO systems in terms of the number of antennas to IRS-aided moderate-scale MIMO, as well as from the existing heterogeneous wireless network to an IRS-aided hybrid network. In contrast to massive MIMO that leverages tens and even hundreds of active antennas to form sharp beams directly, an

IRS-aided MIMO system allows a base station to be equipped with substantially fewer antennas without compromising the users' quality-of-experience by exploiting the large aperture of IRS to create fine-grained reflect beams through smart passive reflection. To do so, the system hardware cost and energy consumption can be significantly lowered, especially for wireless systems migrating to higher frequency bands. On the other hand, although existing wireless networks rely on a heterogeneous multi-tier architecture consisting of macro, micro, and small base stations, remote radio heads, relays, distributed antennas, etc., they are all active nodes that generate signals, thus requiring sophisticated interference coordination and cancelation. This traditional approach inevitably aggravates the network operation overhead and, therefore, may not be able to sustain the wireless network capacity growth cost-effectively. In contrast, integrating IRSs into the wireless network will shift the existing heterogeneous network with active components only to a new hybrid architecture comprising both active and passive components. Since IRSs are of much lower cost as compared with their active counterparts, they can be more densely deployed in a wireless network at even lower cost, yet without bringing interference thanks to their passive reflection and resultant local coverage. By optimally setting the ratios between active nodes and passive IRSs in the hybrid network, a sustainable, green, low-cost network capacity scaling can be achieved [Wu et al., 2021].

Figure 7.1 illustrates several promising applications of IRS-aided wireless transmission [Wu and Zhang, 2020]. The first application shows a dead spot where the direct link between a user and its serving base station is severely blocked by an obstacle, e.g. a reinforced concrete building. In this case, deploying an IRS that has strong links with both the base station and the user can bypass the obstacle via intelligent signal reflection, thereby creating a virtual Line-Of-Sight (LOS) link. This is particularly helpful for the coverage extension in mmWave and THz communications, which are highly susceptible to blockage. The second application focuses on a cell-edge user suffering from both high signal attenuation from its serving base station and significant co-channel interference from a neighboring base station. Deploying an IRS at the cell edge can improve the desired signal strength while simultaneously suppressing the inter-cell interference by properly designing its passive beamforming, thus forming a signal hot spot as well as an interference-free zone in its vicinity. The third application considers the use of IRS to assist the implementation of Simultaneous Wireless Information and Power Transfer (SWIPT). In the deployment scenario of massive low-power or passive devices within an Internet-of-Things (IoT) network, the large aperture of IRS is employed to compensate for the significant power attenuation over a long distance via reflect beamforming to improve the efficiency of wireless power transfer. Last but not least, the fourth application provides a general description of artificially manipulating the channel statistics by adding additional signal paths toward the

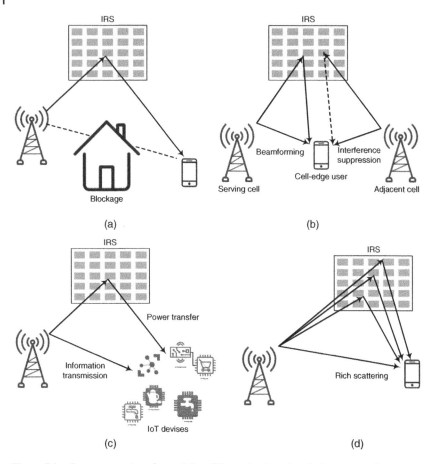

Figure 7.1 Some examples of promising IRS applications in wireless networks. (a) Virtual LOS. (b) Beamforming and interference suppression. (c) SWIPT. (d) Refining channel statistics.

desired direction so as to, e.g. improve the channel rank condition or transform a Rayleigh-faded channel to a Rician-faded channel.

7.2 IRS-Aided Single-Antenna Transmission

In this section, we study the fundamentals of the basic point-to-point IRS-aided wireless transmission in terms of its signal and channel models. Consider a three-node system consisting of a base station, a user, and an IRS with N passive reflecting elements on a planar surface, denoted by $\mathbb{B}$, $\mathbb{U}$, and $\mathbb{I}$, respectively. For the sake of simplicity, we first assume both the base station and the user equipment are equipped with a single antenna, while the signal bandwidth B_s is narrow at a given carrier frequency f_c, with $B_s \ll f_c$. However, the more general cases, i.e. multi-antenna, multi-user, multi-cell, and multi-carrier broadband systems, will be introduced in the subsequent sections.

7.2.1 Signal Model

Mathematically, the baseband equivalent complex-valued transmit signal is denoted by $s_b(t)$. After the up-conversion of the RF chain, the transmit antenna feeds a passband signal

$$s(t) = \Re\left[s_b(t)e^{j2\pi f_c t}\right] \tag{7.1}$$

into the wireless channel, where $\Re[\cdot]$ denotes the real part of a complex number and j is the imaginary unit $j^2 = -1$. Without losing generality, we first focus on the downlink signal transmission from the base station to the user via a particular reflecting element of the IRS, denoted by n, where $n \in \{1, 2, \dots, N\}$. The impulse response for the multi-path fading channel between the base station and the nth reflecting element can be modeled by

$$h_n(\tau) = \sum_{l=1}^{L} \alpha_{n,l}\delta(\tau - \tau_{n,l}), \tag{7.2}$$

where L expresses the total number of resolvable signal paths, $\alpha_{n,l}$ and $\tau_{n,l}$ denote the attenuation and propagation delay of path l, assuming that the channel response is time-invariant during the transmission of $s(t)$.

As a result, the impinging signal on the nth reflecting element of the IRS is expressed as

$$
\begin{aligned}
r_n(t) &= h_n(\tau) * s(t) \\
&= \sum_{l=1}^{L} \alpha_{n,l}s(t - \tau_{n,l}) \\
&= \sum_{l=1}^{L} \alpha_{n,l}\Re\left[s_b(t - \tau_{n,l})e^{j2\pi f_c(t-\tau_{n,l})}\right] \\
&= \Re\left[\sum_{l=1}^{L} \alpha_{n,l}s_b(t - \tau_{n,l})e^{j2\pi f_c(t-\tau_{n,l})}\right] \\
&= \Re\left[\sum_{l=1}^{L} \alpha_{n,l}e^{-j2\pi f_c\tau_{n,l}}s_b(t - \tau_{n,l})e^{j2\pi f_c t}\right] \\
&= \Re\left[(h_n^b(\tau) * s_b(t))e^{j2\pi f_c t}\right],
\end{aligned}
\tag{7.3}
$$

where $*$ stands for the linear convolution, and the baseband equivalent impulse response

$$h_n^b(\tau) = \sum_{l=1}^{L} \alpha_{n,l} e^{-j2\pi f_c \tau_{n,l}} \delta(\tau - \tau_{n,l}). \tag{7.4}$$

We write $\beta_n \in [0,1]$ and τ_n to denote the amplitude attenuation and delay induced by the nth reflecting element. By ignoring hardware impairments, e.g. phase noise and circuit nonlinearity, the reflected signal by the nth reflecting element can be expressed as

$$
\begin{aligned}
x_n(t) &= \beta_n r_n(t - \tau_n) \\
&= \beta_n \Re\left[\left(h_n^b(\tau) * s_b(t - \tau_n) \right) e^{j2\pi f_c(t - \tau_n)} \right] \\
&\approx \Re\left[\left(h_n^b(\tau) * s_b(t) \right) \beta_n e^{-j2\pi f_c \tau_n} e^{j2\pi f_c t} \right] \\
&= \Re\left[\left(h_n^b(\tau) * s_b(t) \right) c_n e^{j2\pi f_c t} \right].
\end{aligned}
\tag{7.5}
$$

The condition $s_b(t - \tau_n) \approx s_b(t)$ is easy to be satisfied since the symbol period is far greater than the physical delay induced by the reflecting element, i.e. $T_s = 1/B_s \gg \tau_n$, in a narrowband wireless system. We write $c_n = \beta_n e^{j\theta_n}$ to denote the reflection coefficient of the nth reflecting element, where $\theta_n = -2\pi f_c \tau_n \in [0, 2\pi)$ is the phase shift induced by the reflecting element, and this phase shift is periodic with respect to 2π.

Similarly, the impulse response for the multi-path fading channel between the nth reflecting element and the user can be modeled by

$$g_n(\tau) = \sum_{l=1}^{\mathfrak{L}} \alpha_{n,l} \delta(\tau - \tau_{n,l}), \tag{7.6}$$

where $\mathfrak{L}$ expresses the total number of resolvable signal paths, $\alpha_{n,l}$ and $\tau_{n,l}$ denote the attenuation and propagation delay of path l, assuming that the channel response is time-invariant during the transmission of $x_n(t)$. Consequently, the received signal at the user due to the nth reflecting element is given by

$$
\begin{aligned}
y_n(t) &= g_n(\tau) * x_n(t) \\
&= \sum_{l=1}^{\mathfrak{L}} \alpha_{n,l} x_n(t - \tau_{n,l}) \\
&= \sum_{l=1}^{\mathfrak{L}} \alpha_{n,l} \Re\left[\left(h_n^b(\tau) * s_b(t - \tau_{n,l}) \right) \beta_n e^{-j2\pi f_c \tau_n} e^{j2\pi f_c(t - \tau_{n,l})} \right] \\
&= \Re\left[\sum_{l=1}^{\mathfrak{L}} \alpha_{n,l} e^{-j2\pi f_c \tau_{n,l}} \left(h_n^b(\tau) * s_b(t - \tau_{n,l}) \right) c_n e^{j2\pi f_c t} \right].
\end{aligned}
\tag{7.7}
$$

Defining the baseband equivalent impulse response of $g_n(\tau)$ as

$$g_n^b(\tau) = \sum_{l=1}^{\mathfrak{L}} \alpha_{n,l} e^{-j2\pi f_c \tau_{n,l}} \delta(\tau - \tau_{n,l}), \tag{7.8}$$

Equation (7.7) can be rewritten as

$$y_n(t) = \Re\left[\left(g_n^b(\tau) * h_n^b(\tau) * s_b(t)\right) c_n e^{j2\pi f_c t}\right]. \tag{7.9}$$

Denoting $y_{n,b}(t)$ by the baseband equivalent received signal, the passband received signal can be also expressed as

$$y_n(t) = \Re\left[y_{n,b}(t) e^{j2\pi f_c t}\right]. \tag{7.10}$$

Comparing Eq. (7.10) with Eq. (7.9), we get

$$
\begin{aligned}
y_{n,b}(t) &= \left(g_n^b(\tau) * h_n^b(\tau) * s_b(t)\right) \beta_n e^{-j2\pi f_c \tau_n} \\
&= g_n^b(\tau) * c_n * h_n^b(\tau) * s_b(t).
\end{aligned} \tag{7.11}
$$

Until now, the baseband equivalent impulse response of the cascaded channel from the base station to the user via the nth reflecting element can be modeled by

$$v_n(t) = g_n^b(\tau) * c_n * h_n^b(\tau). \tag{7.12}$$

Further, we can know that the discrete-time baseband equivalent channel model in a narrowband system can be given by

$$v_n = g_n h_n c_n, \tag{7.13}$$

which is a cascaded channel denoted by a multiplication of three terms, i.e. the channel coefficient between the base station and the reflecting element, the reflecting coefficient, and the channel coefficient between the reflecting element and the user.

The narrowband model is based on the fact that a single tap is sufficient to express a frequency-flat channel, where h_n and g_n stand for the channel coefficients between the base station to the nth reflecting element and the nth reflecting element to the user, respectively. Generally, h_n is circularly symmetric complex Gaussian random variables with zero mean and variance σ_h^2, denoted by $h_n \sim \mathcal{CN}(0, \sigma_h^2)$, as well as $g_n \sim \mathcal{CN}(0, \sigma_g^2)$. It is worth mentioning that the reflected link from the base station to user via the IRS is also referred to as dyadic backscatter channel or pinhole channel in the literature [Wu and Zhang, 2019], with quite distinct behaviors from the direct link. To be specific, each reflecting element on the IRS behaves like a pinhole, which combines all the received multi-path signals at a single physical point, and re-scatters the combined signal as if from a point source.

Suppose there is no signal coupling in the reflection by neighboring IRS elements, i.e. all IRS elements reflect the incident signals independently. Due to the substantial path loss, we only consider signals reflected by the IRS for the first time and ignore those reflected multiple times. As such, the received signal from all reflecting elements can be modeled as a superposition of their respective reflected signals. Based on the reflecting model in Eq. (7.11), therefore, the discrete-time baseband signal model accounting for all the N reflecting elements is computed by

$$y = \left(\sum_{n=1}^{N} g_n c_n h_n\right) \sqrt{P_t} s + z, \tag{7.14}$$

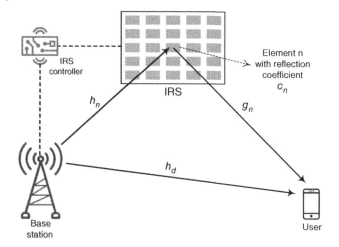

Figure 7.2 Illustration of a discrete-time baseband equivalent model of IRS-aided single-user single-antenna wireless transmission in a frequency-flat channel.

where y denotes the received symbol, s is a normalized transmitted symbol satisfying $\mathbb{E}\left[|s|^2\right] = 1$, P_t expresses the transmitted power of the base station, and n is additive white Gaussian noise (AWGN) with zero mean and variance σ_n^2, i.e. $z \sim C\mathcal{N}(0, \sigma_z^2)$. Let $\mathbf{h} = [h_1, h_2, \ldots, h_N]^T$, $\mathbf{g} = [g_1, g_2, \ldots, g_N]^T$, and $\Theta = \text{diag}(c_1, c_2, \ldots, c_N)$, we can get the vector form of Eq. (7.14) as

$$y = \mathbf{g}^T \Theta \mathbf{h} \sqrt{P_t} s + z. \tag{7.15}$$

It is noted that the IRS is assumed to perform a linear mapping from the incident signals to the reflected signals. If there is signal coupling or joint processing over the reflecting elements, the reflecting matrix Θ would not be diagonal any more. Furthermore, since the receiver can overhear the signals from both the reflected link denoted by $\mathbb{B} - \mathbb{I} - \mathbb{U}$ and the direct link denoted by $\mathbb{B} - \mathbb{U}$, the observation at the user can be expressed as

$$y = \left(\sum_{n=1}^{N} g_n c_n h_n + h_d \right) \sqrt{P_t} s + z = \left(\mathbf{g}^T \Theta \mathbf{h} + h_d \right) \sqrt{P_t} s + z, \tag{7.16}$$

where $h_d \sim C\mathcal{N}(0, \sigma_d^2)$ is the channel coefficient of the direct link. A discrete-time baseband equivalent model of IRS-aided wireless signal transmission is illustrated in Figure 7.2.

7.2.2 Passive Beamforming

The IRS signal model is applicable for either wireless power transfer where the harvested energy is generally modeled as a concave and increasing function of the received signal power or the information transfer where the achievable rate is a logarithmic function of the receive Signal-to-Noise Ratio (SNR). Thus, the main task of the IRS-aided transmission is to perform passive beamforming or reflect beamforming by judiciously adjusting its reflection coefficients, such that the signals reflected by IRS can be added either constructively with those via other signal paths to increase the desired signal strength, or destructively to mitigate the co-channel interference.

Accordingly, the received SNR at the user is given by

$$
\gamma = \frac{P_t | \sum_{n=1}^{N} g_n c_n h_n + h_d |^2}{\sigma_z^2}
$$

$$
= \frac{P_t | \mathbf{g}^T \boldsymbol{\Theta} \mathbf{h} + h_d |^2}{\sigma_z^2}. \tag{7.17}
$$

The channel capacity or maximum achievable rate of the basic point-to-point IRS-aided transmission is computed by $R = \log(1 + \gamma)$.

To maximize the channel capacity, the reflecting coefficients $c_n, \forall n = 1, 2, \ldots, N$ are optimized with respect to the instantaneous channel conditions. Neglecting the constant terms and employing continuous reflection amplitudes and phase shifts, the optimization problem can be formulated as

$$
\max_{\boldsymbol{\beta}, \boldsymbol{\theta}} \left| \sum_{n=1}^{N} g_n c_n h_n + h_d \right|^2
$$

$$
\text{s.t.} \quad \beta_n \in [0, 1], \quad \forall n = 1, 2, \ldots, N \tag{7.18}
$$

$$
\theta_n \in [0, 2\pi), \quad \forall n = 1, 2, \ldots, N,
$$

with $\boldsymbol{\beta} = [\beta_1, \beta_2, \ldots, \beta_N]^T$ and $\boldsymbol{\theta} = [\theta_1, \theta_2, \ldots, \theta_N]^T$.

As we know, the signal strength is maximized when all the incoming signals are coherently combined at the receiver by aligning their phases. Consequently, the optimal phase shift for reflecting element n should be [Wu et al., 2021]

$$
\theta_n^\star = \text{mod} \left[\psi_d - (\phi_{h,n} + \phi_{g,n}), 2\pi \right], \tag{7.19}
$$

where $\phi_{h,n}$, $\phi_{g,n}$, and ψ_d stand for the phases of h_n, g_n, and h_d, respectively, and mod $[\cdot]$ is the modulo operation. This equation implies that the reflecting coefficient of each element compensates for the phase rotation induced by the 𝔹-𝕀 and 𝕀-𝕌 links, resulting in a residual phase aligning with that of the direct link, so as to achieve the coherent combining. Given a blockage in the direct link, h_d approaches to zero, and therefore ψ_d in the aforementioned equation can be replaced with an arbitrary phase value without changing the maximized outcome.

This optimization solution relies on the fact that the values of β_n do not affect the optimality in coherent combining since different signals are co-phased.

Applying the optimal phases in Eq. (7.19), the optimization problem in Eq. (7.18) is simplified to

$$\max_{\beta} \left| \sum_{n=1}^{N} |g_n||h_n|\beta_n + |h_d| \right|^2 \tag{7.20}$$

$$\text{s.t.} \quad \beta_n \in [0,1], \quad \forall n = 1, 2, \ldots, N,$$

following the derivation

$$\left| \sum_{n=1}^{N} g_n c_n h_n + h_d \right|^2 = \left| \sum_{n=1}^{N} |g_n||h_n|\beta_n e^{j(\theta_n^* + \phi_{h,n} + \phi_{g,n})} + h_d \right|^2$$

$$= \left| \sum_{n=1}^{N} |g_n||h_n|\beta_n e^{j\psi_d} + |h_d|e^{j\psi_d} \right|^2$$

$$= \left| \sum_{n=1}^{N} |g_n||h_n|\beta_n + |h_d| \right|^2 |e^{j\psi_d}|^2$$

$$= \left| \sum_{n=1}^{N} |g_n||h_n|\beta_n + |h_d| \right|^2. \tag{7.21}$$

It is easily known that the optimal reflection amplitudes are given by $\beta_n^\star = 1$, $\forall n = 1, 2, \ldots, N$ since maximizing the signal strength in each reflected path achieves the largest receive power under the condition of coherent combining. An interesting observation is that the optimal reflecting coefficient for each IRS element is determined according to the knowledge of its corresponding cascaded channel as a whole, i.e. $g_n h_n$, without the need of knowing g_n and h_n individually. This property can be leveraged to substantially lower the complexity of channel estimation. Then, the maximal received SNR is expressed as

$$\gamma_{\max} = \frac{P_t \left| \sum_{n=1}^{N} |g_n||h_n| + |h_d| \right|^2}{\sigma_z^2}. \tag{7.22}$$

A fundamental question regarding the achievable performance of the IRS-aided signal transmission is how the received SNR grows in terms of the number of reflecting elements N. Under the assumption of independent identically distributed (*i.i.d.*) Rayleigh channels, $\gamma_{\max}$ is a non-central Chi-square random variable with one degree of freedom. When N becomes sufficiently large, the reflected links are more dominant and the direct link can be neglected, i.e. taking the value of $|h_d| = 0$ in Eq. (7.22) to get

$$\gamma_{\max} = \frac{P_t \left| \sum_{n=1}^{N} |g_n||h_n| \right|^2}{\sigma_z^2}. \tag{7.23}$$

According to the central limit theorem [Basar, 2019], we have

$$\gamma_{\max} \approx N^2 \frac{P_t \pi^2 \sigma_h^2 \sigma_g^2}{16\sigma_z^2}, \tag{7.24}$$

implying that the employment of IRS brings an SNR gain on the order of magnitude N^2. That is because the IRS achieves the passive beamforming gain of N in the I-U link, and simultaneously captures an additional aperture gain of N in the B-I link.

7.2.3 Product-Distance Path Loss

Last but not least, it is worth mentioning that the IRS cascaded channel undergoes *product-distance path loss* [Wu et al., 2021] in contrast to the conventional *sum-distance path loss* in typical specular reflection. As we know, the channel coefficients h_n and g_n are decided by large-scale fading (i.e. distance-related path loss and shadowing), and small-scale fading due to multi-path propagation. In particular, the path loss of an IRS-reflected channel captures its average power and is thus essential to the link budget analysis and performance evaluation. Under the far-field propagation environment, where the IRS is located sufficiently far from both the base station and user, the distances of the B-I and I-U links can be denoted by d_h and d_g, respectively, regardless of the slight difference among individual reflecting elements. For the sake of simplicity, the power gain of the B-I channel can be denoted by

$$\mathbb{E}\left[|h_n|^2\right] = \sigma_h^2 \propto \frac{1}{d_h^{\alpha_h}}, \tag{7.25}$$

and that of the I-U channel is

$$\mathbb{E}\left[|g_n|^2\right] = \sigma_g^2 \propto \frac{1}{d_g^{\alpha_g}}, \tag{7.26}$$

with the path loss components α_h and α_g.

Then, the cascaded channel corresponds to

$$\mathbb{E}\left[|h_n g_n|^2\right] = \sigma_h^2 \sigma_g^2 \propto \frac{1}{d_h^{\alpha_h} d_g^{\alpha_g}}. \tag{7.27}$$

indicating that the IRS cascaded channel undergoes double path loss, referred to as the product-distance path loss model. Hence, a large number of IRS reflecting elements are required in practice to compensate for such severe power loss due to double attenuation, by jointly designing their reflection coefficients to achieve high passive beamforming gain.

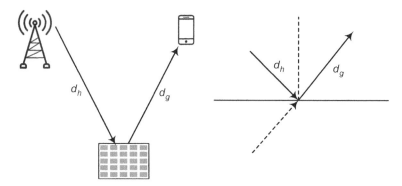

Figure 7.3 Comparison of the product-distance path loss model in the IRS context and the sum-distance path loss model in the conventional specular reflection.

In contrast, the received signal power via an infinitely large perfect electric conductor is inversely proportional to the sum distance of the two-hop links, as shown in Figure 7.3, i.e.

$$\frac{1}{(d_h + d_g)^\alpha}, \tag{7.28}$$

where α is the path loss component. In such a sum-distance path loss model, the received power of a reflected signal is equivalent to the case where a transmitter is located at the image point of the original transmitter, with the same propagation distance $d_h + d_g$.

7.3 IRS-Aided Multi-Antenna Transmission

This section further studies the point-to-point IRS-aided transmission in a narrow-band frequency-flat channel but with multiple antennas at the base station, where the downlink transmission becomes a Multi-Input Single-Output (MISO) system, and the uplink transmission is a Single-Input Multi-Output (SIMO) system. Consequently, the active beamforming at the base station and the passive beamforming at the IRS must be jointly optimized to maximize achievable spectral efficiency. For simplicity, the discussion only focuses on the downlink MISO transmission, but the results are also applicable to the uplink SIMO transmission. Moreover, it becomes an IRS-aided MIMO system if the user terminal is equipped with multiple antennas. After the discussion of the IRS-aided MISO joint beamforming, a brief introduction of the IRS-aided MIMO system will be followed.

7.3.1 Joint Active and Passive Beamforming

Consider a single-user MISO communication system comprising an N_b-antenna base station, a single-antenna user, and an IRS with N passive reflecting elements.

To characterize the theoretical performance gain brought by the IRS, we assume the channel state information of all involved channels is perfectly known and follows quasi-static frequency-flat fading. At the base station, linear beamforming denoted by a transmit vector $\mathbf{w} \in \mathbb{C}^{N_b \times 1}$ is applied, with $\|\mathbf{w}\|^2 \leqslant P_t$, where $\| \cdot \|$ represents the Euclidean norm of a complex vector. Then, the discrete-time baseband equivalent signal received by the user is expressed as

$$y = \left(\sum_{n=1}^{N} g_n c_n \mathbf{h}_n^T + \mathbf{h}_d^T \right) \mathbf{w}s + z = (\mathbf{g}^T \mathbf{\Theta} \mathbf{H} + \mathbf{h}_d^T) \mathbf{w}s + z, \qquad (7.29)$$

where $\mathbf{h}_d = [h_{d,1}, h_{d,2}, \ldots, h_{d,N_b}]^T \in \mathbb{C}^{N_b \times 1}$ is the channel vector from N_b base-station antennas to the user, $\mathbf{h}_n = [h_{n,1}, h_{n,2}, \ldots, h_{n,N_b}]^T \in \mathbb{C}^{N_b \times 1}$ is the channel vector from N_b base-station antennas to the nth reflecting element, $\mathbf{g} = [g_1, g_2, \ldots, g_N]^T$, $\mathbf{\Theta} = \text{diag}(c_1, c_2, \ldots, c_N)$, and $\mathbf{H} \in \mathbb{C}^{N \times N_b}$ denotes the channel matrix from the base station to the IRS, where the nth row of this matrix equals to $\mathbf{h}_n^T$.

By jointly designing the active beamforming $\mathbf{w}$ and passive reflection coefficients $\mathbf{\Theta}$, the spectral efficiency

$$R = \log \left(1 + \frac{|(\mathbf{g}^T \mathbf{\Theta} \mathbf{H} + \mathbf{h}_d^T) \mathbf{w}|^2}{\sigma_z^2} \right) \qquad (7.30)$$

can be maximized, resulting in the following optimization problem

$$\begin{aligned}
\max_{\mathbf{\Theta}, \mathbf{w}} \quad & |(\mathbf{g}^T \mathbf{\Theta} \mathbf{H} + \mathbf{h}_d^T) \mathbf{w}|^2 \\
\text{s.t.} \quad & \|\mathbf{w}\|^2 \leqslant P_t \\
& \beta_n \in [0, 1], \quad \forall n = 1, 2, \ldots, N \\
& \theta_n \in [0, 2\pi), \quad \forall n = 1, 2, \ldots, N.
\end{aligned} \qquad (7.31)$$

As revealed in the previous section, the optimal value of reflecting attenuation is $\beta_n = 1, \forall n = 1, 2, \ldots, N$. Then, the aforementioned optimization problem is simplified to

$$\begin{aligned}
\max_{\theta, \mathbf{w}} \quad & |(\mathbf{g}^T \mathbf{\Theta} \mathbf{H} + \mathbf{h}_d^T) \mathbf{w}|^2 \\
\text{s.t.} \quad & \|\mathbf{w}\|^2 \leqslant P_t \\
& \theta_n \in [0, 2\pi), \quad \forall n = 1, 2, \ldots, N,
\end{aligned} \qquad (7.32)$$

with $\mathbf{\Theta} = \text{diag}\left(e^{j\theta_1}, e^{j\theta_2}, \ldots, e^{j\theta_N} \right)$. Unfortunately, it is still a non-convex optimization problem because its objective function is not jointly concave with respect to θ and $\mathbf{w}$. In fact, the coupling between the active and passive beamforming optimization variables is the major challenge in their joint optimization. Two solutions have been proposed to tackle this problem [Wu and Zhang, 2019], i.e. Semi-Definite Relaxation (SDR) that is applied to obtain a high-quality approximation as well as a performance bound, and Alternating Optimization

(AO), which proposes to alternately optimize the transmit beamforming and reflecting coefficients in an iterative manner.

7.3.1.1 SDR Solution

One can observe that if the reflecting coefficients θ are fixed, the optimization problem becomes a normal method to determine the optimal beamforming vector in a typical multi-antenna transmission system. Thus, the optimal beamformer could be a matched filter, a.k.a. maximum-ratio transmission, which can maximize the strength of a desired signal, we have

$$\mathbf{w}^\star = \sqrt{P_t} \frac{(\mathbf{g}^T \mathbf{\Theta H} + \mathbf{h}_d^T)^H}{\|\mathbf{g}^T \mathbf{\Theta H} + \mathbf{h}_d^T\|}. \tag{7.33}$$

Substituting $\mathbf{w}^\star$ into Eq. (7.32), it is reduced to an optimization problem with respect to θ only, i.e.

$$\begin{aligned}
\max_{\theta} \quad & \|\mathbf{g}^T \mathbf{\Theta H} + \mathbf{h}_d^T\|^2 \\
\text{s.t.} \quad & \theta_n \in [0, 2\pi), \quad \forall n = 1, 2, \ldots, N.
\end{aligned} \tag{7.34}$$

Denoting $\mathbf{q} = [q_1, q_2, \ldots, q_N]^H$ with $q_n = e^{j\theta_n}$, the constraints in Eq. (7.34) are equivalent to the unit-modulus constraints $|q_n| = 1$, $\forall n$. According to the transformation given by Wu and Zhang [2019], i.e. $\mathbf{g}^T \mathbf{\Theta H} = \mathbf{q}^H \mathbf{\Phi}$, where $\mathbf{\Phi} = \text{diag}(\mathbf{g}^T)\mathbf{H} \in \mathbb{C}^{N \times N_b}$, we have

$$\|\mathbf{g}^T \mathbf{\Theta H} + \mathbf{h}_d^T\|^2 = \|\mathbf{q}^H \mathbf{\Phi} + \mathbf{h}_d^T\|^2. \tag{7.35}$$

Then, Eq. (7.34) is equivalent to

$$\begin{aligned}
\max_{\mathbf{q}} \quad & \mathbf{q}^H \mathbf{\Phi} \mathbf{\Phi}^H \mathbf{q} + \mathbf{q}^H \mathbf{\Phi} \mathbf{h}_d^* + \mathbf{h}_d^T \mathbf{\Phi}^H \mathbf{q} + \|\mathbf{h}_d\|^2 \\
\text{s.t.} \quad & |q_n| = 1, \quad \forall n = 1, 2, \ldots, N.
\end{aligned} \tag{7.36}$$

The simplified optimization problem is a non-convex quadratically constrained quadratic program, which is generally NP-hard to solve. In particular, the unit-modulus constraints are non-convex and challenging to handle, imposing another challenge in the optimization algorithm design. Therefore, various methods have been proposed in the literature to obtain high-quality sub-optimal solutions, including SDR with Gaussian randomization [Wu and Zhang, 2019], AO, where each of the phase shifts is optimized in closed-form while the others being fixed iteratively to guarantee a locally optimal, and a Branch-and-Bound (BoB) algorithm to achieve a globally optimal solution [Yu et al., 2020], to name a few. The following part will introduce the SDR solution as an example, while interested readers may refer to the corresponding literature for the other solutions.

By introducing an auxiliary variable t, Eq. (7.36) can be rewritten as

$$\max_{\mathbf{q}} \quad \bar{\mathbf{q}}^H \mathbf{R} \bar{\mathbf{q}} + \|\mathbf{h}_d\|^2$$

$$\text{s.t.} \quad |q_n| = 1, \quad \forall n = 1, 2, \ldots, N + 1, \tag{7.37}$$

with

$$\mathbf{R} = \begin{pmatrix} \mathbf{\Phi}\mathbf{\Phi}^H & \mathbf{\Phi}\mathbf{h}_d^* \\ \mathbf{h}_d^T\mathbf{\Phi}^H & 0 \end{pmatrix} \tag{7.38}$$

and

$$\bar{\mathbf{q}} = \begin{bmatrix} \mathbf{q} \\ t \end{bmatrix}. \tag{7.39}$$

Defining $\mathbf{Q} = \bar{\mathbf{q}}\bar{\mathbf{q}}^H$, we have $\bar{\mathbf{q}}^H \mathbf{R} \bar{\mathbf{q}} = \text{tr}(\mathbf{R}\bar{\mathbf{q}}\bar{\mathbf{q}}^H) = \text{tr}(\mathbf{R}\mathbf{Q})$, which needs to satisfy $\mathbf{Q} \succeq 0$ and rank($\mathbf{Q}$) = 1. As a result, the optimization problem is finally simplified to

$$\max_{\mathbf{Q}} \quad \text{tr}(\mathbf{R}\mathbf{Q})$$

$$\text{s.t.} \quad |q_n| = 1, \quad \forall n = 1, 2, \ldots, N + 1, \tag{7.40}$$

$$\mathbf{Q} \succeq 0,$$

which becomes a convex semi-definite program. Decompose $\mathbf{Q}$ as $\mathbf{Q} = \mathbf{U}\mathbf{\Sigma}\mathbf{U}^H$, where $\mathbf{U}$ is a unitary matrix and $\mathbf{\Sigma}$ is a diagonal matrix, both with the size $(N + 1) \times (N + 1)$. According to Wu and Zhang [2018], a sub-optimal solution for the optimization problem in Eq. (7.37) is given by

$$\bar{\mathbf{q}} = \mathbf{U}\mathbf{\Sigma}^{1/2}\mathbf{r}, \tag{7.41}$$

where $\mathbf{r}$ is a random vector generated according to $\mathbf{r} \in \mathcal{CN}(\mathbf{0}, \mathbf{I}_{N+1})$. Finally, the solution to the optimization problem in Eq. (7.36) can be determined as

$$\mathbf{q} = e^{j \arg\left(\left[\frac{\bar{\mathbf{q}}}{\bar{q}_{N+1}}\right]_{1:N}\right)}, \tag{7.42}$$

where $[\cdot]_{1:N}$ denotes a sub-vector extracting the first N elements.

Although joint active and passive beamforming using the SDR solution achieves good performance, it requires global channel state information, imposing prohibitive channel estimation operations and signaling exchange overhead, especially when the number of antennas at the base station and the number of reflecting elements at the IRS are large. In addition, the calculation of the optimal beamforming vector and reflecting coefficients brings a high computational burden, especially in fast-fading environments where the channels vary quickly.

7.3.1.2 Alternating Optimization

This technique provides an efficient algorithm to lower the complexity of the previous SDR method. The key idea is to optimize the transmit beamforming and reflecting coefficients alternatively (rather than jointly) in an iterative manner until the convergence is achieved.

Given a known transmit beamforming vector $\mathbf{w}_0$, the optimization problem in Eq. (7.32) is simplified to

$$\max_{\theta} \quad |(\mathbf{g}^T\boldsymbol{\Theta}\mathbf{H} + \mathbf{h}_d^T)\mathbf{w}_0|^2$$
$$\text{s.t.} \quad \theta_n \in [0, 2\pi), \quad \forall n = 1, 2, \ldots, N. \tag{7.43}$$

The optimization equation is still non-convex but it can enable a closed-form solution by exploiting the following inequality

$$|(\mathbf{g}^T\boldsymbol{\Theta}\mathbf{H} + \mathbf{h}_d^T)\mathbf{w}_0| \leqslant |\mathbf{g}^T\boldsymbol{\Theta}\mathbf{H}\mathbf{w}_0| + |\mathbf{h}_d^T\mathbf{w}_0|. \tag{7.44}$$

The equality achieves if and only if $\arg\left(\mathbf{g}^T\boldsymbol{\Theta}\mathbf{H}\mathbf{w}_0\right) = \arg\left(\mathbf{h}_d^T\mathbf{w}_0\right) \triangleq \varphi_0$, where $\arg(\cdot)$ stands for the phase of a complex number or the component-wise phase of a complex vector.

Denoting $\mathbf{q} = [q_1, q_2, \ldots, q_N]^H$ with $q_n = e^{j\theta_n}$ and $\chi = \text{diag}(\mathbf{g}^T)\mathbf{H}\mathbf{w}_0 \in \mathbb{C}^{N\times 1}$, we have $\mathbf{g}^T\boldsymbol{\Theta}\mathbf{H}\mathbf{w}_0 = \mathbf{q}^H\chi \in \mathbb{C}$. Then, by ignoring the constant term $|\mathbf{h}_d^T\mathbf{w}_0|$, Eq. (7.43) is transformed to

$$\max_{\mathbf{q}} \quad |\mathbf{q}^H\chi|$$
$$\text{s.t.} \quad |q_n| = 1, \quad \forall n = 1, 2, \ldots, N, \tag{7.45}$$
$$\arg(\mathbf{q}^H\chi) = \varphi_0.$$

It is easy to obtain the optimal solution to maximize the objective function, i.e.

$$\mathbf{q}^* = e^{j(\varphi_0 - \arg(\chi))} = e^{j(\varphi_0 - \arg(\text{diag}(\mathbf{g}^T)\mathbf{H}\mathbf{w}_0))}. \tag{7.46}$$

Accordingly, the optimal phase shift for the nth reflecting element is given by

$$\theta_n^\star = \varphi_0 - \arg(g_n\mathbf{h}_n^T\mathbf{w}_0)$$
$$= \varphi_0 - \arg(g_n) - \arg(\mathbf{h}_n^T\mathbf{w}_0), \tag{7.47}$$

where $\mathbf{h}_n^T\mathbf{w}_0 \in \mathbb{C}$ can be regarded as the effective Single-Input Single-Output (SISO) channel perceived by the nth reflecting element combining the effects of the transmit beamforming $\mathbf{w}_0$ and the channel from the base station to the reflecting element $\mathbf{h}_n$, and g_n denotes the channel coefficient from the nth reflecting element to the user. In this regard, Eq. (7.47) implies that the optimal phase shift should be tuned such that the phase of the signal that passes through the $\mathbb{B} - \mathbb{I}$ and $\mathbb{I} - \mathbb{U}$ links is compensated, and the residual phase is aligned with that of the signal over the direct link to achieve coherent combining at the receiver.

Once the reflecting coefficients are known, the AO solution is alternated to optimize $\mathbf{w}$ given $\theta_n^\star$, or equivalently the knowledge of the overall channel $\mathbf{g}^T\mathbf{\Theta H} + \mathbf{h}_d^T$. The optimal beamformer could be a matched filter, which can maximize the strength of a desired signal, i.e.

$$\mathbf{w}^\star = \sqrt{P_t}\frac{(\mathbf{g}^T\mathbf{\Theta H} + \mathbf{h}_d^T)^H}{\|\mathbf{g}^T\mathbf{\Theta H} + \mathbf{h}_d^T\|}. \tag{7.48}$$

Next, the AO solution is alternated again to optimize θ given $\mathbf{w}^\star$. This process iterates until the convergence is achieved. This alternating optimization approach is practically meaningful since both the transmit beamforming and phase shifts are obtained in closed-form expressions.

7.3.2 Joint Precoding and Reflecting

The previous discussions mainly focused on SISO or MISO systems where the user equipment has only a single antenna. In a more general IRS-aided MIMO case with multiple antennas at both the base station and user, the joint optimization of the IRS reflecting coefficients and MIMO transmit covariance matrix or precoding matrix is required. It is more difficult than the conventional MIMO system without the IRS. The system model and optimization formulation of the IRS-aided MIMO communications will be introduced as follows.

Consider a point-to-point MIMO communication system comprising an N_b-antenna base station, a single user with N_u antennas, and an IRS with N passive reflecting elements. Assume quasi-static blocking fading in frequency-flat channels, and one particular fading block is considered where the channel state information is perfectly known and remains approximately constant. Then, as illustrated in Figure 7.4, the discrete-time baseband equivalent system model can be expressed as

$$\mathbf{y} = \left(\sum_{n=1}^{N}\mathbf{g}_n e^{j\theta_n}\mathbf{h}_n^T + \mathbf{H}_d\right)\mathbf{x} + \mathbf{z}, \tag{7.49}$$

where $\mathbf{y} = [y_1, y_2, \dots, y_{N_u}]^T \in \mathbb{C}^{N_u \times 1}$ denotes the received symbol vector at the user, $\mathbf{H}_d \in \mathbb{C}^{N_u \times N_b}$ is the channel matrix of the direct link from the base station to the user, $\mathbf{h}_n = [h_{n,1}, h_{n,2}, \dots, h_{n,N_b}]^T \in \mathbb{C}^{N_b \times 1}$ is the channel vector from N_b base-station antennas to the nth reflecting element, $\mathbf{g}_n = [g_{n,1}, g_{n,2}, \dots, g_{n,N_u}]^T \in \mathbb{C}^{N_u \times 1}$ expresses the channel vector from the nth reflecting element to N_u user antennas, θ_n is the induced phase rotation on the nth reflecting element (assuming the maximal amplitude $\beta_n = 1$), $\mathbf{z} = [z_1, z_2, \dots, z_{N_u}]^T \in \mathbb{C}^{N_u \times 1}$ denotes the vector of AWGN, satisfying $\mathbf{z} \sim \mathcal{CN}\left(\mathbf{0}, \sigma_z^2 \mathbf{I}_{N_u}\right)$, and $\mathbf{x} = [x_1, x_2, \dots, x_{N_b}]^T \in \mathbb{C}^{N_b \times 1}$ denotes the transmitted symbol vector, with the transmit power constraint $\mathbb{E}[\mathbf{x}^H\mathbf{x}] \leqslant P_t$.

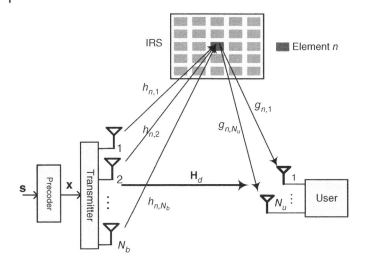

Figure 7.4 Schematic diagram of an IRS-aided MIMO system.

The transmit covariance matrix is defined by $\mathbf{Q} = \mathbb{E}[\mathbf{x}\mathbf{x}^H] \in \mathbb{C}^{N_b \times N_b}$, and the power constraint can also be expressed as $\mathrm{tr}(\mathbf{Q}) \leqslant P_t$.

Equivalently, the system model can also be expressed in matrix form as

$$\mathbf{y} = (\mathbf{G}\mathbf{\Theta}\mathbf{H} + \mathbf{H}_d)\mathbf{x} + \mathbf{z}, \tag{7.50}$$

where the diagonal reflecting matrix is written as $\mathbf{\Theta} = \mathrm{diag}\left\{e^{j\theta_1}, e^{j\theta_2}, \ldots, e^{j\theta_N}\right\}$, and $\mathbf{H} \in \mathbb{C}^{N \times N_b}$ denotes the channel matrix from the base station to the IRS, where the nth row of this matrix equals to $\mathbf{h}_n^T$, and $\mathbf{G} \in \mathbb{C}^{N_u \times N}$ denotes the channel matrix from the IRS to the user, expressed by $\mathbf{G} = [\mathbf{g}_1, \mathbf{g}_2, \ldots, \mathbf{g}_N]$. Treating $\tilde{\mathbf{H}} = \mathbf{G}\mathbf{\Theta}\mathbf{H} + \mathbf{H}_d$ as an effective channel matrix in the conventional MIMO system, yields the channel capacity of the IRS-aided MIMO system, which is given by

$$
\begin{aligned}
C &= \log_2 \det\left(\mathbf{I}_{N_u} + \frac{\tilde{\mathbf{H}}\mathbf{Q}\tilde{\mathbf{H}}^H}{\sigma_z^2}\right) \\
&= \log_2 \det\left(\mathbf{I}_{N_u} + \frac{\left(\mathbf{G}\mathbf{\Theta}\mathbf{H} + \mathbf{H}_d\right)\mathbf{Q}\left(\mathbf{G}\mathbf{\Theta}\mathbf{H} + \mathbf{H}_d\right)^H}{\sigma_z^2}\right).
\end{aligned} \tag{7.51}
$$

Differing from the conventional MIMO channel without the IRS, for which the capacity is solely determined by the channel matrix $\mathbf{H}_d$, the capacity for the IRS-aided MIMO channel is also dependent on the IRS reflection matrix $\mathbf{\Theta}$, since it influences the effective channel matrix $\tilde{\mathbf{H}} = \mathbf{G}\mathbf{\Theta}\mathbf{H} + \mathbf{H}_d$ as well as the resultant optimal transmit covariance matrix $\mathbf{Q}$. To maximize the capacity of an IRS-aided MIMO channel, we need to jointly optimize the IRS reflection matrix and the transmit covariance matrix, subject to uni-modular constraints on the reflection coefficients and a sum power constraint at the transmitter. As a result, the optimization problem is formulated as

$$\max_{\Theta, Q} \quad \log_2 \det \left(I_{N_u} + \frac{\left(G\Theta H + H_d\right) Q \left(G\Theta H + H_d\right)^H}{\sigma_z^2} \right)$$

$$\text{s.t.} \quad \theta_n \in [0, 2\pi), \quad \forall n = 1, 2, \ldots, N,$$

$$\text{tr}(Q) \leqslant P_t. \tag{7.52}$$

It is a non-convex optimization problem since the objective function can be shown to be non-concave over the reflection matrix, and the uni-modular constraint on each reflection coefficient is also non-convex. Moreover, the transmit covariance matrix Q is coupled with Θ in the objective function, which makes this optimization more difficult to solve. In Zhang and Zhang [2020], an alternating algorithm has been proposed for solving this optimization problem. Specifically, the objective function is first transformed into a more tractable form in terms of the optimization variables Q and θ_n, $\forall n$, based on which we then solve two sub-problems, for optimizing respectively the transmit covariance matrix Q or one reflection coefficient θ_n, $\forall n$ with all the other variables being fixed. The optimal solutions to both subproblems are closed-form, enabling an efficient alternating optimization algorithm to obtain a locally optimal solution by iteratively solving these subproblems.

In addition to optimizing the transmit covariance matrix Q, there is an alternative approach by maximizing the performance through joint reflecting and precoding design. Let $x = Ps$, where $s \in \mathbb{C}^{N_s \times 1}$ denotes the vector of N_s information symbols transmitted over the channel simultaneously, satisfying $\mathbb{E}\left[ss^H\right] = I_{N_s}$, and $P \in \mathbb{C}^{N_t \times N_s}$ denotes the precoding matrix used to encode N_s information symbols into N_b transmitted symbols. Then, the system model can be rewritten as

$$y = \left(\sum_{n=1}^{N} g_n e^{j\theta_n} h_n^T + H_d \right) Ps + z = (G\Theta H + H_d)Ps + z. \tag{7.53}$$

In this case, the optimization problem becomes the process to find the optimal precoding matrix and reflecting coefficient matrix. Therefore, the optimization problem can be rewritten as

$$\max_{\Theta, P} \quad \log_2 \det \left(I_{N_u} + \frac{\left(G\Theta H + H_d\right) PP^H \left(G\Theta H + H_d\right)^H}{\sigma_z^2} \right)$$

$$\text{s.t.} \quad \theta_n \in [0, 2\pi), \quad \forall n = 1, 2, \ldots, N,$$

$$\text{tr}(PP^H) \leqslant P_t. \tag{7.54}$$

In addition to maximizing the channel capacity, the transmit covariance matrix or precoding matrix can be optimized to improve other performance metrics. For example, Ye et al. [2020] proposes a joint optimization of precoding and reflecting to minimize symbol error rate for IRS-aided MIMO communications.

7.4 Dual-Beam Intelligent Reflecting Surface

Making use of hybrid beamforming, a Base Station (BS) can generate a pair of independent beams toward the IRS and User Equipment (UE), respectively. As a result, the optimal reflecting phases are directly calculated from the estimated Channel State Information (CSI), independently with the active beamforming. Hence, the optimization of passive and active beamforming is decoupled, resulting in a simplified system design. In contrast to the joint passive and active beamforming optimization, the high computational burden and high latency due to iterative optimization can be avoided. Moreover, the coverage is optimized in terms of the cell-edge and cell-center performance.

7.4.1 Dual Beams Over Hybrid Beamforming

There are three major beamforming structures: digital, analog, and hybrid digital-analog. Implementing digital beamforming over a large-scale array, e.g. in a mmWave or THz transceiver, needs a lot of RF components, e.g. high power amplifiers, leading to high hardware cost and power consumption. This constraint has driven the application of analog beamforming, using only a single RF chain. Analog beamforming is implemented as the *de-facto* approach for indoor mmWave systems. However, it only supports single-stream transmission and suffers from the hardware impairment of analog phase shifters. As a consequence, hybrid beamforming [Zhang et al., 2019] has been proposed as an efficient approach to support multi-stream transmission with only a few RF chains and a phase-shifter network. Compared with analog beamforming, hybrid beamforming supports spatial multiplexing, diversity, and spatial-division multiple access. It achieves spectral efficiency comparable to digital beamforming with much lower hardware complexity and cost.

Hybrid beamforming can be implemented with different forms of circuit networks, resulting in two basic structures:

- *Fully Connected Hybrid Beamforming*
  The transmit data is first precoded in the baseband into M data streams, and each stream is processed by an independent RF chain. Each RF chain is connected to all N antennas via an analog network, where $N \gg M$. Hence, a total of MN analog phase shifters are required.

- *Partially Connected Hybrid Beamforming*
  Each RF chain is only connected to a subset of all antenna elements called a sub-array. This structure is preferred from the perspective of practice since it substantially lowers hardware complexity (as well as power consumption) by dramatically reducing the number of analog phase shifters from MN to N.

Without loss of generality, we focus on the partially connected hybrid beamforming in this part for simple analysis but it is also applicable to the fully connected (FC) hybrid beamforming. Mathematically, the output of the baseband precoder is denoted by $s_m[t]$, $1 \leq m \leq M$. After the digital-to-analog conversion and up-conversion, the mth RF chain feeds

$$\Re \left[s_m[t] e^{j2\pi f_0 t} \right], \quad 1 \leq m \leq M, \tag{7.55}$$

into the analog network, where $\Re[\cdot]$ denotes the real part of a complex number, and f_0 stands for the carrier frequency. In the partially connected beamforming, an array is divided into several sub-arrays, and each antenna is allocated to only one RF chain. Each sub-array contains $N_s = N/M$ elements (suppose N_s is an integer). Denote the weighted phase shift on the nth antenna of the mth sub-array by ψ_{nm}, where $1 \leq n \leq N_s$ and $1 \leq m \leq M$ to denote, and thus the transmit signals of the mth sub-array are expressed by

$$\mathbf{s}_m(t) = \left[\Re \left[s_m[t] e^{j2\pi f_0 t} e^{j\psi_{1m}} \right], \ldots, \Re \left[s_m[t] e^{j2\pi f_0 t} e^{j\psi_{N_s m}} \right] \right]^T. \tag{7.56}$$

Radiating a plane wave into a homogeneous media in the direction indicated by the angle of departure θ, the time difference between a typical element n of the mth sub-array and the reference point is denoted by $\tau_{nm}(\theta)$. Within a flat-fading wireless channel, the received passband signal is

$$\begin{aligned}
y(t) &= \sum_{m=1}^{M} \sum_{n=1}^{N_s} \Re \left[h_m(t) s_m[t] e^{j2\pi f_0(t-\tau_{nm}(\theta))} e^{j\psi_{nm}} \right] + n(t) \\
&= \Re \left[\left(\sum_{m=1}^{M} h_m(t) s_m[t] \sum_{n=1}^{N_s} e^{-j2\pi f_0 \tau_{nm}(\theta)} e^{j\psi_{nm}} \right) e^{j2\pi f_0 t} \right] + n(t),
\end{aligned} \tag{7.57}$$

where $h_m(t)$ represents the channel response between the mth sub-array and receiver, and $n(t)$ the noise [Jiang and Schotten, 2022].

Denoting the steering vector of sub-array m as

$$\mathbf{a}_m(\theta) = \left[e^{-j2\pi f_0 \tau_{1m}(\theta)}, e^{-j2\pi f_0 \tau_{2m}(\theta)}, \ldots, e^{-j2\pi f_0 \tau_{N_s m}(\theta)} \right]^T, \tag{7.58}$$

and its weighting vector (due to analog phase shift) as

$$\mathbf{w}_m = \left[e^{j\psi_{1m}}, e^{j\psi_{2m}}, \ldots, e^{j\psi_{N_s m}} \right]^T, \tag{7.59}$$

Eq. (7.57) can be rewritten as

$$r(t) = \Re \left[\left(\sum_{m=1}^{M} h_m(t) s_m[t] \mathbf{a}_m^T(\theta) \mathbf{w}_m \right) e^{j2\pi f_0 t} \right] + n(t),$$

$$= \Re \left[\left(\sum_{m=1}^{M} h_m(t) s_m[t] B_m(\theta, t) \right) e^{j2\pi f_0 t} \right] + n(t), \tag{7.60}$$

with the beam pattern of the mth sub-array:

$$B_m(\theta, t) = \mathbf{a}_m^T(\theta) \mathbf{w}_m = \sum_{n=1}^{N_s} e^{-j2\pi f_0 \tau_{nm}(\theta)} e^{j\psi_{nm}}. \tag{7.61}$$

After the down-conversion and sampling at the receiver, the baseband equivalent received signal can be simplified into (neglecting the time index for simplicity)

$$r = \sum_{m=1}^{M} h_m B_m(\theta) s_m + n, \tag{7.62}$$

where r is the received symbol, s_m is the modulated symbol for the mth RF chain, and h_m denotes the channel coefficient between the reference antenna of the mth sub-array and the receiver.

Without losing generality, we use the partially connected structure with two branches hereinafter to present and analyze the dual-beam IRS scheme. As shown in Figure 7.5, a pair of beams with patterns $B_1(\theta)$ and $B_2(\theta)$ over the

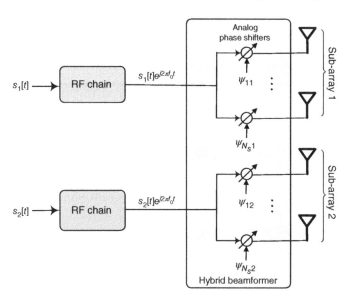

Figure 7.5 Block diagrams of two-branch partially connected hybrid beamforming.

hybrid beamformer are obtained. After the down-conversion and sampling at the receiver, the baseband received signal can be simplified into

$$r = h_1 B_1(\theta_1)s_1 + h_2 B_2(\theta_2)s_2 + n, \tag{7.63}$$

where h_1 and h_1 denote the channel coefficients from the reference antenna of the first and second sub-arrays to the UE, and θ_1 and θ_2 stand for the departure-of-angles (DOA) of the transmitted signals over sub-arrays 1 and 2, respectively.

7.4.2 Dual-Beam IRS

The fundamental of DB-IRS is to steer one beam toward the user, like a conventional beamforming system, while the other beam is used to focus energy on the IRS, as demonstrated in Figure 7.6. Since both the locations of the IRS and BS are deliberately selected and fixed, there is usually no blockage in-between and the DOA information θ_I is deterministic. The beam toward the IRS is fixed, which substantially simplifies the system implementation. The inter-antenna spacing in hybrid beamforming is small, typically half wavelength $d = \lambda/2$. As a result, the antennas are highly correlated. Without losing generality, we assume the first sub-array serves the IRS, and the second sub-array toward the UE. The channel vector from the first sub-array to the nth reflecting element can be expressed as $\mathbf{h}_n = h_n \mathbf{a}_1(\theta_I) \in \mathbb{C}^{N_s \times 1}$, where h_n denotes the channel response from the reference antenna of this sub-array to the nth reflecting element. Likewise, the channel vector from the second sub-array to the UE is given by $\mathbf{h}_d = h_d \mathbf{a}_2(\theta_d)$, where h_d denotes the channel response from the reference antenna of the second sub-array to the UE, and θ_d is the DOA of the user. We write $\mathbf{w}_1$ and $\mathbf{w}_2$ to denote the weighting vectors of two sub-arrays, respectively. Meanwhile, we assume two

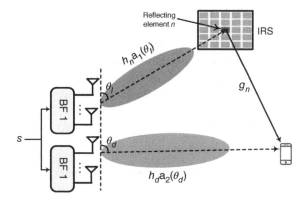

Figure 7.6 Illustration of the dual-beam IRS using hybrid beamforming.

sub-arrays have a common transmitted symbol s with $\mathbb{E}\left[|s|^2\right] = 1$. The observation at the user in a DB-IRS system can be given by

$$
\begin{aligned}
r &= \sqrt{P_t}\left(\sum_{n=1}^{N} g_n c_n \mathbf{h}_n^T \mathbf{w}_1 + \mathbf{h}_d^T \mathbf{w}_2\right)s + n \\
&= \sqrt{P_t}\left(\sum_{n=1}^{N} g_n c_n h_n \mathbf{a}_1^T(\theta_I)\mathbf{w}_1 + h_d \mathbf{a}_2^T(\theta_d)\mathbf{w}_2\right)s + n \\
&= \sqrt{P_t}\left(\sum_{n=1}^{N} g_n c_n h_n B_r(\theta_I) + h_d B_d(\theta_d)\right)s + n,
\end{aligned}
\tag{7.64}
$$

with the beam patterns corresponding to the IRS and UE $B_r(\theta) = \mathbf{a}_1^T(\theta)\mathbf{w}_1$ and $B_d(\theta) = \mathbf{a}_2^T(\theta)\mathbf{w}_2$, respectively.

7.4.3 Optimization Design

Accordingly, the received SNR at the user is given by

$$
\gamma = \frac{P_t\left|\sum_{n=1}^{N} g_n c_n h_n B_r(\theta_I) + h_d B_d(\theta_d)\right|^2}{\sigma_n^2}.
\tag{7.65}
$$

The aim of the optimization design is to maximize the spectral efficiency $R = \log(1 + \gamma)$ through selecting the optimal weighting vectors and reflecting coefficients, resulting in the following optimization problem

$$
\begin{aligned}
\max_{\Theta,\mathbf{w}_m} \quad & \left|\sum_{n=1}^{N} g_n c_n h_n B_r(\theta_I) + h_d B_d(\theta_d)\right|^2 \\
\text{s.t.} \quad & \|\mathbf{w}_m\|^2 \leqslant 1, \quad m = 1, 2, \\
& \phi_n \in [0, 2\pi), \quad \forall n = 1, 2, \ldots, N.
\end{aligned}
\tag{7.66}
$$

Mostly importantly, Θ and $\mathbf{w}_m$, $m = 1, 2$ in this optimization problem is decoupled thanks to the dual-beam approach. Thus, Θ and $\mathbf{w}_m$, $m = 1, 2$ can be optimized independently, in contrast to the joint optimization in (7.32). In addition, the determination of $\mathbf{w}_1$ and $\mathbf{w}_2$ is also independent.

To get the optimal beam toward the IRS, we need to solve

$$
\begin{aligned}
\max_{\mathbf{w}_1} \quad & |\mathbf{a}_1^T(\theta_I)\mathbf{w}_1|^2 \\
\text{s.t.} \quad & \|\mathbf{w}_1\|^2 \leqslant 1.
\end{aligned}
\tag{7.67}
$$

What we need to know is the DOA information, which can be estimated through classical algorithms efficiently, such as MUltiple SIgnal Classification (MUSIC) and ESPRINT [Gardner, 1988]. Given the estimated θ_I, the optimal weighting vector is $\mathbf{w}_1 = \sqrt{1/N_s}\mathbf{a}_1^*(\theta_I)$, resulting in

$$B_r(\theta_I) = \sqrt{\frac{1}{N_s}}\mathbf{a}_1^T(\theta_I)\mathbf{a}_1^*(\theta_I) = \sqrt{N_s}. \tag{7.68}$$

Since the BS and IRS are fixed, θ_d is easier to obtain and keeps constant in a long-term basis. Similarly, the direct beam gain $B_d(\theta_d) = \sqrt{N_s}$ if the weighting vector is set to

$$\mathbf{w}_2 = \sqrt{\frac{1}{N_s}}\mathbf{a}_2^*(\theta_d). \tag{7.69}$$

For simplicity, we assume that the sidelobes of two beam patterns in the other's direction are negligible, i.e. $|B_d(\theta_I)| = 0$ and $|B_r(\theta_d)| = 0$.

Afterward, the optimization problem in (7.66) is reduced to

$$\max_{\Theta} \quad N_s \left| \sum_{n=1}^{N} g_n c_n h_n + h_d \right|^2 \tag{7.70}$$

$$\text{s.t.} \quad \phi_n \in [0, 2\pi), \quad \forall n = 1, 2, \dots, N,$$

which is equivalent to an IRS system with a single-antenna BS. Therefore, the optimal phase shift for reflecting element n equals to

$$\phi_n^\star = \text{mod}\,[\phi_d - (\phi_{h,n} + \phi_{g,n}), 2\pi], \tag{7.71}$$

where $\phi_{h,n}$, $\phi_{g,n}$, and ϕ_d stand for the phases of h_n, g_n, and h_d, respectively, and mod $[\cdot]$ is the modulo operation. Given the optimal beams in (7.68) and (7.69) and the optimal phase shifts, the received SNR of the dual-beam IRS system, as given by (7.65), is maximized, equaling to

$$\gamma_{max} = \frac{N_s P_t \left| \sum_{n=1}^{N} |g_n||h_n| + |h_d| \right|^2}{\sigma_n^2}, \tag{7.72}$$

implying an active beamforming gain of $N_s = N_b/2$ and a passive beamforming gain of N^2.

Compared with the joint active and passive beamforming optimization, which needs an iterative optimization process, the calculation of dual-beam IRS parameters is substantially simple. In a nutshell, a DB-IRS system only needs to estimate the DOA of the user, and then steers a beam toward it. The beam toward the IRS is fixed, which needs an operation in a long-term basis since the locations of the IRS and BS are fixed and known. Meanwhile, the optimal phase shifts for the IRS are obtained in terms of the acquired CSI directly, as (7.19).

To provide an insight into this technique, Figure 7.7 shows a performance comparison of the receive SNR among several transmission setups: (1) Alternating optimization for the joint active and passive beamforming, as presented in Section

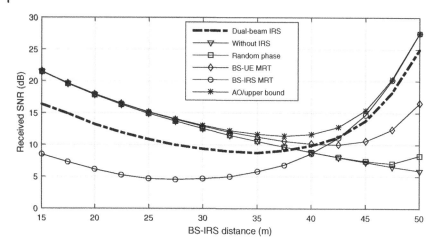

Figure 7.7 Simulation results of the received SNR versus the horizontal distance between the BS and IRS.

7.3.1.2. The number of iterations is set to three, which is enough for well convergence. Since AO achieves the optimal results as the same as the SDR scheme or the upper performance bound, the other two are not shown in the figure for simplicity. (2) BS-UE maximum-ratio transmission (MRT), which sets $\mathbf{w}^{\star} = \mathbf{h}_d^{*}/\|\mathbf{h}_d\|$ to achieve matched filtering in terms of the $\mathbb{B} - \mathbb{U}$ direct link; (3) BS-IRS MRT that sets $\mathbf{w}^{\star} = \mathbf{h}_n^{*}/\|\mathbf{h}_n\|$ to achieve matched filtering based on the $\mathbb{B} - \mathbb{I}$ link, while the optimal reflecting phases are computed accordingly. Since this link is LOS, or a rank-one channel, $\mathbf{h}_n$ can be any row of $\mathbf{H}$. (4) Random phase shifts, where θ_n, $\forall n$ is randomly selected and then perform MRT at the BS based on the combined channel, i.e. $\mathbf{w}^{\star} = \frac{(\mathbf{g}^T\boldsymbol{\Theta}\mathbf{H}+\mathbf{h}_d^T)^{*}}{\|\mathbf{g}^T\boldsymbol{\Theta}\mathbf{H}+\mathbf{h}_d^T\|}$. (5) A benchmark scheme in a MISO system without the aid of IRS by setting $\mathbf{w}^{\star} = \mathbf{h}_d^{*}/\|\mathbf{h}_d\|$.

In a conventional system without IRS, the cell-edge user suffers from low SNR due to severe propagation loss. The cell-edge problem is alleviated by deploying an IRS, as shown in Figure 7.7, where the IRS-aided wireless networks exhibit comparable cell-edge performance as the cell center. This is because the user farther away from the BS is closer to the IRS and thus it is able to get strong reflected signals. It implies that the system coverage can be effectively extended by deploying passive IRS in addition to a BS or active relay. Specially, the BS-UE MRT scheme performs near-optimally when the UE is at the cell center, whereas suffering from a considerable SNR loss at the cell edge. That is because the received signal is dominated by the direct link at the cell center whereas the reflected link is dominant at the cell edge. Moreover, it can be observed that BS-IRS MRT behaves oppositely as the UE moves away from the BS toward the IRS. It also reveals the significance of accurate phase shifts since random phase shifts overwhelm the potential of IRS. The dual-beam IRS scheme can achieve a good balance at the cell center and cell edge. Specifically, its cell-edge performance approaches to the optimal result but the implementation complexity is reduced.

7.5 IRS-Aided Wideband Communications

Previous sections merely discussed the IRS-aided communications in frequency-flat fading (narrow-band) channels, which can be simply modeled by a single channel tap, e.g. $h \sim C\mathcal{N}(0, \sigma_h^2)$. However, one of the technological trends in the wireless transmission is that the signal bandwidth becomes increasingly wider, from tens of kHz in the first-generation systems to hundreds of MHz in the fifth-generation systems, aiming to support a higher transmission rate. As a consequence, most of wireless communications systems nowadays are broadband with signal bandwidths far wider than the *coherence bandwidth*, leading to severe frequency selectivity.

7.5.1 Cascaded Frequency-Selective Channel

A multipath fading channel can be described by the response at time t to an impulse at time $t - \tau$, namely

$$h(\tau, t) = \sum_{l=1}^{L} a_l(t)\delta(\tau - \tau_l(t)), \tag{7.73}$$

where $a_l(t)$ and $\tau_l(t)$ denote the attenuation and propagation delay of the lth path at time t, respectively, and L is the total number of resolvable paths. In a particular situation where the transmitter, receiver, and surrounding environment are all stationary, the attenuation and propagation delays do not change over time. Hence, we obtain the linear time-invariant channel model with an impulse response

$$h(\tau) = \sum_{l=1}^{L} a_l \delta(\tau - \tau_l). \tag{7.74}$$

Practical wireless communications are passband transmission that is carried out in a bandwidth at carrier frequency f_c. However, most of the signal processing in wireless communications, such as channel coding, modulation, detection, and channel estimation, are usually implemented at the baseband. Hence, it makes sense to obtain a complex baseband equivalent model, which is given by Tse and Viswanath [2005]

$$h_b(\tau, t) = \sum_{l=1}^{L} a_l(t)e^{-2\pi j f_c \tau_l(t)}\delta(\tau - \tau_l(t)). \tag{7.75}$$

Following the sampling theorem, we can create a more useful discrete-time channel model by figuring out the ζth tap of the channel filter at (discrete) time n, i.e.

$$h_\zeta[n] = \sum_{l=1}^{L} a_l(nT_s)e^{-2\pi jf_c\tau_l(nT_s)}\text{sinc}\left(\zeta - \frac{\tau_l(nT_s)}{T_s}\right), \quad \zeta = 0, 1, \ldots, Z - 1,$$

(7.76)

where $T_s = 1/B_s$ stands for the sampling period with the signal bandwidth B_s, and the *sinc* function is defined as

$$\text{sinc}(t) := \frac{\sin(\pi t)}{\pi t}.$$

(7.77)

In a special case, where the gains and delays of the paths are time-invariant, Eq. (7.76) is simplified to

$$h_\zeta = \sum_{l=1}^{L} a_l e^{-2\pi jf_c\tau_l}\text{sinc}(\zeta - \frac{\tau_l}{T_s}).$$

(7.78)

For the exposition of the system model of IRS-aided OFDM transmission, we assume both the base station and user are equipped with a single antenna. As derived in Section 7.2.1, the baseband equivalent impulse response of the cascaded IRS channel from the base station to the user via the nth reflecting element can be modeled by

$$v_n(t) = g_n^b(\tau) * c_n * h_n^b(\tau).$$

(7.79)

In other words, the impulse response of the cascaded channel is the linear convolution of the impulse response of the base station to element n channel, the reflection coefficient, and the impulse response of the element n to user channel. According to Eq. (7.78), the discrete-time baseband equivalent channel from the base station to the nth reflecting element in a wideband system can be expressed by

$$\mathbf{h}_n = [h_n^1, h_n^2, \ldots, h_n^\zeta, \ldots, h_n^{Z_n^{BI}}]^T,$$

(7.80)

with the number of delayed taps for reflecting element n in the 𝔹-𝟙 link Z_n^{BI}. Similarly, the discrete-time baseband equivalent channel from the nth reflecting element to the user is denoted by

$$\mathbf{g}_n = [g_n^1, g_n^2, \ldots, g_n^\zeta, \ldots, g_n^{Z_n^{IU}}]^T,$$

(7.81)

where Z_n^{IU} is the number of delayed taps for reflecting element n in the 𝟙-𝕌 link.

Hence, referring to Eq. (7.79), the cascaded channel from the base station to user via the nth reflecting element can be modeled as

$$\mathbf{v}_n = \mathbf{g}_n * c_n * \mathbf{h}_n = c_n(\mathbf{g}_n * \mathbf{h}_n).$$

(7.82)

We can also write the cascaded channel as

$$\mathbf{v}_n = [v_n^1, v_n^2, \dots, v_n^\zeta, \dots, v_n^{Z_n^{\text{BU}}}]^T, \tag{7.83}$$

with the number of delayed taps $Z_n^{\text{BU}} = Z_n^{\text{BI}} + Z_n^{\text{IU}} - 1$. Last but not least, the discrete-time baseband equivalent channel of the direct link can be written as

$$\mathbf{h}_d = [h_d^1, h_d^2, \dots, h_d^\zeta, \dots, h_d^{Z_d}]^T, \tag{7.84}$$

using Z_d to denote the number of delayed taps in the direct link.

7.5.2 IRS-Aided OFDM System

Due to its ability to cope with multipath frequency-selective fading without the need for complex equalization and simple implementation through the use of the digital Fourier transform, OFDM or orthogonal frequency-division multiplexing has become the most dominant modulation technique for wired and wireless communication systems over the past two decades. It has been extensively applied in a wide variety of well-known standards, e.g. Long-Term Evolution (LTE)-Advanced and 5G NR. It is envisioned that it will serve as a key technology in the forthcoming 6G system [Jiang et al., 2021] either in the conventional sub-6 GHz band and high frequency bands. Therefore, it is worth paying strong attention on the modeling and designing of IRS-aided OFDM transmission.

The data transmission in an OFDM system is organized in block-wise in a block fading channel, where the channel remains constant per OFDM symbol. We write

$$\tilde{\mathbf{x}}[t] = [\tilde{x}_0[t], \dots, \tilde{x}_m[t], \dots, \tilde{x}_{M-1}[t]]^T \tag{7.85}$$

to denote the frequency-domain transmission block of the base station on the tth OFDM symbol. (The tilde ~ marks frequency-domain variables.) Transform $\tilde{\mathbf{x}}[t]$ into a time-domain sequence

$$\mathbf{x}[t] = [x_0[t], \dots, x_{m'}[t], \dots, x_{M-1}[t]]^T \tag{7.86}$$

through an M-point Inverse Discrete Fourier Transform (IDFT), i.e.

$$x_{m'}[t] = \frac{1}{M} \sum_{m=0}^{M-1} \tilde{x}_m[t] e^{\frac{2\pi j m' m}{M}}, \tag{7.87}$$

$\forall m' = 0, 1, \dots, M - 1$. Defining the Discrete Fourier Transform (DFT) matrix

$$\mathbf{F} = \begin{bmatrix} \omega_M^{0\cdot 0} & \cdots & \omega_M^{0\cdot(M-1)} \\ \vdots & \ddots & \vdots \\ \omega_M^{(M-1)\cdot 0} & \cdots & \omega_M^{(M-1)\cdot(M-1)} \end{bmatrix} \tag{7.88}$$

with a primitive Mth root of unity $\omega_M^{m \cdot m'} = e^{2\pi j m m'/M}$, the OFDM modulation can also be written in matrix form as

$$\mathbf{x}[t] = \mathbf{F}^{-1} \tilde{\mathbf{x}}[t] = \frac{1}{M} \mathbf{F}^* \tilde{\mathbf{x}}[t]. \tag{7.89}$$

To avoid Inter-Symbol Interference (ISI) and preserve the orthogonality of sub-carriers, a guard interval known as Cyclic Prefix (CP) or called cyclic extension originally is added between two consecutive blocks. CP insertion implies that the last part of an OFDM symbol is copied and inserted at the beginning of this OFDM symbol. Thus, the OFDM symbol with the insertion of CP is expressed by

$$\mathbf{x}^{cp}[t] = [x_{M-N_{cp}}[t], \dots, x_{M-1}[t], x_0[t], \dots, x_{M-1}[t]]^T. \tag{7.90}$$

The ISI can be completely eliminated if the length of CP is no less than the length of any channel filter, i.e.

$$N_{cp} \geqslant \max\left(Z_1^{\mathbb{BU}}, \dots, Z_N^{\mathbb{BU}}, Z_d\right). \tag{7.91}$$

The transmitted signal $\mathbf{x}^{cp}[t]$ goes through the direct channel $\mathbf{h}_d$ to reach the user, resulting in a received signal component $\mathbf{x}^{cp}[t] * \mathbf{h}_d$. Meanwhile, the channel from the base station to user via the nth reflecting element corresponds to a signal component $\mathbf{x}^{cp}[t] * \mathbf{v}_n = c_n \mathbf{x}^{cp}[t] * (\mathbf{g}_n * \mathbf{h}_n)$. Thus, the received symbol vector at the user is computed by

$$
\begin{aligned}
\mathbf{y}^{cp}[t] &= \sum_{n=1}^{N} \mathbf{v}_n * \mathbf{x}^{cp}[t] + \mathbf{h}_d * \mathbf{x}^{cp}[t] + \mathbf{z}[t] \\
&= \sum_{n=1}^{N} c_n (\mathbf{g}_n * \mathbf{h}_n) * \mathbf{x}^{cp}[t] + \mathbf{h}_d * \mathbf{x}^{cp}[t] + \mathbf{z}[t],
\end{aligned} \tag{7.92}
$$

with a vector of additive noise $\mathbf{z}[t]$. Removing the CP, we get

$$
\begin{aligned}
\mathbf{y}[t] &= \sum_{n=1}^{N} \ddot{\mathbf{v}}_n \otimes \mathbf{x}[t] + \ddot{\mathbf{h}}_d \otimes \mathbf{x}[t] + \mathbf{z}[t] \\
&= \left(\sum_{n=1}^{N} \ddot{\mathbf{v}}_n + \ddot{\mathbf{h}}_d\right) \otimes \mathbf{x}[t] + \mathbf{z}[t],
\end{aligned} \tag{7.93}
$$

where $\otimes$ stands for *the cyclic convolution* [Jiang and Kaiser, 2016], $\ddot{\mathbf{v}}_n$ is an M-point channel filter formed by padding zeros at the tail of $\mathbf{v}_n$, i.e.

$$\ddot{\mathbf{v}}_n = [v_n^1, v_n^2, \dots, v_n^{\zeta}, \dots, v_n^{Z_n^{\mathbb{BU}}}, \underbrace{0, \dots, 0}_{\text{Zero padding}}]^T, \tag{7.94}$$

and

$$\ddot{\mathbf{h}}_d = [h_d^1, h_d^2, \dots, h_d^{\zeta}, \dots, h_d^{Z_d}, \underbrace{0, \dots, 0}_{\text{Zero padding}}]^T. \tag{7.95}$$

Denoting the effective channel grasping all paths between the base station and user by $\ddot{\mathbf{h}} = \sum_{n=1}^{N} \ddot{\mathbf{v}}_n + \ddot{\mathbf{h}}_d$, the system model in Eq. (7.93) is simplified to

$$\mathbf{y}[t] = \ddot{\mathbf{h}} \otimes \mathbf{x}[t] + \mathbf{z}[t]. \tag{7.96}$$

Afterward, the DFT demodulator outputs the frequency-domain received signal

$$\tilde{\mathbf{y}}[t] = \mathbf{F}\mathbf{y}[t]. \tag{7.97}$$

Substituting (7.89) and (7.93) into (7.97), and applying *the convolution theorem* for DFT, we have

$$
\begin{aligned}
\tilde{\mathbf{y}}[t] &= \mathbf{F}\left(\ddot{\mathbf{h}} \otimes \mathbf{x}[t]\right) + \mathbf{F}\mathbf{z}[t] \\
&= \sum_{n=1}^{N} \mathbf{F}(\ddot{\mathbf{v}}_n \otimes \mathbf{x}[t]) + \mathbf{F}(\ddot{\mathbf{h}}_d \otimes \mathbf{x}[t]) + \mathbf{F}\mathbf{z}[t] \\
&= \sum_{n=1}^{N} \tilde{\mathbf{v}}_n \odot \tilde{\mathbf{x}}[t] + \tilde{\mathbf{h}}_d \odot \tilde{\mathbf{x}}[t] + \tilde{\mathbf{z}}[t],
\end{aligned} \tag{7.98}
$$

where $\odot$ represents the Hadamard product (i.e. element-wise multiplication), and $\tilde{\mathbf{v}}_n$, $\tilde{\mathbf{h}}_d$, and $\tilde{\mathbf{z}}[t]$ denote frequency-domain channel responses and noise, respectively, which are computed by

$$\tilde{\mathbf{v}}_n = \mathbf{F}\ddot{\mathbf{v}}_n, \quad \tilde{\mathbf{h}}_d = \mathbf{F}\ddot{\mathbf{h}}_d, \quad \tilde{\mathbf{z}}[t] = \mathbf{F}\mathbf{z}[t]. \tag{7.99}$$

Alternatively, we can get the overall frequency-domain channel response as

$$\tilde{\mathbf{h}} = \mathbf{F}\ddot{\mathbf{h}} = \mathbf{F}\left(\sum_{n=1}^{N} \ddot{\mathbf{v}}_n + \ddot{\mathbf{h}}_d\right) = \sum_{n=1}^{N}\mathbf{F}\ddot{\mathbf{v}}_n + \mathbf{F}\ddot{\mathbf{h}}_d = \sum_{n=1}^{N}\tilde{\mathbf{v}}_n + \tilde{\mathbf{h}}_d. \tag{7.100}$$

In this notation, the Channel Frequency Response (CFR) at a typical OFDM subcarrier m can be expressed as

$$\tilde{h}_m = \mathbf{f}_m\left(\sum_{n=1}^{N}\ddot{\mathbf{v}}_n + \ddot{\mathbf{h}}_d\right) = \sum_{n=1}^{N}\mathbf{f}_m\ddot{\mathbf{v}}_n + \mathbf{f}_m\ddot{\mathbf{h}}_d = \sum_{n=1}^{N}\tilde{v}_{m,n} + \tilde{h}_{m,d}, \tag{7.101}$$

where $\mathbf{f}_m$ is the mth row of the DFT matrix $\mathbf{F}$, and $\tilde{h}_m$ denotes the mth entry of $\tilde{\mathbf{h}}$, and $\tilde{v}_{m,n}$ and $\tilde{h}_{m,d}$ are the mth element of $\tilde{\mathbf{v}}_n$ and $\tilde{\mathbf{h}}_d$, respectively. Then, the system model in Eq. (7.98) can be rewritten as

$$\tilde{\mathbf{y}}[t] = \tilde{\mathbf{h}} \odot \tilde{\mathbf{x}}[t] + \tilde{\mathbf{z}}[t], \tag{7.102}$$

which is equivalent to the form of the conventional OFDM system.

At last any frequency-selective channel in the IRS-aided OFDM system is transformed into a set of M independent frequency-flat subcarriers. The signal transmission on the mth subcarrier can be modeled by

$$
\begin{aligned}
\tilde{y}_m[t] &= \tilde{h}_m\tilde{x}_m[t] + \tilde{z}_m[t] \\
&= \left(\sum_{n=1}^{N}\tilde{v}_{m,n} + \tilde{h}_{m,d}\right)\tilde{x}_m[t] + \tilde{z}_m[t], \quad m = 0, 1, \ldots, M - 1.
\end{aligned} \tag{7.103}
$$

Then, we can observe that the signal model of the IRS-aided OFDM transmission over each subcarrier is equivalent to that of the IRS-aided narrowband transmission as given in Eq. (7.16).

7.5.3 Rate Maximization

Neglecting the bandwidth loss due to the insertion of CP, the achievable rate (i.e. spectral efficiency in bps/Hz) of the IRS-aided OFDM system can be computed by

$$
R = \sum_{m=0}^{M-1} \log_2 \left(1 + \frac{P_m \left| \sum_{n=1}^{N} \tilde{v}_{m,n} + \tilde{h}_{m,d} \right|^2}{\sigma_z^2/M} \right)
$$

$$
= \sum_{m=0}^{M-1} \log_2 \left(1 + \frac{P_m \left| \sum_{n=1}^{N} \mathbf{f}_m \ddot{\mathbf{v}}_n + \mathbf{f}_m \ddot{\mathbf{h}}_d \right|^2}{\sigma_z^2/M} \right),
\tag{7.104}
$$

where P_m stands for the transmit power assigned to the mth subcarrier with the constraint $\sum_{m=0}^{M-1} P_m \leqslant P_t$.

To maximize the achievable rate, the reflecting phase shifts need to cater to the frequency-selective channels across different subcarriers, or equivalently, the time-domain channels at different delayed taps. Moreover, $\boldsymbol{\Theta}$ needs to be jointly optimized with the transmit power allocations $\mathbf{p} = [P_0, P_1, \ldots, P_{M-1}]^T$ over the M sub-carriers, formulating the following optimization problem

$$
\max_{\boldsymbol{\Theta}, \mathbf{p}} \; \sum_{m=0}^{M-1} \log_2 \left(1 + \frac{P_m \left| \sum_{n=1}^{N} \mathbf{f}_m \ddot{\mathbf{v}}_n + \mathbf{f}_m \ddot{\mathbf{h}}_d \right|^2}{\sigma_z^2/M} \right)
$$

$$
\text{s.t.} \quad \theta_n \in [0, 2\pi), \quad \forall n = 1, 2, \ldots, N,
$$

$$
P_m \geqslant 0, \quad \forall m = 0, 1, \ldots, M-1,
$$

$$
\sum_{m=0}^{M-1} P_m \leqslant P_t,
\tag{7.105}
$$

which is more difficult to solve as compared with the narrow-band case. To tackle this problem, an efficient Successive Convex Approximation (SCA) based algorithm was proposed in Yang et al. [2020] by approximating the non-concave rate function in this objective function using its concave lower bound based on the first-order Taylor expansion. The SCA-based algorithm is guaranteed to converge to a stationary point of the joint IRS reflection coefficients and transmit power optimization problem, and requires only polynomial complexity over N and M.

To further lower the complexity, Zheng and Zhang [2020] proposed a simplified algorithm where the IRS phase shifts are designed to only align with the time-domain channel with the strongest path strength, thus termed as the strongest Channel Impulse Response (CIR) maximization. Assuming equal power allocation across the OFDM subcarriers, Eq. (7.104) is rewritten as

$$R = \sum_{m=0}^{M-1} \log_2 \left(1 + \frac{P_t \left| \sum_{n=1}^{N} \tilde{v}_{m,n} + \tilde{h}_{m,d} \right|^2}{\sigma_z^2} \right),$$ (7.106)

applying $P_m = P_t/M$. It is non-concave over Θ and thus difficult to maximize optimally. Alternatively, it is considered to maximize the rate upper bound with the aid of the Jensen's inequality, which is given by

$$R \leqslant \log_2 \left(1 + \frac{1}{M} \sum_{m=0}^{M-1} \frac{P_t \left| \sum_{n=1}^{N} \tilde{v}_{m,n} + \tilde{h}_{m,d} \right|^2}{\sigma_z^2} \right).$$ (7.107)

Neglecting the constant terms for simplicity, the following optimization problem is formulated

$$\max_{\Theta} \quad \sum_{m=0}^{M-1} \left| \sum_{n=1}^{N} \tilde{v}_{m,n} + \tilde{h}_{m,d} \right|^2$$ (7.108)

$$\text{s.t.} \quad \theta_n \in [0, 2\pi), \quad \forall n = 1, 2, \ldots, N.$$

By exploiting the time-domain property, a low-complexity method referred to as the strongest CIR maximization was applied to solve this optimization problem sub-optimally, which can refer to Zheng and Zhang [2020].

Last but not least, it is worth noting that the IRS phase shifts θ_n, $\forall n$ affect the channel response at each OFDM subcarrier *identically*. In other words, any reflecting element can only induce the same phase rotation across all OFDM subcarriers at a given time instant, and cannot realize frequency-specific phase rotation due to the restriction of hardware implementation. Wu et al. [2021] shows the upper bound of the achievable rate by assuming that (ideally) different IRS reflection coefficients can be designed for different subcarriers, thus making the reflection design *frequency-specific*. It is observed that this rate upper bound outperforms the SCA-based solution with the practical frequency-flat IRS reflection quite substantially, and the rate gap increases with the number of subcarriers. This thus reveals that a fundamental limit of IRS-aided OFDM systems lies in the lack of frequency-specific IRS reflection due to its passive operation.

7.6 Multi-User IRS Communications

Due to its disruptive capability in smartly reconfiguring the wireless propagation environment and hardware constraints, the integration of IRS brings some fundamental particularities in the coordination of multi-user signal transmission. For instance, the lack of frequency-selective reflection leads to the performance loss of frequency-division approaches. Hence, it is worth paying particular attention to the impact of IRS on multi-user signal transmission.

7.6.1 Multiple Access Model

Figure 7.8 shows an IRS-assisted multi-user MIMO downlink communications system, where an intelligent surface with N reflecting elements is deployed to assist the transmission from an N_b-antenna BS to K single-antenna UE. The IRS is a passive device, where TDD is usually adopted to simplify channel estimation. The users send pilot signals in uplink training so that the BS can estimate uplink CSI, which is used for optimizing downlink data transmission due to channel reciprocity. To characterize the theoretical analysis, assume that the CSI of all involved channels is perfectly known at the BS. In addition, the channels follow frequency-flat block fading. Since the direct paths from either the BS or the IRS to UEs may be blocked, the corresponding small-scale fading follows Rayleigh distribution. Consequently, the channel gain between antenna element $n_b \in \{1, 2, \ldots, N_b\}$ and user $k \in \mathcal{K} \triangleq \{1, 2, \ldots, K\}$ is a circularly symmetric complex Gaussian random variable with zero mean and variance σ_f^2, i.e. $f_{kN_b} \sim \mathcal{CN}(0, \sigma_f^2)$. Thus, the channel vectors from the BS and the IRS to the kth UE are denoted by

$$\mathbf{f}_k = [f_{k1}, f_{k2}, \ldots, f_{kN_b}]^T, \tag{7.109}$$

and

$$\mathbf{g}_k = [g_{k1}, g_{k2}, \ldots, g_{kN}]^T, \tag{7.110}$$

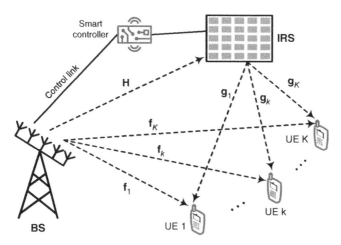

Figure 7.8 Schematic diagram of an IRS-aided multi-user MIMO system, consisting of a multi-antenna BS, K single-antenna UE, and a reflecting surface with N IRS elements.

respectively, where $g_{kn} \sim \mathcal{CN}(0, \sigma_g^2)$ is the channel gain between IRS element $n \in \mathcal{N} \triangleq \{1, 2, \ldots, N\}$ and user k. We write

$$\mathbf{h}_n = [h_{n1}, h_{n2}, \ldots, h_{nN_b}]^T \tag{7.111}$$

to denote the channel vector between the BS and the nth reflecting element. So the channel matrix from the BS to the IRS is expressed as $\mathbf{H} \in \mathbb{C}^{N \times N_b}$, where the nth row of $\mathbf{H}$ equals to $\mathbf{h}_n^T$. In contrast to randomly distributed and moving UEs, a favorable location is deliberately selected for the IRS to exploit a LOS path of the fixed BS without any blockage, resulting in Rician fading, i.e.

$$\mathbf{H} = \sqrt{\frac{\Gamma \sigma_h^2}{\Gamma + 1}} \mathbf{H}_{\text{LOS}} + \sqrt{\frac{\sigma_h^2}{\Gamma + 1}} \mathbf{H}_{\text{NLOS}}, \tag{7.112}$$

with the Rician factor Γ, the LOS component $\mathbf{H}_{\text{LOS}}$, the multipath component $\mathbf{H}_{\text{NLOS}}$ consisting of independent entries that follow $\mathcal{CN}(0, 1)$, and the BS-IRS path loss σ_h^2.

The reflecting surface is equipped with a smart controller, which can adaptively adjust the phase shift of each IRS element in terms of the CSI acquired through periodic channel estimation. We write $c_{nk} = \beta_{nk} e^{j\phi_{nk}}$ to denote the reflection coefficient of the nth IRS element for user k, with the induced phase shift $\phi_{nk} \in [0, 2\pi)$ and the amplitude attenuation $\beta_{nk} \in [0, 1]$. As revealed by Wu and Zhang [2019], the optimal value of reflecting attenuation is $\beta_{nk} = 1, \forall n, k$ to maximize the received power and simplify the hardware implementation. Due to the high path loss, the signals that are reflected by the IRS twice or more are negligible. By ignoring hardware impairments such as quantified phase shifts and phase noise, the kth UE observes the received signal

$$r_k = \sqrt{P_d} \left(\sum_{n=1}^{N} g_{kn} e^{j\phi_{kn}} \mathbf{h}_n^T + \mathbf{f}_k^T \right) \mathbf{s} + n_k, \tag{7.113}$$

where $\mathbf{s}$ denotes the $N_b \times 1$ vector of transmitted signals over the BS antenna array, P_d expresses the power constraint of the BS, n_k is AWGN with zero mean and variance σ_n^2, namely, $n_k \sim \mathcal{CN}(0, \sigma_n^2)$. Define $\mathbf{\Theta}_k = \text{diag}\{e^{j\phi_{1k}}, \ldots, e^{j\phi_{Nk}}\}$, Eq. (7.113) can be rewritten in matrix form as

$$r_k = \sqrt{P_d} (\mathbf{g}_k^T \mathbf{\Theta}_k \mathbf{H} + \mathbf{f}_k^T) \mathbf{s} + n_k. \tag{7.114}$$

7.6.2 Orthogonal Multiple Access

This part analyzes the achievable spectral efficiency of two typical Orthogonal Multiple Access (OMA) schemes in an IRS-aided multi-user MIMO system and presents an alternating method to optimize active beamforming at the BS and passive reflection at the IRS jointly.

7.6.2.1 Time-Division Multiple Access

This scheme divides the signaling dimensions along the time axis into orthogonal portions called time slots. Each user transmits over the entire bandwidth but cyclically accesses its assigned slot. It implies non-continuous transmission, which simplifies the system design since some processing such as channel estimation can be performed during the time slots of other users. Another advantage is that TDMA is able to assign multiple time slots for a single user, increasing system flexibility. Mathematically, a radio frame is orthogonally divided into K time slots, where the CSI keeps constant. At the kth slot, the BS applies linear beamforming $\mathbf{w}_k \in \mathbb{C}^{N_b \times 1}$, where $\|\mathbf{w}_k\|^2 \leqslant 1$, to send the information-bearing symbol s_k with zero mean and unit variance, i.e. $\mathbb{E}\left[|s_k|^2\right] = 1$, intended for a general user k.

Substituting $\mathbf{s} = \mathbf{w}_k s_k$ into Eq. (7.114), we obtain

$$r_k = \sqrt{P_d}(\mathbf{g}_k^T \boldsymbol{\Theta}_k \mathbf{H} + \mathbf{f}_k^T)\mathbf{w}_k s_k + n_k. \tag{7.115}$$

By jointly optimizing active beamforming $\mathbf{w}_k$ and reflection $\boldsymbol{\Theta}_k$, the instantaneous SNR of user k, i.e.

$$\gamma_k = \frac{P_d |(\mathbf{g}_k^T \boldsymbol{\Theta}_k \mathbf{H} + \mathbf{f}_k^T)\mathbf{w}_k|^2}{\sigma_n^2} \tag{7.116}$$

can be maximized, formulating the following optimization

$$\begin{aligned}
\max_{\boldsymbol{\Theta}_k, \, \mathbf{w}_k} \quad & |(\mathbf{g}_k^T \boldsymbol{\Theta}_k \mathbf{H} + \mathbf{f}_k^T)\mathbf{w}_k|^2 \\
\text{s.t.} \quad & \|\mathbf{w}_k\|^2 \leqslant 1 \\
& \phi_{nk} \in [0, 2\pi), \quad \forall n = 1, \dots, N, \ \forall k = 1, \dots, K,
\end{aligned} \tag{7.117}$$

which is non-convex because the objective function is not jointly concave with respect to $\boldsymbol{\Theta}_k$ and $\mathbf{w}_k$. To solve this problem, we can apply alternating optimization that alternately optimizes $\boldsymbol{\Theta}_k$ and $\mathbf{w}_k$ in an iterative manner [Wu and Zhang, 2019]. Given an initialized transmit vector $\mathbf{w}_k^{(0)}$, Eq. (7.117) is simplified to

$$\begin{aligned}
\max_{\boldsymbol{\Theta}_k} \quad & |(\mathbf{g}_k^T \boldsymbol{\Theta}_k \mathbf{H} + \mathbf{f}_k^T)\mathbf{w}_k^{(0)}|^2 \\
\text{s.t.} \quad & \phi_{nk} \in [0, 2\pi), \quad \forall n = 1, \dots, N, \ \forall k = 1, \dots, K.
\end{aligned} \tag{7.118}$$

The objective function is still non-convex but it enables a closed-form solution through applying the well-known triangle inequality

$$|(\mathbf{g}_k^T \boldsymbol{\Theta}_k \mathbf{H} + \mathbf{f}_k^T)\mathbf{w}_k^{(0)}| \leqslant |\mathbf{g}_k^T \boldsymbol{\Theta}_k \mathbf{H} \mathbf{w}_k^{(0)}| + |\mathbf{f}_k^T \mathbf{w}_k^{(0)}|. \tag{7.119}$$

The equality achieves if and only if

$$\arg\left(\mathbf{g}_k^T \boldsymbol{\Theta}_k \mathbf{H} \mathbf{w}_k^{(0)}\right) = \arg\left(\mathbf{f}_k^T \mathbf{w}_k^{(0)}\right) \triangleq \varphi_{0k}. \tag{7.120}$$

Define $\mathbf{q}_k = [q_{1k}, q_{2k}, \ldots, q_{Nk}]^H$ with $q_{nk} = e^{j\phi_{nk}}$ and $\chi_k = \mathrm{diag}(\mathbf{g}_k^T)\mathbf{H}\mathbf{w}_k^{(0)} \in \mathbb{C}^{N \times 1}$, we have $\mathbf{g}_k^T \boldsymbol{\Theta}_k \mathbf{H}\mathbf{w}_k^{(0)} = \mathbf{q}_k^H \chi_k \in \mathbb{C}$. Ignore the constant term $|\mathbf{f}_k^T \mathbf{w}_k^{(0)}|$, Eq. (7.118) is transformed to

$$\max_{\mathbf{q}_k} \quad |\mathbf{q}_k^H \chi_k|$$

$$\text{s.t.} \quad |q_{nk}| = 1, \quad \forall n = 1, \ldots, N, \ \forall k = 1, \ldots, K, \tag{7.121}$$

$$\arg(\mathbf{q}_k^H \chi_k) = \varphi_{0k}.$$

The solution for Eq. (7.121) can be derived as

$$\mathbf{q}_k^{(1)} = e^{j(\varphi_{0k} - \arg(\chi_k))} = e^{j\left(\varphi_{0k} - \arg\left(\mathrm{diag}(\mathbf{g}_k^T)\mathbf{H}\mathbf{w}_k^{(0)}\right)\right)}. \tag{7.122}$$

Accordingly,

$$\begin{aligned}
\phi_{nk}^{(1)} &= \varphi_{0k} - \arg\left(g_{nk}\mathbf{h}_n^T\mathbf{w}_k^{(0)}\right) \\
&= \varphi_{0k} - \arg\left(g_{nk}\right) - \arg\left(\mathbf{h}_n^T\mathbf{w}_k^{(0)}\right),
\end{aligned} \tag{7.123}$$

where $\mathbf{h}_n^T\mathbf{w}_k^{(0)} \in \mathbb{C}$ can be regarded as an effective SISO channel perceived by the nth reflecting element, combining the effects of transmit beamforming $\mathbf{w}_k^{(0)}$ and channel response $\mathbf{h}_n$. In this regard, Eq. (7.123) implies that an IRS reflector should be tuned such that the phase of the reflected signal through the cascaded link is compensated, and the residual phase is aligned with that of the signal over the direct link, to achieve coherent combining at the receiver. Once the reflecting phases at the first iteration, i.e. $\boldsymbol{\Theta}_k^{(1)} = \mathrm{diag}\{e^{j\phi_{1k}^{(1)}}, e^{j\phi_{2k}^{(1)}}, \ldots, e^{j\phi_{Nk}^{(1)}}\}$ are determined, the optimization is alternated to update $\mathbf{w}_k$. The BS can apply matched filtering to maximize the strength of a desired signal, resulting in

$$\mathbf{w}_k^{(1)} = \frac{(\mathbf{g}_k^T\boldsymbol{\Theta}_k^{(1)}\mathbf{H} + \mathbf{f}_k^T)^H}{\|\mathbf{g}_k^T\boldsymbol{\Theta}_k^{(1)}\mathbf{H} + \mathbf{f}_k^T\|}. \tag{7.124}$$

After the completion of the first iteration, the BS gets $\boldsymbol{\Theta}_k^{(1)}$ and $\mathbf{w}_k^{(1)}$, which serve as the initial input for the second iteration to derive $\boldsymbol{\Theta}_k^{(2)}$ and $\mathbf{w}_k^{(2)}$. This process iterates until the convergence is achieved with the optimal beamformer $\mathbf{w}_k^\star$ and optimal reflection $\boldsymbol{\Theta}_k^\star$. Substituting $\mathbf{w}_k^\star$ and $\boldsymbol{\Theta}_k^\star$ into Eq. (7.116), we can derive the achievable spectral efficiency of user k as

$$R_k = \frac{1}{K}\log\left(1 + \frac{P_d|(\mathbf{g}_k^T\boldsymbol{\Theta}_k^\star\mathbf{H} + \mathbf{f}_k^T)\mathbf{w}_k^\star|^2}{\sigma_n^2}\right). \tag{7.125}$$

Thereby, the sum rate of the TDMA IRS system can be computed by

$$R_{\text{TDMA}} = \sum_{k=1}^{K} R_k. \tag{7.126}$$

7.6.2.2 Frequency-Division Multiple Access

In FDMA, the system bandwidth is divided along the frequency axis into K orthogonal subchannels. Each user occupies a dedicated subchannel over the entire time. The BS employs linear beamforming $\mathbf{w}_k$ to transmit s_k over the kth subchannel with equally allocated transmit power P_d/K. Thus, the achievable spectral efficiency of user k is

$$R_k = \frac{1}{K} \log \left(1 + \frac{P_d/K \left| (\mathbf{g}_k^T \mathbf{\Theta}_k \mathbf{H} + \mathbf{f}_k^T) \mathbf{w}_k \right|^2}{\sigma_n^2/K} \right). \tag{7.127}$$

In contrast to TDMA, where the IRS phase shifts can be dynamically adjusted in different slots, the surface can be optimized only for a particular user, whereas other users suffer from phase-unaligned reflection. That is because the hardware limitation of IRS passive elements, which can be fabricated in *time-selective* rather than *frequency-selective*.

Without losing generality, we suppose the FDMA system optimizes the IRS to aid the signal transmission of user $\hat{k}$, the optimal parameters $\mathbf{\Theta}_{\hat{k}}^{\star}$ and $\mathbf{w}_{\hat{k}}^{\star}$ can be derived using the same alternating optimization as that of TDMA. Once the phase shifts of the surface are completely adjusted for $\hat{k}$, what the remaining $K - 1$ users, denoted by $\{i | i = 1, 2, \ldots, K, \ i \neq \hat{k}\}$, can do is to realize partial optimization (instead of joint optimization) by updating their respective active beamforming based on the combined channel gain $\mathbf{g}_i^T \mathbf{\Theta}_{\hat{k}}^{\star} \mathbf{H} + \mathbf{f}_i^T$. For user i, the beamformer can be optimized as

$$\mathbf{w}_i^{\star} = \frac{(\mathbf{g}_i^T \mathbf{\Theta}_{\hat{k}}^{\star} \mathbf{H} + \mathbf{f}_i^T)^H}{\| \mathbf{g}_i^T \mathbf{\Theta}_{\hat{k}}^{\star} \mathbf{H} + \mathbf{f}_i^T \|}. \tag{7.128}$$

Then, the sum rate of the FDMA IRS system can be calculated by

$$R_{FDMA} = \frac{1}{K} \log \left(1 + \frac{P_d | (\mathbf{g}_{\hat{k}}^T \mathbf{\Theta}_{\hat{k}}^{\star} \mathbf{H} + \mathbf{f}_{\hat{k}}^T) \mathbf{w}_{\hat{k}}^{\star} |^2}{\sigma_n^2} \right) \tag{7.129}$$

$$+ \sum_i \frac{1}{K} \log \left(1 + \frac{P_d | (\mathbf{g}_i^T \mathbf{\Theta}_{\hat{k}}^{\star} \mathbf{H} + \mathbf{f}_i^T) \mathbf{w}_i^{\star} |^2}{\sigma_n^2} \right).$$

7.6.3 Non-Orthogonal Multiple Access

Although the inter-user interference among orthogonally multiplexed users is mitigated to facilitate low-complexity multi-user detection at the receiver, it is widely recognized that OMA cannot achieve the sum-rate capacity of a multi-user wireless system. Superposition coding and successive interference cancelation (SIC) make it possible to reuse each orthogonal resource unit by more than one user. At the transmitter, all the individual information symbols are superimposed into a single waveform, while the SIC at the receiver decodes the signals iteratively until it gets the desired signal.

Mathematically, the BS superimposes K information-bearing symbols into a composite waveform

$$\mathbf{s} = \sum_{k=1}^{K} \sqrt{\alpha_k} \mathbf{w}_k s_k, \tag{7.130}$$

where α_k represents the power allocation coefficient subjecting to $\sum_{k=1}^{K} \alpha_k \leqslant 1$. The challenge is to decide how to allocate the power among the users, which is critical for interference cancelation at the receiver. That is why NOMA is regarded as a kind of power-domain multiple access. Generally, more power is allocated to the users with smaller channel gain, e.g. located farther from the BS, to improve the received SNR, so that high detection reliability can be guaranteed. Despite of less power assigned to a user with a stronger channel gain, e.g. close to the BS, it is capable of detecting its signal correctly with reasonable SNR. As an example, illustration of a downlink IRS-aided NOMA system consisting of a BS, a surface, and $K = 2$ users are given in Figure 7.9. Substitute Eq. (7.130) into Eq. (7.114) to

Figure 7.9 Illustration of a downlink IRS-aided NOMA system consisting of a BS, a surface, a far user, and a near user.

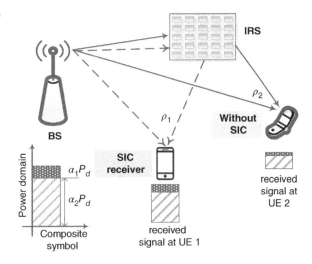

yield the observation of user k as

$$r_k = \sqrt{P_d}(\mathbf{g}_k^T\boldsymbol{\Theta}_k\mathbf{H} + \mathbf{f}_k^T)\sum_{k'=1}^{K}\sqrt{\alpha_{k'}}\mathbf{w}_{k'}s_{k'} + n_k$$

$$= \underbrace{\sqrt{\alpha_k P_d}(\mathbf{g}_k^T\boldsymbol{\Theta}_k\mathbf{H} + \mathbf{f}_k^T)\mathbf{w}_k s_k}_{\text{Desired signal}} \tag{7.131}$$

$$+ \underbrace{\sqrt{P_d}(\mathbf{g}_k^T\boldsymbol{\Theta}_k\mathbf{H} + \mathbf{f}_k^T)\sum_{k'=1,k'\neq k}^{K}\sqrt{\alpha_{k'}}\mathbf{w}_{k'}s_{k'} + n_k}_{\text{Multi-user interference}}.$$

Due to the hardware limit, the IRS can only assist one user while other users have to share the common phase shifts that are not favorable for them. As FDMA, we suppose the system optimizes the IRS to aid the signal transmission of user $\hat{k}$.

The optimal parameters $\boldsymbol{\Theta}_{\hat{k}}^\star$ and $\mathbf{w}_{\hat{k}}^\star$ can be derived using the same alternating optimization as the TDMA. Once the phase shifts of the surface are completely adjusted for $\hat{k}$, a user $k \neq \hat{k}$ can partially optimize its transmission by deriving its active beamforming given the combined channel gain $\mathbf{g}_k^T\boldsymbol{\Theta}_{\hat{k}}^\star\mathbf{H} + \mathbf{f}_k^T$. Similar to Eq. (7.128), the beamformer for user k is figured out as

$$\mathbf{w}_k^\star = \frac{(\mathbf{g}_k^T\boldsymbol{\Theta}_{\hat{k}}^\star\mathbf{H} + \mathbf{f}_k^T)^H}{\|\mathbf{g}_k^T\boldsymbol{\Theta}_{\hat{k}}^\star\mathbf{H} + \mathbf{f}_k^T\|}. \tag{7.132}$$

The same signal $\mathbf{x}$ that contains all information symbols is delivered to all users. The optimal order of interference cancelation is detecting the user with the most power allocation (the weakest channel gain) to the user with the least power allocation (the strongest channel gain). We write $\rho_k = (\mathbf{g}_k^T\boldsymbol{\Theta}_{\hat{k}}^\star\mathbf{H} + \mathbf{f}_k^T)\mathbf{w}_k^\star, \forall k$ to denote the effective gain of the combined channel for user k. Without loss of generality, assume that user 1 has the largest combined channel gain, and user K is the weakest, i.e.

$$\|\rho_1\|^2 \geqslant \|\rho_2\|^2 \geqslant \cdots \geqslant \|\rho_K\|^2. \tag{7.133}$$

With this order, each NOMA user decodes s_K first, and then subtracts its resultant component from the received signal. As a result, a typical user k after the first SIC iteration gets

$$\tilde{r}_k = r_k - \rho_k\sqrt{\alpha_K P_d}s_K = \rho_k\sum_{k=1}^{K-1}\sqrt{\alpha_k P_d}s_k + n_k, \tag{7.134}$$

assuming error-free detection and perfect channel knowledge. In the second iteration, the user decodes s_{K-1} using the remaining signal $\tilde{r}_k$. The cancelation iterates until each user gets the symbol intended for it. Particularly, the weakest user decodes its own signal directly without SIC since it is allocated the most power. Treating the multi-user interference as noise, the SNR for user K can be written as

$$\gamma_K = \frac{\|\rho_K\|^2 \alpha_K P_d}{\|\rho_K\|^2 \sum_{k=1}^{K-1} \alpha_k P_d + \sigma_n^2}. \tag{7.135}$$

In general, user k successfully cancels the signals from user $k + 1$ to K but suffering from the interference from user 1 to $k - 1$. Consequently, the received SNR for user k is

$$\gamma_k = \frac{\|\rho_k\|^2 \alpha_k P_d}{\|\rho_k\|^2 \sum_{k'=1}^{k-1} \alpha_{k'} P_d + \sigma_n^2}, \tag{7.136}$$

resulting in the achievable rate of $R_k = \log(1 + \gamma_k)$. The sum rate of downlink IRS-aided NOMA transmission is computed by

$$R_{\text{NOMA}} = \sum_{k=1}^{K} \log\left(1 + \frac{\|\rho_k\|^2 \alpha_k P_d}{\|\rho_k\|^2 \sum_{k'=1}^{k-1} \alpha_{k'} P_d + \sigma_n^2}\right). \tag{7.137}$$

7.7 Channel Aging and Prediction

To fully realize the potential of the IRS in boosting power and spectral efficiencies, a base station needs to know instantaneous channel state information of the direct link and the cascaded link. In a Frequency-Division Duplex (FDD) system, the CSI is estimated at the user and fed back to the base station through a limited feedback channel. Due to the delay of feedback, the available CSI at the base station may become aged before its actual usage. Although the feedback delay can be avoided in a TDD system exploiting channel reciprocity, it is still possible that the CSI is aged due to a processing delay. That is, we should consider the time taken by the calculation of optimal reflecting phases and the configuration of the IRS elements, especially in high-mobility or high-frequency environments. In practice, the system performance is vulnerable to such feedback and processing delays since the knowledge of CSI is outdated quickly. This phenomenon is referred to as *channel aging*, subjected to channel fading and hardware impairments.

In addition to the IRS-aided systems, it has been extensively recognized that channel aging severely degrades the performance of a wide variety of adaptive wireless communications systems, including MIMO precoding [Zheng and Rao, 2008], multi-user MIMO [Wang et al., 2014], massive MIMO, cell-free massive MIMO [Jiang and Schotten, 2021a], beamforming, opportunistic relay selection [Jiang and Schotten, 2021b], interference alignment, closed-loop transmit diversity, transmit antenna selection [Yu et al., 2017], orthogonal frequency-division multiple access, Coordinated Multi-Point (CoMP) transmission, physical layer security, mobility management, to name a few.

To tackle the problem of channel aging, many mitigation algorithms and protocols have been proposed in the literature. These techniques either passively compensate for the performance loss with a price of scarce radio resources or aim to achieve only the full performance potential partially by designing a system under the assumption of outdated CSI. In contrast, an alternative method known as *channel prediction* opens a new way for an efficient and effective approach to directly improve the CSI accuracy without wasting radio resources, attracting a lot of attention. Two model-based predictive techniques, namely, Auto-Regressive (AR) model [Baddour and Beaulieu, 2005] and Parametric Model [Adeogun et al., 2014], have been applied by statistically modeling wireless channels. The AR prediction method models a wireless channel as an auto-regressive process and derives a next-time channel state by utilizing a weighted linear combination of the past and current channel states. Although the AR model is simple, it is vulnerable to noise and error propagation, making it unattractive in long-range prediction [Jiang et al., 2020]. The parametric model assumes that a fading channel is the superposition of some complex sinusoids, and its parameters, e.g. attenuation, phases, spatial angels, the Doppler shifts, and the number of scatters, change much slowly relative to channels' fading rate and can be assessed accurately. In addition to the tedious estimation process, the estimated parameters will expire soon in a time-varying channel and therefore need to be re-estimated iteratively, leading to high computational complexity [Jiang and Schotten, 2019*b*].

In March 2016 when AlphaGo, a computer program developed by Google Deep-Mind [Silver et al., 2016], achieved an overwhelming victory versus a human champion in the game of Go, the passion for exploring Artificial Intelligence (AI) technology was sparked almost in all scientific and engineering branches. Actually, the wireless research community started to apply AI techniques to solve communication problems long ago. Exploiting the capability of time-series prediction [Connor et al., 1994], a classical AI technique called Recurrent Neural Network (RNN) has been applied to form channel predictors for single-antenna frequency-flat fading channels [Jiang and Schotten, 2018*a,b*] and then further extended to MIMO frequency-selective fading channels [Jiang and Schotten, 2019c]. Recently, the feasibility and effectiveness of applying a deep neural network based on Long-Short Term Memory (LSTM) and Gated Recurrent Unit (GRU) to predict fading channels have also been investigated [Jiang and Schotten, 2020b,c]. This section will provide the readers a comprehensive view of channel aging, channel prediction, and the fundamentals of using deep learning to predict fading channels.

7.7.1 Outdated Channel State Information

Due to the feedback and processing delays, there is a gap between the time instant when pilot signals probe the uplink channels and the time instant when the downlink data transmission happens with the phase shifts of the IRS elements tuned in terms of the measured CSI. The measured CSI may be outdated under the fluctuation of channels raised mainly by *Doppler shifts* (due to *user mobility* and *high frequency*), as well as *phase noise* at the transceiver.

Recall that the channel tap gain

$$h_l[t] = \sum_i a_i(tT_s)e^{-j2\pi f_c \tau_i(tT_s)}\mathrm{sinc}\left[l - \frac{\tau_i(tT_s)}{T_s}\right] \tag{7.138}$$

for $l = 0, \ldots, L - 1$, with the carrier frequency f_c, sampled attenuation $a_i(tT_s)$ and delay $\tau_i(tT_s)$ of the ith signal path, the sampling period T_s, and $\mathrm{sinc}(x) \triangleq \frac{\sin(x)}{x}$ for $x \neq 0$. To model how fast the taps $h_l[t]$ evolve with time t, a statistical quantity known as the tap gain auto-correlation function is defined as

$$R_l[\tau] = \mathbb{E}[h_l^*[t]h_l[t + \tau]]. \tag{7.139}$$

With the classical Doppler spectrum of the Jakes model, the auto-correlation function takes the value

$$R_l[\tau] = J_0(2\pi f_d \tau), \tag{7.140}$$

where f_d means the maximal Doppler shift, τ stands for the delay between the outdated and actual CSI, and $J_0(\cdot)$ represents the *zeroth* order Bessel function of the first kind. In particular, the maximal Doppler frequency can be computed by

$$f_d = \frac{f_c v}{c} = \frac{v}{\lambda}, \tag{7.141}$$

where v denotes the velocity of a moving object, c is the speed of the light in the free space, and λ represents the wavelength of carrier frequency. To provide a concrete view, the auto-correlation values of fading channels with Doppler shifts of 50, 100, and 200 Hz are demonstrated in Figure 7.10.

7.7.1.1 Doppler Shift

For the sake of simplicity, we ignore time indices and denote an actual channel realization of a MIMO flat-fading channel by $\mathbf{H} = [h_{n_r n_t}]_{N_r \times N_t}$ and its outdated version $\mathbf{H}' = [h'_{n_r n_t}]_{N_r \times N_t}$. In the context of IRS, this MIMO channel can model the links between a multi-antenna BS to a multi-antenna UE, between a multi-antenna BS to an IRS, or between an IRS to a multi-antenna UE. A correlation coefficient is used to quantify the inaccuracy of the outdated CSI in independent and identically distributed (*i.i.d.*) channels, i.e.

$$\rho = \frac{|cov(h_{n_r n_t}, h'_{n_r n_t})|}{\mu_h \mu_{h'}}, \tag{7.142}$$

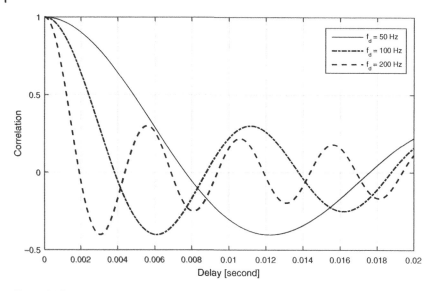

Figure 7.10 Auto-correlation of fading channels in terms of different Doppler shifts: 50, 100, and 200 Hz.

where $cov(\cdot)$ means the covariance of two random variables, μ stands for the standard deviation, and $h_{n_t n_r}$ denotes the channel gain between transmit antenna n_t and receive antenna n_r or the n_r^{th} reflecting element. Due to *i.i.d.* elements in $\mathbf{H}$ and $\mathbf{H}'$, ρ is independent of n_r and n_t. Thus, Eq. (7.142) can be simplified into

$$\rho = \frac{|cov(h, h')|}{\mu_h \mu_{h'}}. \tag{7.143}$$

Since the elements of $\mathbf{H}$ and $\mathbf{H}'$ are both zero mean circularly symmetric Gaussian distributed, according to Ramya and Bhashyam [2009], their relationship can be given by

$$\mathbf{H}' = \rho \mathbf{H} + \sqrt{1 - \rho^2} \mathbf{E}, \tag{7.144}$$

where $\mathbf{E} = [\varepsilon_{n_r n_t}]_{N_r \times N_t}$ is a matrix consisting of normalized Gaussian random variables, i.e. $\varepsilon_{n_r n_t} \sim \mathcal{CN}(0, 1)$. For frequency-flat fading in a narrowband system, a single tap and its outdated version are denoted by h and h', respectively. Accordingly, we have

$$h' = \sigma_{h'} \left(\frac{\rho}{\sigma_h} h + \varepsilon \sqrt{1 - \rho^2} \right), \tag{7.145}$$

where ε is a random variable with standard normal distribution $\varepsilon \sim \mathcal{CN}(0, 1)$ and $\sigma_{h'}^2$ is the variance of h'.

Assuming the Jakes' model, $\mathbf{H}$ and $\mathbf{H}'$ follow joint complex Gaussian distribution, where the correlation coefficient takes the value $\rho = J_0(2\pi f_d \tau)$. Thus, $\mathbf{H}$ conditioned on $\mathbf{H}'$ follows Gaussian distribution, which is modeled by

$$\mathbf{H}|\mathbf{H}' \sim \mathcal{CN}(\rho \mathbf{H}', 1 - \rho^2).$$

Due to the assumption of a normalized channel gain $\mathbb{E}[|h|^2] = 1$, the average SNR $\bar{\gamma} = \mathbb{E}\left[\frac{|h|^2 P_t}{\sigma_n^2}\right]$ is simplified to $\bar{\gamma} = P_t/\sigma_n^2$. Thus, an instantaneous SNR $\gamma = \frac{||\mathbf{H}||^2 P_t}{\sigma_n^2}$ can be rewritten as $\gamma = ||\mathbf{H}||^2 \bar{\gamma}$. Conditioned on its outdated version $\gamma' = ||\mathbf{H}'||^2 \bar{\gamma}$, γ follows a noncentral Chi-square distribution with two degrees of freedom. Its Probability Density Function (PDF) is expressed as [Jiang et al., 2016]

$$f_{\gamma|\gamma'}(\gamma|\gamma') = \frac{1}{\bar{\gamma}(1 - \rho^2)} e^{-\frac{\gamma + \rho^2 \gamma'}{\bar{\gamma}(1-\rho^2)}} J_0\left(\frac{2\sqrt{\rho^2 \gamma \gamma'}}{\bar{\gamma}(1 - \rho^2)}\right). \tag{7.146}$$

7.7.1.2 Phase Noise

Due to imperfect oscillators at the transmitter, the transmitted signals suffer from phase noise during the up-conversion processing from baseband to passband signals, and *vice versa* at the receiver. Such phase noise is not only random but also time-varying, leading to the outdated CSI that is equivalent to the effect of Doppler shift. Utilizing a well-established Wiener process [Khanzadi et al., 2016], the phase noise of the BS and the UE at *discrete-time* instant t can be modeled as

$$\begin{cases} \Delta\phi_t = \phi_t - \phi_{t-1}, & \Delta\phi_t \sim \mathcal{CN}(0, \sigma_\phi^2) \\ \Delta\varphi_t = \varphi_t - \varphi_{t-1}, & \Delta\varphi_t \sim \mathcal{CN}(0, \sigma_\varphi^2), \end{cases} \tag{7.147}$$

where the increment variances are given by $\sigma_i^2 = 4\pi^2 f_c c_i T_s$, $\forall i = \phi, \varphi$ with symbol period T_s and oscillator-dependent constant c_i. Since the passive reflecting elements in IRS have no RF components, there is no phase noise in a reflecting surface.

7.7.2 Impact of Channel Aging on IRS

From a practical point of view, the estimated CSI used to calculate the optimal reflecting phases may substantially differ from the actual CSI at the instant of using the selected phases to reflect the signals. Utilizing an outdated version of the CSI rather than the actual CSI may severely deteriorate the system performance of an IRS-enhanced system.

Considering the effect of Doppler shifts and phase noise, we can write

$$\mathfrak{h}_d = h_d e^{j(\phi_u + \varphi_u)} \tag{7.148}$$

to denote the overall channel gain of the direct link between the BS and UE during the signal transmission denoted by time u, where ϕ_u and φ_u denote the corresponding phase noise of the BS and UE at instant u, respectively. Its outdated version acquired during the channel-estimation process (time p) is given by

$$
\begin{aligned}
\mathfrak{h}'_d &= h'_d e^{j(\phi_p + \varphi_p)} \\
&= \left(\rho h_d + \varepsilon \sqrt{1 - \rho^2} \right) e^{j(\phi_u - \Delta\phi_u + \varphi_u - \Delta\psi_u)},
\end{aligned}
\tag{7.149}
$$

where ϕ_p and φ_p denote the corresponding phase noises of the BS and UE at instant p, respectively. Similarly, we can get the actual CSI between the BS and the nth IRS element during the signal transmission

$$
\mathfrak{h}_n = h_n e^{j\phi_u},
\tag{7.150}
$$

since the IRS elements are passive. The corresponding outdated CSI at instant p is given by

$$
\begin{aligned}
\mathfrak{h}'_n &= h'_n e^{j\phi_p} \\
&= \left(\rho h_n + \varepsilon \sqrt{1 - \rho^2} \right) e^{j(\phi_u - \Delta\phi_u)}.
\end{aligned}
\tag{7.151}
$$

The actual and outdated CSI between the nth IRS element and the UE are expressed as

$$
\mathfrak{g}_n = g_n e^{j\psi_u}.
\tag{7.152}
$$

and

$$
\begin{aligned}
\mathfrak{g}'_n &= g'_n e^{j\psi_p} \\
&= \left(\rho g_n + \varepsilon \sqrt{1 - \rho^2} \right) e^{j(\psi_u - \Delta\psi_u)},
\end{aligned}
\tag{7.153}
$$

respectively. Under good conditions where the channels exhibit slow fading if Doppler shifts are small and the quality of oscillators is high, the effect of channel aging is not explicit and the performance loss is possible to be negligible. Otherwise, the impact should be serious either in fast fading environments or low-cost hardware utilization.

The received signal at the UE in an IRS-enhanced system can be expressed as

$$
y = \left(\sum_{n=1}^{N} \mathfrak{g}_n c_n \mathfrak{h}_n + \mathfrak{h}_d \right) \sqrt{P_t} s + z.
\tag{7.154}
$$

However, the BS only has the outdated channel information $\mathfrak{g}'_n$, $\mathfrak{h}'_n$, and $\mathfrak{h}'_d$. The optimal phase shift for reflecting element n is set to

$$
\theta_n^\star = \operatorname{mod} \left[\psi_d - (\phi_{h,n} + \phi_{g,n}), 2\pi \right],
\tag{7.155}
$$

where $\phi_{h,n}$, $\phi_{g,n}$, and ψ_d stand for the phases of $\mathfrak{h}'_n$, $\mathfrak{g}'_n$, and $\mathfrak{h}'_d$, respectively, and mod [·] is the modulo operation. Then, the maximal received SNR is expressed as

$$\gamma_{\max} = \frac{P_t \left| \sum_{n=1}^{N} |g_n| |h_n| e^{j\phi_e} + |h_d| e^{j\psi_e} \right|^2}{\sigma_z^2}, \tag{7.156}$$

where we use $e^{j\phi_e}$ and $e^{j\psi_e}$ to denote the residual phases (errors) due to the channel aging, which destroys the desired coherent combining and definitely degrades the system performance.

7.7.3 Classical Channel Prediction

Based on typical statistical methods, predictive models can be built to approximate the dynamics of a fading channel using a set of propagation parameters. With the knowledge of the current and past channel information, these parameters are derived and then the CSI at the next step can be extrapolated. Existing model-based channel prediction is mainly differentiated into two major categories, i.e. auto-regressive and parametric models [Jiang and Schotten, 2019a]. The basic principles, parameter estimation methods, and constraints of these two models are presented as follows.

7.7.3.1 Autoregressive Model

Exploiting the auto-correlation of a time-varying channel in the time domain, this technique models the channel impulse response as an AR process and estimate its coefficients by using a Kalman Filter (KF). Then, a linear predictor is formed to extrapolate the future channel coefficient by combining weighted current and past channel coefficients. We write AR(p) to denote a complex AR process with an order of p, which can be denoted by a time-domain recursion [Baddour and Beaulieu, 2005], i.e.

$$x[n] = \sum_{k=1}^{p} a_k x[n-k] + w[n], \tag{7.157}$$

where $w[n]$ is complex Gaussian noise with zero mean and variance σ_p^2, and $\{a_1, a_2, \ldots, a_p\}$ denote the AR coefficients. The corresponding Power Spectral Density (PSD) of AR(p) has a rational form

$$S_{xx}(f) = \frac{\sigma_p^2}{\left| 1 + \sum_{k=1}^{p} a_k e^{-2\pi jfk} \right|^2}. \tag{7.158}$$

In a Rayleigh fading channel, the theoretical PSD associated with either in-phase or quadrature parts of a fading signal has a well-known U-shaped band-limited form, i.e.

$$S(f) = \begin{cases} \dfrac{1}{\pi f_d \sqrt{1-\left(\frac{f}{f_d}\right)^2}}, & |f| \leq f_d \\ 0, & f > f_d \end{cases}, \tag{7.159}$$

where f_d is the maximum Doppler shift in Hertz. The corresponding discrete-time auto-correlation function is given by

$$R[n] = J_0(2\pi f_m |n|), \tag{7.160}$$

with the normalized maximal Doppler shift $f_m = f_d T_s$. An arbitrary spectrum can be closely approximated by an AR model with sufficiently large order. The basic relationship between a desired autocorrelation function $R[n]$ and an AR(p) model parameters can be given in matrix form by

$$\mathbf{v} = \mathbf{Ra}, \tag{7.161}$$

where

$$\mathbf{R} = \begin{bmatrix} R[0] & R[-1] & \cdots & R[-p+1] \\ R[1] & R[0] & \cdots & R[-p+2] \\ \vdots & \vdots & \ddots & \vdots \\ R[p-1] & R[p-2] & \cdots & R[0] \end{bmatrix}, \tag{7.162}$$

$$\mathbf{a} = \begin{bmatrix} a_1 & a_2 & \cdots & a_p \end{bmatrix}^T, \tag{7.163}$$

$$\mathbf{v} = \begin{bmatrix} R[1] & R[2] & \cdots & R[p] \end{bmatrix}^T, \tag{7.164}$$

and

$$\sigma_p^2 = R[0] + \Sigma_{k=1}^{p} a_k R[k]. \tag{7.165}$$

Substituting Eqs. (7.162)–(7.164) into Eq. (7.161), yields the AR coefficients. Then, we can build a KF predictor for a single-antenna frequency-flat fading channel as

$$\hat{h}[t+1] = \sum_{k=1}^{p} a_k h[t-k+1]. \tag{7.166}$$

Treating a MIMO channel as a set of independent sub-channels, a KF predictor for a multi-antenna system can also be given by

$$\hat{\mathbf{H}}[t+1] = \sum_{k=1}^{p} a_k \mathbf{H}[t-k+1]. \tag{7.167}$$

This scheme is not optimal since it only takes advantage of temporal correlation of individual sub-channels, whereas ignoring spatial and frequency correlations among multiple antennas in a MIMO channel. Moreover, this scheme is vulnerable to noise [Jiang and Schotten, 2019b] and suffers from error propagation, limiting its significance in practice.

7.7.3.2 Parametric Model

This scheme models a fading channel as the superposition of a finite number of complex sinusoids, each of which has its respective amplitude, Doppler shift, and phase [Adeogun et al., 2014]. The rationale is based on an observation that multi-path parameters change slowly in comparison with the fading rate of channels, and future CSI within a certain range can be extrapolated if these parameters are known.

Following a commonly used sum of sinusoids model, a MIMO channel is expressed as the superposition of P scattering sources

$$\mathbf{H}(t) = \sum_{p=1}^{P} \alpha_p \mathbf{a}_r(\theta_p) \mathbf{a}_t^T(\phi_p) e^{j\omega_p t}, \tag{7.168}$$

where α_p is the amplitude of the pth scattering source, ω_p denotes its Doppler shift, θ_p and ϕ_p stands for the angles of arrival and departure, respectively, $\mathbf{a}_r$ represents the response vector of the receive antenna array, while $\mathbf{a}_t$ for the transmit antenna array. Using a uniform linear array with M equally spaced elements as an example, its steering vector can be formulated as

$$\mathbf{a}(\psi) = \left[1, e^{-j\frac{2\pi}{\lambda} d \sin(\psi)}, \dots, e^{-j\frac{2\pi}{\lambda}(M-1)d \sin(\psi)} \right]^T, \tag{7.169}$$

where ψ stands for the angle of arrival or departure, d is the antenna spacing, and λ denotes the wavelength of carrier frequency. Prediction of a MIMO channel in terms of this model is essentially a problem of parameter estimation, in which the number of scattering sources, the amplitude and Doppler shift for each path, as well as its angles of arrival and departure, need to be estimated. In other words, the main work of building a parametric model is to figure out $\hat{P}$ and $\{\hat{\alpha}_p, \hat{\theta}_p, \hat{\phi}_p, \hat{\omega}_p\}_{p=1}^{\hat{P}}$ with the knowledge of a number of discrete-time channel gain samples $\{\mathbf{H}[k]|k = 1, \dots, K\}$.

The procedure of parameters' estimation for the parametric model is divided into the following stages [Jiang and Schotten, 2019a]:

i. Using the K available channel matrices to form a sufficiently large matrix exhibiting the required translational invariance structure in all dimensions. Therefore, forming an $N_r Q \times N_t L$ block-Hankel matrix, which can be written as

$$\hat{\mathbf{D}} = \begin{bmatrix} \mathbf{H}[1] & \mathbf{H}[2] & \cdots & \mathbf{H}[S] \\ \mathbf{H}[2] & \mathbf{H}[3] & \cdots & \mathbf{H}[S+1] \\ \vdots & \vdots & \ddots & \vdots \\ \mathbf{H}[Q] & \mathbf{H}[Q+1] & \cdots & \mathbf{H}[K] \end{bmatrix}, \tag{7.170}$$

where Q is the size of Hankel matrix and $S = K - Q + 1$.

ii. From the transformed data, a covariance matrix containing the temporal and spatial correlation is calculated. The spatio-temporal covariance matrix $\hat{\mathbf{C}}$ is then derived as $\hat{\mathbf{C}} = \hat{\mathbf{D}}\hat{\mathbf{D}}^H/(N_t S)$, where $(\cdot)^H$ denotes the Hermitian conjugate transpose.

iii. Then, the number of dominant scattering sources can be estimated using the Minimum Description Length (MDL) criterion as

$$\hat{P} = \arg\min_{p=1,\ldots,N_r Q-1}\left[S\log(\lambda_p) + \frac{1}{2}(p^2 + p)\log S\right], \tag{7.171}$$

where λ_p is the pth eigenvalue of $\hat{\mathbf{C}}$.

iv. The invariance structure in $\hat{\mathbf{C}}$ is exploited to jointly estimate the structural parameters. Making full use of classical estimation algorithms, such as MUSIC and Estimation of Signal Parameters by Rotational Invariance Techniques (ESPRIT) [Gardner, 1988], the angles of arrival and departure, as well as the Doppler shifts, i.e. $\{\hat{\theta}_p, \hat{\phi}_p, \hat{\omega}_p\}_{p=1}^{\hat{P}}$, can be computed.

v. Obtained the estimated structural parameters $\{\hat{\theta}_p, \hat{\phi}_p, \hat{\omega}_p\}_{p=1}^{\hat{P}}$, in together with $\hat{P}$, the complex amplitudes $\{\hat{\alpha}_p\}_{p=1}^{\hat{P}}$ then can be calculated.

vi. Once all parameters have been determined, the channel prediction is conducted as follows

$$\hat{\mathbf{H}}(\tau) = \sum_{p=1}^{\hat{P}}\hat{\alpha}_p\mathbf{a}_r(\hat{\theta}_p)\mathbf{a}_t^T(\hat{\phi}_p)e^{j\hat{\omega}_p\tau}, \tag{7.172}$$

where τ denotes a time range for which the CSI is to be predicted.

As we can see, the process of estimating parameters is tedious, leading to high computational complexity. More importantly, the estimated parameters become invalid quickly with the change of mobile propagation environments, especially in a fast fading channel. That means these parameters need to periodically estimate, which is unattractive from the practical viewpoint. To overcome the weakness of the conventional channel prediction, some machine learning techniques exhibit high potential. The following part will introduce the principles of machine learning-based channel prediction, including the fundamentals of recurrent neural network, long-short term memory, and gated recurrent unit in both shallow and deep networks.

7.7.4 Recurrent Neural Network

Recurrent neural network is a class of machine learning that has shown great potential in the field of time-series prediction [Connor et al., 1994]. Unlike a feed-forward network that only learns from training data, a recurrent neural network can also use its memory of past states to process sequences of inputs.

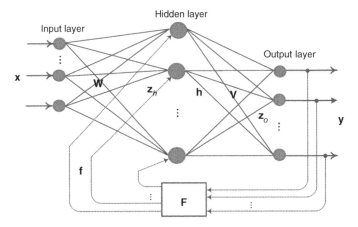

Figure 7.11 The structure of a typical RNN network.

RNN has several variants, among which a Jordan network is currently used to build a channel predictor. Basically, a simple network consists of three layers: an input layer with N_i neurons, a hidden layer with N_h neurons, and a layer having N_o outputs, as shown in Figure 7.11. Each connection between the activation of a neuron in the predecessor layer and the input of a neuron in the successor layer is assigned a weight. Let w_{ln} denote the weight connecting the nth input and the lth hidden neuron, while v_{ol} is the weight for hidden neuron l and output o, where $1 \leq n \leq N_i$, $1 \leq l \leq N_h$, and $1 \leq o \leq N_o$. Constructing a $N_h \times N_i$ weight matrix **W** as

$$\mathbf{W} = \begin{bmatrix} w_{11} & \cdots & w_{1N_i} \\ \vdots & \ddots & \vdots \\ w_{N_h1} & \cdots & w_{N_hN_i} \end{bmatrix}, \tag{7.173}$$

and denoting the activation vector of the input layer and the recurrent component (feedback) at time step t as

$$\mathbf{x}(t) = \left[x_1(t), x_2(t), \ldots, x_{N_i}(t) \right]^T \tag{7.174}$$

and

$$\mathbf{f}(t) = \left[f_1(t), f_2(t), \ldots, f_{N_h}(t) \right]^T, \tag{7.175}$$

respectively, the input for the hidden layer is expressed in matrix form by

$$\mathbf{z}_h(t) = \mathbf{W}\mathbf{x}(t) + \mathbf{f}(t) + \mathbf{b}_h, \tag{7.176}$$

where $\mathbf{b}_h = \left[b_1^h, \ldots, b_{N_h}^h \right]^T$ denotes the vector of biases in the hidden layer. Using a matrix **F** to represent the mapping from the output at the previous time step, i.e.

$\mathbf{y}(t-1) = \left[y_1(t-1), \dots, y_{N_o}(t-1)\right]^T$, to the recurrent component, we have

$$\mathbf{f}(t) = \mathbf{F}\mathbf{y}(t-1). \tag{7.177}$$

The behavior of a neural network depends on activation functions, typically falling into the following categories: linear, rectified linear, threshold, sigmoid, and tangent. In general, a sigmoid function is employed to deal with nonlinearity, which is defined as

$$S(x) = \frac{1}{1+e^{-x}}. \tag{7.178}$$

Substituting Eq. (7.177) into Eq. (7.178), the activation vector of the hidden layer is thus

$$\mathbf{h}(t) = S(\mathbf{z}_h(t)) = S(\mathbf{W}\mathbf{x}(t) + \mathbf{F}\mathbf{y}(t-1) + \mathbf{b}_h), \tag{7.179}$$

where $S(\mathbf{z}_h)$ means an element-wise operation for simplicity, i.e.

$$S(\mathbf{z}_h) = \left[S(z_1), S(z_2), \dots, S(z_{N_h})\right]^T. \tag{7.180}$$

In analogous to Eq. (7.173), another weight matrix $\mathbf{V}$ having a dimension of $N_o \times N_h$ with entries $\{v_{ol}\}$ is introduced. Then, the input for the output layer is $\mathbf{z}_o(t) = \mathbf{V}\mathbf{h}(t) + \mathbf{b}_y$, where $\mathbf{b}_y$ is the vector of biases in the output layer, resulting in an output vector:

$$\mathbf{y}(t) = S\left(\mathbf{z}_o(t)\right) = S\left(\mathbf{V}\mathbf{h}(t) + \mathbf{b}_y\right). \tag{7.181}$$

Like other data-driven AI techniques, the operation of a RNN is categorized into two phases: training and predicting. The training of a neural network is typically based on a fast algorithm known as Back-Propagation (BP). Provided a training dataset, the network feeds forward input data and compares the resulting output $\mathbf{y}$ against the desired value $\mathbf{y}_0$. Measured by a cost function, e.g. $C = \|\mathbf{y}_0 - \mathbf{y}\|^2$, prediction errors are propagated back through the network, causing iteratively updating of weights and biases until a certain convergence condition reaches. To provide an initial impression of this process, the BP algorithm in combination with gradient descent learning for a feed forward network is briefly depicted:

Start from an initial network state where $\{\mathbf{W}, \mathbf{V}, \mathbf{b}_h, \mathbf{b}_y\}$ are randomly set.

1. Input a training example $(\mathbf{x}, \mathbf{y}_0)$.
2. *Feed-Forward*: For the hidden layer, its input $\mathbf{z}_h$ and activation $\mathbf{h}$ can be computed using Eqs. (7.176) and (7.179), respectively. For simplicity, the BP algorithm applied in a feed forward network without a recurrent component is illustrated. Hence, the exact equation to calculate the input is $\mathbf{z}_h(t) = \mathbf{W}\mathbf{x}(t) + \mathbf{b}_h$. Also, $\mathbf{z}_o$ and $\mathbf{y}$ in Eq. (7.181) for the output layer are obtained.

3. Compute the output error $\mathbf{e}^y = \left[e_1^y, e_2^y, \ldots, e_{N_o}^y \right]^T$ in terms of

$$\mathbf{e}^y = \nabla_y C \odot S'(\mathbf{z}_o), \tag{7.182}$$

where ∇ notates a vector whose entries are partial derivatives, namely, $\nabla_y C = \left[\frac{\partial C}{\partial y_1}, \ldots, \frac{\partial C}{\partial y_{N_o}} \right]^T$. In addition, $S'(\mathbf{z}_o)$ stands for the derivative of the activation function with respect to its corresponding input $\mathbf{z}_o$. Given a sigmoid function in Eq. (7.178), for instance, we have

$$S'(\mathbf{z}_o) = \frac{\partial S(\mathbf{z}_o)}{\partial \mathbf{z}_o} = S(\mathbf{z}_o)(1 - S(\mathbf{z}_o)). \tag{7.183}$$

4. **BP**: $\mathbf{e}^y$ is propagated back to the hidden layer to derive the error vector there, i.e.

$$\mathbf{e}^h = \mathbf{V}^T \mathbf{e}^y \odot S'(\mathbf{z}_h). \tag{7.184}$$

5. **Gradient Descent**: With the back-propagated errors, the weights and biases are able to be updated according to the following rules:

$$\begin{cases} \mathbf{W} = \mathbf{W} - \eta \mathbf{e}^h \mathbf{x}^T \\ \mathbf{V} = \mathbf{V} - \eta \mathbf{e}^y \mathbf{h}^T \\ \mathbf{b}_h = \mathbf{b}_h - \eta \mathbf{e}^h \\ \mathbf{b}_y = \mathbf{b}_y - \eta \mathbf{e}^y \end{cases}, \tag{7.185}$$

where η stands for the learning rate.

The weights and biases are iteratively updated until the cost function is below a predefined threshold or the number of epochs reaches its maximal value. Once the training process completed, the trained network can be used to process upcoming samples. The training of a RNN typically use a variant of the BP algorithm, known as Back-Propagation Through Time (BPTT). It requires to unfold a recurrent neural network in time steps to form a pseudo feed-forward network, where the BP algorithm is applicable. Upon this, other more advanced or efficient approaches such as real-time recurrent learning and extended Kalman filtering have been designed [Jiang and Schotten, 2019a].

7.7.5 RNN-Based Channel Prediction

Observing the MIMO channel model and the structure of a neural network, high similarity that both have multiple inputs and outputs with fully weighted connections can be found. A neural network well suits to process MIMO channels by adapting the number of input and output neurons with respect to the number of transmit and receive antennas. A RNN predictor is quite flexibly to be configured to forecast channel response or envelope on demand in either frequency-flat or frequency-selective fading channels. The following discussion starts from the simplest case that applies a RNN to predict a flat fading channel in a single-antenna system, and then extends step by step until a frequency-domain predictor for frequency-selective MIMO channels.

7.7.5.1 Flat-Fading Channel Prediction

To begin with, consider a discrete-time baseband equivalent model for a SISO flat-fading channel:

$$r[t] = h[t]s[t] + n[t]. \tag{7.186}$$

The aim of RNN predictor is to get a predicted value $\hat{h}[t + \tau]$ that is as close as possible to its actual value $h[t + \tau]$. To deal with complex-valued channel gains, a network with complex-valued weights called a complex-valued RNN hereinafter is needed. At time t, $h[t]$ is obtained through channel estimation, while a series of d past values $h[t - 1], h[t - 2], \ldots, h[t - d]$ can be memorized simply through a tapped delay line. These $d + 1$ channel gains are fed into the RNN as the input, i.e.

$$\mathbf{x}[t] = \left[h[t], h[t - 1], \ldots, h[t - d] \right]^T. \tag{7.187}$$

In together with the delayed feedback, the prediction of a future channel gain $\mathbf{y}[t] = \left[\hat{h}[t + 1] \right]^T$ is obtained.

The extension of this predictor to flat-fading MIMO channels is straightforward. To adapt to the input layer of a RNN, channel matrices are required to be vectorized into a $N_r N_t \times 1$ vector, as follows:

$$\mathbf{h}[t] = \vec{\mathbf{H}}[t] = \left[h_{11}[t], h_{12}[t], \ldots, h_{N_r N_t}[t] \right]. \tag{7.188}$$

Together with a number of d past values $\mathbf{H}[t - 1], \ldots, \mathbf{H}[t - d]$, the input of RNN this case is

$$\mathbf{x}[t] = [\mathbf{h}[t], \mathbf{h}[t - 1], \ldots, \mathbf{h}[t - d]]^T, \tag{7.189}$$

resulting in a predictive value $\mathbf{y}[t] = \hat{\mathbf{h}}^T[t + 1]$, which can be transformed to a predicted matrix $\hat{\mathbf{H}}[t + 1]$.

In comparison with a complex-valued RNN, a recurrent neural network with real-valued weights called a real-valued RNN has lower complexity and higher prediction accuracy, whereas it can only deal with real-valued data. Fortunately, a complex-valued channel gain can be decomposed into two real values, namely, $h = h^r + jh^i$. Hence, a real-valued RNN was proposed in Jiang and Schotten [2018b] to build a simpler predictor with higher accuracy by means of decoupling the real and imaginary parts. Without a necessity of using two RNNs, the real and imaginary parts can be processed jointly in a single predictor. In this case, the input of the network is

$$\mathbf{x}[t] = \left[h^r[t], h^i[t], \dots, h^r[t-d], h^i[t-d] \right]^T, \tag{7.190}$$

generating an output $\mathbf{y}[t] = \left[\hat{h}^r[t+1], \hat{h}^i[t+1] \right]^T$ that synthesizes to a predicted channel gain $\hat{h}[t+1] = \hat{h}^r[t+1] + j\hat{h}^i[t+1]$. Similarly, $\mathbf{H}[t]$ is decomposed into

$$\mathbf{H}[t] = \mathbf{H}_R[t] + j\mathbf{H}_I[t], \tag{7.191}$$

where $\mathbf{H}_R = \mathfrak{R}(\mathbf{H}) = [h^r_{n_r n_t}]_{N_r \times N_t}$ denotes a matrix composed by the real parts of channel gains and $\mathbf{H}_I = \mathfrak{I}(\mathbf{H}) = [h^i_{n_r n_t}]_{N_r \times N_t}$ is the imaginary counterpart. Like Eq. (7.188), these matrices are vectorized as

$$\mathbf{h}_r[t] = \vec{\mathbf{H}}_R[t] = \left[h^r_{11}[t], h^r_{12}[t], \dots, h^r_{N_r N_t}[t] \right]. \tag{7.192}$$

Feeding

$$\mathbf{x}[t] = [\mathbf{h}_r[t], \mathbf{h}_i[t], \dots, \mathbf{h}_r[t-d], \mathbf{h}_i[t-d]]^T \tag{7.193}$$

into the network, the resulting output is written as $\mathbf{y}[t] = \left[\hat{\mathbf{h}}_r[t+1], \hat{\mathbf{h}}_i[t+1] \right]^T$, which can be transformed into $\hat{\mathbf{H}}_R[t+D]$ and $\hat{\mathbf{H}}_I[t+D]$. Then, a predicted matrix is reaped simply by $\hat{\mathbf{H}}[t+1] = \hat{\mathbf{H}}_R[t+1] + j\hat{\mathbf{H}}_I[t+1]$.

Many adaptive transmission systems only need to know the envelope of channel response $|h|$, rather than a complex-valued gain h. Therefore, a real-valued RNN can be directly applied, which in turn can lower complexity, speed up training process, and improve prediction accuracy, in comparison with predicting channel gains. The channel envelope at time t denoted by $|h[t]|$ is known, with a number of d past values $|h[t-1]|, |h[t-2]|, \dots, |h[t-d]|$, the input in this case is written as

$$\mathbf{x}[t] = \left[|h[t]|, |h[t-1]|, \dots, |h[t-d]| \right]^T, \tag{7.194}$$

which generate $|\hat{h}[t+1]|$ through the network. Further, let $\mathbf{Q}[t] = \left[|h_{n_r n_t}[t]| \right]_{N_r \times N_t}$ denotes a matrix, in which the $(n_r, n_t)^{\text{th}}$ entry is the envelope of $h_{n_r n_t}[t]$ in $\mathbf{H}[t]$. Likewise, $\mathbf{Q}[t]$ is vectorized as

$$\mathbf{q}[t] = \vec{\mathbf{Q}}[t] = \left[|h_{11}[t]|, |h_{12}[t]|, \dots, |h_{N_r N_t}[t]| \right]. \tag{7.195}$$

With the input $\mathbf{x}[t] = [\mathbf{q}[t], \mathbf{q}[t-1], \dots, \mathbf{q}[t-d]]^T$, the prediction $\mathbf{y}[t] = \hat{\mathbf{q}}[t+1]$ is got and further transformed into $\hat{\mathbf{Q}}[t+1]$.

7.7.5.2 Frequency-Selective Fading Channel Prediction

To begin with, let us consider the discrete-time model for a frequency-selective SISO system:

$$r[t] = \sum_{l=0}^{L-1} h_l[t]s[t-l] + n[t], \tag{7.196}$$

where s and r denote the transmitted and received symbol, respectively, $h_l[t]$ stands for the lth tap for a time-varying channel filter at time t, and n is additive noise. Dropped time index for simplicity, a frequency-selective channel is modeled as a linear L-tap filter

$$\mathbf{h} = \left[h_0, h_1, \ldots, h_{L-1} \right]^T. \tag{7.197}$$

It can be converted into a set of N orthogonal narrow-band channels known as sub-carriers through the OFDM modulation, which is represented by

$$\tilde{r}_n[t] = \tilde{h}_n[t]\tilde{s}_n[t] + \tilde{n}_n[t], \quad n = 0, 1, \ldots, N-1, \tag{7.198}$$

where $\tilde{s}_n[t]$, $\tilde{r}_n[t]$, and $\tilde{n}_n[t]$ stand for the transmitted signal, received signal, and noise, respectively, at sub-carrier n. According to the picket fence effect in discrete Fourier transform, the frequency response of the channel filter denoted by $\tilde{\mathbf{h}} = [\tilde{h}_0, \tilde{h}_1, \ldots, \tilde{h}_{N-1}]^T$ is the DFT of

$$\mathbf{h}' = \left[h_0, h_1, \ldots, h_{L-1}, 0, \ldots, 0 \right]^T \tag{7.199}$$

that pads $\mathbf{h}$ in Eq. (7.197) with $N - L$ zeros at the tail.

Extending Eq. (7.198) to a multi-antenna system is straightforward though a MIMO-OFDM system that is modeled as

$$\tilde{\mathbf{r}}_n[t] = \tilde{\mathbf{H}}_n[t]\tilde{\mathbf{s}}_n[t] + \tilde{\mathbf{n}}_n[t], \quad n = 0, 1, \ldots, N-1, \tag{7.200}$$

where $\tilde{\mathbf{s}}_n[t]$ represents $N_t \times 1$ transmit symbol vector on sub-carrier n at time t, $\tilde{\mathbf{r}}_n[t]$ is $N_r \times 1$ received symbol vector, and $\tilde{\mathbf{n}}[t]$ is the vector of additive noise. The subchannel between transmit antenna n_t and receive antenna n_r is equivalent to a frequency-selective SISO channel, denoted by a channel filter

$$\mathbf{h}^{n_r n_t} = [h_0^{n_r n_t}, h_1^{n_r n_t}, \ldots, h_{L-1}^{n_r n_t}]^T. \tag{7.201}$$

Likewise, the frequency response of this filter can be obtained by conducting DFT, i.e.

$$\tilde{\mathbf{h}}^{n_r n_t} = [\tilde{h}_0^{n_r n_t}, \tilde{h}_1^{n_r n_t}, \ldots, \tilde{h}_{N-1}^{n_r n_t}]^T. \tag{7.202}$$

Then, the channel matrix on sub-carrier n can be notated as

$$\tilde{\mathbf{H}}_n[t] = \left[\tilde{h}_n^{n_r n_t}[t] \right]_{N_r \times N_t}. \tag{7.203}$$

The main idea of frequency-domain channel prediction is to convert a frequency-selective channel into a set of orthogonal flat fading sub-carriers, and then utilize a frequency-domain predictor to forecast the frequency response on each sub-carrier [Jiang and Schotten, 2019c]. At time t over sub-carrier n, $\tilde{\mathbf{H}}_n[t]$, as well as its d-step delays $\tilde{\mathbf{H}}_n[t-1], \ldots, \tilde{\mathbf{H}}_n[t-d]$, are fed into the RNN. Dropping the time index for simplicity, these matrices are vectorized as

$$\tilde{\mathbf{h}}_n = \text{vec}\left(\tilde{\mathbf{H}}_n \right) = [\tilde{h}_n^{11}, \tilde{h}_n^{12}, \ldots, \tilde{h}_n^{N_r N_t}]. \tag{7.204}$$

The RNN outputs a D-step prediction, i.e. $\hat{\mathbf{h}}_n[t+D] = [\hat{h}_n^{11}[t+D], \ldots, \hat{h}_n^{N,N_t}[t+D]]^T$, transforming into a predicted matrix $\hat{\mathbf{H}}_n[t+D]$ via a vector-to-matrix module.

Although the prediction is conducted at sub-carrier level, we do not need to deal with all N sub-carriers taking into account channel's frequency correlation. Integrated with a pilot-assisted system, only predicting the CSI on sub-carriers carrying pilot symbols is enough. Suppose one pilot is inserted uniformly every N_P sub-carriers, amounts to a total of $P = \left\lfloor \frac{N}{N_P} \right\rfloor$ pilot sub-carriers, where $\lfloor \cdot \rfloor$ denotes a ceiling function. Given predicted values $\hat{\mathbf{H}}_p[t+D], p = 1, \ldots, P$, frequency-domain interpolation can be applied to get the predicted values on all sub-carriers $\hat{\mathbf{H}}_n[t+D], n = 0, \ldots, N-1$.

7.7.6 Long-Short Term Memory

RNN is good at processing data sequences through storing indefinite historical information in its internal state, exhibiting great potential in time-series prediction. Nevertheless, it suffers from the gradient exploding and vanishing problems with the gradient-based BPTT training technique, where a back-propagated error signal is apt to be very large, leading to oscillating weights, or tends to zero that implies a prohibitively long training time or training does not work at all.

To this end, Hochreiter and Schmidhuber designed an elegant RNN structure - long short-term memory – in 1997 in their pioneer work of [Hochreiter and Schmidhuber, 1997]. The key innovation of LSTM to deal with long-term dependency is the introduction of special units called memory cells in the recurrent hidden layer and multiplicative gates that regulate the information flow. In the original structure of LSTM, each memory block contains two gates: an *input* gate protecting the memory contents stored in the cell from perturbation by irrelevant interference, and an *output* gate that controls the extent to which the memory information applied to generate the output activation. To address a weakness of LSTM, namely, the internal state grows indefinitely and eventually cause the network to break down when processing continual input streams that are not segmented into subsequences, a *forget* gate was added [Gers et al., 2000]. It scales the internal state of the memory cell before cycling back through self-recurrent connections. Although its history is not long, LSTM has been applied successfully to sequence prediction and labeling tasks. It has already gotten the state-of-the-art technological results in many fields such as machine translation, speech recognition, and handwriting recognition, and has also achieved a great commercial success, justified by many unprecedented intelligent services such as Google Translate and Apple Siri.

Like a deep RNN consisting of multiple recurrent hidden layers, a deep LSTM network is built by stacking multiple LSTM layers. Without loss of generality,

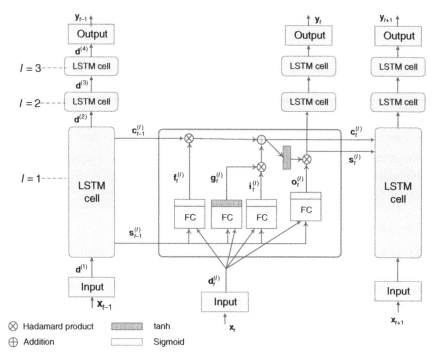

Figure 7.12 Illustration of a three-hidden-layer deep LSTM network. Source: Adapted from Jiang and Schotten [2020a].

Figure 7.12 shows an example of a deep LSTM network that consists of an input layer, three hidden layers, and an output layer. At an arbitrary time step, a data vector $\mathbf{x}$ goes through the input feedforward layer to get $\mathbf{d}^{(1)}$, which is the activation for memory cells in the first hidden layer. Along with the recurrent unit feeding back from the previous time step, $\mathbf{d}^{(2)}$ is generated and then forwarded to the second hidden layer. This recursive process continues until the output layer gets $\mathbf{y}$ according to $\mathbf{d}^{(4)}$. Unrolling the network through time, the memory block at the lth hidden layer has two internal states at time step $t-1$, i.e. the short-term state $\mathbf{s}_{t-1}^{(l)}$ and the long-term state $\mathbf{c}_{t-1}^{(l)}$. Traversing the memory cells from the left to the right, $\mathbf{c}_{t-1}^{(l)}$ first throws away some old memories at the forget gate, integrates new information selected by the input gate, and then sends out as the current long-term state $\mathbf{c}_t^{(l)}$. The input vector $\mathbf{d}_t^{(l)}$ and the previous short-term memory $\mathbf{s}_{t-1}^{(l)}$ are fed into four different Fully Connected (FC) layer, generating the activation vectors of gates as

$$\mathbf{f}_t^{(l)} = \sigma_g\left(\mathbf{W}_f^{(l)}\mathbf{d}_t^{(l)} + \mathbf{U}_f^{(l)}\mathbf{s}_{t-1}^{(l)} + \mathbf{b}_f^{(l)}\right), \tag{7.205}$$

$$\mathbf{i}_t^{(l)} = \sigma_g \left(\mathbf{W}_i^{(l)} \mathbf{d}_t^{(l)} + \mathbf{U}_i^{(l)} \mathbf{s}_{t-1}^{(l)} + \mathbf{b}_i^{(l)} \right), \qquad (7.206)$$

$$\mathbf{o}_t^{(l)} = \sigma_g \left(\mathbf{W}_o^{(l)} \mathbf{d}_t^{(l)} + \mathbf{U}_o^{(l)} \mathbf{s}_{t-1}^{(l)} + \mathbf{b}_o^{(l)} \right), \qquad (7.207)$$

where $\mathbf{W}$ and $\mathbf{U}$ are weight matrices for the FC layers, $\mathbf{b}$ stands for bias vectors, the subscripts f, i, and o associate with the forget, input, and output gate, respectively, and σ_g represents the *sigmoid* activation function

$$\sigma_g(x) = \frac{1}{1 + e^{-x}}. \qquad (7.208)$$

Dropping some old memories at the forget gate, and adding some new information selected from current memory input that is defined as

$$\mathbf{g}_t^{(l)} = \sigma_h \left(\mathbf{W}_g^{(l)} \mathbf{d}_t^{(l)} + \mathbf{U}_g^{(l)} \mathbf{s}_{t-1}^{(l)} + \mathbf{b}_g^{(l)} \right), \qquad (7.209)$$

the previous long-term memory $\mathbf{c}_{t-1}^{(l)}$ is thus transformed into

$$\mathbf{c}_t^{(l)} = \mathbf{f}_t^{(l)} \otimes \mathbf{c}_{t-1}^{(l)} + \mathbf{i}_t^{(l)} \otimes \mathbf{g}_t^{(l)}, \qquad (7.210)$$

where $\otimes$ denotes the Hadamard product (element-wise multiplication) for matrices, and σ_h is the *hyperbolic tangent* function denoted by tanh, defining by

$$\sigma_h(x) = \frac{e^{2x} - 1}{e^{2x} + 1}. \qquad (7.211)$$

In addition to sigmoid and tanh, there are other commonly used activation functions, e.g. the *Rectified Linear Unit* (ReLU) that can be written as

$$\sigma_r(x) = \max(0, x), \qquad (7.212)$$

which returns 0 if it receives any negative input, but for any positive value x it returns that value back. Further, $\mathbf{c}_t$ passes through the *tanh* function and then is filtered by the output gate to produce the current short-term memory, as well as the output for this memory block, i.e.

$$\mathbf{s}_t^{(l)} = \mathbf{d}_t^{(l+1)} = \mathbf{o}_t^{(l)} \otimes \sigma_h \left(\mathbf{c}_t^{(l)} \right). \qquad (7.213)$$

Since the advent of LSTM, its original structure continues to evolve. Cho et al. [2014] presents a simplified version with fewer parameters in 2014, known as gated recurrent unit or GRU, which exhibits even better performance over LSTM on certain smaller data sets. In a GRU memory block, the short- and long-term states are merged into a single one, and a single gate $\mathbf{z}_t^{(l)}$ is used to replace the forget and input gates, namely,

$$\mathbf{z}_t^{(l)} = \sigma_g \left(\mathbf{W}_z^{(l)} \mathbf{d}_t^{(l)} + \mathbf{U}_z^{(l)} \mathbf{s}_{t-1}^{(l)} + \mathbf{b}_z^{(l)} \right). \qquad (7.214)$$

The output gate is removed, but an intermediate state $\mathbf{r}_t^{(l)}$ is newly introduced, i.e.

$$\mathbf{r}_t^{(l)} = \sigma_g \left(\mathbf{W}_r^{(l)} \mathbf{d}_t^{(l)} + \mathbf{U}_r^{(l)} \mathbf{s}_{t-1}^{(l)} + \mathbf{b}_r^{(l)} \right). \tag{7.215}$$

Likewise, the hidden state at the previous time step transverses the memory cells, drops some old memory, and loads some now information, resulting in the current state:

$$\begin{aligned} \mathbf{s}_t^{(l)} = {} & (1 - \mathbf{z}_t^{(l)}) \otimes \mathbf{s}_{t-1}^{(l)} \\ & + \mathbf{z}_t^{(l)} \otimes \sigma_h \left(\mathbf{W}_s^{(l)} \mathbf{d}_t^{(l)} + \mathbf{U}_s^{(l)} (\mathbf{r}_t^{(l)} \otimes \mathbf{s}_{t-1}^{(l)}) + \mathbf{b}_s^{(l)} \right). \end{aligned} \tag{7.216}$$

7.7.7 Deep Learning-Based Channel Prediction

To shed light on the operation of the downlink (DL) predictor, a point-to-point flat-fading MIMO system with N_t transmit and N_r receive antennas is studied. Its transmission is modeled as

$$\mathbf{r}[t] = \mathbf{H}[t]\mathbf{s}[t] + \mathbf{n}[t], \tag{7.217}$$

where $\mathbf{r}[t]$ and $\mathbf{s}[t]$ denote the received and transmitted signal vectors at time step t, respectively, $\mathbf{n}[t]$ is the vector of additive noise, and $\mathbf{H}[t]$ represents an $N_r \times N_t$ channel matrix, whose (n_r, n_t)-entry $h_{n_r n_t}$ is the complex-valued gain of the channel between transmit antenna n_t and receive antenna n_r. The transmitter requires CSI feedback so as to adapt its transmission parameters to a time-varying channel. Due to feedback delay τ, when the transmitter uses $\mathbf{H}[t]$ to choose parameters, the instantaneous channel gain already changes to $\mathbf{H}[t + \tau]$. It is probably that $\mathbf{H}[t] \neq \mathbf{H}[t + \tau]$, especially in high mobility environment. Outdated CSI imposes severe performance loss on a wide variety of adaptive wireless techniques. Hence, it is worth conducting channel prediction at the receiver to obtain predicted CSI $\hat{\mathbf{H}}[t + D]$, where $D \geqslant \tau$, to counteract the effect of feedback delay, or equivalently, performing prediction at the transmitter.

The instantaneous channel matrix $\mathbf{H}[t]$ is estimated at the receiver, which is fed into the predictor, rather than being fed back to the transmitter directly, as does a typical adaptive MIMO system. Based on an observation that the magnitude of channel gain, i.e. $|h_{n_r n_t}|$, is already enough for most of adaptive tasks, the prediction is applied on such real-valued channel data. In order to adapt the input layer of a neural network, we add a data pre-processing layer in the predictor, where a channel matrix is transferred into a vector of channel magnitudes, like:

$$\mathbf{H}[t] \rightarrow \left[|h_{11}[t]|, |h_{12}[t]|, \ldots, |h_{N_r N_t}[t]| \right]^T. \tag{7.218}$$

Replacing $\mathbf{x}_t$ in Figure 7.12 with this channel vector and going through a number of hidden layers, the output layer generates $\hat{\mathbf{H}}[t + D]$, which is the D-step prediction.

In addition to the magnitude, the predictor can also be applied to process complex-valued channel gains. To this end, the predictor generally needs to be built on a complex-valued deep neural network, which is not well implemented in current AI software tools. Instead, a real-valued network is applied to predict the real and imaginary parts of channel gains, where the data pre-processing layer transforms $H[t]$ into

$$[\Re(h_{11}[t]), \dots, \Re(h_{N_r N_t}[t]), \Im(h_{11}[t]), \dots, \Im(h_{N_r N_t}[t])]^T. \tag{7.219}$$

where $\Re(\cdot)$ and $\Im(\cdot)$ denote the real and imaginary units, respectively. Feeding this vector into the predictor at time t, $\hat{H}[t + D]$ can be obtained after simply combining the predicted real and imaginary units.

In addition to flat fading channels, the predictor can also be applied for frequency-selective channels simply by means of converting it into a set of N narrow-band sub-carriers through, for example, the OFDM modulation [Jiang and Kaiser, 2016]. At the nth sub-carrier, the signal transmission is represented by

$$\tilde{r}_n[t] = \tilde{H}_n[t]\tilde{s}_n[t] + \tilde{n}_n[t], \quad n = 0, 1, \dots, N-1, \tag{7.220}$$

where $\tilde{r}_n[t]$ represents N_r received symbols for sub-carrier n at time t, $\tilde{s}_n[t]$ corresponds to N_t transmit symbols, $\tilde{n}_n[t]$ is a vector of additive noise. We write $\tilde{H}_n[t] = [\tilde{h}_n^{n_r n_t}[t]]_{N_r \times N_t}$ to denote the frequency-domain channel matrix, where $1 \leqslant n_r \leqslant N_r$, $1 \leqslant n_t \leqslant N_t$, and $\tilde{h}_n^{n_r n_t} \in \mathbb{C}^{1 \times 1}$ stands for the CFR on sub-carrier n between transmit antenna n_t and receive antenna n_r. The prediction process on each sub-carrier is as same as that of the flat fading channel.

7.8 Summary

Using a large number of small, passive, and low-cost reflecting elements, IRS can *proactively* form a smart and programmable wireless environment. Thus, it provides a new degree of freedom to the design of 6G wireless systems, enabling sustainable capacity and performance growth with affordable cost, low complexity, and low energy consumption. This chapter introduced the system model and the signal transmission of IRS-aided systems with either single-antenna or multi-antenna base stations in both frequency-flat and frequency-selective fading channels. Joint optimization of active and passive beamforming and dual-beam IRS transmission were also presented. In addition, the impact of channel aging on the IRS and machine learning-based channel prediction were given. This chapter can serve as a tutorial for the readers as a starting point to further carrier out research works on this promising topic.

References

Adeogun, R. O., Teal, P. D. and Dmochowski, P. A. [2014], 'Extrapolation of MIMO mobile-to-mobile wireless channels using parametric-model-based prediction', *IEEE Transactions on Vehicular Technology* **64**(10), 4487–4498.

Baddour, K. and Beaulieu, N. [2005], 'Autoregressive modeling for fading channel simulation', *IEEE Transactions on Wireless Communications* **4**, 1650–1662.

Basar, E. [2019], Transmission through large intelligent surfaces: A new frontier in wireless communications, *in* 'Proceedings of 2019 European Conference on Networks and Communications (EuCNC)', Valencia, Spain, pp. 112–117.

Cho, K., van Merrienboer, B., Gulcehre, C., Bahdanau, D., Bougares, F., Schwenk, H., Bengio, Y. [2014], 'Learning phrase representations using RNN encoder-decoder for statistical machine translation', *preprint arXiv:1406.1078*.

Connor, J., Martin, R. and Atlas, L. (1994], 'Recurrent neural networks and robust time series prediction', *IEEE Transactions on Neural Networks* **5**(2), 240–254.

Gardner, W. [1988], 'Simplification of MUSIC and ESPRIT by exploitation of cyclostationarity', *Proceedings of the IEEE* **76**(7), 845–847.

Gers, F. A., Schmidhuber, J., Cummins, F. [2000], 'Learning to forget: Continual prediction with LSTM', *Neural Computation* **12**(10), 2451–2471.

Hochreiter, S. and Schmidhuber, J. [1997], 'Long short-term memory', *Neural Computation* **9**(8), 1735–1780.

Hu, S., Rusek, F. and Edfors, O. [2018], 'Beyond massive MIMO: The potential of data transmission with large intelligent surfaces', *IEEE Transactions on Signal Processing* **66**(10), 2746–2758.

Jiang, W. and Kaiser, T. [2016], From OFDM to FBMC: Principles and Comparisons, *in* F. L. Luo and C. Zhang, eds, '*Signal Processing for 5G: Algorithms and Implementations*', John Wiley&Sons and IEEE Press, United Kingdom, Chapter 3.

Jiang, W. and Schotten, H. [2018*a*], Neural network-based channel prediction and its performance in multi-antenna systems, *in* 'Proceedings of 2018 IEEE Vehicular Technology Conference (VTC-Fall)', Chicago, USA.

Jiang, W. and Schotten, H. D. [2018*b*], Multi-antenna fading channel prediction empowered by artificial intelligence, *in* 'Proceedings of 2018 IEEE Vehicular Technology Conference (VTC-Fall)', Chicago, USA.

Jiang, W. and Schotten, H. [2019*a*], 'Neural network-based fading channel prediction: A comprehensive overview', *IEEE Access* **7**, 118112–118124.

Jiang, W. and Schotten, H. D. [2019*b*], A comparison of wireless channel predictors: Artificial Intelligence versus Kalman filter, *in* 'Proceedings of 2019 IEEE International Communications Conference (ICC)', Shanghai, China.

Jiang, W. and Schotten, H. D. [2019*c*], Recurrent neural network-based frequency-domain channel prediction for wideband communications, *in* 'Proceedings of 2019 IEEE Vehicular Technology Conference (VTC)', Kuala Lumpur, Malaysia.

Jiang, W. and Schotten, H. D. [2020*a*], 'Deep learning for fading channel prediction', *IEEE Open Journal of the Communications Society* **1**, 320–332.

Jiang, W. and Schotten, H. D. [2020*b*], A deep learning method to predict fading channel in multi-antenna systems, *in* 'Proceedings of 2020 IEEE Vehicular Technology Conference (VTC-Spring)', Antwerp, Belgium.

Jiang, W. and Schotten, H. D. [2020*c*], Recurrent neural networks with long short-term memory for fading channel prediction, *in* 'Proceedings of 2018 IEEE Vehicular Technology Conference (VTC-Spring)', Antwerp, Belgium.

Jiang, W. and Schotten, H. D. [2021*a*], 'Cell-free massive MIMO-OFDM transmission over frequency-selective fading channels', *IEEE Communications Letters* **25**(8), 2718–2722.

Jiang, W. and Schotten, H. D. [2021*b*], 'A simple cooperative diversity method based on deep-learning-aided relay selection', *IEEE Transactions on Vehicular Technology* **70**(5), 4485–4500.

Jiang, W. and Schotten, H. D. [2022], Initial access for millimeter-wave and terahertz communications with hybrid beamforming, *in* 'Proceedings of 2022 IEEE International Communications Conference (ICC)', Seoul, South Korea.

Jiang, W., Kaiser, T. and Vinck, A. J. H. [2016], 'A robust opportunistic relaying strategy for co-operative wireless communications', *IEEE Transactions on Wireless Communications* **15**(4), 2642–2655.

Jiang, W., Strufe, M. and Schotten, H. [2020], Long-range fading channel prediction using recurrent neural network, *in* 'Proceedings of 2020 IEEE Consumer Communications and Networking Conference (CCNC)', Las Vegas, USA.

Jiang, W., Han, B., Habibi, M. A. and Schotten, H. D. [2021], 'The road towards 6G: A comprehensive survey', *IEEE Open Journal of the Communications Society* **2**, 334–366.

Khanzadi, M. R., Krishnan, N., Wu, Y. i., Amat, A. G., Eriksson, T. and Schober, R. [2016], 'Linear massive MIMO precoders in the presence of phase noise – a large-scale analysis', *IEEE Transactions on Vehicular Technology* **65**(5), 3057–3071.

Ramya, T. R. and Bhashyam, S. [2009],'Using delayed feedback for antenna selection in MIMO systems', *IEEE Transactions on Wireless Communications* **8**(12), 6059–6067.

Silver, D., Huang, A., Maddison, C. J., Guez, A., Sifre, L., van den Driessche, G., Schrittwieser, J., Antonoglou, I., Panneershelvam, V., Lanctot, M., Dieleman, S., Grewe, D., Nham, J., Kalchbrenner, N., Sutskever, I., Lillicrap, T., Leach, M., Kavukcuoglu, K., Graepel, T. and Hassabis, D. [2016], 'Mastering the game of Go with deep neural networks and tree search', *Nature* **529**, 484–489.

Tang, W., Chen, M. Z., Dai, J. Y., Zeng, Y., Zhao, X., Jin, S., Cheng, Q. and Cui, T. J. [2020], 'Wireless communications with programmable metasurface: New paradigms, opportunities, and challenges on transceiver design', *IEEE Wireless Communications* **27**(2), 180–187.

Tse, D. and Viswanath, P. [2005], *Fundamentals of Wireless Communication*, Cambridge University Press, Cambridge, United Kingdom.

Wang, Q., Greenstein, L. J., Cimini, L. J., Chan, D. S. and Hedayat, A. [2014], 'Multi-user and single-user throughputs for downlink MIMO channels with outdated channel state information', *IEEE Wireless Communications Letters* **3**, 321–324.

Wu, Q. and Zhang, R. [2018], Intelligent reflecting surface enhanced wireless network: Joint active and passive beamforming design, Abu Dhabi, United Arab Emirates.

Wu, Q. and Zhang, R. [2019], 'Intelligent reflecting surface enhanced wireless network via joint active and passive beamforming', *IEEE Transactions on Wireless Communications* **18**(11), 5394–5409.

Wu, Q. and Zhang, R. [2020], 'Towards smart and reconfigurable environment: Intelligent reflecting surface aided wireless network', *IEEE Communications Magazine* **58**(1), 106–112.

Wu, Q., Zhang, S., Zheng, B., You, C. and Zhang, R. [2021], 'Intelligent reflecting surface-aided wireless communications: A tutorial', *IEEE Transactions on Communications* **69**(5), 3313–3351.

Yang, Y., Zheng, B., Zhang, S. and Zhang, R. [2020], 'Intelligent reflecting surface meets OFDM: Protocol design and rate maximization', *IEEE Transactions on Communications* **68**(7), 4522–4535.

Ye, J., Guo, S. and Alouini, M.-S. [2020], 'Joint reflecting and precoding designs for SER minimization in reconfigurable intelligent surfaces assisted MIMO systems', *IEEE Transactions on Wireless Communications* **19**(8), 5561–5574.

Yu, X., Xu, W., Leung, S.-H. and Wang, J. [2017], 'Unified performance analysis of transmit antenna selection with OSTBC and imperfect CSI over Nakagami-m fading channels', *IEEE Transactions on Vehicular Technology* **67**, 494–508.

Yu, X., Xu, D. and Schober, R. [2020], Optimal beamforming for MISO communications via intelligent reflecting surfaces, *in* 'Proceedings of 2020 IEEE 21st International Workshop on Signal Processing Advances in Wireless Communications (SPAWC)', Atlanta, USA.

Yuan, X., Zhang, Y.-J. A., Shi, Y., Yan, W. and Liu, H. [2021], 'Reconfigurable-intelligent-surface empowered wireless communications: Challenges and opportunities', *IEEE Wireless Communications* **28**(2), 136–143.

Zhang, S. and Zhang, R. [2020], 'Capacity characterization for intelligent reflecting surface aided MIMO communication', *IEEE Journal on Selected Areas in Communications* **38**(8), 1823–1838.

Zhang, J., Yu, X. and Letaief, K. B. [2019], 'Hybrid beamforming for 5G and beyond millimeter-wave systems: A holistic view', *IEEE Open Journal of the Communications Society* **1**, 77–91.

Zheng, B. and Zhang, R. [2020], 'Intelligent reflecting surface-enhanced OFDM: Channel estimation and reflection optimization', *IEEE Wireless Communications Letters* **9**(4), 518–522.

Zheng, J. and Rao, B. D. [2008], 'Capacity analysis of MIMO systems using limited feedback transmit precoding schemes', *IEEE Transactions on Signal Processing* **56**, 2886–2901.

8

Multiple Dimensional and Antenna Techniques for 6G

Wireless communications suffer from deep fades due to the destructive super-position of multiple signal components that arrive at a receiver from different propagation paths. It brings a series of continuous erroneous symbols, which are the primary source of poor performance in wireless transmission. This motivates us to exploit various diversity techniques that carry identical information through independently faded paths, and reliable communications are achieved as long as one of the paths is good. Due to the restriction of time and frequency resources, time and frequency diversity are not the optimal selection. Therefore, spatial diversity, also known as antenna diversity, which is realized by simply adding an antenna array at the transmitter or the receiver without any loss of precious radio resources, becomes attractive. Spatial diversity can be further divided into different forms: *receive combining* that employs multiple antennas at the receiver to tackle independent faded signals, *transmit diversity* utilizing multiple transmit antennas to carry identical information in the space-time domain, and *transmit antenna selection*, which opportunistically selects the best channel. When an antenna array has multiple antennas with small inter-antenna spacing and no polarization, the signal paths corresponding to different antennas are highly correlated. In this case, spatial diversity is not available. Therefore, beamforming can be applied to achieve a power gain by steering the beam to concentrate the energy toward the desired direction or mitigate interfering signals. Under rich scattering environments, using multiple antennas at both the transmitter and receiver enables an additional degree of freedom by *spatially multiplexing* parallel data streams. The capacity of such a MIMO channel increases linearly with the number of antennas, whereas the channel capacity of spatial diversity or beamforming increases only on a logarithmic scale.

6G Key Technologies: A Comprehensive Guide, First Edition. Wei Jiang and Fa-Long Luo.

This chapter will focus on the fundamentals of multi-antenna transmission, consisting of

- The basics of spatial diversity and its particular advantages.
- Combining multiple spatial signals at the receiver for receive diversity through maximal-ratio combining, selection combining, and equal-gain combining.
- The design of space-time coding, including Space-Time Trellis Codes, Alamouti codes, and Space-Time Block Codes, for transmit diversity.
- Conventional beamforming over highly correlated antenna arrays and single-stream precoding over low-correlated arrays for power gain or interference suppression.
- The principle and advantages of transmit antenna selection.
- The fundamentals of point-to-point MIMO or single-user MIMO to realize a spatial-multiplexing gain. MIMO precoding at the transmitter and typical MIMO detection methods (i.e. linear decoding and successive interference cancelation).

8.1 Spatial Diversity

Compared with an Additive White Gaussian Noise (AWGN) channel, a wireless channel suffers from deep fades due to the destructive combination of signal replicas arriving from different propagation paths. A deep fade brings successive symbol and bit errors due to a very low Signal-to-Noise Ratio (SNR), which is the primary source of poor performance in wireless communications. This motivates us to exploit various diversity techniques to improve the performance. The basic idea of diversity is transmitting signals that carry identical information through multiple signal paths, each of which fades independently. It ensures that the receiver obtains multiple independent signal replicas, and reliable communication is achieved as long as one of the replicas is strong.

There are many approaches to achieving diversity. Diversity over time slots, referring to as *time diversity*, can be obtained via coding and interleaving. Information bits are coded, and the coded symbols are dispersed across multiple coherence periods through interleaving so that different parts of a codeword experience independent fades. Analogously, one can also exploit *frequency diversity* if a channel is wide enough to exhibit frequency selectivity. Techniques, such as single-carrier equalization, direct-sequence spread-spectrum, and Orthogonal Frequency Division Multiplexing (OFDM), are conventional ways to improve performance through exploiting frequency diversity. Alternatively, we can explore *spatial diversity* or *antenna diversity* by using multiple transmit or receive antennas if spaced sufficiently or polarized. In a cellular network, *macro diversity* or *multi-user diversity* can be exploited by the fact that the channels between a base station and multiple users are different. A wireless system typically employs several different types of diversity to achieve better performance.

In wireless communications, some radio resources are precious, i.e. time and frequency, which are limited and hard to produce in a specific location artificially. Time diversity wastes time resources by repeating identical information at multiple time slots. Moreover, interleaving and coding across several coherence periods increase system delay, which might be unacceptable for delay-sensitive applications when the channel coherence time is large. The identical drawback happens in frequency diversity, e.g. a narrow-band signal occupies a wide bandwidth in spread-spectrum techniques. Spatial diversity can be obtained by adding an antenna array at the transmitter or the receiver at a price of hardware cost and power consumption without any loss of precious radio resources. If the inter-antenna distance is sufficiently large, the received signals corresponding to different antennas have a low mutual correlation. In other words, different antennas fade more or less independently, resulting in independent signal paths. The antenna distance required for low fading correlation depends on wavelength and local scattering environment. A mobile terminal on the ground is generally surrounded by a lot of scatterers. In this context, the channel decorrelates over shorter distances, and a typical antenna separation of the order of only half a wavelength is sufficient to achieve a relatively low correlation. For a typical macro-cell base station mounted on a high tower, a more considerable inter-antenna distance of several to tens of wavelengths is typically required to ensure a low fading correlation. Another method of realizing low inter-antenna correlation is by using different polarization with vertically and horizontally polarized waves. Although their average received powers are approximately the same, the possibility of two paths falling into deep fades simultaneously is low since the scattering angle relative to either polarization is random. It is sometimes also called *polarization diversity*. Spatial diversity can be further divided into several forms:

- *Receive Diversity* employs multiple antennas at the receiver to form independent faded paths over Single Input Multiple Output (SIMO) channels. In addition to a diversity gain, this scheme has a power gain.
- *Transmit Diversity* utilizes multiple transmit antennas to transmit signals carrying identical information over a Multiple Input Single Output (MISO) channel. Interesting coding problems arise, leading to the necessity of designing space-time codes. It is attractive for the downlink of a cellular system, allowing for low-complexity, low-cost, lightweight mobile terminals.

Channels with multiple transmit and receive antennas, so-called Multi-Input Multi-Output (MIMO) channels, provide even more high order diversity. Besides the provisioning of spatial diversity, MIMO channels also offer an additional degree of freedom to transmit multiple data streams parallelly, referred to as *spatial multiplexing*, which will be introduced in the subsequent section.

- *Transmit Antenna Selection (TAS)*, which selects a single antenna from multiple transmit antennas to transmit a signal. A diversity order equaling the number of all transmit antennas can be achieved if the selected antenna has the best path with the strongest SNR. This technique can substantially lower the complexity of implementation, lower hardware cost, and improve power efficiency.

8.2 Receive Combining

The most commonly form of spatial diversity applied historically is the use of an antenna array at the receiver side to achieve receive diversity. It combines independent faded paths to obtain a resultant signal that is then detected to recover the original symbol. Different combining schemes have different complexity and performance. Linear combining techniques are usually employed, where the resulting signal is just a weighted sum of the received signals at all branches.

Figure 8.1 illustrates the structure of a linear combination of received signals $r_1, r_2, \ldots, r_{N_r}$ from N_r different antennas. If antennas are sufficiently far away or polarized, each channel is assumed to experience independent and identically distributed (*i.i.d.*) frequency-flat Rayleigh fading. If the signal bandwidth is greater than channel coherence bandwidth, a wireless channel suffers from frequency-selective fading. Multi-carrier techniques such as OFDM can convert a frequency-selective fading channel into a magnitude of frequency-flat fading channels. Therefore, this chapter only employs frequency-flat fading channels, whereas the subsequent chapter will discuss multi-antenna techniques over

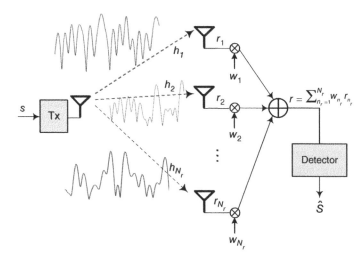

Figure 8.1 Schematic diagram of multi-antenna receive diversity with a linear combiner, where the signal from a single transmit antenna arrives N_r receive antennas through independently faded frequency-flat channels.

frequency-selective fading channels. A channel including the effects of the transmitter, the wireless link, and the receiver is modeled by a complex multiplicative distortion consisting of a magnitude gain and a phase shift. Usually, the channel coefficient between the transmit antenna and the n_r^{th} receive antenna can be expressed by a complex circularly symmetric Gaussian random variable $h_{n_r} \sim \mathcal{CN}(0, 1)$, $n_r = 1, 2, \ldots, N_r$. A typical channel suffers from a phase rotation θ_{n_r} and a corresponding gain a_{n_r}, namely

$$h_{n_r} = a_{n_r} e^{j\theta_{n_r}}, \tag{8.1}$$

where θ_{n_r} and a_{n_r} are real scalar values. Thus, the received signal at a typical antenna n_r is

$$r_{n_r} = h_{n_r} s + n_{n_r}, \tag{8.2}$$

where s is the transmitted symbol with the average power P, and n_{n_r} denotes AWGN with variance σ_n^2, i.e. $n \sim \mathcal{CN}(0, \sigma_n^2)$. All branches have an identical average SNR, which is denoted by

$$\bar{\gamma}_n = \frac{P}{\sigma_n^2}. \tag{8.3}$$

The primary purpose of multiplexing a weight at each antenna is to compensate for the corresponding channel phase, ensuring that the phases of the received signals are aligned to maximize the signal strength. Without this phase alignment, the signals cannot add up coherently in the combiner, leading to a resulting signal that still exhibits significant fading due to the constructive and destructive addition of all received signals. Consequently, a typical complex-valued weight w_{n_r} would contain a phase value $-\theta_{n_r}$ for compensation. The output signal of the linear combiner is

$$r = \sum_{n_r=1}^{N_r} w_{n_r} r_{n_r}. \tag{8.4}$$

We write γ to denote the instantaneous SNR of the combined signal r. It varies randomly, where the distribution is a function of the number of diversity paths, the fading distribution on each path, and the combining scheme. Given γ and its statistics, two metrics are usually applied to measure diversity performance, i.e. the *average error probability*

$$\bar{P} = \int_0^{+\infty} P(\gamma) f_\gamma(\gamma) d\gamma, \tag{8.5}$$

where $P(\gamma)$ is the error probability in AWGN, and $f_\gamma(\gamma)$ denotes the Power Density Function (PDF) of the instantaneous SNR, and the *outage probability*

$$P_{\text{out}} = \mathbb{P}(\gamma \leq \gamma_0) = \int_0^{\gamma_0} f_\gamma(\gamma) d\gamma, \tag{8.6}$$

where $\mathbb{P}$ notates mathematical probability, and γ_0 expresses a target SNR value.

In the following parts we will study three typical combining techniques:

- *Selection Combining (SC)* chooses the path with the highest SNR, and performs detection based on the signal from the selected path.
- *Maximal Ratio Combining (MRC)* makes decisions based on an optimal linear combination (a matched filter) of the path signals.
- *Equal-Gain Combining (EGC)* simply adds the path signals together after they have been co-phased.

8.2.1 Selection Combining

In SC, the received signal from the antenna with the largest SNR $\max(\gamma_{n_r})$, $n_r = 1, 2, \ldots, N_r$ is chosen for processing at the receiver. The SC receiver needs only one RF chain that is switched into the selected receive antenna, bringing the advantages of low hardware cost, low complexity, and low power consumption. Moreover, the co-phasing among multiple branches is not required, and therefore this technique can be applied for either coherent detection or differential modulation.

Mathematically, the weights for SC are determined by

$$w_{n_r} = \begin{cases} 1, & \text{if } n_r = \arg\max_{n_r}(\gamma_{n_r}) \\ 0, & \text{otherwise} \end{cases}. \tag{8.7}$$

The instantaneous SNR of the output of the linear combiner is then

$$\gamma_{SC} = \max_{n_r}(\gamma_{n_r}). \tag{8.8}$$

Derive the Cumulative Distribution Function (CDF) of γ_{SC} as

$$\begin{aligned} F_{\gamma_{SC}}(\gamma) &= \mathbb{P}(\gamma_{SC} < \gamma) \\ &= \mathbb{P}\left(\max_{n_r}(\gamma_{n_r}) < \gamma\right) \\ &= \prod_{n_r=1}^{N_r} \mathbb{P}(\gamma_{n_r} < \gamma) = \prod_{n_r=1}^{N_r} F_{\gamma_{n_r}}(\gamma), \end{aligned} \tag{8.9}$$

where $F_{\gamma_{n_r}}(\gamma)$ denotes the CDF of the instantaneous SNR at a typical receive antenna n_r. With the assumption of *i.i.d.* Rayleigh fading, we know that γ_{n_r} follows the exponential distribution. Thus,

$$F_{\gamma_{n_r}}(\gamma) = 1 - e^{-\gamma/\bar{\gamma}_{n_r}}, \tag{8.10}$$

with the average SNR at the n_r^{th} receive antenna $\bar{\gamma}_{n_r}$, and this equation is further simplified to

$$F_{\gamma_{n_r}}(\gamma) = 1 - e^{-\gamma/\bar{\gamma}_n}, \tag{8.11}$$

applying $\bar{\gamma}_{n_r} = \bar{\gamma}_n$, for any $n_r = 1, 2, \ldots, N_r$.

Substituting Eq. (8.11) into Eq. (8.9) yields

$$F_{\gamma_{SC}}(\gamma) = \left[1 - e^{-\gamma/\bar{\gamma}_n}\right]^{N_r}, \tag{8.12}$$

implying that even all branches follow Rayleigh fading, the combined signal is no longer Rayleigh distributed.

Then, the outage probability of SC for the target SNR γ_0 is obtained by substituting $\gamma = \gamma_0$ into Eq. (8.12):

$$P_{\text{out}}^{SC}(\gamma_0) = \mathbb{P}(\gamma_{SC} \leq \gamma_0) = F_{\gamma_{SC}}(\gamma_0) = \left[1 - e^{-\gamma_0/\bar{\gamma}_n}\right]^{N_r}. \tag{8.13}$$

Differentiating Eq. (8.12) relative to γ gets the PDF for γ_{SC}:

$$f_{\gamma_{SC}}(\gamma) = \frac{\partial F_{\gamma_{SC}}(\gamma)}{\partial \gamma} = \frac{N_r}{\bar{\gamma}_n}\left[1 - e^{-\gamma/\bar{\gamma}_n}\right]^{N_r-1}e^{-\gamma/\bar{\gamma}_n}. \tag{8.14}$$

In AWGN, error probability depends on the received SNR. But the received signal power in a wireless channel varies randomly due to multipath fading. Therefore, we consider a fading channel as an AWGN with a variable gain, and the average Bit Error Rate (BER) and Symbol Error Rate (SER) can be computed by integrating the error probability in AWGN over the fading distribution. The BERs and SERs of several typical digital modulations in AWGN are listed in Table 8.1. The average probability is computed using Eq. (8.5). Closed-form expressions do not exist for most of modulation schemes, except for Differential Phase Shift Keying (DPSK), as given by

$$\begin{aligned}
\bar{P}_b &= \int_0^\infty \frac{1}{2}e^{-\gamma}f_{\gamma_{SC}}(\gamma)d\gamma \\
&= \int_0^\infty \frac{1}{2}e^{-\gamma}\frac{N_r}{\bar{\gamma}_n}\left[1 - e^{-\gamma/\bar{\gamma}_n}\right]^{N_r-1}e^{-\gamma/\bar{\gamma}_n}d\gamma \\
&= \frac{N_r}{2}\sum_{n_r=0}^{N_r-1}(-1)^{n_r}\frac{\binom{N_r-1}{n_r}}{1 + n_r + \bar{\gamma}_n}.
\end{aligned} \tag{8.15}$$

The more favorable distribution for the overall received SNR results in a faster drop of error probability or outage probability due to the benefit of diversity. In particular, we use a metric called **diversity order** to reflect how quickly the error probability decays in terms of the average SNR. The diversity order equals N_r, when the error performance can be expressed in the form $c\bar{\gamma}_n^{-N_r}$, where c is a constant that depends on the specific modulation and coding, and $\bar{\gamma}_n$ is the average SNR per branch. The performance gain increases with the diversity order, but not linearly. The most significant gain is obtained by going from a single antenna (i.e. no diversity) to two antennas. Increasing the number of diversity branches from two to three will give much less gain than going from one to

Table 8.1 Approximate symbol and bit error probabilities for coherent modulations.

Modulation	Symbol error rate	Bit error rate
BPSK	—	$P_b(\gamma_b) = Q(\sqrt{2\gamma_b})$
QPSK	$P_s(\gamma_s) \approx 2Q(\sqrt{\gamma_s})$	$P_b(\gamma_b) = Q(\sqrt{2\gamma_b})$
DPSK	$P_s(\gamma_s) = \frac{1}{2}e^{-\gamma_s}$	$P_b(\gamma_b) = \frac{1}{2}e^{-\gamma_b}$
MPSK	$P_s(\gamma_s) \approx 2Q\left(\sqrt{2\gamma_s}\sin\left(\frac{\pi}{M}\right)\right)$	$P_b(\gamma_b) = \frac{2}{\log_2 M}Q\left(\sqrt{2\gamma_b\log_2 M}\sin\left(\frac{\pi}{M}\right)\right)$
MQAM	$P_s(\gamma_s) \approx 4Q\left(\sqrt{\frac{3\overline{\gamma}_s}{M-1}}\right)$	$P_b(\gamma_b) \approx \frac{4}{\log_2 M}Q\left(\sqrt{\frac{3\overline{\gamma}_b\log_2 M}{M-1}}\right)$

Note: $Q(x) = \frac{1}{\sqrt{2\pi}}\int_x^\infty \exp\left(-\frac{u^2}{2}\right)du$

Source: Goldsmith [2005]/With permission of Cambridge University Press.

two, and in general, increasing N_r yields a diminishing gain [Goldsmith, 2005]. This rule can also be proved by the average SNR of the combined signal

$$
\begin{aligned}
\overline{\gamma}_{SC} &= \int_0^\infty \gamma f_{\gamma_{SC}}(\gamma)d\gamma \\
&= \int_0^\infty \gamma \frac{N_r}{\overline{\gamma}_n}\left[1 - e^{-\gamma/\overline{\gamma}_n}\right]^{N_r-1} e^{-\gamma/\overline{\gamma}_n}d\gamma \\
&= \overline{\gamma}_n \sum_{n_r=1}^{N_r}\frac{1}{n_r},
\end{aligned}
\tag{8.16}
$$

which grows with the number of N_r but the incremental step is substantially diminishing.

According to Eq. (8.15), Figure 8.2 shows the average BER of SC with Quadrature Phase Shift Keying (QPSK) modulation as a function of the average SNR per branch. As an example, the SNR gain is approximately 12 dB decibel when the diversity order is increased from one to two at the BER level of 10^{-3}. However, going from two-branch to four-branch results in an additional gain of about 6 dB. Doubling the number of diversity further to 8 and 16 results in an additional reduction of approximately 3 dB and less than 2 dB, respectively.

8.2.2 Maximal Ratio Combining

The MRC receiver is also a matched filter that can maximize the SNR of the combined signal by weighting the received signal in each branch in proportion

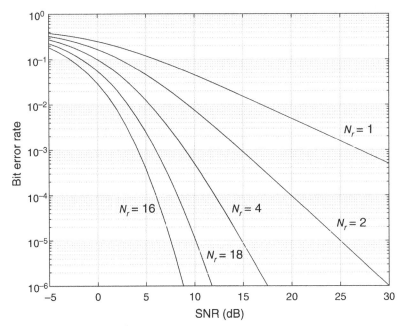

Figure 8.2 Average BER performance of selection combining with QPSK under *i.i.d.* Rayleigh fading channels.

to the signal strength and also align the phases of received signals at different antennas to compensate for the corresponding channel phases. Assume the perfect Channel State Information (CSI) h_{n_r}, $n_r = 1, 2, \ldots, N_r$ is obtained by the n_r^{th} receive antenna through channel estimation. Using the weights

$$w_{n_r} = h_{n_r}^*, \quad n_r = 1, 2, \ldots, N_r, \tag{8.17}$$

where $(\cdot)^*$ denotes the complex conjugate, the resultant signal of the linear combiner is obtained by substituting Eq. (8.17) into Eq. (8.4)

$$r_{\text{MRC}} = \sum_{n_r=1}^{N_r} h_{n_r}^* r_{n_r} = \sum_{n_r=1}^{N_r} h_{n_r}^* (h_{n_r} s + n_{n_r})$$

$$= \sum_{n_r=1}^{N_r} |h_{n_r}|^2 s + \sum_{n_r=1}^{N_r} h_{n_r}^* n_{n_r}. \tag{8.18}$$

The instantaneous SNR is

$$\gamma_{\text{MRC}} = \sum_{n_r=1}^{N_r} |h_{n_r}|^2 \frac{P}{\sigma_n^2} = \left(\sum_{n_r=1}^{N_r} |h_{n_r}|^2 \right) \overline{\gamma}_n, \tag{8.19}$$

applying $\overline{\gamma}_n = P/\sigma_n^2$. Under *i.i.d.* Rayleigh distribution $h_{n_r} \sim \mathcal{CN}(0,1)$, $n_r = 1$, $2, \dots, N_r$, the item

$$\sum_{n_r=1}^{N_r} |h_{n_r}|^2 \tag{8.20}$$

is a sum of the squares of $2N_r$ independent real Gaussian random variables. Consequently, the overall received SNR γ_{MRC} is Chi-square distributed with expected value $\overline{\gamma}_{\mathrm{MRC}} = N_r \overline{\gamma}_n$ and variance $2N_r \overline{\gamma}_n$. Thus, its PDF is given by

$$f_{\gamma_{\mathrm{MRC}}}(\gamma) = \frac{\gamma^{N_r-1} e^{-\gamma/\overline{\gamma}_n}}{\overline{\gamma}_n^{N_r}(N_r - 1)!}, \quad \gamma \geq 0, \tag{8.21}$$

where the operator ! means the factorial of a non-negative integer, e.g.

$$n! = n \cdot (n - 1) \cdot (n - 2) \cdots 2 \cdot 1. \tag{8.22}$$

Then, substitute Eq. (8.21) back in Eq. (8.6) to obtain the outage probability of MRC:

$$\begin{aligned}
P_{\mathrm{out}}^{\mathrm{MRC}}(\gamma_0) &= \mathbb{P}(\gamma_{\mathrm{MRC}} \leq \gamma_0) \\
&= \int_0^{\gamma_0} f_{\gamma_{\mathrm{MRC}}}(\gamma) d\gamma \\
&= 1 - e^{-\gamma_0/\overline{\gamma}_n} \sum_{n_r=1}^{N_r} \frac{(\gamma_0/\overline{\gamma}_n)^{n_r-1}}{(n_r - 1)!}
\end{aligned} \tag{8.23}$$

Meanwhile, the average error probability can be computed by applying Eq. (8.5). For example, the BER of QPSK modulation with *i.i.d.* Rayleigh fading is

$$\begin{aligned}
\overline{P}_b &= \int_0^\infty Q\left(\sqrt{2\gamma}\right) f_{\gamma_{\mathrm{MRC}}}(\gamma) d\gamma \\
&= \left(\frac{1-F}{2}\right)^{N_r} \sum_{n_r=0}^{N_r-1} \binom{N_r + n_r - 1}{n_r} \left(\frac{1+F}{2}\right)^{n_r},
\end{aligned} \tag{8.24}$$

where

$$F = \sqrt{\frac{\overline{\gamma}_n}{1 + \overline{\gamma}_n}}. \tag{8.25}$$

At high SNR, apply the Taylor series expansion in $1/\overline{\gamma}_n$ to get the approximations:

$$\frac{1-F}{2} \approx \frac{1}{4\overline{\gamma}_n}, \quad \text{and} \quad \frac{1+F}{2} \approx 1. \tag{8.26}$$

Moreover,

$$\sum_{n_r=0}^{N_r-1} \binom{N_r + n_r - 1}{n_r} = \binom{2N_r - 1}{N_r}. \tag{8.27}$$

Applying Eq. (8.26) and Eq. (8.27), Eq. (8.24) can be approximated to

$$\overline{P}_b \approx \binom{2N_r - 1}{N_r} \left(\frac{1}{4\overline{\gamma}_n}\right)^{N_r}, \tag{8.28}$$

implying that the error probability decays at a rate of N_r^{th} power of SNR and the MRC receiver achieves the diversity order of N_r. The maximal diversity order of a system with N_r antennas is N_r, and therefore the MRC receiver achieves full diversity order at high SNR.

8.2.3 Equal-Gain Combining

A simpler combining technique is equal-gain combining, where each signal branch is weighted equally, irrespective of their channel gains. It only phase rotates the signals of different receive antennas to ensure phase alignment when added together.

Applying the weights

$$w_{n_r} = e^{-j\theta_{n_r}}, \quad n_r = 1, 2, \ldots, N_r, \tag{8.29}$$

resulting in the output signal of the linear combiner

$$\begin{aligned}
r_{\text{EGC}} &= \sum_{n_r=1}^{N_r} w_{n_r} r_{n_r} = \sum_{n_r=1}^{N_r} w_{n_r} \left(h_{n_r} s + n_{n_r} \right) \\
&= \sum_{n_r=1}^{N_r} e^{-j\theta_{n_r}} \left(|h_{n_r}| e^{j\theta_{n_r}} s + n_{n_r} \right) \\
&= \sum_{n_r=1}^{N_r} |h_{n_r}| s + \sum_{n_r=1}^{N_r} e^{-j\theta_{n_r}} n_{n_r}.
\end{aligned} \tag{8.30}$$

The instantaneous SNR is

$$\gamma_{\text{EGC}} = \frac{1}{N_r} \left(\sum_{n_r=1}^{N_r} |h_{n_r}| \right)^2 \frac{P}{\sigma_n^2} = \frac{1}{N_r} \left(\sum_{n_r=1}^{N_r} |h_{n_r}| \right)^2 \overline{\gamma}_n. \tag{8.31}$$

The general PDF and CDF of γ_{EGC} with an arbitrary number of N_r do not exist in closed-form. For $N_r = 2$ under *i.i.d.* Rayleigh fading, an expression for the CDF can be given by Goldsmith [2005]

$$F_{\gamma_{\text{EGC}}}(\gamma) = 1 - e^{-2\gamma/\overline{\gamma}_n} - e^{-\gamma/\overline{\gamma}_n} \sqrt{\frac{\pi\gamma}{\overline{\gamma}_n}} \left[1 - 2Q\left(\sqrt{\frac{2\gamma}{\overline{\gamma}_n}}\right) \right]. \tag{8.32}$$

Let $\gamma = \gamma_0$ in Eq. (8.32) to obtain the outage probability

$$\begin{aligned}
P_{\text{out}}^{\text{EGC}}(\gamma_0) &= \mathbb{P}(\gamma_{\text{EGC}} \leq \gamma_0) = F_{\gamma_{\text{EGC}}}(\gamma_0) \\
&= 1 - e^{-2\gamma_0/\overline{\gamma}_n} - e^{-\gamma_0/\overline{\gamma}_n} \sqrt{\frac{\pi\gamma_0}{\overline{\gamma}_n}} \left[1 - 2Q\left(\sqrt{\frac{2\gamma_0}{\overline{\gamma}_n}}\right) \right].
\end{aligned} \tag{8.33}$$

Differentiating Eq. (8.32) relative to γ obtains the PDF

$$f_{\gamma_{EGC}}(\gamma) = \frac{\partial F_{\gamma_{EGC}}(\gamma)}{\partial \gamma} \tag{8.34}$$

$$= \frac{1}{\bar{\gamma}_n} e^{-2\gamma/\bar{\gamma}_n} + \sqrt{\pi} e^{-\gamma/\bar{\gamma}_n} \left(\frac{1}{\sqrt{4\gamma\bar{\gamma}_n}} - \frac{1}{\bar{\gamma}_n}\sqrt{\frac{\gamma}{\bar{\gamma}_n}} \right)$$

$$\times \left[1 - 2Q\left(\sqrt{\frac{2\gamma_0}{\bar{\gamma}_n}} \right) \right].$$

Similarly, the average BER for QPSK with the EGC receiver can be obtained as

$$\bar{P}_b = \int_0^\infty Q\left(\sqrt{2\gamma}\right) f_{\gamma_{EGC}}(\gamma) d\gamma = \frac{1}{2}\left(1 - \sqrt{1 - \left(\frac{1}{1+\bar{\gamma}_n}\right)^2} \right). \tag{8.35}$$

The performance of EGC is quite close to that of MRC, typically exhibiting a power penalty of less than 1 dB decibel, with the price paid for low complexity of using equal gains.

8.3 Space-Time Coding

In most scattering environments, receive diversity is a practical, effective, and, therefore, a commonly applied technique for alleviating the effect of multipath fading. However, receive diversity is not suitable for some deployment scenarios, e.g. the downlink of a cellular system where a mobile terminal is hard to integrate an antenna array, especially when operating at sub-6 GHz frequencies. As an alternative, spatial diversity can also be achieved by applying multiple antennas at the transmitter, referred to as *transmit diversity*. Multiple antennas are primarily attractive for base stations, which provide more space, sufficient power supply, and strong processing capability. It offers an equivalent spatial diversity gain without the need for additional receive antennas and corresponding RF chains at mobile terminals. As a consequence, a cellular system becomes more economical by using receive diversity in the uplink and simultaneously transmit diversity in the downlink.

Transmit diversity design relies on whether the CSI is known at the transmitter. When the CSI is available, also known as CSI at the transmitter (CSIT), the system is very similar to receive diversity. By multiplying a complex weight $h_{n_t}^*$ with the transmitted signal of transmit antenna n_t, these signals co-phased combine in the air, resulting in a signal that is equivalent to the MRC combined signal in receive diversity. Transmit diversity with CSIT achieves not only spatial diversity but also a power gain. It is also called transmit *beamforming* (over low-correlation antennas) or *precoding*, which will be discussed in detail in Section 8.5. This part focuses on the case where the transmitter does not know CSI.

8.3.1 Repetition Coding

If no knowledge of downlink channels is available at the transmitter, multiple transmit antennas cannot provide beamforming but only spatial diversity. Low mutual correlation among different channels is required, achieved by means of sufficiently large inter-antenna distances or different antenna polarization. Under such an antenna configuration, we still need an approach to realize the spatial diversity offered by multiple transmit antennas. One can simply apply a repetition code that repeats the same symbol N_t times to transmit over N_t transmit antennas simultaneously. In contrast to the receive diversity where each antenna receives a signal experienced an independently fading channel, transmitting an identical signal over multiple transmit antennas cannot naturally form multiple independent paths. Figure 8.3 demonstrates a repetition-coding strategy, where two transmit antennas send the same signal $s(t)e^{j2\pi f_c t}$ and equally divide the transmit energy.

The received signal is then

$$r(t) = \sum_{l=1}^{L_1} \frac{\alpha_l}{\sqrt{2}} s(t - \tau_l) e^{j2\pi f_c(t-\tau_l)} + \sum_{l=1}^{L_2} \frac{\alpha_l}{\sqrt{2}} s(t - \tau_l) e^{j2\pi f_c(t-\tau_l)}, \qquad (8.36)$$

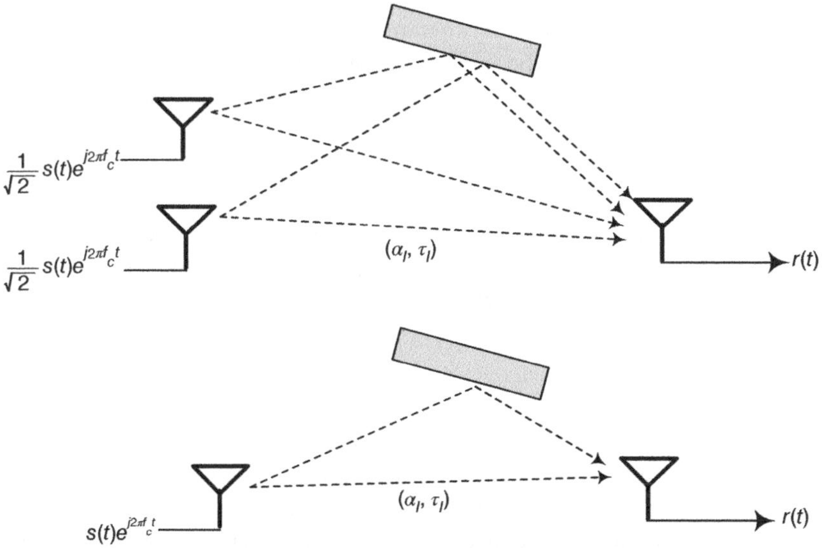

Figure 8.3 Comparison between single-antenna transmission and multi-antenna transmission with an identical transmitted signal per antenna.

where L_1 and L_2 denote the total number of propagation paths for transmit antennas 1 and 2, respectively, α_l and τ_l represent the amplitude gain and delay of path l. Under the narrow-band assumption, the signal bandwidth is far less than the carrier frequency, so that the baseband signal keeps almost constant over the interval of $\tau_n(\theta)$, i.e. the approximation $s(t) \approx s(t - \tau_n(\theta))$ holds. Then, Eq. (8.36) becomes

$$r(t) = \sum_{l=1}^{L_1+L_2} \frac{\alpha_l}{\sqrt{2}} s(t) e^{j2\pi f_c(t-\tau_l)} = \left(\sum_{l=1}^{L_1+L_2} \frac{\alpha_l}{\sqrt{2}} e^{-j2\pi f_c \tau_l} \right) s(t) e^{j2\pi f_c t}. \tag{8.37}$$

Recalling the transmit signal $s(t)e^{j2\pi f_c t}$, we gets the channel response

$$h(\tau) = \sum_{l=1}^{L_1+L_2} \frac{\alpha_l}{\sqrt{2}} e^{-j2\pi f_c \tau_l}. \tag{8.38}$$

As we know, the channel response for the signal-antenna transmission is expressed by

$$h(\tau) = \sum_{l=1}^{L} \alpha_l e^{-j2\pi f_c \tau_l}. \tag{8.39}$$

Except for more propagation paths, the channel of two-antenna transmission with an identical signal exhibit no difference from the signal-antenna case from the perspective of a receiver. Therefore, no diversity is available.

Alternatively, we can derive the same conclusion through a baseband equivalent model. Using a repetition code over two transmit antennas, the transmitted symbol on either antenna is $\frac{1}{\sqrt{2}}s$ for the same total transmit power relative to the signal antenna. Applying channel gains $h_i \sim \mathcal{CN}(0,1), i = 1, 2$, the received signal is then

$$r = \frac{1}{\sqrt{2}} \left(h_1 + h_2 \right) s + n. \tag{8.40}$$

The effective channel $\frac{1}{\sqrt{2}} \left(h_1 + h_2 \right)$ is the sum of two complex Gaussian random variables, and is thus a complex Gaussian random variable with *zero mean* and *unit* variance, i.e. $\frac{1}{\sqrt{2}} \left(h_1 + h_2 \right) \sim \mathcal{CN}(0,1)$. It is equivalent to the single-antenna transmission where

$$r = hs + n \tag{8.41}$$

with $h \sim \mathcal{CN}(0,1)$. In other words, a system cannot reap spatial diversity gain when applying a repetition code over multiple transmit antennas.

An approach to implement transmit diversity with full diversity order is simply sending the same symbol over the N_t transmit antennas during N_t symbol periods. That is a repetition code in the time domain, rather than in the spatial domain. At only one time, only a single antenna is activated while the others keep silent. This repetition code is quite wasteful of degrees of freedom. Inevitably, we need to design codes specifically for transmit diversity systems, leading to many efforts in the design of *space-time coding*. Space-time coding refers to the set of schemes aimed at realizing joint encoding of multiple transmit antennas. In these schemes, a number of coded symbols equaling to the number of transmit antennas are generated and transmitted simultaneously, one symbol per antenna. A space-time encoder generates these symbols such that the diversity gain is maximized by using the appropriate signal processing and decoding procedure at the receiver [Gesbert et al., 2003]. The development of the concept for space-time coding was initially revealed by Tarokh et al. [1998] in the form of trellis codes, referred to as Space-Time Trellis Code (STTC). Later, the simplest and yet one of the most elegant space-time codes called the Alamouti scheme [Alamouti, 1998] was proposed, which has been successfully applied in the Universal Mobile Telecommunications System (UMTS) and subsequent systems such as Long-Term Evolution (LTE) and LTE-Advanced. The Alamouti scheme is designed specifically for two transmit antennas, while the generalization of space-time coding to any number of antennas is possible. Such generalized space-time codes with linear combination for any number of transmit antennas are often known as Space-Time Block Codes (STBC) [Tarokh et al., 1999].

8.3.2 Space-Time Trellis Codes

The first attempt to design a space-time code was presented by Seshadri and Winters [1993]. However, the key milestone of developing space-time coding was initially completed by Tarokh *et al.* by introducing STTCs in the late 1990s [Tarokh et al., 1998]. Consider a wireless communication system that employs N_t transmit antennas at the transmitter and a single receive antenna where the channels are quasi-static and frequency-flat. The encoding for these trellis codes depends on the current state of the encoder, and the input symbol. At each time t, an input symbol s_t is encoded by selecting a transition path labeled by

$$\mathbf{c}_t = [c_1^t, c_2^t, \dots, c_{N_t}^t]^T. \tag{8.42}$$

Then, these coded symbols are transmitted by N_t antennas simultaneously where antenna n_t sends $c_{n_t}^t$, $n_t = 1, 2, \dots, N_t$.

Figure 8.4 demonstrates an example for STTCs, we provide a QPSK four-state trellis code design for two transmit antennas. The constellation of QPSK and the labeling of the trellis description is also given in the figure. Each row of the matrix represents the edge labels for transitions from the corresponding states. The edge label $c_1 c_2$ means that symbol c_1 is transmitted over the first antenna while symbol c_2 is transmitted over the second antenna simultaneously. At the beginning and

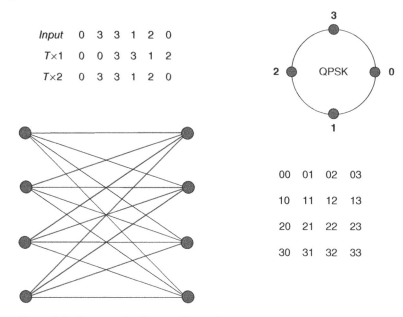

Figure 8.4 An example of space-time trellis coding with four states, QPSK constellation, and two transmit antennas.

the end of each block, the encoder is required to be in the zero state. The label of the first selected branch is *00* since the input symbol is *0* at the first symbol period and the initial state is *0*. Then, the label of the second selected branch becomes *03* to send symbol *3*. The third selected branch is labeled by *33*, indicating that sending symbol *3* from state *3*. This process iterates until the whole block of input symbols is transmitted.

Suppose the length of block is T and then the code vector sequence is

$$\mathbf{C} = \{\mathbf{c}_1, \mathbf{c}_2, \dots, \mathbf{c}_T\}. \tag{8.43}$$

We write r_t to denote the received symbol at time t, the branch metric for a transition labeled $\mathbf{c}_t$ is given by

$$\left| r_t - \mathbf{h}_t^T \mathbf{c}_t \right|^2 \tag{8.44}$$

with the channel vector between N_t transmit antennas and the receive antenna $\mathbf{h}_t = [h_1, h_2, \dots, h_{N_t}]^T$. The Viterbi algorithm [Viterbi, 2006] is then used to compute the path with the lowest accumulated metric. Consider the probability that the decoder decides erroneously in favor of the legitimate code vector sequence

$$\tilde{\mathbf{C}} = \{\tilde{\mathbf{c}}_1, \tilde{\mathbf{c}}_2, \dots, \tilde{\mathbf{c}}_T\}, \tag{8.45}$$

an error matrix is defined as

$$\mathbf{A}[\mathbf{C}, \tilde{\mathbf{C}}] = \sum_{t=1}^{T} (\mathbf{c}_t - \tilde{\mathbf{c}}_t)(\mathbf{c}_t - \tilde{\mathbf{c}}_t)^*. \tag{8.46}$$

The probability of transmitting $\mathbf{C}$ and deciding in favor of $\tilde{\mathbf{C}}$ is upper bounded for a Rayleigh fading channel by

$$\mathbb{P}(\mathbf{C} \to \tilde{\mathbf{C}}) \leq \left(\prod_{i=1}^{r} \beta_i \right)^{-N_t} \left(\frac{E_s}{4N_0} \right)^{-rN_t}, \tag{8.47}$$

where E_s is the symbol energy, N_0 denotes the noise spectral density, r is the rank of the error matrix $\mathbf{A}$, and β_i, $i = 1, 2, \ldots, r$ stand for the nonzero eigenvalues of $\mathbf{A}$ [Gesbert et al., 2003].

8.3.3 Alamouti Coding

STTCs provide a diversity gain equaling the number of transmit antennas and a coding gain that depends on the complexity of the code (i.e. the number of states in the trellis) without any loss in bandwidth efficiency. But it requires a multi-dimensional Viterbi algorithm at the receiver for decoding. When the number of antennas is fixed, the decoding complexity of STTC (measured by the number of trellis states at the decoder) rises exponentially as a function of the diversity level and transmission rate.

In addressing the issue of decoding complexity, Siavash M. Alamouti created a remarkable space-time coding scheme for transmission with two antennas [Alamouti, 1998] in 1998. The Alamouti scheme supports Maximum-Likelihood (ML) detection based only on linear processing at the receiver. It is the unique complex-symbol STBC with full transmit diversity at a full symbol rate. The scheme works over two consecutive symbol periods where it is assumed that fading is constant across this time span. The input symbols are divided into groups of two symbols each. At a given symbol period, the two symbols in each group s_1, s_2 are transmitted simultaneously from the two antennas. The signal transmitted from antenna 1 is s_1, and the signal transmitted from antenna 2 is s_2. During the next symbol period, the signal $-s_2^*$ is transmitted from antenna 1, and the signal s_1^* is transmitted from antenna 2, as shown in Figure 8.5. Let h_1 and h_2 be channel coefficients from the first and second transmit antennas to the single receive antenna, respectively. Then, the received symbols at two symbol periods can then be expressed as

$$\begin{cases} r_1 = h_1 s_1 + h_2 s_2 + n_1 \\ r_2 = -h_1 s_2^* + h_2 s_1^* + n_2 \end{cases}. \tag{8.48}$$

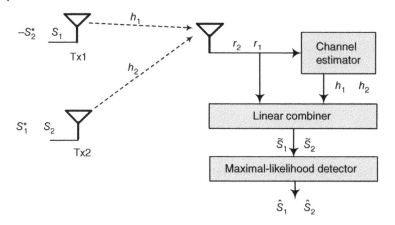

Figure 8.5 The Alamouti scheme over two-branch transmit diversity system with one receiver.

The combiner builds two combined symbols that are sent to the ML detector:

$$\tilde{s}_1 = h_1^* r_1 + h_2 r_2^*$$
$$\tilde{s}_2 = h_2^* r_1 - h_1 r_2^*. \tag{8.49}$$

Substituting Eq. (8.48) into Eq. (8.49), we obtain

$$\tilde{s}_1 = \left(|h_1|^2 + |h_2|^2\right) s_1 + h_1^* n_1 + h_2 n_2^*$$
$$\tilde{s}_2 = \left(|h_1|^2 + |h_2|^2\right) s_s - h_1 n_2^* + h_2^* n_1, \tag{8.50}$$

which are equivalent to that obtained from two-branch maximal ratio combining. Thus, the instantaneous SNR of the received signal is

$$\gamma_{\text{ala}} = \left(|h_1|^2 + |h_2|^2\right) \frac{P}{\sigma_n^2}. \tag{8.51}$$

Therefore, the Alamouti scheme realizes a diversity order of 2 (full diversity) without any loss of transmission rate (full rate). Furthermore, it does not need the knowledge of CSI at the transmitter.

Alternatively, the coding and decoding process can be expressed in vector form. A pair of symbols $\mathbf{s} = [s_1, s_2]^T$ are precoded as follows:

$$\begin{pmatrix} s_1 \\ s_2 \end{pmatrix} \xrightarrow{\text{Precoding}} \begin{pmatrix} s_1 & -s_2^* \\ s_2 & s_1^* \end{pmatrix}. \tag{8.52}$$

The row of the precoded matrix corresponds to the spatial domain, i.e. different antennas, while the column stands for the temporal domain.

Building a vector of received symbols as $\mathbf{r} = [r_1, r_2^*]^T$, a noise vector $\mathbf{n} = [n_1, n_2^*]^T$, and a composite channel matrix

$$\mathbf{H} = \begin{pmatrix} h_1 & h_2 \\ h_2^* & -h_1^* \end{pmatrix}. \tag{8.53}$$

Then, Eq. (8.48) is rewritten in matrix form as

$$\mathbf{r} = \mathbf{Hs} + \mathbf{n}. \tag{8.54}$$

The transmitted symbols can be detected by simply employing Zero-Forcing (ZF) decoding

$$\hat{\mathbf{s}} = \mathbf{H}^H \mathbf{r} = \mathbf{H}^H \mathbf{Hs} + \mathbf{H}^H \mathbf{n}, \tag{8.55}$$

to get

$$\hat{\mathbf{s}} = \begin{pmatrix} |h_1|^2 + |h_2|^2 & 0 \\ 0 & |h_1|^2 + |h_2|^2 \end{pmatrix} \mathbf{s} + \mathbf{H}^H \mathbf{n}. \tag{8.56}$$

Meanwhile, we can apply the Minimum Mean-Square Error (MMSE) method to get better detection performance

$$\hat{\mathbf{s}} = \left(\mathbf{H}^H \mathbf{H} + \sigma_n^2 \mathbf{I} \right)^{-1} \mathbf{H}^H \mathbf{r}, \tag{8.57}$$

where σ_n^2 is the noise variance, and $\mathbf{I}$ denotes the identity matrix.

8.3.4 Space-Time Block Codes

The Alamouti scheme is appealing in terms of simplicity and performance, which motivated the development of similar coding techniques. By applying the theory of orthogonal design, STBCs for more than two transmit antennas have been created. In STBC, data symbols are encoded and split into N_t streams, which are simultaneously transmitted using N_t transmit antennas. The received signal is a linear superposition of the N_t transmitted signals distorted by channel fading and perturbed by noise. STBCs are designed to achieve the maximum diversity order for a given number of transmit antennas subject to the constraint of having a simple decoding algorithm. The code orthogonality enables ML decoding based only on linear processing at the receiver, rather than joint detection like the Viterbi algorithm.

A generalization of orthogonal methods is shown to provide STBCs for both real-valued and complex-valued constellations for any number of transmit antennas. These codes achieve the maximum possible transmission rate for any number of transmit antennas using any arbitrary real-valued constellation such as Pulse-Amplitude Modulation (PAM). A real orthogonal design of size N_t is an $N_t \times N_t$ matrix with entries the indeterminates $\pm s_1, \pm s_1, \dots, \pm s_{N_t}$. The simplest orthogonal design are a 2×2 matrix

$$\mathcal{O}_2 = \begin{pmatrix} s_1 & s_2 \\ -s_2 & s_1 \end{pmatrix}, \tag{8.58}$$

satisfying

$$\mathcal{O}_2^T \mathcal{O}_2 = \begin{pmatrix} s_1^2 + s_2^2 & 0 \\ 0 & s_1^2 + s_2^2 \end{pmatrix} = (s_1^2 + s_2^2)\mathbf{I}_2, \tag{8.59}$$

where $\mathbf{I}_2$ is the identity matrix with size 2. Given an orthogonal design, one can negate certain columns of $\mathcal{O}$ to get another orthogonal design where all the entries of the first row have positive signs. Thus, the 4×4 orthogonal design is

$$\mathcal{O}_4 = \begin{pmatrix} s_1 & s_2 & s_3 & s_4 \\ -s_2 & s_1 & -s_4 & s_3 \\ -s_3 & s_4 & s_1 & -s_2 \\ -s_4 & -s_3 & s_2 & s_1 \end{pmatrix}. \tag{8.60}$$

STBCs can achieve a full rate and full diversity by encoding N_t real-valued symbols $s_1, s_2, \ldots, s_{N_t}$ into $\mathcal{O}_{N_t}\left(s_1, s_2, \ldots, s_{N_t}\right)$ and then transmitting each row of $\mathcal{O}_{N_t}$ over N_t transmit antennas simultaneously.

For an arbitrary complex-valued constellation such as PSK and QAM, STBCs achieve $1/2$ of the maximum possible transmission rate for any number of transmit antennas. Moreover, it can achieve the full, $3/4$, and $3/4$ of maximum transmission rate for the specific cases of two, three, and four transmit antennas, respectively, using arbitrary complex-valued constellations [Tarokh et al., 1999]. For instance, rate $1/2$ codes for transmission using four transmit antennas are given by

$$\mathcal{G}_4 = \begin{pmatrix} s_1 & s_2 & s_3 & s_4 \\ -s_2 & s_1 & -s_4 & s_3 \\ -s_3 & s_4 & s_1 & -s_2 \\ -s_4 & -s_3 & s_2 & s_1 \\ s_1^* & s_2^* & s_3^* & s_4^* \\ -s_2^* & s_1^* & -s_4^* & s_3^* \\ -s_3^* & s_4^* & s_1^* & -s_2^* \\ -s_4^* & -s_3^* & s_2^* & s_1^* \end{pmatrix}, \tag{8.61}$$

where s_1, s_2, s_3, s_4 denotes complex-valued constellation points, whereas s_1, s_2, s_3, s_4 in Eq. (8.60) are real-valued symbols. It implies that four complex symbols are transmitted by four antennas over eight symbols periods.

It is desired to get higher rates than $1/2$ for STBCs with complex linear processing. There are only specific designs for $N_t < 5$. The Alamouti scheme achieve the full rate at $N_t = 2$. For $N_t = 3$ and $N_t = 4$, the maximal rate is limited to $3/4$ through orthogonal design such as

$$
\mathcal{G}_4 = \begin{pmatrix}
s_1 & s_2 & \dfrac{s_3}{\sqrt{2}} & \dfrac{s_3}{\sqrt{2}} \\[2mm]
-s_2^* & s_1^* & \dfrac{s_3}{\sqrt{2}} & -\dfrac{s_3}{\sqrt{2}} \\[2mm]
\dfrac{s_3^*}{\sqrt{2}} & \dfrac{s_3^*}{\sqrt{2}} & \dfrac{-s_1 - s_1^* + s_2 - s_2^*}{2} & \dfrac{s_1 - s_1^* - s_2 - s_2^*}{2} \\[2mm]
\dfrac{s_3^*}{\sqrt{2}} & -\dfrac{s_3^*}{\sqrt{2}} & \dfrac{s_1 - s_1^* + s_2 + s_2^*}{2} & -\dfrac{s_1 + s_1^* + s_2 - s_2^*}{2}
\end{pmatrix}
\tag{8.62}
$$

for $N_t = 4$.

8.4 Transmit Antenna Selection

A high diversity order is only possible if space-time codes are employed for a large number of transmit antennas. However, the design and implementation of high-dimensional space-time codes are challenging. STBCs can achieve a full transmission rate with complex constellations only for two transmit antennas. For STTCs, the code design for a large number of transmit antennas is computationally challenging, and ML decoding becomes very complex in an exponential scale. Moreover, the number of RF chains in a conventional multiple-antenna system equals to the number of antennas, imposing high complexity, high hardware cost, and high power consumption. Therefore, a low-cost, low-complexity, and power-efficient alternative to realize high diversity order is attractive [Sanayei and Nosratinia, 2004].

TAS is similar to SC in receive diversity except for a feedback channel from the receiver to the transmitter. At any time, a single (or a few) antennae with the highest SNR are chosen to transmit the signal. The number of required RF chains is drastically reduced, bringing a significant benefit in terms of hardware cost and size, implementation complexity, and power consumption. Interestingly, TAS can achieve the full diversity order that equals the number of all transmit antennae participating in the selection, rather than the number of antennae transmitting signals simultaneously [Yu et al., 2018].

Figure 8.6 shows the principle of TAS in a multi-antenna system with N_t transmit antennae and only a single receive antenna for simplicity. Multiple receive antennae are optional for such a system but it is applicable when receive combining is jointly applied. Relying on antenna-specific pilot symbols or reference

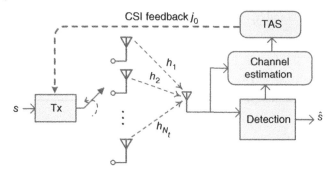

Figure 8.6 Schematic diagram of transmit antenna selection where the best antenna with the largest instantaneous SNR are activated for single-stream transmission.

signals inserted in transmitted signals, instantaneous CSI can be estimated at the receiver accurately. Consider a frequency-flat fading channel, a $1 \times N_t$ channel vector at time t denoted by

$$\mathbf{h}[t] = [h_1[t], h_2[t], \cdots, h_{N_t}[t]] \tag{8.63}$$

is available at the receiver, where $h_{n_t} \in \mathbb{C}$ represents the complex channel coefficient between the n_t^{th} transmit antenna and the receive antenna. Assuming L out of N_t transmit antennas are selected, the total number of possible choices is a combination of n *choose* k notated by $\binom{N_t}{L}$. Regarding the jth choice, where $1 \leqslant j \leqslant \binom{N_t}{L}$, using $\mathbf{h}_j[t]$ with a dimension of $1 \times L$ to indicate the channel vector from L potential transmit antennas, which is a subset of $\mathbf{h}[t]$. With the knowledge of CSI, namely, $\mathbf{h}[t]$ in this case, the receiver finds out the best choice that has the largest overall channel gain:

$$j_0 = \arg\max{}_{1 \leqslant j \leqslant \binom{N_t}{L}} \left\| \mathbf{h}_j[t] \right\|^2, \tag{8.64}$$

where $\| \cdot \|$ denotes the Frobenius norm of a matrix or a vector [Chen et al., 2005]. The receiver feeds the index of the selected choice j_0 back to the transmitter through a feedback channel. Once received the feedback, the transmitter activates the antennas belonging to choice j_0 to transmit signals.

Without loss of generality, we use a single antenna selection, namely, $L = 1$, for analyzing the performance. Therefore, the best antenna is selected in terms of

$$n_b = \arg\max{}_{1 \leqslant n_t \leqslant N_t} \left\{ |h_{n_t}[t]|^2 \right\}. \tag{8.65}$$

Then, the instantaneous SNR of the receive signal equals

$$\gamma_b = \max_{n_t = 1, \ldots, N_t} (\gamma_{n_t}), \tag{8.66}$$

where γ_{n_t} denotes the instantaneous SNR of the signal propagated from the n_t^{th} transmit antenna to the receive antenna. The CDF for γ_b is derived as

$$
\begin{aligned}
F_{\gamma_b}(\gamma) &= \mathbb{P}(\gamma_b < \gamma) \\
&= \mathbb{P}\left(\max_{n_t}(\gamma_{n_t}) < \gamma\right) \\
&= \prod_{n_t=1}^{N_t} \mathbb{P}(\gamma_{n_t} < \gamma) = \prod_{n_t=1}^{N_t} F_{\gamma_{n_t}}(\gamma).
\end{aligned}
\tag{8.67}
$$

Under the assumption of *i.i.d.* Rayleigh fading, Eq. (8.67) can be rewritten into

$$
F_{\gamma_b}(\gamma) = \left[1 - e^{-\frac{\gamma}{\bar{\gamma}_n}}\right]^{N_t},
\tag{8.68}
$$

applying

$$
F_{\gamma_{n_t}}(\gamma) = 1 - e^{-\frac{\gamma}{\bar{\gamma}_n}}.
\tag{8.69}
$$

Differentiating Eq. (8.68) relative to γ gets the PDF for γ_b, i.e.

$$
f_{\gamma_b}(\gamma) = \frac{\partial F_{\gamma_b}(\gamma)}{\partial \gamma} = \frac{N_t}{\bar{\gamma}_n}\left[1 - e^{-\frac{\gamma}{\bar{\gamma}_n}}\right]^{N_t-1} e^{-\frac{\gamma}{\bar{\gamma}_n}}.
\tag{8.70}
$$

Given the PDF of the received SNR, the average BER can be computed using Eq. (8.5). For example, the closed-form expression of average BER for DPSK modulation is given by

$$
\begin{aligned}
\bar{P}_b &= \int_0^\infty \frac{1}{2} e^{-\gamma} f_{\gamma_b}(\gamma) d\gamma \\
&= \int_0^\infty \frac{1}{2} e^{-\gamma} \frac{N_t}{\bar{\gamma}_n}\left[1 - e^{-\gamma/\bar{\gamma}_n}\right]^{N_t-1} e^{-\gamma/\bar{\gamma}_n} d\gamma \\
&= \frac{N_t}{2} \sum_{n_t=0}^{N_t-1} (-1)^{n_t} \frac{\binom{N_t-1}{n_t}}{1 + n_t + \bar{\gamma}_n}.
\end{aligned}
\tag{8.71}
$$

Then, the outage probability of a transmit-antenna-selection system given a target SNR γ_0 is obtained by substituting $\gamma = \gamma_0$ into Eq. (8.68):

$$
P_{\text{out}}^{\text{TAS}}(\gamma_0) = \mathbb{P}(\gamma_b \leq \gamma_0) = F_{\gamma_b}(\gamma_0) = \left[1 - e^{-\frac{\gamma_0}{\bar{\gamma}_n}}\right]^{N_t}.
\tag{8.72}
$$

To provide an insight into the achievable diversity, an asymptotic performance analysis is further made. Applying the Taylor series expansion yields

$$
e^{-\frac{\gamma_0}{\bar{\gamma}_n}} = \sum_{m=0}^\infty \frac{\left(-\frac{\gamma_0}{\bar{\gamma}_n}\right)^m}{m!} = 1 + \left(-\frac{\gamma_0}{\bar{\gamma}_n}\right) + \frac{\left(-\frac{\gamma_0}{\bar{\gamma}_n}\right)^2}{2!} + \cdots.
\tag{8.73}
$$

At high SNR, we have

$$1 - e^{-\frac{\gamma_0}{\bar{\gamma}_n}} \approx \frac{\gamma_0}{\bar{\gamma}_n}, \tag{8.74}$$

resulting in an asymptotic approximation of Eq. (8.72), i.e.

$$P_{\text{out}}^{\text{TAS}}(\gamma_0) = \left[1 - e^{-\frac{\gamma_0}{\bar{\gamma}_n}}\right]^{N_t} \approx \left(\frac{\gamma_0}{\bar{\gamma}_n}\right)^{N_t}. \tag{8.75}$$

It implies that TAS with a single selected transmit antenna and a single receive antenna can obtain full diversity order equaling to the number of all transmit antennae N_t.

8.5 Beamforming

When an antenna array has multiple antennas with small inter-antenna spacing and no polarization, the signal paths corresponding to different antennas are highly correlated. As a result, there is no spatial diversity, and only beamforming can be applied to achieve a power gain. Beamforming at the transmitter is called transmit beamforming or called receive beamforming at the receiver. In addition to power gain, beamforming can mitigate interfering signals in particular directions.

The mutual correlation among antennae is low if a transmitter has multiple transmit antennae with large inter-antenna spacing or polarized antennas. As discussed in the previous section, one can use space-time coding to exploit independently faded paths for spatial diversity. In addition, coherent beamforming can also be applied over a low-correlation antenna array, aiming at transmit diversity and power gain. This technique is also called transmitter-side beamforming, MIMO beamforming, precoding, or single-stream precoding. To distinguish these two forms, we use the term *classical beamforming* to refer the beamforming over a high-correlation antenna array, while the term *single-stream precoding* to refer the beamforming over a low-correlation antenna array. In contrast to highly correlated antennas, which merely exhibit phase difference, both the phases and instantaneous gains of the signals corresponding to different low-correlated antennas may differ. As a consequence, classical beamforming only adjusts signal phases, whereas precoding controls both phases and amplitudes.

8.5.1 Classical Beamforming

The term *beamforming* was derived from early spatial filters designed to form pencil beams to receive a signal radiating from a specific direction and attenuate signals from other directions [Veen and Buckley, 1988]. Forming a beam seems to indicate energy radiation at the transmitter; however, beamforming can be applied to provide directional beams for either radiation or reception of signals. The directionality can concentrate the signal energy in a narrow direction, resulting in high signal power, suppressed co-channel interference to others, and reduced multi-path delay spread. Wireless systems designed to receive spatially propagating signals often encounter the presence of interference signals. A desired signal and interference usually originate from different spatial locations. This spatial separation can be exploited to form a high-gain beam (beamforming) toward the direction of the desired signal while a high-attenuation null (null-forming) toward the direction of interference.

The conventional beamforming is fully digital, where the desired beam is formed by simply multiplying the baseband signal with a weighting vector. However, *digital beamforming* requires an RF chain for each antenna element, leading to unaffordable energy consumption and hardware cost for a large-scale array. Therefore, another form that can lower implementation complexity, called *analog beamforming*, has been adopted. By employing analog phase-shifters to adjust the phases of signals, analog beamforming needs only a single RF chain to steer the beam, leading to low hardware cost and energy consumption. However, since an analog circuit can only partly adjust the phases of signals, it is difficult to adapt a beam to a particular channel condition appropriately, and this leads to a considerable performance loss. As a result, hybrid analog-digital beamforming to balance the benefits of fully digital and fully analog beamforming was recognized, especially for mmWave transmission [Zhang et al., 2019]. *Hybrid beamforming* can significantly reduce the number of RF chains, resulting in lower hardware cost and energy consumption, while achieving a comparable performance relative to digital beamforming.

Consider an array of N omnidirectional elements (neglecting the difference of N_t at the transmitter and N_r at the receiver), indexed by $n = 1, \ldots, N$, radiating (or acquiring) signals into a homogeneous media in the far field of uncorrelated sinusoidal point sources of frequency f_0. As illustrated in Figure 8.7, the time taken by a plane wave propagating from the nth transmit element to a receive antenna located in the direction indicated by the angle of departure θ is

$$\tau_n(\theta) = \frac{\mathbf{r}_n \cdot \mathbf{u}(\theta)}{c}, \tag{8.76}$$

where $\mathbf{r}_n$ denotes the position vector of the nth element relative to the reference point, $\mathbf{u}(\theta)$ is the unit vector in angle θ, c is the speed of propagation of the plane wave front, and $\cdot$ denotes the inner product.

The receiver observes the signal transmitted by the reference element expressed in complex notation as

$$s(t)e^{j2\pi f_0 t} \tag{8.77}$$

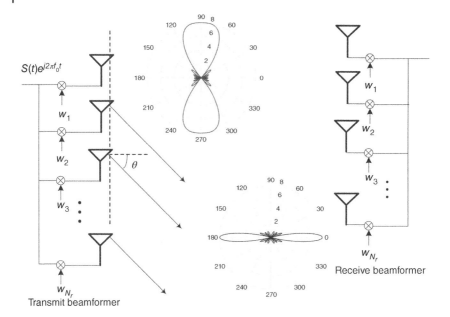

Figure 8.7 Schematic diagram of a transmit beamformer and a receive beamformer with two examples of possible patterns. Beamforming can be implemented in digital-beamforming form by multiplexing a complex weighting coefficient on each branch or in analog-beamforming form by a phase-shifter network.

with $s(t)$ stands for the complex baseband signal. The wavefront on the nth element arrives with a time difference $\tau_n(\theta)$ compared with that of the reference element. Thus, the signal induced at the receive antenna due to the nth element can be expressed as

$$s(t)e^{j2\pi f_0[t-\tau_n(\theta)]}. \tag{8.78}$$

This expression is based on *the narrow-band assumption* for array signal processing, assuming that the signal bandwidth is narrow enough and the array dimension is small enough, so that the baseband signal keeps almost constant over the interval of $\tau_n(\theta)$, i.e. the approximation $s(t) \approx s(t - \tau_n(\theta))$ holds.

The overall received signal induced due to all N elements is

$$y(t) = \sum_{n=1}^{N} s(t)e^{j2\pi f_0[t-\tau_n(\theta)]} + n(t), \tag{8.79}$$

where $n(t)$ indicates white Gaussian noise at the receive antenna. The principle of narrow-band beam-former is to impose a phase shift of the signal on each element by multiplexing a complex weight $w_n(t)$ with the baseband signal, or alternatively shifting the signal phases directly $w_n(t) = e^{j\theta_n(t)}$. Thus, the received signal with beamforming becomes

$$y(t) = \sum_{n=1}^{N} w_n^*(t)s(t)e^{j2\pi f_0[t-\tau_n(\theta)]} + n(t). \tag{8.80}$$

For a Uniform Linear Array (ULA) with equally element spacing d aligned with a direction such that the first element is situated at the origin, Eq. (8.76) can be rewritten as [Jiang and Yang, 2012]

$$\tau_n(\theta) = \frac{d}{c}(n-1)\sin\theta. \tag{8.81}$$

In an AWGN channel, the received signal is obtained by substituting Eq. (8.81) into Eq. (8.80), yielding

$$\begin{aligned}
y(t) &= \sum_{n=1}^{N} w_n^*(t)s(t)e^{j2\pi f_0 t}e^{-j\frac{2\pi}{\lambda}(n-1)d\sin\theta} + n(t) \\
&= \left(\sum_{n=1}^{N} w_n^*(t)e^{-j\frac{2\pi}{\lambda}(n-1)d\sin\theta} \right) s(t)e^{j2\pi f_0 t} + n(t) \\
&= g(\theta, t)s(t)e^{j2\pi f_0 t} + n(t),
\end{aligned} \tag{8.82}$$

where $g(\theta, t)$ is the effect of beamforming, referred to as the beam pattern. Defining the weighting vector

$$\mathbf{w}(t) = \left[w_1(t), w_2(t), \ldots, w_N(t) \right]^T, \tag{8.83}$$

and the steering vector of ULA

$$\mathbf{a}(\theta) = \left[1, e^{-j\frac{2\pi}{\lambda}d\sin\theta}, e^{-j\frac{2\pi}{\lambda}2d\sin\theta}, \ldots, e^{-j\frac{2\pi}{\lambda}(N-1)d\sin\theta} \right]^T, \tag{8.84}$$

the beam pattern of ULA can be calculated by

$$g(\theta, t) = \mathbf{w}^H(t)\mathbf{a}(\theta). \tag{8.85}$$

Figure 8.7 illustrates two formed beam patterns over an eight-element ULA. The radiated energy can be concentrated in a particular direction with a power gain equaling to the number of antenna elements. In other words, an eight-element ULA brings a power gain of 8 (in 0° and 180° of receive beam while 90° and 270° of transmit beam) relative to an omnidirectional antenna with the same power. Through changing weighting vectors or adjusting phases, a beam can be steered toward any particular direction in terms of the angular information of a mobile user, which can be estimated by using some classical algorithms such as MUltiple SIgnal Classification (MUSIC) and Estimation of Signal Parameters via Rational Invariance Techniques (ESPRIT) [Gardner, 1988]. Suppose the angle of a mobile user is θ_0, letting

$$\mathbf{w} = \mathbf{a}(\theta_0) = \left[1, e^{-j\frac{2\pi}{\lambda}d\sin\theta_0}, \ldots, e^{-j\frac{2\pi}{\lambda}(N-1)d\sin\theta_0} \right]^T \tag{8.86}$$

would form a beam pointing to the desired angle. Then, the beam pattern is

$$g(\theta) = \mathbf{w}^H(t)\mathbf{a}(\theta) = \sum_{n=1}^{N} e^{-j\frac{2\pi d}{\lambda}(n-1)[\sin\theta - \sin\theta_0]}, \tag{8.87}$$

which achieve the peak amplitude of

$$g(\theta_0) = \sum_{n=1}^{N} e^{j0} = N \tag{8.88}$$

in the desired angle θ_0, equivalent to a power gain of $|g(\theta_0)|^2/N = N$. In a nutshell, beamforming brings a power gain equaling to the number of elements in the antenna array N. The principle of receive beamforming at the receiver is equivalent to the transmit beamforming. The transmitter and receiver can jointly apply transmit and receive beamforming to obtain a higher power gain.

8.5.2 Single-Stream Precoding

Low mutual correlation usually implies either a sufficiently large antenna separation or different polarization directions. The principle of single-stream precoding under low correlation is similar to that of classical beamforming with high correlation. Each signal to be transmitted on its respective antenna is multiplied by a complex weight. However, in contrast to classical beamforming that only adjusts signal phases, precoding should consider both the phases and amplitudes of transmitted signals. This reflects the fact that both the phases and instantaneous gains of different channels differ due to low correlation.

Another key difference between classical beamforming and precoding is that the latter needs the knowledge of channels, i.e. CSIT. The adjustment of the precoding weights is thus typically performed on a relatively short time scale to follow the fading variations, especially in a fast-fading environment. For example, in the case of Frequency-Division Duplex (FDD), where uplink and downlink transmission takes place in different frequency carriers, their fading is generally uncorrelated between the downlink and uplink channels. In this context, a receiver needs to acquire knowledge of the downlink channel and then report it to its transmitter through a feedback channel. Alternatively, a receiver can select the best precoding vector from a limited set of possible vectors in a predefined codebook. It is efficient since only the index of the selected vector is transferred. Still, there is a performance penalty due to the difference between the predefined vector and the optimal precoding vector. On the other hand, there is typically a high correlation between the downlink and uplink in Time-Division Duplex (TDD), where uplink and downlink transmission perform in the same frequency carrier but separately non-overlapping time slots. In this case, the transmitter could, at least in theory, determine the instantaneous downlink fading from measurements on the uplink, thus avoiding the need for any feedback.

Suppose the transmitter has N_t antennas, the complex weight for the n_t^{th} transmit antenna is denoted by v_{n_t}, $n_t = 1, 2, \ldots, N_t$. The precoder encodes a transmit symbol s to a series of N_t coded symbols

$$s_{n_t} = v_{n_t} s, \quad n_t = 1, 2, \ldots, N_t, \tag{8.89}$$

which are transmitted over N_t antennas simultaneously, as shown in Figure 8.8. The precoding can also be expressed in vector notation by using a precoding vector $\mathbf{v} = [v_1, v_2, \ldots, v_{N_t}]^T$ as

$$\mathbf{s} = \mathbf{v} s \tag{8.90}$$

with the transmit symbol vector $\mathbf{s} = [s_1, s_2, \ldots, s_{N_t}]^T$.

Assuming that the signals transmitted from different antennas are only subject to frequency-flat fading h_{n_t}, $n_t = 1, \ldots, N_t$. In order to maximize the received signal power, the precoding weights should be selected according to

$$v_{n_t} = \frac{h_{n_t}^*}{\sqrt{\sum_{n_t=1}^{N_t} |h_{n_t}|^2}}, \tag{8.91}$$

which is the complex conjugate of the corresponding channel coefficient with normalization to ensure a fixed overall transmit power [Dahlman et al., 2011]. The purpose of the precoding vector is threefold:

- Phase rotates the transmitted signals to compensate for the instantaneous channel phase and ensure that the received signals are received phase-aligned;

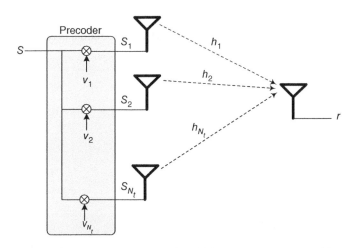

Figure 8.8 Schematic diagram of multiple transmit antennas with precoding.

- Allocates power to different antennas proportional to the instantaneous channel power gain, namely, more power being allocated to antennas with good channel conditions;
- Keep an overall constant transmit power.

Assume the receiver has a single antenna, the received signal can be expressed by

$$r_{\text{pre}} = \sum_{n_t=1}^{N_t} h_{n_t} v_{n_t} s + n = \sum_{n_t=1}^{N_t} \frac{h_{n_t}^* h_{n_t} s}{\sqrt{\sum_{n_t=1}^{N_t} |h_{n_t}|^2}} + n$$

$$= \left(\sqrt{\sum_{n_t=1}^{N_t} |h_{n_t}|^2} \right) s + n,$$

through which the receiver can directly get the transmit symbol. The instantaneous SNR of the received signal is

$$\gamma_{\text{pre}} = \sum_{n_t=1}^{N_t} |h_{n_t}|^2 \frac{P}{\sigma_n^2} = \left(\sum_{n_t=1}^{N_t} |h_{n_t}|^2 \right) \overline{\gamma}_n, \tag{8.92}$$

which is equivalent to the received SNR of maximal ratio combining, see Eq. (8.19). It implies that the precoding in terms of the instantaneous CSI over multiple low-correlation antennas can achieve full diversity order equaling to the number of transmit antennas. Its performance is comparable with that of MRC at the receiver.

It is noted that classical beamforming can also be modeled according to Eq. (8.90) with the constraint that the weights are limited to unit gain and only provide phase shifts. We write h_1 to denote the channel response of the reference antenna. Then, the channel response for an antenna array is $h_1 \mathbf{a}(\theta_0)$, namely, $h_{n_t} = h_1 e^{-j\frac{2\pi}{\lambda}(n_t-1)d\sin\theta_0}$ for antenna n_t. Applying Eq. (8.91), we obtain a precoding weight

$$v_{n_t} = \frac{h_{n_t}^*}{\sqrt{\sum_{n_t=1}^{N_t} |h_{n_t}|^2}} = \frac{h_1^* e^{j\frac{2\pi}{\lambda}(n_t-1)d\sin\theta_0}}{\sqrt{N_t |h_1|^2}}, \tag{8.93}$$

resulting in a received signal

$$r_{\text{bf}} = \sum_{n_t=1}^{N_t} h_{n_t} v_{n_t} s + n$$

$$= \sum_{n_t=1}^{N_t} \frac{\left(h_1^* e^{j\frac{2\pi}{\lambda}(n_t-1)d\sin\theta_0} \right) \left(h_1 e^{-j\frac{2\pi}{\lambda}(n_t-1)d\sin\theta_0} \right)}{\sqrt{N_t |h_1|^2}} s + n$$

$$= \sqrt{N_t |h_1|^2} s + n. \tag{8.94}$$

The received SNR for beamforming is

$$\gamma_{bf} = N_t |h_1|^2 \frac{P}{\sigma_n^2}. \tag{8.95}$$

Compared with a single antenna (the reference antenna) transmission with equal total power, which has a received SNR

$$\gamma_1 = |h_1|^2 \frac{P}{\sigma_n^2}, \tag{8.96}$$

the beamforming achieves a power gain of N_t.

Consider multiple antennas at the receiver, the channel response is then modeled by a channel matrix $\mathbf{H}$. The precoding methodology is similar but becomes more complex, especially when multiple parallel data streams are transmitted simultaneously. It belongs to the precoding of *Spatial Multiplexing* for higher channel capacity, rather than spatial diversity, which will be introduced in the subsequent section.

8.6 Spatial Multiplexing

The previous two sections studied the use of multiple transmit antennas or receive antennas for spatial diversity or power gains (sometimes called array gains) through coherent combining, space-time coding, antenna selection, or beamforming. Higher-order diversity and power gains are available by simultaneously applying multiple transmit **and** receive antennas at the transmitter and receiver, respectively. Spatial diversity can be seen as an approach to improve the SNR against channel fading. This gain is more significant in the low SNR regime where a wireless system is power-limited but becomes marginal in the high SNR regime that exhibits bandwidth-limited. In this section, we will study a novel approach to exploit a MIMO channel having multiple transmit antennae and multiple receive antennae. Under rich scattering environments, a MIMO channel enables an additional spatial dimension for signal transmission, yielding a degree-of-freedom gain [Tse and Viswanath, 2005]. These degrees of freedom can be exploited by *spatially multiplexing* parallel data streams, resulting in a linear increase in the channel capacity. In other words, the capacity of such a MIMO channel is proportional to the number of antennas. In contrast, the channel capacity of spatial diversity or beamforming increases only in a logarithmic scale with the number of antennas (diversity order). This section will introduce the basic principle of spatially multiplexing over a point-to-point MIMO system, also known as Single-User Multi-Input Multi-Output (SU-MIMO), and then discuss its practical implementation via MIMO precoding and detection.

8.6.1 Single-User MIMO

Consider a narrowband wireless channel with N_t transmit antennas and N_r receive antennas, which can be described by an $N_r \times N_t$ matrix $\mathbf{H}$. As discussed in the previous section, the use of spatial diversity and beamforming can improve the quality of the received signal with a maximal available SNR gain in proportion to $N_t \times N_r$. The channel capacity can be expressed by

$$C = \log_2 \left(1 + N_t N_r \frac{P}{\sigma_n^2} \right). \tag{8.97}$$

Observe that

$$\log_2(1 + \gamma) \approx \gamma \log_2(e), \tag{8.98}$$

when the SNR is small ($\gamma \to 0$), implying that the capacity grows approximately linearly with the diversity order at the low-power regime. If the system is power-limited, every 3 dB decibel increase (or, doubling) in the power doubles the capacity [Tse and Viswanath, 2005]. However, the gain of spatial diversity diminishes quickly and saturates if the received SNR falls into the bandwidth-limited region. At high SNR $\gamma \gg 1$, we have

$$\log_2(1 + \gamma) \approx \log_2(\gamma), \tag{8.99}$$

implying that the channel capacity grows only logarithmically with the diversity order. In other words, every 3 dB decibel increase in the power yields only one additional bit for channel capacity.

The baseband-equivalent channel model for a single-user MIMO channel can be expressed by

$$\mathbf{r} = \mathbf{Hs} + \mathbf{n}, \tag{8.100}$$

where $\mathbf{r} = \left[r_1, r_2, \ldots, r_{N_r} \right]^T \in C^{N_r \times 1}$ stands for the received signal vector, and $\mathbf{s} = \left[s_1, s_2, \ldots, s_{N_t} \right]^T \in C^{N_t \times 1}$ denotes the transmitted signal vector, with the constraint of a total power P, i.e. $\mathbb{E}[\mathbf{s}^H \mathbf{s}] \leqslant P$. Equivalently, since $\mathbf{s}^H \mathbf{s} = \mathrm{tr}(\mathbf{ss}^H)$, and commuting expectation and trace, we get

$$\mathrm{tr}\left(\mathbb{E}\left[\mathbf{ss}^H \right] \right) \leqslant P. \tag{8.101}$$

This second form of the power constraint is more useful in some discussions. We use $\mathbf{n} = \left[n_1, n_2, \ldots, n_{N_r} \right]^T \in C^{N_r \times 1}$ to represent the vector of additive complex Gaussian noise. The noise corrupting different received signals are usually independent, zero-mean, and equal variance σ_n^2, i.e. $\mathbb{E}[\mathbf{nn}^H] = \sigma_n^2 \mathbf{I}_{N_r}$, which can be interchangeably expressed by $\mathbf{n} \in \mathcal{CN}(0, \sigma_n^2 \mathbf{I}_{N_r})$. With the assumption of flat fading, a MIMO channel can be modeled into an $N_r \times N_t$ matrix, i.e.

$$
\mathbf{H} = \begin{pmatrix} h_{11} & h_{12} & \cdots & h_{1N_t} \\ h_{21} & h_{22} & \cdots & h_{2N_t} \\ \vdots & \vdots & \ddots & \vdots \\ h_{N_r 1} & h_{N_r 2} & \cdots & h_{N_r N_t} \end{pmatrix}
\tag{8.102}
$$

with a typical entry $h_{n_r n_t}$ on the n_r^{th} row and the n_t^{th} column, denoting the channel coefficient between transmit antenna n_t and receive antenna n_r, where $n_r = 1, 2, \ldots, N_r$ and $n_t = 1, 2, \ldots, N_t$, as illustrated in Figure 8.9. $\mathbf{H}$ is a complex-valued random matrix follows a particular probability distribution, and each use of the channel corresponds to an independent realization of $\mathbf{H}$. In general, each entry is a complex circularly symmetric Gaussian random variable with zero mean, independent real and imaginary parts with variance $1/2$, denoted by $h \sim \mathcal{CN}(0, 1)$. This choice models a Rayleigh fading environment with enough separation within the receive antennas and the transmit antennas such that the fades for each transmit-receive antenna pair are independent. In all cases, the realization of $\mathbf{H}$ is deterministic, and the receiver obtains the channel output consisting of the pair $(\mathbf{r}, \mathbf{H})$ [Telatar, 1999]. Throughout this chapter and the whole book, bold lower-case and upper-case letters denote vectors and matrices, respectively. For their operation, $(\cdot)^*$, $(\cdot)^T$, and $(\cdot)^H$ notate the conjugate, transpose, and Hermitian transpose, respectively.

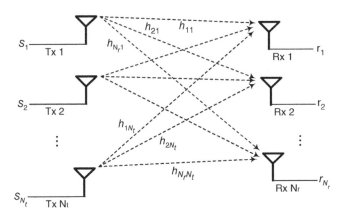

Figure 8.9 Schematic diagram of a MIMO channel having N_t transmit antennas and N_r receive antennas, where the channel coefficient between transmit antenna n_t and receive antenna n_r is denoted by $h_{n_r n_t}$.

As we know, the Shannon capacity for a single-antenna system in a flat fading channel expressed in bps/Hz is given by

$$C = \log_2\left(1 + |h|^2 \frac{P}{\sigma_n^2}\right). \tag{8.103}$$

Deploying multiple receive antennas, we have a SIMO system with the improved capacity

$$C = \log_2\left(1 + \frac{P}{\sigma_n^2}\sum_{n_r=1}^{N_r}|h_{n_r}|^2\right). \tag{8.104}$$

Similarly, if we opt for transmit diversity without CSI at the transmitter, we have a MISO system with the capacity

$$C = \log_2\left(1 + \frac{P}{N_t\sigma_n^2}\sum_{n_t=1}^{N_t}|h_{n_t}|^2\right), \tag{8.105}$$

where the normalization by N_t ensures a fixed total transmit power and implies the absence of array (power) gain unlike the case of multiple receive antennae. Either the capacity of a SIMO or MISO channel, as given in Eqs. (8.104) or (8.105), respectively, exhibits a logarithmic relationship with respect to the number of antennas.

Telatar [1999] and Foschini and Gans [1998] demonstrated that the capacity of a MIMO channel can grow linearly with

$$N_m = \min(N_t, N_r) \tag{8.106}$$

rather than logarithmically. The formula for generalized capacity is given by

$$C = \log_2 \det\left[\mathbf{I}_{N_r} + \frac{P}{N_t\sigma_n^2}\mathbf{H}\mathbf{H}^H\right], \tag{8.107}$$

where the operator $\det[\cdot]$ means the determinant of a matrix. This is a vector Gaussian channel, and therefore its capacity can be figured out by decomposing the vector channel into a set of parallel, independent scalar Gaussian subchannels.

From basic linear algebra, a linear transformation can be expressed as a composition of three operators, i.e. a rotation operation, a scaling operation, and another rotation operation. As a consequence, any matrix can be decomposed by the Singular Value Decomposition (SVD) as

$$\mathbf{H} = \mathbf{U}\mathbf{\Sigma}\mathbf{V}^H, \tag{8.108}$$

where $\mathbf{U} \in C^{N_r \times N_r}$ and $\mathbf{V} \in C^{N_t \times N_t}$ are unitary, satisfying

$$\mathbf{U}^H\mathbf{U} = \mathbf{U}\mathbf{U}^H = \mathbf{I}_{N_r} \tag{8.109}$$

$$\mathbf{V}^H\mathbf{V} = \mathbf{V}\mathbf{V}^H = \mathbf{I}_{N_t}, \tag{8.110}$$

and $\Sigma \in \mathcal{R}^{N_r \times N_t}$ is a rectangular matrix whose diagonal elements are non-negative real numbers while off-diagonal elements are zero. The diagonal elements are generally the ordered singular values of $\mathbf{H}$ denoted by $\lambda_1 \geq \lambda_2 \geq \ldots \geq \lambda_{N_m}$. This matrix likes

$$\Sigma = \begin{pmatrix} \lambda_1 & 0 & \cdots & 0 & 0 & \cdots & 0 \\ 0 & \lambda_2 & \cdots & 0 & 0 & \cdots & 0 \\ \vdots & \vdots & \ddots & \vdots & & \vdots & \\ 0 & 0 & \cdots & \lambda_{N_m} & 0 & \cdots & 0 \end{pmatrix}, \tag{8.111}$$

when $N_r < N_t$, or

$$\Sigma = \begin{pmatrix} \lambda_1 & 0 & \cdots & 0 \\ 0 & \lambda_2 & \cdots & 0 \\ \vdots & \vdots & \ddots & \vdots \\ 0 & 0 & \cdots & \lambda_{N_m} \\ 0 & 0 & \cdots & 0 \\ \vdots & \vdots & & \vdots \\ 0 & 0 & \cdots & 0 \end{pmatrix}, \tag{8.112}$$

when $N_r > N_t$ [Yang, 2016].

Define a matrix

$$\mathbf{W} = \begin{cases} \mathbf{HH}^H, & N_r \leq N_t \\ \mathbf{H}^H\mathbf{H}, & N_r > N_t \end{cases}. \tag{8.113}$$

These singular values are also the nonzero eigenvalues of $\mathbf{W}$, the columns of $\mathbf{U}$ are the eigenvectors of $\mathbf{HH}^H$, and the columns of $\mathbf{V}$ are the eigenvectors of $\mathbf{H}^H\mathbf{H}$. Thus, we can rewrite Eq. (8.100) as

$$\mathbf{r} = \mathbf{U}\Sigma\mathbf{V}^H\mathbf{s} + \mathbf{n}. \tag{8.114}$$

If the transmitter knows $\mathbf{r}$, it can precode the information symbol vector $\mathbf{x} = \left[x_1, x_2, \ldots, x_{N_t}\right]^T$ as

$$\mathbf{s} = \mathbf{Vx}, \tag{8.115}$$

resulting in

$$\mathbf{r} = \mathbf{U}\Sigma\mathbf{V}^H\mathbf{Vx} + \mathbf{n} = \mathbf{U}\Sigma\mathbf{x} + \mathbf{n}, \tag{8.116}$$

since $\mathbf{V}^H\mathbf{V} = \mathbf{I}_{N_t}$. As illustrated in Figure 8.10, if the receive employs $\mathbf{U}^H$ to decode the received signal, we have

$$\mathbf{y} = \mathbf{U}^H\mathbf{r} = \mathbf{U}^H\mathbf{U}\Sigma\mathbf{x} + \mathbf{U}^H\mathbf{n} = \Sigma\mathbf{x} + \mathbf{z}. \tag{8.117}$$

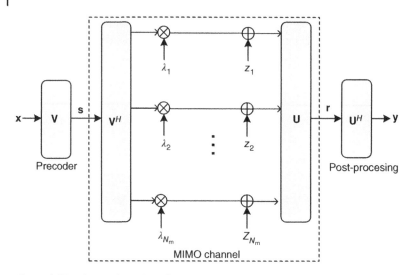

Figure 8.10 Converting a MIMO channel into parallel subchannels through the singular value decomposition of the channel matrix.

The precoding does not change the power constraint because

$$\mathbb{E}[\mathbf{x}^H\mathbf{x}] = \mathbb{E}[\mathbf{s}^H\mathbf{V}^H\mathbf{V}\mathbf{s}] = \mathbb{E}[\mathbf{s}^H\mathbf{s}] \leqslant P. \tag{8.118}$$

Similarly, we know $\mathbf{z} = \mathbf{U}^H\mathbf{n}$ has the same distribution as $\mathbf{n}$, i.e. $\mathbf{z} \sim \mathcal{CN}(0, \sigma_n^2\mathbf{I}_{N_r})$. Expanding $\mathbf{y} = \Sigma\mathbf{x} + \mathbf{z}$, we obtain (only for the case $N_r < N_t$)

$$\begin{pmatrix} y_1 \\ y_2 \\ \vdots \\ y_{N_r} \end{pmatrix} = \begin{pmatrix} \lambda_1 & 0 & \cdots & 0 & 0 & \cdots & 0 \\ 0 & \lambda_2 & \cdots & 0 & 0 & \cdots & 0 \\ \vdots & \vdots & \ddots & \vdots & & \vdots & \\ 0 & 0 & \cdots & \lambda_{N_m} & 0 & \cdots & 0 \end{pmatrix} \begin{pmatrix} x_1 \\ x_2 \\ x_3 \\ \vdots \\ x_{N_t} \end{pmatrix} + \begin{pmatrix} z_1 \\ z_2 \\ \vdots \\ z_{N_r} \end{pmatrix}, \tag{8.119}$$

implying that we get N_m parallel Gaussian channels:

$$y_n = \lambda_n x_n + z_n, \quad \text{for } n = 1, \dots, N_m. \tag{8.120}$$

Its capacity is computed by directly summing the individual capacity of all parallel subchannels, which is expressed by

$$C = \sum_{n=1}^{N_m} \log_2\left(1 + \lambda_n^2 \frac{P}{N_t\sigma_n^2}\right), \tag{8.121}$$

assuming equal power allocation among transmit antennas.

Alternatively, the capacity can be derived by substituting $\mathbf{H} = \mathbf{U}\Sigma\mathbf{V}^H$ into Eq. (8.107):

$$C = \log_2 \det\left[\mathbf{I}_{N_r} + \frac{P}{N_t\sigma_n^2}\mathbf{U}\Sigma\mathbf{V}^H\mathbf{V}\Sigma^H\mathbf{U}^H\right]$$

$$= \log_2 \det\left[\mathbf{I}_{N_r} + \frac{P}{N_t\sigma_n^2}\mathbf{U}\Sigma\Sigma^H\mathbf{U}^H\right]$$

$$= \log_2 \det\left[\mathbf{U}\left(\mathbf{I}_{N_r} + \frac{P}{N_t\sigma_n^2}\Sigma\Sigma^H\right)\mathbf{U}^H\right]$$

$$= \log_2 \det\left[\mathbf{I}_{N_r} + \frac{P}{N_t\sigma_n^2}\Sigma\Sigma^H\right]. \tag{8.122}$$

Applying Eq. (8.112) yields

$$\Sigma\Sigma^H = \begin{pmatrix} \lambda_1^2 & 0 & \cdots & 0 & 0 & \cdots & 0 \\ 0 & \lambda_2^2 & \cdots & 0 & 0 & \cdots & 0 \\ \vdots & \vdots & \ddots & \vdots & & \vdots & \\ 0 & 0 & \cdots & \lambda_{N_m}^2 & 0 & \cdots & 0 \\ 0 & 0 & \cdots & 0 & 0 & \cdots & 0 \\ \vdots & \vdots & \ddots & \vdots & & \vdots & \\ 0 & 0 & \cdots & 0 & 0 & \cdots & 0 \end{pmatrix}, \tag{8.123}$$

Thus, Eq. (8.122) is further figured out as

$$C = \sum_{n=1}^{N_m} \log_2\left(1 + \lambda_n^2 \frac{P}{N_t\sigma_n^2}\right), \tag{8.124}$$

which is exactly identical to Eq. (8.121). Each subchannel with a nonzero eigenvalue can support a data stream, so the MIMO channel can support the spatial multiplexing of multiple streams. The channel capacity now grows linearly with the number of antennas, and the maximal number of parallel subchannels equals the minimum of transmit and receive antennas' number.

There are two key parameters to determine the performance. The *rank* means the number of nonzero singular values of a channel matrix, we have

$$r \leqslant \min(N_t, N_r). \tag{8.125}$$

It indicates how many parallel subchannels a MIMO channel provides, and with **full rank**, a MIMO channel provides $N_m = \min(N_t, N_r)$ spatial degrees of freedom. Another parameter is called the *condition number* of $\mathbf{H}$, which is defined as the ratio between the maximal singular value and the minimal singular value

$$c = \frac{\max_{n=1,2,\ldots,N_m}(\lambda_n)}{\min_{n=1,2,\ldots,N_m}(\lambda_n)}. \tag{8.126}$$

A channel matrix is said to be *well-conditioned* if the condition number approaches to 1; otherwise, it is *ill-conditioned*.

8.6.2 MIMO Precoding

Although the benefits of MIMO are realizable when the receiver alone has the knowledge of channels, better performance can be achieved when the transmitter knows the CSI, known as CSIT. For instance, in a four-transmit two-receive antenna system with *i.i.d.* Rayleigh fading channels, exploiting CSI at the transmitter can double the capacity at the SNR of -5 dB and increase 1.5 bps/Hz additional capacity at the SNR of 5 dB [Paulraj, 2007]. To this end, the transmitter needs to precode the transmit signals in terms of instantaneous CSI prior to transmission. This subsection studies the precoding in a single-user MIMO system with either full CSIT or limited CSIT.

8.6.2.1 Full CSI at the Transmitter

In a TDD system, where the channel reciprocity is exploited, or in an FDD system with slow fading, it may be possible to track the channel variations at the transmitter. Despite of CSI inaccuracy, the precoding with the perfect CSIT is also essential theoretically. As discussed in the previous section, the SVD precoding can realize the capacity-achieving performance assume the channel matrix is full rank. The precoder is simply $\mathbf{V}$, and the water-filling strategy is applied to allocate power among subchannels, i.e.

$$P_n = \left(\mu - \frac{N_0}{\lambda_n^2} \right)^+,$$
(8.127)

with μ chosen so that the total power constraint is satisfied

$$\mathbb{E}\left[\sum_{n=1}^{N_m} \left(\mu - \frac{N_0}{\lambda_n^2} \right)^+ \right] \leqslant P.$$
(8.128)

However, $\mathbf{H}$ is not always a full-rank matrix, and the number of spatial degrees of freedom is equal to the rank of $\mathbf{H}$. We write N_L to denote the number of transmission layers, also referred to as the transmission rank, which should be no greater than the rank of the instantaneous channel matrix, i.e. $N_L \leqslant r \leqslant N_m$. Note that the transmission rank is not necessarily selected to equal the matrix rank. For example, one can transmit two data streams over a four-rank channel.

If $N_L = r \leqslant N_m$, the channel matrix contains $N_L \leqslant N_m$ nonzero singular values. Taking advantage of the effect of the rows or columns with all zeros in Σ, the SVD formula given in Eq. (8.108) can be rewritten in an equivalent form

$$\mathbf{H} = \mathbf{U}\Sigma\mathbf{V}^H = \tilde{\mathbf{U}}\tilde{\Sigma}\tilde{\mathbf{V}}^H,$$
(8.129)

where $\tilde{\mathbf{U}} \in C^{N_r \times N_L}$ denotes the first N_L columns of $\mathbf{U} \in C^{N_r \times N_r}$ (because the entries at the remaining $N_r - N_L$ columns will multiplex with zeros at the last rows of Σ), $\tilde{\mathbf{V}}^H \in C^{N_L \times N_t}$ stands for the first N_L rows of $\mathbf{V}^H \in C^{N_t \times N_t}$, and $\tilde{\Sigma} \in C^{N_L \times N_L}$ equals

$$
\tilde{\Sigma} = \begin{pmatrix} \lambda_1 & 0 & \cdots & 0 \\ 0 & \lambda_2 & \cdots & 0 \\ \vdots & \vdots & \ddots & \vdots \\ 0 & 0 & \cdots & \lambda_{N_L} \end{pmatrix}. \tag{8.130}
$$

As shown in Figure 8.11, channel-coded information symbols are demultiplexed into N_L streams, denoted by $\mathbf{x} = [x_1, x_2, \ldots, x_{N_L}]^T$, and each stream is mapped upon one layer of the precoder. For the SVD precoding with a partial rank, the precoding matrix $\mathbf{P} = \tilde{\mathbf{V}}$. Then, the vector of transmitted symbols $\mathbf{s} = [s_1, s_2, \ldots, s_{N_t}]^T$ is formed by

$$
\mathbf{s} = \mathbf{P}\mathbf{x}. \tag{8.131}
$$

Applying a linear detector $\mathbf{D} = \tilde{\mathbf{U}}^H$, the receiver computes

$$
\mathbf{y} = \mathbf{D}\mathbf{r} = \tilde{\mathbf{U}}^H \mathbf{r} = \tilde{\mathbf{U}}^H (\tilde{\mathbf{U}}\tilde{\Sigma}\tilde{\mathbf{V}}^H \tilde{\mathbf{V}}\mathbf{x} + \mathbf{n}) = \tilde{\Sigma}\mathbf{x} + \tilde{\mathbf{n}}, \tag{8.132}
$$

which is transformed into a general detection problem in an AWGN channel.

Figure 8.11 illustrates a typical structure of linear precoding in a single-user MIMO system. It implies that linear processing by means of a size $N_t \times N_L$ precoding matrix applied at the transmitter. In the case of a single layer, the linear precoding for spatial multiplexing falls back to precoding-based beamforming. In the single-stream precoding (or transmit beamforming over low-correlation antenna arrays), as studied in Section 8.5, an identical signal is emitted from each of transmit antennas with appropriate weighting, adjusting both the phase and gain, to maximize the SNR at the receiver. In a SU-MIMO system, precoding

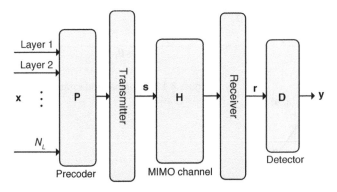

Figure 8.11 Schematic diagram of linear precoding in a single-user MIMO system.

means that multiple data streams are simultaneously emitted from the transmit antennas with independent and appropriate weighting such that the link throughput is maximized. The precoding serves two purposes:

- When the number of transmission layers to be spatially multiplexed equals the number of transmit antennas ($N_L = N_t$), it can be used to *orthogonalize* the parallel signals, allowing for improved signal isolation at the receiver.
- When the number of transmission layers to be spatially multiplexed is less than the number of transmit antennas ($N_L < N_t$), it also provides the mapping of the N_L spatially multiplexed signals to the N_t transmit antennas, including the combination of spatially multiplexing and beamforming.

The SVD precoding can effectively transform a MIMO channel into a set of parallel subchannels. However, the error performance is dominated by that subchannel with the smallest singular value. Significant improvement can be achieved by jointly coding over pairs of subchannels as long as the pairs are appropriately chosen. The motivation for the subchannel pairing arises from the idea of rotation coding, which is depicted by a rotation matrix

$$\begin{pmatrix} \cos\theta & \sin\theta \\ -\sin\theta & \cos\theta \end{pmatrix}. \tag{8.133}$$

It generates a rotation of the original constellation points with an angle of θ, as demonstrated in Figure 8.12.

The pairing of subchannels is achieved by X-codes and Y-codes proposed in Mohammed et al. [2011], where a pairing matrix **G** is applied to pair different subchannels in order to improve the overall diversity order. For instance, the X-code structure for $N_L = 6$ is given by

$$\mathbf{G} = \begin{pmatrix} \cos\theta_1 & & & & & \sin\theta_1 \\ & \cos\theta_2 & & & \sin\theta_2 & \\ & & \cos\theta_3 & \sin\theta_3 & & \\ & & -\sin\theta_3 & \cos\theta_3 & & \\ & -\sin\theta_2 & & & \cos\theta_2 & \\ -\sin\theta_1 & & & & & \cos\theta_1 \end{pmatrix}. \tag{8.134}$$

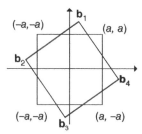

Figure 8.12 Original and rotated constellation.

Then, the precoding matrix becomes

$$\mathbf{P} = \tilde{\mathbf{V}}\mathbf{G}. \tag{8.135}$$

X-codes have better error performance than the standard SVD precoding. Nevertheless, its performance degrades when the subchannel pairs are ill-conditioned. This degradation along with the motivation of further lower the complexity results in Y-codes. The details of its structure and performance can refer to Mohammed et al. [2011].

8.6.2.2 Limited CSI at the Transmitter

In practice, unfortunately, time-varying channel response, especially in fast-fading environments, makes it difficult and often expensive to obtain CSI accurately and timely at the transmitter. Limited feedback resources, associated feedback delays, transmission errors, and scheduling lags degrade the system performance of CSIT with short channel coherence time in FDD systems. In TDD forms, antenna calibration errors and switching time lags between downlink and uplink transmission again restrict CSIT accuracy. Therefore, we often only have imperfect, partial instantaneous CSI for precoding at the transmitter.

In the context of limited feedback, an effective solution is to define a limited set of precoding matrices, referred to as the codebook, which is known by both the transmitter and receiver. Codebook-based precoding allows a wireless system to dispense with full CSIT, considerably reducing the feedback overhead. Based on measurements on reference signals, the receiver selects a suitable transmission rank and corresponding precoding matrix. Information containing the chosen rank and precoding matrix is then reported to the transmitter in the form of, e.g. a Rank Indicator (RI) and a Precoding-Matrix Indicator (PMI) as defined in 3GPP LTE specifications. The RI and PMI are only recommendations, and the transmitter does not necessarily follow the RI/PMI provided by the receiver when selecting the actual transmission rank and precoding matrix. When not following the recommendation, the transmitter must explicitly inform the receiver what precoding matrix is used. On the other hand, if the transmitter uses the recommended precoding matrix, only acknowledge is signaled.

To facilitate a concrete view of the readers, we hereby use the precoding codebook specified by LTE as an example. The system supports multi-antenna transmission using two and four antenna ports. The codebooks are defined for

- Two antenna ports for one and two layers, corresponding to precoding matrices of size 2×1 and 2×2, respectively, as listed in Table 8.2.
- Four antenna ports for one, two, three, and four layers, corresponding to precoding matrices of size 4×1, 4×2, 4×3, and 4×4, respectively, as listed in Table 8.3.

Table 8.2 Precoding codebooks for LTE with two antenna ports.

Number of layers	Codebook index					
	0	1	2	3	4	5
One layer	$\begin{bmatrix} 1 \\ 0 \end{bmatrix}$	$\begin{bmatrix} 0 \\ 1 \end{bmatrix}$	$\frac{1}{\sqrt{2}}\begin{bmatrix} 1 \\ 1 \end{bmatrix}$	$\frac{1}{\sqrt{2}}\begin{bmatrix} 1 \\ -1 \end{bmatrix}$	$\frac{1}{\sqrt{2}}\begin{bmatrix} 1 \\ j \end{bmatrix}$	$\frac{1}{\sqrt{2}}\begin{bmatrix} 1 \\ -j \end{bmatrix}$
Two layers	$\frac{1}{\sqrt{2}}\begin{bmatrix} 1 & 0 \\ 0 & 1 \end{bmatrix}$	$\frac{1}{2}\begin{bmatrix} 1 & 1 \\ 1 & -1 \end{bmatrix}$	$\frac{1}{2}\begin{bmatrix} 1 & 1 \\ j & -j \end{bmatrix}$			

Source: Adapted from 3GPP TS36.211 [2018].

Table 8.3 Precoding codebooks for LTE with four antenna ports.

Codebook Index	$\mathbf{u}_n$	Number of layers N_L			
		1	2	3	4
0	$\mathbf{u}_0 = [1, -1, -1, 1]^T$	$\mathbf{W}_0^{(1)}$	$\frac{1}{\sqrt{2}}\mathbf{W}_0^{(14)}$	$\frac{1}{\sqrt{3}}\mathbf{W}_0^{(124)}$	$\frac{1}{2}\mathbf{W}_0^{(1234)}$
1	$\mathbf{u}_1 = [1, -j, 1, j]^T$	$\mathbf{W}_1^{(1)}$	$\frac{1}{\sqrt{2}}\mathbf{W}_1^{(12)}$	$\frac{1}{\sqrt{3}}\mathbf{W}_1^{(123)}$	$\frac{1}{2}\mathbf{W}_1^{(1234)}$
2	$\mathbf{u}_2 = [1, 1, -1, 1]^T$	$\mathbf{W}_2^{(1)}$	$\frac{1}{\sqrt{2}}\mathbf{W}_2^{(12)}$	$\frac{1}{\sqrt{3}}\mathbf{W}_2^{(123)}$	$\frac{1}{2}\mathbf{W}_2^{(3214)}$
3	$\mathbf{u}_3 = [1, j, 1, -j]^T$	$\mathbf{W}_3^{(1)}$	$\frac{1}{\sqrt{2}}\mathbf{W}_3^{(12)}$	$\frac{1}{\sqrt{3}}\mathbf{W}_3^{(123)}$	$\frac{1}{2}\mathbf{W}_3^{(3214)}$
4	$\mathbf{u}_4 = \left[1, \frac{-1-j}{\sqrt{2}}, -j, \frac{1-j}{\sqrt{2}}\right]^T$	$\mathbf{W}_4^{(1)}$	$\frac{1}{\sqrt{2}}\mathbf{W}_4^{(14)}$	$\frac{1}{\sqrt{3}}\mathbf{W}_4^{(124)}$	$\frac{1}{2}\mathbf{W}_4^{(1234)}$
5	$\mathbf{u}_5 = \left[1, \frac{1-j}{\sqrt{2}}, j, \frac{-1-j}{\sqrt{2}}\right]^T$	$\mathbf{W}_5^{(1)}$	$\frac{1}{\sqrt{2}}\mathbf{W}_5^{(14)}$	$\frac{1}{\sqrt{3}}\mathbf{W}_5^{(124)}$	$\frac{1}{2}\mathbf{W}_5^{(1234)}$
6	$\mathbf{u}_6 = \left[1, \frac{1+j}{\sqrt{2}}, -j, \frac{-1+j}{\sqrt{2}}\right]^T$	$\mathbf{W}_6^{(1)}$	$\frac{1}{\sqrt{2}}\mathbf{W}_6^{(13)}$	$\frac{1}{\sqrt{3}}\mathbf{W}_6^{(134)}$	$\frac{1}{2}\mathbf{W}_6^{(1324)}$
7	$\mathbf{u}_7 = \left[1, \frac{-1+j}{\sqrt{2}}, j, \frac{1+j}{\sqrt{2}}\right]^T$	$\mathbf{W}_7^{(1)}$	$\frac{1}{\sqrt{2}}\mathbf{W}_7^{(13)}$	$\frac{1}{\sqrt{3}}\mathbf{W}_7^{(134)}$	$\frac{1}{2}\mathbf{W}_7^{(1324)}$
8	$\mathbf{u}_8 = [1, -1, 1, 1]^T$	$\mathbf{W}_8^{(1)}$	$\frac{1}{\sqrt{2}}\mathbf{W}_8^{(12)}$	$\frac{1}{\sqrt{3}}\mathbf{W}_8^{(124)}$	$\frac{1}{2}\mathbf{W}_8^{(1234)}$
9	$\mathbf{u}_9 = [1, -j, -1, -j]^T$	$\mathbf{W}_9^{(1)}$	$\frac{1}{\sqrt{2}}\mathbf{W}_9^{(14)}$	$\frac{1}{\sqrt{3}}\mathbf{W}_9^{(134)}$	$\frac{1}{2}\mathbf{W}_9^{(1234)}$
10	$\mathbf{u}_{10} = [1, 1, 1, -1]^T$	$\mathbf{W}_{10}^{(1)}$	$\frac{1}{\sqrt{2}}\mathbf{W}_{10}^{(13)}$	$\frac{1}{\sqrt{3}}\mathbf{W}_{10}^{(123)}$	$\frac{1}{2}\mathbf{W}_{10}^{(1324)}$
11	$\mathbf{u}_{11} = [1, j, -1, j]^T$	$\mathbf{W}_{11}^{(1)}$	$\frac{1}{\sqrt{2}}\mathbf{W}_{11}^{(13)}$	$\frac{1}{\sqrt{3}}\mathbf{W}_{11}^{(134)}$	$\frac{1}{2}\mathbf{W}_{11}^{(1324)}$
12	$\mathbf{u}_{12} = [1, -1, -1, 1]^T$	$\mathbf{W}_{12}^{(1)}$	$\frac{1}{\sqrt{2}}\mathbf{W}_{12}^{(12)}$	$\frac{1}{\sqrt{3}}\mathbf{W}_{12}^{(123)}$	$\frac{1}{2}\mathbf{W}_{12}^{(1234)}$
13	$\mathbf{u}_{13} = [1, -1, 1, -1]^T$	$\mathbf{W}_{13}^{(1)}$	$\frac{1}{\sqrt{2}}\mathbf{W}_{13}^{(13)}$	$\frac{1}{\sqrt{3}}\mathbf{W}_{13}^{(123)}$	$\frac{1}{2}\mathbf{W}_{13}^{(1324)}$
14	$\mathbf{u}_{14} = [1, 1, -1, -1]^T$	$\mathbf{W}_{14}^{(1)}$	$\frac{1}{\sqrt{2}}\mathbf{W}_{14}^{(13)}$	$\frac{1}{\sqrt{3}}\mathbf{W}_{14}^{(123)}$	$\frac{1}{2}\mathbf{W}_{14}^{(3214)}$
15	$\mathbf{u}_{15} = [1, 1, 1, 1]^T$	$\mathbf{W}_{15}^{(1)}$	$\frac{1}{\sqrt{2}}\mathbf{W}_{15}^{(12)}$	$\frac{1}{\sqrt{3}}\mathbf{W}_{15}^{(123)}$	$\frac{1}{2}\mathbf{W}_{15}^{(1234)}$

Source: 3GPP TS36.211 [2018]/ETSI.

Unlike Table 8.2 that explicitly provides the precoding matrices for two antenna ports, the precoding matrices in Table 8.3 need an interpretation. For a typical row $n, n = 0, 1, \ldots, 15$ the corresponding $\mathbf{W}_n$ is calculated by

$$\mathbf{W}_n = \mathbf{I} - \frac{2\mathbf{u}_n\mathbf{u}_n^H}{\mathbf{u}_n^H\mathbf{u}_n}, \tag{8.136}$$

and $\mathbf{W}_n^{(i_a i_b)}$ denotes a submatrix of $\mathbf{W}_n$ consisting of columns i_a and i_b. For example,

$$\mathbf{W}_0 = \frac{1}{2}\begin{bmatrix} 1 & 1 & 1 & 1 \\ 1 & 1 & -1 & -1 \\ 1 & -1 & 1 & -1 \\ 1 & -1 & -1 & 1 \end{bmatrix}, \tag{8.137}$$

and

$$\mathbf{W}_0^{(1)} = \frac{1}{2}\begin{bmatrix} 1 \\ 1 \\ 1 \\ 1 \end{bmatrix}, \quad \mathbf{W}_0^{(14)} = \frac{1}{2}\begin{bmatrix} 1 & 1 \\ 1 & -1 \\ 1 & -1 \\ 1 & 1 \end{bmatrix}. \tag{8.138}$$

The aforementioned mechanism is also known as *closed-loop* codebook-based precoding for multi-antenna transmission. A variant is called **open-loop precoding** that does not rely on any detailed precoding matrix recommendation from the receiver and does not require any explicit signaling of the actual precoding matrix used for the transmission. Instead, a precoding matrix is selected in a predefined and deterministic way known to both the transmitter and receiver in advance. One use of the open-loop precoding is in high-mobility scenarios where accurate feedback is difficult to achieve due to the latency in the PMI reporting.

In addition to codebook-based precoding, another variant referred to as **non-codebook-based precoding** was introduced in LTE Release 9. Despite the ambiguity raised by the term, it still can use the precoding matrices given by a predefined codebook. The major difference compared with codebook-based precoding is the presence of demodulation reference signals before the precoding. The transmission of precoded reference signals allows for demodulation and recovery of the transmitted layers at the receiver without *explicit knowledge of the precoding* applied at the transmitter. Channel estimation based on precoded reference signals will reflect the experienced channel, including the effect of precoding, and can therefore be used directly for coherent demodulation. There is thus no need to signal any precoding-matrix information to the receiver, which only needs to know the transmission rank. Consequently, the transmitter can apply an arbitrary precoding matrix, and there is no need for explicit signaling [Dahlman et al., 2011].

8.6.3 MIMO Detection

When the transmitter knows the channel, the SVD architecture enables the transmitter to send parallel data streams through the channel to arrive orthogonally at the receiver without interference between the streams. This is achieved by pre-rotating the data so that parallel streams can be sent along the eigenmodes of the channel. When the transmitter does not know the channel, precoding is not possible. However, the full degrees of freedom can be attained if proper MIMO detection algorithms are applied. Detection of spatially multiplexed signals is one of the crucial receiver functions in a MIMO wireless communication system. That is because, in addition to additive noise and channel fading, a received signal contains spatial interference due to simultaneous transmission from multiple transmit antennas. Consequently, the design, analysis, and implementation of efficient single-processing algorithms to detect received signals in the presence of this spatial interference are attractive.

We start with a system model

$$\mathbf{r} = \mathbf{Hs} + \mathbf{n}, \tag{8.139}$$

where the channel matrix $\mathbf{H}$ is perfectly known at the receiver but unknown at the transmitter. There is no precoding at the transmitter, i.e. $\mathbf{P} = \mathbf{I}$, implying that independent data streams are transmitted on the different transmit antennas. Note that both the receiver and transmitter know the precoding matrix in the case of CSIT, and the effect of $\mathbf{P}$ can be regarded as part of the overall channel response. The detection of precoded signals with a known precoding matrix at the receiver is similar to detecting independently transmitted signals. Therefore, this part focuses on the MIMO detection without CSIT but is applicable to the latter.

In the rest of this part, several well-known MIMO detection algorithms, including the optimal ML detection, linear detection (matched filter, ZF, and MMSE), and nonlinear detection (successive interference cancelation, SIC), are introduced.

8.6.3.1 Maximum-Likelihood Detection

The receiver generates an estimate $\hat{\mathbf{s}}$ of the transmitted symbol vector $\mathbf{s}$, based on its knowledge of the channel matrix $\mathbf{H}$, and the observation $\mathbf{r}$, i.e.

$$\hat{\mathbf{s}} = f(\mathbf{H}, \mathbf{r}). \tag{8.140}$$

The optimal detector from the point of view of minimizing the average error probability is realized by the ML algorithm. The ML detector solves a nonlinear optimization problem of minimizing the squared Euclidean distance between the observation $\mathbf{r}$ and the hypothesized received signal $\mathbf{Hs}$. Thus, Eq. (8.140) becomes

$$\hat{\mathbf{s}} = \arg\min_{\mathbf{s} \in \mathcal{X}^{N_t}} \|\mathbf{r} - \mathbf{Hs}\|^2, \tag{8.141}$$

applying the Euclidean norm $\| \cdot \|$, which is defined as

$$\|\mathbf{x}\| = \sqrt{|x_1|^2 + |x_2|^2 + \cdots + |x_N|^2} \tag{8.142}$$

for an N-dimensional complex space $\mathbb{C}^N$. The minimization is over $\mathbf{s} \in \mathcal{X}^{N_t}$, which is the set of all possible transmitted vectors. To be specific, there are a total of $|\mathcal{X}|^{N_t}$ vectors in the sample space, where $\mathcal{X}$ is the set of all constellation points and $|\cdot|$ denotes the cardinality of a set. Computing the exact solution to this optimization problem through an exhaustive search imposes an exponential complexity regarding the number of transmit antennas. It is only possible for small N_t but prohibitive when N_t becomes large. Knowing the exact ML solution is desired since it is a benchmark to evaluate various detection algorithms. To this end, low-complexity bounds on the ML performance are usually applied.

8.6.3.2 Linear Detection

Linear detection gets a soft estimate (marked by the *tilde* operator) of the transmitted symbol vector by means of a linear transformation of the received symbol vector, i.e.

$$\tilde{\mathbf{s}} = \phi(\mathbf{r}) = \mathbf{D}\mathbf{r} \tag{8.143}$$

with the detection matrix $\mathbf{D}$. It spatially decouples the effect of the channel, also known as MIMO equalization. The complexity of linear detectors is the same order of magnitude as that of inverting or factorizing a matrix of dimensions $N_r \times N_t$, which is therefore very attractive. A hard estimate (marked by the *hat* operator) is then obtained by mapping each entry of $\tilde{\mathbf{s}}$ to its closest constellation point, according to

$$\hat{s}_n = \arg \min_{s \in \mathcal{X}} |\tilde{s}_n - s| \tag{8.144}$$

with $\hat{\mathbf{s}}_n = [\hat{s}_1, \hat{s}_2, \dots, \hat{s}_{N_t}]^T$.

Matched Filter We write $\mathbf{h}_n, n = 1, 2, \dots, N_t$ to denote the nth column of the channel matrix $\mathbf{H}$, Eq. (8.139) can be rewritten as

$$\mathbf{r} = \mathbf{H}\mathbf{s} + \mathbf{n} = \sum_{n=1}^{N_t} \mathbf{h}_n s_n + \mathbf{n}. \tag{8.145}$$

We utilize s_m to denote the desired symbol sent from transmit antenna m. To detect s_m, the detector can focus only on transmit antenna m and simply treat the signals from other antennas (inter-antenna interference) as noise. Thus, the detector has a view of

$$\mathbf{r} = \underbrace{\mathbf{h}_m s_m}_{\text{the desired signal}} + \underbrace{\sum_{n=1, n \neq m}^{N_t} \mathbf{h}_n s_n + \mathbf{n}}_{\text{noise}}, \tag{8.146}$$

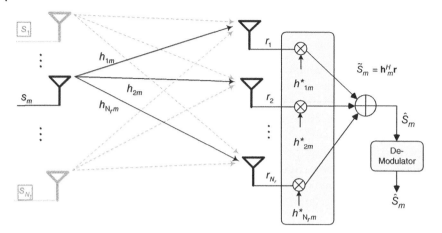

Figure 8.13 Principle of MIMO detection based on a matched filter, where the transmitted signals except the desired symbol denoted by s_m are treated as noise. Ignoring other transmit antennas, it seems like a SIMO channel and applies maximal ratio combining to form a soft estimate $\tilde{s}_m$ at the receiver, where $\mathbf{h}_m$ represents the mth column of $\mathbf{H}$.

where the first term is the desired component for detecting s_m, and the second term is noise. Ignoring other transmit antennas, a MIMO channel is transformed into a SIMO channel, as modeled by

$$\mathbf{r} = \mathbf{h}s + \mathbf{n}. \tag{8.147}$$

Similar to the matched filter in a SIMO channel, the receiver then applies the maximal ratio combining to form a soft estimate $\tilde{s}_m$, as demonstrated in Figure 8.13. A total of N_t parallel, independent decorrelators detect N_t transmitted symbols individually, forming a *Matched-Filter* (MF) detector for a MIMO channel.

A soft estimate of s_m is obtained by

$$\tilde{s}_m = \frac{\mathbf{h}_m^H}{\|\mathbf{h}_m\|^2}\mathbf{r} = s_m + \frac{\sum_{n=1,n\neq m}^{N_t}\mathbf{h}_m^H\mathbf{h}_n s_n + \mathbf{n}}{\|\mathbf{h}_m\|^2}. \tag{8.148}$$

Subsequently, a hard estimate $\hat{s}_m$ is obtained by mapping $\tilde{s}_m$ to its nearest constellation point in terms of the Euclidean distance according to Eq. (8.144). The MF detection can also be expressed in vector form

$$\tilde{\mathbf{s}} = \mathbf{H}^H\mathbf{r}, \tag{8.149}$$

implying a detection matrix

$$\mathbf{D} = \mathbf{H}^H. \tag{8.150}$$

It achieves a sub-optimal performance when $N_t \ll N_r$, whereas its performance severely deteriorates with the increasing number of transmit antennas due to a raised level of inter-antenna interference.

Zero Forcing In contrast to maximizing the strength of a desired single in the MF detection, ZF aims to cancel the interference from other streams completely (a.k.a. interference nulling) through a linear transformation of the received signal vector. Based on the observation and channel knowledge $(\mathbf{r}, \mathbf{H})$, solve the optimization problem

$$\tilde{\mathbf{s}} = \arg\min_{\mathbf{s}} \left\| \mathbf{r} - \mathbf{H}\mathbf{s} \right\|^2, \tag{8.151}$$

which removes the constellation constrains on $\mathbf{s}$ compared with Eq. (8.141) because the output is a soft estimate rather than a hard estimate. This can substantially lower the complexity.

When $\mathbf{H}$ is a square invertible matrix, the solution is given by

$$\tilde{\mathbf{s}} = \mathbf{H}^{-1}\mathbf{r}, \tag{8.152}$$

where $(\cdot)^{-1}$ denotes the inverse of a matrix.

If $\mathbf{H}$ is a square matrix but not invertible or $\mathbf{H}$ is not square, the pseudo inverse (a.k.a. the Moore–Penrose pseudo inverse)

$$\mathbf{H}^{\dagger} = (\mathbf{H}^H\mathbf{H})^{-1}\mathbf{H}^H \tag{8.153}$$

is employed.

We write $\mathbf{h}_k^{\dagger}$, $k = 1, 2, \ldots, N_t$ to denote the kth row of $\mathbf{H}^{\dagger} \in \mathbb{C}^{N_t \times N_r}$. Since $\mathbf{H}^{\dagger}\mathbf{H} = \mathbf{I} \in \mathbb{C}^{N_t \times N_t}$, we know that $\mathbf{h}_k^{\dagger}\mathbf{H}$ is a $1 \times N_t$ row vector with all zeros except for a unit in the kth entry. Rewriting Eq. (8.152) in a symbol wise, a soft estimate of a typical symbol s_k is given by

$$\tilde{s}_k = \mathbf{h}_k^{\dagger}\mathbf{r} = \mathbf{h}_k^{\dagger}\mathbf{H}\mathbf{s} + \mathbf{h}_k^{\dagger}\mathbf{n} = s_k + \mathbf{h}_k^{\dagger}\mathbf{n}. \tag{8.154}$$

Afterward, a hard estimate $\hat{s}_k$ is obtained by mapping $\tilde{s}_k$ to its nearest constellation point in terms of the Euclidean distance according to Eq. (8.144).

ZF can amplify the noise if the minimum singular value of $\mathbf{H}$ is too small, demonstrated by the SNR of $\tilde{s}_k$

$$\gamma_{\tilde{s}_k} = \frac{|s_k|^2}{\left\| \mathbf{h}_k^{\dagger} \right\|^2 \sigma_n^2}, \tag{8.155}$$

where the noise variance is enhanced by a factor of $\|\mathbf{h}_k^{\dagger}\|^2$. At low SNRs, the amplified noise is dominant, and the performance of a ZF detector may be worse than an MF detector.

Expressing the ZF detection in vector form, the detection matrix is then

$$\mathbf{D} = \mathbf{H}^{\dagger} = (\mathbf{H}^H\mathbf{H})^{-1}\mathbf{H}^H, \tag{8.156}$$

or simply

$$\mathbf{D} = \mathbf{H}^{-1} \tag{8.157}$$

if **H** is an invertible matrix. The soft estimate of the transmitted vector equals

$$\tilde{\mathbf{s}} = \mathbf{H}^{\dagger}\mathbf{r} = (\mathbf{H}^H\mathbf{H})^{-1}\mathbf{H}^H(\mathbf{H}\mathbf{s} + \mathbf{n})$$
$$= \mathbf{s} + (\mathbf{H}^H\mathbf{H})^{-1}\mathbf{H}^H\mathbf{n}$$
$$= \mathbf{s} + \tilde{\mathbf{n}}, \tag{8.158}$$

where inter-stream interference is completely canceled, hence the name *ZF*. The complexity of computing $\mathbf{H}^{\dagger}$ is roughly cubic in N_t, which is one order of magnitude more than that of an MF detector.

Minimum Mean-Squared Error To reduce the effect of noise amplification in ZF detection, a linear detector that minimizes the mean squared error between the transmitted symbol vector and the estimated vector is applied. That is, an optimal detection matrix **D** is chosen to satisfy

$$\min_{\mathbf{D}} \mathbb{E}\left[\|\mathbf{s} - \mathbf{D}\mathbf{r}\|^2\right]. \tag{8.159}$$

Equivalently, it solves an optimization equation expressed by

$$\tilde{\mathbf{s}} = \arg\min_{\mathbf{s}} \left\|\mathbf{r} - \mathbf{H}\mathbf{s}\right\|^2 + \lambda\|\mathbf{s}\|^2, \quad \text{for } \lambda > 0, \tag{8.160}$$

which adds a regularization term in Eq. (8.151) to alleviate the effect of noise amplification. The solution is given by

$$\mathbf{D} = (\mathbf{H}^H\mathbf{H} + \sigma_n^2\mathbf{I})^{-1}\mathbf{H}^H. \tag{8.161}$$

The MMSE detector outperforms both the MF and ZF detectors. At high SNRs, it approaches to the ZF detector since the impact of the second term inside the inverse operation is negligible when σ_n^2 is small [Chockalingam and Rajan, 2014]. At low SNRs, it behaves like the MF detector due to the prominence of the diagonal elements of $\mathbf{H}^H\mathbf{H}$.

8.6.3.3 Successive Interference Cancelation

Compared with the optimal ML detection, the linear detectors (MF, ZF, and MMSE) are simpler to implement, but their error performance is much inferior. A class of *nonlinear* detectors based on interference cancelation, which iteratively removes the interference of the detected streams, can remarkably improve the performance of detecting the remaining streams. Typical interference cancelation techniques include SIC and Parallel Interference Cancelation (PIC). The former is more attractive for its low complexity. Initially, the symbol in the strongest data stream is first detected employing a linear detector. Hence, this technique is also called MF-SIC, ZF-SIC, and MMSE-SIC if the applied detector is MF, ZF, and MMSE, respectively. Once a data stream is successfully retrieved, we can use

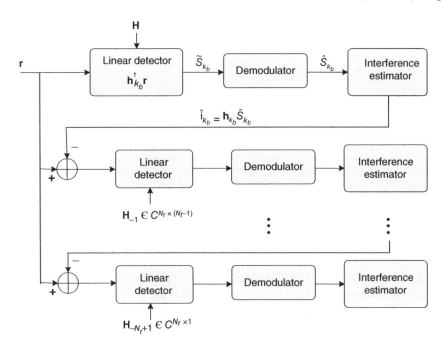

Figure 8.14 Schematic diagram of successive interference cancelation: a bank of linear detectors, each estimating one of the data streams, with estimated interference successively canceled from the received vector at each stage.

the detected symbol and the known channel matrix to estimate its corresponding interference. Then, this interference is subtracted from the received signal vector to reduce the overall interference of the remaining data streams. If the first stream is successfully decoded, there are only $N_t - 1$ streams from the viewpoint of the second detector. This process iterates until the last detector does not encounter any interference from the other data streams (assuming successful subtraction in each preceding stage).

The early breakthrough for MIMO techniques was due to the successful demonstration of Vertical Bell Laboratories Layered Space-Time (V-BLAST) architecture [Wolniansky et al., 1998; Foschini, 1996]. In V-BLAST, the ZF-SIC detection was demonstrated, which is summarized as follows:

1. *Symbol Detection*: Build $\mathbf{H}^{\dagger} = (\mathbf{H}^H \mathbf{H})^{-1} \mathbf{H}^H$ and determine the strongest stream in terms of $\|\mathbf{h}_k^{\dagger}\|^2$, where $\mathbf{h}_k^{\dagger}, k = 1, 2, \ldots, N_t$ denotes the kth row of $\mathbf{H}^{\dagger}$. We write k_b to denote the index of the strongest stream, and get a soft estimate applying $\tilde{s}_{k_b} = \mathbf{h}_{k_b}^{\dagger} \mathbf{r}$. Then, a hard estimate $\hat{s}_{k_b}$ is obtained from $\tilde{s}_{k_b}$, as illustrated in Figure 8.14.

2. *Interference Estimation*: Estimate the vector of inter-stream interference due to the k_b^{th} symbol through

$$\hat{\mathbf{i}}_{k_b} = \mathbf{h}_{k_b}\hat{s}_{k_b}, \tag{8.162}$$

where $\mathbf{h}_{k_b}$ stands for the k_b^{th} column of the channel matrix $\mathbf{H}$.

3. *Interference Cancelation*: Subtract this interference to form a new received vector

$$\mathbf{r}_{-1} = \mathbf{r} - \hat{\mathbf{i}}_{k_b}, \tag{8.163}$$

where $\mathbf{r}_{-1} \in \mathbb{C}^{N_r \times 1}$. Meanwhile, removing the k_b^{th} column from $\mathbf{H}$ yields a sub-matrix $\mathbf{H}_{-1} \in \mathbb{C}^{N_r \times (N_t - 1)}$, and removing the k_b^{th} entry from $\mathbf{s}$ yields $\mathbf{s}_{-1} \in \mathbb{C}^{(N_t - 1) \times 1}$, we obtain

$$\mathbf{r}_{-1} = \mathbf{H}_{-1}\mathbf{s}_{-1} + \mathbf{n}. \tag{8.164}$$

It is equivalent to a MIMO system with $N_t - 1$ transmit antennas and N_r received antennas, without the k_b^{th} transmit antenna.

4. *Successive Operation*: Let $\mathbf{r} = \mathbf{r}_{-1}$ and $\mathbf{H} = \mathbf{H}_{-1}$, and go to Step 1 until all symbols are successfully detected.

For MMSE-SIC and MF-SIC, the corresponding detection matrix in Step 1 becomes $(\mathbf{H}^H\mathbf{H} + \sigma_n^2\mathbf{I})^{-1}\mathbf{H}^H$ and $\mathbf{H}^H$, respectively, whereas other processing keeps identical.

Let's utilize a simple MIMO system with three transmit antennas and two receive antennas as an example to provide the readers with a deep understanding. Transmitting a symbol vector $\mathbf{s} = [s_1, s_2, s_3]^T$ through a channel

$$\mathbf{H} = \begin{bmatrix} h_{11} & h_{12} & h_{13} \\ h_{21} & h_{22} & h_{23} \end{bmatrix} \tag{8.165}$$

generates a received symbol vector

$$\mathbf{r} = \begin{bmatrix} r_1 \\ r_2 \end{bmatrix} = \begin{bmatrix} h_{11}s_1 + h_{12}s_2 + h_{13}s_3 \\ h_{21}s_1 + h_{22}s_2 + h_{23}s_3 \end{bmatrix}. \tag{8.166}$$

Suppose s_3 is first detected correctly, namely, $\hat{s}_3 = s_3$, its corresponding interference is estimated by

$$\hat{\mathbf{i}}_3 = \mathbf{h}_3\hat{s}_3 = \begin{bmatrix} h_{13}s_3 \\ h_{23}s_3 \end{bmatrix}. \tag{8.167}$$

Cancel $\hat{\mathbf{i}}_3$ from $\mathbf{r}$ to get

$$\mathbf{r}_{-1} = \mathbf{r} - \hat{\mathbf{i}}_3 = \begin{bmatrix} h_{11}s_1 + h_{12}s_2 \\ h_{21}s_1 + h_{22}s_2 \end{bmatrix}. \tag{8.168}$$

It is equivalent to a MIMO system with two transmit antennas and two receive antennas, transmitting $\mathbf{s} = [s_1, s_2]^T$ through a 2×2 MIMO channel

$$\mathbf{H} = \begin{bmatrix} h_{11} & h_{12} \\ h_{21} & h_{22} \end{bmatrix}. \tag{8.169}$$

8.7 Summary

This chapter explored the fundamentals of multi-antenna transmission, including spatial diversity, beamforming, and spatial multiplexing. Spatial diversity can effectively combat the deep fades in wireless channels through receive combining (e.g. MRC, EGC, and SC), space-time coding (e.g. STTC, Alamouti, and STBC), and TAS. Conventional beamforming over high-correlated arrays and single-stream beamforming (a.k.a. transmit precoding) over low-correlated arrays bring a power gain. In addition, beam steering enables the capability of suppressing the interference in particular directions. A more attractive multi-antenna technique is MIMO, which uses multiple antennas at both the transmitter and receiver to achieve an additional degree of freedom by *spatially multiplexing* parallel data streams. The capacity of a MIMO channel increases linearly with the number of antennas under rich scattering environments, whereas spatial diversity and beamforming only bring a logarithmic increase. However, the implementation of spatial multiplexing in a single-user MIMO system is restricted by the hardware limit of terminals and is vulnerable to channels conditions. The next chapter will study multi-user MIMO and massive MIMO, which can further unleash the potential of multi-antenna transmission.

References

3GPP TS36.211 [2018], Evolved universal terrestrial radio access (E-UTRA); physical channels and modulation (release 14), Technical specification, 3GPP.

Alamouti, S. [1998], 'A simple transmit diversity technique for wireless communications', *IEEE Journal on Selected Areas in Communications* **16**(8), 1451–1458.

Chen, Z., Yuan, J. and Vucetic, B. [2005], 'Analysis of transmit antenna selection/maximal-ratio combining in Rayleigh fading channels', *IEEE Transactions on Vehicular Technology* **54**(4), 1312–1321.

Chockalingam, A. and Rajan, B. S. [2014], *Large MIMO Systems*, Cambridge University Press, New York, USA.

Dahlman, E., Parkvall, S. and Sköld, J. [2011], *4G LTE/LTE-Advanced for Mobile Broadband*, Academic Press, Elsevier, Oxford, The United Kingdom.

Foschini, G. [1996], 'Layered space-time architecture for wireless communication in a fading environment when using multi-element antennas', *Bell Labs Technical Journal* **1**(2), 41–59.

Foschini, G. and Gans, M. [1998], 'On limits of wireless communications in a fading environment when using multiple antennas', *Wireless Personal Communications* **6**, 311–335.

Gardner, W. [1988], 'Simplification of MUSIC and ESPRIT by exploitation of cyclostationarity', *Proceedings of the IEEE* **76**(7), 845–847.

Gesbert, D., Shafi, M., shan Shiu, D., Smith, P. and Naguib, A. [2003], 'From theory to practice: An overview of MIMO space-time coded wireless systems', *IEEE Journal on Selected Areas in Communications* **21**(3), 281–302.

Goldsmith, A. [2005], *Wireless Communications*, Cambridge University Press, Stanford University, California.

Jiang, W. and Yang, X. [2012], An enhanced random beamforming scheme for signal broadcasting in multi-antenna systems, *in* 'Proceedings of IEEE 23rd International Symposium on Personal, Indoor and Mobile Radio Communications (PIMRC)', Sydney, Australia, pp. 2055–2060.

Mohammed, S. K., Viterbo, E., Hong, Y. and Chockalingam, A. [2011], 'MIMO precoding with X- and Y-codes', *IEEE Transactions on Information Theory* **57**(6), 3542–3566.

Paulraj, M. V. A. [2007], 'MIMO wireless linear precoding', *IEEE Signal Processing Magazine* **24**(5), 86–105.

Sanayei, S. and Nosratinia, A. [2004] , 'Antenna selection in MIMO systems', *IEEE Communications Magazine* **42**(10), 68–73.

Seshadri, N. and Winters, J. [1993], Two signaling schemes for improving the error performance of frequency-division-duplex (FDD) transmission systems using transmitter antenna diversity, in '*Proceedings of IEEE 43rd Vehicular Technology Conference (VTC)*', Secaucus, USA.

Tarokh, V., Seshadri, N. and Calderbank, A. [1998], '*Space-time codes for high data rate wireless communication: Performance criterion and code construction*', *IEEE Transactions on Information Theory* **44**(2), 744–765.

Tarokh, V., Jafarkhani, H. and Calderbank, A. [1999], '*Space-time block codes from orthogonal designs*', *IEEE Transactions on Information Theory* **45**(5), 1456–1467.

Telatar, E. [1999], 'Capacity of multi-antenna Gaussian channels', *European Transactions on Telecommunications* **10**(6), 585–595.

Tse, D. and Viswanath, P. [2005], *Fundamentals of Wireless Communication*, Cambridge University Press, Cambridge, United Kingdom.

Veen, B. V. and Buckley, K. [1988], '*Beamforming: A versatile approach to spatial filtering*', *IEEE ASSP Magazine* **5**(2), 4–24.

Viterbi, A. [2006], '*A personal history of the Viterbi algorithm*', *IEEE Signal Processing Magazine* **23**(4), 120–142.

Wolniansky, P., Foschini, G., Golden, G. and Valenzuela, R. [1998], V-BLAST: An architecture for realizing very high data rates over the rich-scattering wireless channel, in '*Proceedings of 1998 International Symposium on Signals, Systems, and Electronics*', Pisa, Italy, pp. 295–300.

Yang, X.-Z. [2016], *Communication Road: From Calculus to 5G (Chinese Edition)*, Electronic Industry Press, Beijing, China.

Yu, X., Xu, W., Leung, S.-H. and Wang, J. [2018], '*Unified performance analysis of transmit antenna selection with OSTBC and imperfect CSI over Nakagami-m fading channels*', *IEEE Transactions on Vehicular Technology* **67**(1), 494–508.

Zhang, J., Yu, X. and Letaief, K. B. [2019], '*Hybrid beamforming for 5G and beyond millimeter-wave systems: A holistic view*', *IEEE Open Journal of the Communications Society* **1**, 77–91.

9

Cellular and Cell-Free Massive MIMO Techniques in 6G

Multi-User Multiple-Input Multiple-Output (MU-MIMO) refers to the deployment scenario, where a base station equipped with multiple antennas serves multiple terminals with a single or only a few antennae. By breaking up spatial-multiplexed streams among multiple terminals, MU-MIMO gains superiority over SU-MIMO with three fundamental advantages. First, it facilitates the use of low-complexity, low-cost, power-saving terminals. Second, it is less vulnerable to propagation environments due to the spatial distribution of terminals, even under line-of-sight conditions. Third, information-theoretic analyses reveal that the sum rate of multi-user transmission is higher than the channel capacity of single-user communication. Nevertheless, the conventional MU-MIMO is still hard to scale up for high-order spatial multiplexing since the capacity-achieving precoding (e.g. dirty-paper coding) and decoding impose exponentially growing complexity. Most seriously, the transmitter requires the knowledge of the downlink channel, and the resources spent acquiring channel state information rise with the number of service antennas. A revolutionary technique called massive MIMO breaks this scalability barrier by not attempting to achieve the full Shannon limit and paradoxically by increasing the size of the system. In massive MIMO, only the base station learns the channel knowledge for precoding in the downlink and decoding in the uplink, whereas the terminals do not need to know it. Taking advantage of the channel reciprocity of Time Division Duplex (TDD) operation, the overhead required to acquire CSI depends on the number of terminals. Therefore, the base station side can install a large-scale antenna array so that the number of service antennas is typically increased to several times the number of active users, while the scale of users keeps small such that the implementation complexity is low. One of the benefits of using an unlimited number of service antennas is *channel hardening* where the effects of uncorrelated receiver noise and fast fading are eliminated completely. However, high performance in collocated MIMO is primarily achieved by users that stay near the cell center. Most users at the cell edge restrict to substantially worse

6G Key Technologies: A Comprehensive Guide, First Edition. Wei Jiang and Fa-Long Luo.

quality of services due to inter-cell interference. Therefore, a distributed massive MIMO system called cell-free massive MIMO, where a large number of service antennas randomly spread over a wide area has been proposed. All antennas cooperate phase-coherently via a fronthaul network and serve all users in the same time-frequency resource. There are no cells or cell boundaries. Since this setup combines the distributed MIMO and massive MIMO concepts, it is expected to reap all benefits from these two systems.

This chapter mainly consists of

- An information-theoretic introduction of MU-MIMO, i.e. the sum capacity analyses of MIMO broadcast channels and MIMO multi-access channels.
- The fundamentals of well-known dirty-paper coding that can achieve the full capacity, and the principles of its sub-optimal, low-complexity counterparts called zero-forcing precoding, and block diagonalization.
- The basic setup of massive MIMO, including the acquisition of channel knowledge, linear precoding in the downlink, and linear detection in the uplink.
- Pilot contamination in multi-cell massive MIMO systems, and the system models of downlink and uplink data transmission.
- The layout of a cell-free massive MIMO network, CSI acquisition via uplink training, and data transmission in the uplink.
- Conjugate beamforming and zero-forcing precoding in a cell-free massive MIMO network, and the performance impact of channel aging.

9.1 Multi-User MIMO

Spatial multiplexing discussed in the previous section is also often referred to as *Multiple-Input Multiple-Output* (MIMO), reflecting the fact that multiple parallel data streams are simultaneously transmitted on the same frequency to a single receiver. Using multiple antennas at both the transmitter and receiver in combination with precoding and detection processing aims to separate spatially multiplexed signals and suppress the interference among different transmission layers. This technique has a more specific term *Single-User MIMO or SU-MIMO* for reasons that will become clear in the following text. As a direct extension of spatial multiplexing, parallel transmission layers formed by multiple transmit antennas can intend for different receivers with a single or a few receiver antennas, and vice versa. In the context of mobile communication systems or wireless local area networks, the term *Multi-User MIMO* (MU-MIMO) refers to the deployment scenario, where a base station or an access point equipped with multiple transmit antennas communicates with multiple terminals. A set of terminals with a single or only a few antennae can form a virtual array to cultivate spatial multiplexing gains along with a multi-antenna base station. Due to relatively large signal-processing capability and sufficient power supply, the base station side bears the burden of spatially separating parallel streams. Thus, the base station performs precoding or transmits beamforming toward multiple users in the downlink and multi-user detection in the uplink. Consequently, a remarkable benefit of MU-MIMO over SU-MIMO is that the spatial-multiplexing gain is preserved even in the case of low-cost terminals with a small number of antennas. This is an imperative requirement for achieving the economy of scale in the mobile industry.

Another fundamental distinction between MU-MIMO and SU-MIMO comes from the difference in the underlying channel. Achieving spatial-multiplexing gain highly relies on well-conditioned channels. In a SU-MIMO system, the decorrelation among the spatial signatures of the antennas demands rich-scattering environments with large inter-antenna spacing or the use of antenna polarization. In an MU-MIMO system, the decorrelation among the spatial signatures of different terminals occurs naturally since the highly distributed nature of these terminals. The users are geographically separated, a signal propagates in different directions even there is limited scattering in the environment. Furthermore, a multi-user diversity gain is available in such a setup in addition to the spatial-multiplexing gain.

Nevertheless, the achievement of the potential of MU-MIMO depends on accurate Channel State Information (CSI) at the transmitter. It has been shown that a small amount of feedback can be very beneficial in steering the power toward the receiver's antennas. More precisely, the accuracy of CSI in SU-MIMO only causes an Signal-to-Noise Ratio (SNR) penalty but does not affect the multiplexing gain. However, the accuracy of CSI available at the transmitter does affect the multiplexing gain of an MU-MIMO system. Therefore, it is essential to provide CSI accurately and timely to the transmitter, which is always challenging due to the constraint of feedback resources or the severity of wireless channels.

9.1.1 Broadcast and Multiple-Access Channels

SU-MIMO is a *symmetric* point-to-point system, which can therefore be described using a transmitter and a receiver, and does not need to distinguish downlink and uplink. In contrast, MU-MIMO is an *asymmetric* system, where the downlink transmission from a base station to several terminals is referred to as *a Gaussian MIMO broadcast channel*, while the uplink transmission from several terminals to the base station is called *a Gaussian MIMO multiple access channel*.

In an MU-MIMO system, K terminals are selected for simultaneous communications with a base station over the same time-frequency resource. A typical terminal k is equipped with $N_k, k = 1, 2, \dots, K$ antennas, and therefore these terminals have a total of $N_u = \sum_{k=1}^{K} N_k$ terminal-side antennas. Assume the base station has N_b antennas, the system forms an $N_u \times N_b$ channel in the downlink of a cellular system while an $N_b \times N_u$ channel in the uplink. The base station side can support up to $N_m = \min(N_b, N_u)$ parallel streams, whereas a typical terminal k is assigned to L_k streams, satisfying $L_k \leqslant \min(N_k, N_b)$ and equivalently $\sum_{k=1}^{K} L_k \leqslant N_m$. Such a multi-user MIMO setting can be denoted as a $([N_1, N_2, \dots, N_K], N_b)$ system for the downlink or $(N_b, [N_1, N_2, \dots, N_K])$ for the uplink.

Let's first look at the uplink transmission where K terminals simultaneously transmit toward the base station. We write $\mathbf{H}_{ul}^{(k)} \in \mathbb{C}^{N_b \times N_k}$ to model the channel matrix from the kth user to the base station. Each entry denotes the channel gain from a transmit antenna at a terminal to a receive antenna at the base station. The active terminals can be assumed to randomly locate within a cell and the antennas of a terminal are sufficiently spaced or polarized, resulting in independent fading channels. In flat Rayleigh-fading channels, a channel coefficient is denoted by a circularly symmetric Gaussian complex random variance with zero mean and unit variance, namely, $h \sim \mathcal{CN}(0, 1)$.

The MIMO multiple access channel can be modeled as

$$\mathbf{r} = \sum_{k=1}^{K} \mathbf{H}_{ul}^{(k)} \mathbf{s}_k + \mathbf{n}, \tag{9.1}$$

where $\mathbf{r} \in \mathbb{C}^{N_b \times 1}$, $\mathbf{s}_k \in \mathbb{C}^{N_k \times 1}$, and $\mathbf{n} \in \mathbb{C}^{N_b \times 1}$ stand for the vector of received symbols, the vector of transmitted symbols at terminal k, and the vector of noise, respectively. The power for transmission at the kth terminal is constrained by $\mathbb{E}[\mathbf{s}_k^H \mathbf{s}_k] \leqslant P_k$, or equivalently $\mathrm{tr}\left(\mathbb{E}\left[\mathbf{s}_k \mathbf{s}_k^H\right]\right) \leqslant P_k$, while the noise per receiver antenna is independent complex Gaussian noise with zero mean and variance σ_n^2, namely, $\mathbf{n} \sim \mathcal{CN}(0, \sigma_n^2 \mathbf{I}_{N_b})$.

We denote by $\mathbf{u}_k \in \mathbb{C}^{L_k \times 1}$ the vector of information symbols from user k, which is transformed into $\mathbf{s}_k$ according to

$$\mathbf{s}_k = \mathbf{T}_k \mathbf{u}_k, \tag{9.2}$$

where $\mathbf{T}_k \in \mathbb{C}^{N_k \times L_k}$ stands for the precoding matrix of user k.

Using $\mathbf{H}_{ul} \in \mathbb{C}^{N_b \times N_u}$ to denote the overall MIMO multiple access channel in an MU-MIMO system, we have

$$\mathbf{H}_{ul} = [\mathbf{H}_{ul}^{(1)}, \mathbf{H}_{ul}^{(2)}, \dots, \mathbf{H}_{ul}^{(K)}]. \tag{9.3}$$

Building a vector to contain all transmit symbols from K terminals

$$\mathbf{s} = \begin{bmatrix} \mathbf{s}_1 \\ \vdots \\ \mathbf{s}_K \end{bmatrix}, \tag{9.4}$$

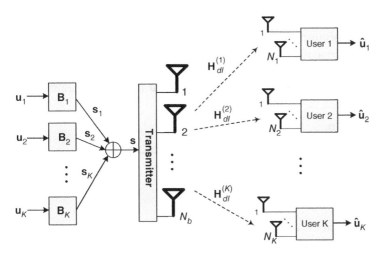

Figure 9.1 Schematic diagram of the downlink for a multi-user MIMO system.

Eq. (9.1) can be rewritten to

$$\mathbf{r} = \sum_{k=1}^{K} \mathbf{H}_{ul}^{(k)} \mathbf{s}_k + \mathbf{n} = \sum_{k=1}^{K} \mathbf{H}_{ul}^{(k)} \mathbf{T}_k \mathbf{u}_k + \mathbf{n} = \mathbf{H}_{ul}\mathbf{s} + \mathbf{n}, \tag{9.5}$$

which is equivalent to the model of a SU-MIMO system with N_u transmit antennae and N_b receive antennae. In other words, a typical terminal k transmits $\mathbf{s}_k$ to the base station, which generates a received component $\mathbf{r}_k = \mathbf{H}_{ul}^{(k)} \mathbf{s}_k$, leading to an overall received signal $\mathbf{r} = \sum_{k=1}^{K} \mathbf{r}_k$.

Then, let's focus on the downlink transmission where the base station sends signals toward K terminals over the same time-frequency resource, as demonstrated in Figure 9.1. The overall system can be modeled as

$$\mathbf{r} = \mathbf{H}_{dl}\mathbf{s} + \mathbf{n}, \tag{9.6}$$

where $\mathbf{H}_{dl} \in \mathbb{C}^{N_u \times N_b}$ denotes the channel matrix between N_b transmit antennae at the base station and N_u receive antennae distributed over K terminals, $\mathbf{r} \in \mathbb{C}^{N_u \times 1}$, $\mathbf{s} \in \mathbb{C}^{N_b \times 1}$, and $\mathbf{n} \in \mathbb{C}^{N_u \times 1}$ stand for the vector of all received symbols, the vector of transmitted symbols at the base station, and the vector of noise, respectively. The power constraint at the base station is expressed by $\mathbb{E}[\mathbf{s}^H\mathbf{s}] \leqslant P$, or equivalently $\mathrm{tr}\left(\mathbb{E}\left[\mathbf{s}\mathbf{s}^H\right]\right) \leqslant P$.

Decomposing Eq. (9.6) as

$$\begin{bmatrix} \mathbf{r}_1 \\ \mathbf{r}_2 \\ \vdots \\ \mathbf{r}_K \end{bmatrix} = \begin{bmatrix} \mathbf{H}_{dl}^{(1)} \\ \mathbf{H}_{dl}^{(2)} \\ \vdots \\ \mathbf{H}_{dl}^{(K)} \end{bmatrix} \mathbf{s} + \begin{bmatrix} \mathbf{n}_1 \\ \mathbf{n}_2 \\ \vdots \\ \mathbf{n}_K \end{bmatrix}, \tag{9.7}$$

where $\mathbf{r}_k \in \mathbb{C}^{N_k \times 1}$ and $\mathbf{n}_k \in \mathbb{C}^{N_k \times 1}$ represent the vector of received symbols and the vector of noise, respectively, at terminal k, and $\mathbf{H}_{dl}^{(k)} \in \mathbb{C}^{N_k \times N_b}$ models the channel from the base station to the kth user, which is a sub-matrix consisting of N_k rows of $\mathbf{H}_{dl}$.

Then, we know that the individual model dedicated to a typical terminal k is given by

$$\mathbf{r}_k = \mathbf{H}_{dl}^{(k)}\mathbf{s} + \mathbf{n}_k, \quad k = 1, 2, \dots, K, \tag{9.8}$$

applying $\mathbf{r} = [\mathbf{r}_1^T, \mathbf{r}_2^T, \dots, \mathbf{r}_K^T]^T$, $\mathbf{n} = [\mathbf{n}_1^T, \mathbf{n}_2^T, \dots, \mathbf{n}_K^T]^T$, and

$$\mathbf{H}_{dl} = \begin{bmatrix} \mathbf{H}_{dl}^{(1)} \\ \vdots \\ \mathbf{H}_{dl}^{(K)} \end{bmatrix}. \tag{9.9}$$

For simplicity, we also use $\mathbf{u}_k \in \mathbb{C}^{L_k \times 1}$ to denote the vector of information symbols intended for user k, which can be precoded individually according to

$$\mathbf{s}_k = \mathbf{B}_k\mathbf{u}_k, \tag{9.10}$$

where $\mathbf{B}_k \in \mathbb{C}^{N_b \times L_k}$ stands for the precoding matrix dedicated to user k at the base station, and $\mathbf{s}_k$ is a component of the overall transmitted signal $\mathbf{s}$ due to user k, satisfying

$$\mathbf{s} = \sum_{k=1}^{K} \mathbf{s}_k = \sum_{k=1}^{K} \mathbf{B}_k\mathbf{u}_k. \tag{9.11}$$

Alternatively, the transmitted signal can be generated by joint precoding

$$\mathbf{s} = \mathbf{B}\mathbf{u}, \tag{9.12}$$

where $\mathbf{B} \in \mathbb{C}^{N_b \times L}$ is the precoding matrix for all information symbols $\mathbf{u} \in \mathbb{C}^{L \times 1} = [\mathbf{u}_1^T, \mathbf{u}_2^T, \dots, \mathbf{u}_K^T]^T$ with the total number of data streams $L = \sum_{k=1}^{K} L_k$. It is not difficult to derive that

$$\mathbf{B} = [\mathbf{B}_1, \mathbf{B}_2, \dots, \mathbf{B}_K]. \tag{9.13}$$

Then, Eq. (9.8) can be also expressed by

$$\mathbf{r}_k = \mathbf{H}_{dl}^{(k)}\mathbf{s} + \mathbf{n}_k = \mathbf{H}_{dl}^{(k)}\mathbf{B}\mathbf{u} + \mathbf{n}_k = \mathbf{H}_{dl}^{(k)}\sum_{k=1}^{K}\mathbf{B}_k\mathbf{u}_k + \mathbf{n}_k. \tag{9.14}$$

Similarly, Eq. (9.6) can be rewritten to

$$\mathbf{r} = \mathbf{H}_{dl}\mathbf{s} + \mathbf{n} = \mathbf{H}_{dl}\mathbf{B}\mathbf{u} + \mathbf{n} = \mathbf{H}_{dl}\sum_{k=1}^{K}\mathbf{B}_k\mathbf{u}_k + \mathbf{n}. \tag{9.15}$$

9.1.2 Multi-User Sum Capacity

In a point-to-point system, the channel capacity provides a measure of the performance limit: reliable communications with an arbitrarily small error probability can be achieved at any rate $R < C$, whereas reliable communications are impossible when $R > C$. For a multi-user system consisting of a base station and K terminals, the concept is extended to a similar performance metric called *a capacity region*. It is characterized by a K-dimensional space $\mathfrak{C} \in \mathbb{R}_+^K$, where $\mathbb{R}_+$ denotes the set of non-negative real-valued numbers, and $\mathfrak{C}$ is the set of all K-tuples $(R_1, R_2, \ldots, R_K)$ such that a generic user k can reliably communicate at rate R_k simultaneously with others. Due to the shared transmission resource, there is a trade-off: if one desires a higher rate, some of other users have to lower their rates. From this capacity region, a performance metric can be derived, i.e. the sum capacity

$$C_{\text{sum}} = \max_{(R_1, R_2, \ldots, R_K) \in \mathfrak{C}} \left(\sum_{k=1}^{K} R_k \right), \tag{9.16}$$

indicating the maximum total throughput that can be achieved.

Let's use a simplest multi-user system consisting of a single-antenna receiver and two users equipped with a single transmit antenna. Users 1 and 2 send the transmitted symbols s_1 and s_2 to a receiver in an uplink AWGN channel. The received symbol is

$$r = s_1 + s_2 + n, \tag{9.17}$$

where the power constrains for s_1 and s_2 are P_1 and P_2, respectively, and $n \sim \mathcal{CN}(0, \sigma_n^2)$ is complex Gaussian noise. The rates of users 1 and 2 are R_1 and R_2, respectively, forming the following capacity region

$$\mathfrak{C} = \left\{ (R_1, R_2) \in \mathbb{R}_+^2 \,\middle|\, \begin{array}{l} R_1 < \log_2 \left(1 + \dfrac{P_1}{\sigma_n^2} \right) \\[2ex] R_2 < \log_2 \left(1 + \dfrac{P_2}{\sigma_n^2} \right) \\[2ex] R_1 + R_2 < \log_2 \left(1 + \dfrac{P_1 + P_2}{\sigma_n^2} \right) \end{array} \right\}. \tag{9.18}$$

The first constraint in Eq. (9.18) implies that the achievable rate of user 1 is bounded by a single-user system where user 2 is absent. Similarly, the second constraint indicates *the single-user bound* for user 2. The third constraint says that the sum rate cannot exceed the capacity of a point-to-point system with the received signal power equal to the sum of the received signal powers of these two users.

It is of particular interest that one user can achieve its single-user bound while another user can simultaneously transmit with a non-zero rate. That is a benefit of a multi-user system over a single-user system. It is achieved by successive interference cancellation or SIC made in two steps. In the first step, the receiver detects the symbol of user 1, treating the signal from user 2 as a colored noise. The achieved rate for user 1 is

$$R_1 = \log_2 \left(1 + \frac{P_1}{P_2 + \sigma_n^2} \right). \tag{9.19}$$

Subtracting s_1 from r, the receiver can then detect s_2 with only the additive white noise. In this case,

$$R_2 = \log_2\left(1 + \frac{P_2}{\sigma_n^2}\right). \tag{9.20}$$

The sum rate equals to

$$C_{\text{sum}} = R_1 + R_2 = \log_2\left(1 + \frac{P_1}{P_2 + \sigma_n^2}\right) + \log_2\left(1 + \frac{P_2}{\sigma_n^2}\right). \tag{9.21}$$

It can be extended to a two-user system where an N_b-antenna base station communicates with two multi-antenna users $k = 1, 2$. User k is equipped with N_k antennas, and simultaneously sends $\mathbf{s}_k \in \mathbb{C}^{N_k \times 1}$ in uplink flat-fading channels, with the constrain of transmit power

$$tr(\mathbf{R}_k) \leqslant P_k, \tag{9.22}$$

where $\mathbf{R}_k = \mathbb{E}[\mathbf{s}_k \mathbf{s}_k^H]$ is the covariance matrix of $\mathbf{s}_k$. We denote by $\mathbf{H}_k \in \mathbb{C}^{N_b \times N_k}$ the channel matrix from user k to the base station.

The capacity region then becomes

$$\mathfrak{C} = \left\{ (R_1, R_2) \in \mathbb{R}_+^2 \,\middle|\, \begin{array}{l} R_1 < \log_2 \det\left[\mathbf{I}_{N_b} + \dfrac{\mathbf{H}_1 \mathbf{R}_1 \mathbf{H}_1^H}{\sigma_n^2}\right] \\[2ex] R_2 < \log_2 \det\left[\mathbf{I}_{N_b} + \dfrac{\mathbf{H}_2 \mathbf{R}_2 \mathbf{H}_2^H}{\sigma_n^2}\right] \\[2ex] C_{sum} < \log_2 \det\left[\mathbf{I}_{N_b} + \dfrac{\mathbf{H}_1 \mathbf{R}_1 \mathbf{H}_1^H + \mathbf{H}_2 \mathbf{R}_2 \mathbf{H}_2^H}{\sigma_n^2}\right] \end{array} \right\}. \tag{9.23}$$

The first two constraints indicate that the achievable rate of either user is bounded by the capacity of a SU-MIMO system, with N_k transmit antennas and N_b receive antennas, where another user is removed. The third constraint implies that the sum capacity of two users equals to a point-to-point system where two active users act as a single user with $N_1 + N_2$ transmit antennas, sending independent signals and subjecting to different power constraints. An example of the capacity region is illustrated in Figure 9.2. The point A corresponds to an optimal case where the receiver detect $\mathbf{s}_1$ first, regarding inter-user interference from user 2 as a colored noise. The maximal rate of user 1 is limited by Tse and Viswanath [2005]

$$R_1 = \log_2 \det\left[\mathbf{I}_{N_b} + \left(\mathbf{I}_{N_b} + \frac{\mathbf{H}_2 \mathbf{R}_2 \mathbf{H}_2^H}{\sigma_n^2}\right)^{-1} \frac{\mathbf{H}_1 \mathbf{R}_1 \mathbf{H}_1^H}{\sigma_n^2}\right], \tag{9.24}$$

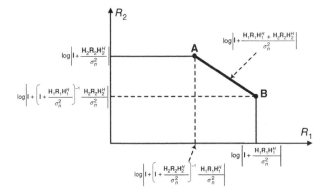

Figure 9.2 Capacity region of a two-user multi-antenna system.

and afterward the receiver can detect $\mathbf{s}_2$ with a rate of

$$R_2 = \log_2 \det \left[\mathbf{I}_{N_b} + \frac{\mathbf{H}_2 \mathbf{R}_2 \mathbf{H}_2^H}{\sigma_n^2} \right], \tag{9.25}$$

as the same as a SU-MIMO system consisting of the base station and only user 2. If we reverse the order of interference cancellation, then the point B is achieved. The other points on the segment AB contains all optimal operating points to maximize the sum capacity. Any point on this segment can be obtained by time-sharing between two cancellation priorities in point A and point B.

For the sake of illustration in Figure 9.2, the discussion has been restricted to a two-user system, but the generalization to K users is natural. The capacity region is now a K-dimensional polyhedron, which can be mathematically described by

$$\mathfrak{C} = \left\{ (R_1, \ldots, R_K) \in \mathbb{R}_+^K \left| \begin{array}{l} R_k < \log_2 \det \left[\mathbf{I}_{N_b} + \frac{\mathbf{H}_k \mathbf{R}_k \mathbf{H}_k^H}{\sigma_n^2} \right], \forall k \\[3mm] C_{\text{sum}} < \log_2 \det \left[\mathbf{I}_{N_b} + \frac{\sum_{k=1}^{K} \mathbf{H}_k \mathbf{R}_k \mathbf{H}_k^H}{\sigma_n^2} \right] \end{array} \right. \right\}. \tag{9.26}$$

It concludes that *the sum capacity of an MU-MIMO system is generally strictly larger than the single-user capacity* of any of the users in this system. That is a particular advantage of applying multi-user transmission.

9.1.3 Dirty Paper Coding

In the uplink of an MU-MIMO system, the received signals result in a non-orthogonal superposition of data streams from different users, thereby meaning that the data detection process can be performed using the well-known multi-user detection techniques. In particular, the optimal receiver architectures are derived based on the maximum-likelihood principle or the Maximum

A Posteriori (MAP) probability criterion [Sanguinetti and Poor, 2009]. They offer performance close to that of an interference-free system but at the price of prohibitive complexity, which increases exponentially with the number of users and data streams. Sub-optimal detection algorithms apply a linear transformation at the receiver in the forms of a matched filter, zero forcing, or minimum Mean Square Error (MSE) detectors, which achieve a reasonable trade-off between performance and complexity. Alternatively, better performance can be obtained employing nonlinear detection with interference cancellation (e.g. zero forcing (ZF)-SIC, matched filter (MF)-SIC, and minimum mean-square error (MMSE)-SIC). Therefore, we skip the uplink and focus only on multi-user precoding or beamforming algorithms in the downlink transmission, where Dirty Paper Coding (DPC) and two linear processing, i.e. zero-forcing precoding and Block Diagonalization (BD) precoding will be studied.

The name of DPC comes from "Writing on Dirty Paper", the title of an article by Costa [1983] on the capacity of a Gaussian channel, as depicted by

$$r = s + i + n, \tag{9.27}$$

where $i \sim \mathcal{N}(0, I)$ is the interference, $n \sim \mathcal{N}(0, \sigma^2)$ is Gaussian noise, the received signal is $r \in \mathbb{R}$, and the transmitted signal $s \in \mathbb{R}$, which is used to transmit u and satisfies $s^2 \leqslant P$. If i is unknown to both transmitter and receiver, the capacity is

$$C = \frac{1}{2}\log_2\left(1 + \frac{P}{\sigma^2 + I}\right). \tag{9.28}$$

Costa presented the surprising result that, if i is perfectly known to the encoder, the capacity of this system is the same as the standard Gaussian channel with an SNR of P/σ^2:

$$C = \frac{1}{2}\log_2\left(1 + \frac{P}{\sigma^2}\right), \tag{9.29}$$

regardless of the interference. The title of Costa's work came by making an analogy to the problem of writing on dirty paper, where the reader cannot nominally distinguish dirt from ink. We can imagine to design a transmitted signal

$$s = u - i \tag{9.30}$$

to realize it, but the optimal transmitter adapts its signals to the interference rather than attempting to cancel it.

DPC has been applied for MIMO broadcast channels to design an optimal capacity-achieving transmission strategy. Subjected to the perfectly known CSI at the transmitter, it can completely alleviate the effect of multi-user interference and achieve the capacity of the AWGN channel, without a power penalty and without requiring the receiver to know the interfering signal.

Consider a system with a multi-antenna transmitter and multiple receiver with a single antenna, the system model in Eq. (9.6) can be adjusted to

$$\mathbf{r} = \mathbf{Hs} + \mathbf{n}, \tag{9.31}$$

where $\mathbf{H} \in \mathbb{C}^{K \times N_b}$ denotes the channel matrix between N_b transmit antennae and K single-antenna receivers, the received vector, transmitted vector, and noise vector are expressed by $\mathbf{r} \in \mathbb{C}^{K \times 1}$, $\mathbf{s} \in \mathbb{C}^{N_b \times 1}$, and $\mathbf{n} \in \mathbb{C}^{K \times 1}$, respectively. The channel knowledge $\mathbf{H}$ is known to the transmitter and to all receivers, and the input is constrained by $\mathbb{E}[\mathbf{s}^H \mathbf{s}] \leqslant P$.

Let $\mathbf{s} = \mathbf{Bu}$, where $\mathbf{B} \in \mathbb{C}^{N_b \times K}$ denote a precoding matrix, and the entries of $\mathbf{u}$ are generated by successive dirty-paper encoding with Gaussian codebooks [Caire and Shamai, 2003]. Equation (9.31) transfers to

$$\mathbf{r} = \mathbf{HBu} + \mathbf{n} = \mathbf{Wu} + \mathbf{n}. \tag{9.32}$$

The precoded channel yields a set of $k = 1, 2, \dots, K$ interference channels

$$r_k = w_{kk} u_k + \sum_{k' < k} w_{kk'} u_{k'} + \sum_{k' > k} w_{kk'} u_{k'} + n_k, \tag{9.33}$$

where $w_{kk'}$ represents the $(k, k')^{\text{th}}$ element of $\mathbf{W}$, r_k, u_k, and n_k denote the kth entry of $\mathbf{r}$, $\mathbf{u}$, and $\mathbf{n}$, respectively, and the transmit power is constrained by $\mathbb{E}[u_k u_k^*] \leqslant P_k$. Given a specific order of interference channels, the encoder considers the interference signal $\sum_{k' < k} w_{kk'} u_{k'}$ caused by users $k' < k$ as known non-causally, and the detector of user k treats the inference signal $\sum_{k' > k} w_{kk'} u_{k'}$ as additional noise. By applying DPC at the transmitter and the minimum Euclidean distance decoding at each receiver, the achievable sum rate is

$$R_{\text{DPC}} = \sum_{k=1}^{K} \log \left(1 + \frac{|w_{kk}|^2 P_k}{\sigma_n^2 + \sum_{k' > k} |w_{kk'}|^2 P_{k'}} \right). \tag{9.34}$$

A choice of the precoding matrix can be obtained by conducting the QR decomposition $\mathbf{H} = \mathbf{LQ}$, where $\mathbf{Q} \in \mathbb{C}^{N_m \times N_b}$ has orthogonal rows, satisfying $\mathbf{QQ}^H = \mathbf{I}$, $\mathbf{L} \in \mathbb{C}^{K \times N_m}$ is lower triangular, and $N_m = \min(K, N_b)$. Letting $\mathbf{B} = \mathbf{Q}^H$, the received vector becomes

$$\mathbf{r} = \mathbf{HBu} + \mathbf{n} = \mathbf{LQQ}^H \mathbf{u} + \mathbf{n} = \mathbf{Lu} + \mathbf{n}, \tag{9.35}$$

corresponding to a set of K interference channels if $N_b \geq K$, i.e.

$$r_k = l_{kk} u_k + \sum_{k' < k} l_{kk'} u_{k'} + n_k, \quad k = 1, \dots, K, \tag{9.36}$$

where $l_{kk'}$ represents the $(k, k')^{\text{th}}$ element of $\mathbf{L}$. The input signals $\mathbf{u}$ are generated by successive dirty-paper coding, where the interference signal $\sum_{k' < k} l_{kk'} u_{k'}$ is non-causally known at the transmitter. There might exist coding schemes such as a typical user k does not see the interference from users $k' < k$. Meanwhile, as observed in Eq. (9.36), the precoding matrix is chosen to force the interference signal from users $k' > k$ to zero. Hence, this scheme is also referred to as Zero-Forcing Dirty-Paper Coding (ZF-DPC).

Then, an MU-MIMO system is transformed into a set of parallel sub-channels, as that of a SU-MIMO system, i.e.

$$r_k = l_{kk} u_k + n_k, \quad k = 1, \dots, K. \tag{9.37}$$

The achievable sum rate is [Caire and Shamai, 2001]

$$R_{\mathrm{DPC}} = \sum_{k=1}^{K} \log \left(1 + \frac{|l_{kk}|^2 P_k}{\sigma_n^2} \right),$$ (9.38)

which can be optimized jointly with respect to the power allocation P_k, $k = 1, 2, \ldots, K$ and the user ordering.

9.1.4 Zero-Forcing Precoding

Despite its significance from the information-theoretic point of view, the implementation of DPC requires prohibitive complexity at both transmitter and receivers, while the practical design of codes close to the capacity is still an open issue. Attempts in this direction were discussed such as a generalization of Tomlinson–Harashima Precoding (THP) to a multidimensional vector quantization scheme. On the other hand, a sub-optimal but simple multi-user transmission strategy is represented by linear precoding also known as linear transmit beamforming.

In the case of single-antenna receives, interference mitigation can be accomplished only at the base station. The most straightforward approach applies *channel inversion* before the transmission, referred to as zero-forcing precoding or ZF linear beamforming. It pre-inverts the channel matrix at the transmitter such that the inter-user interference completely vanishes at all receivers. This approach can be easily applied when the number of users is smaller than the number of transmit antennas ($N_b > K$). At the same time, it applies to the case $N_b < K$ provided that an appropriate user selection algorithm is employed. Albeit the simplicity and easy to implement, such schemes achieve good performance, especially when the number of users is large.

Letting the precoding matrix equal to the pseudo inverse of the channel matrix, i.e.

$$\mathbf{B} = \mathbf{H}^H \left(\mathbf{H}\mathbf{H}^H \right)^{-1},$$ (9.39)

resulting in the received signal

$$\mathbf{r} = \mathbf{H}\mathbf{B}\mathbf{u} + \mathbf{n} = \mathbf{H}\mathbf{H}^H \left(\mathbf{H}\mathbf{H}^H \right)^{-1} \mathbf{u} + \mathbf{n} = \mathbf{u} + \mathbf{n},$$ (9.40)

where the inter-user interference is completely suppressed. In doing so, a generic receiver k observes

$$r_k = u_k + n_k,$$ (9.41)

which is equivalent to an AWGN channel. The sum throughput of the ZF precoding system is thus given by

$$R_{\mathrm{ZF}} = \sum_{k=1}^{K} \log \left(1 + \frac{P_k}{\sigma_n^2} \right).$$ (9.42)

Analytical and numerical results reveal that the simplicity of the ZF precoding comes at the price of a non-negligible loss in terms of sum rate with respect to the optimal DPC technique especially when $K \leqslant N_b$. The main reason for this penalty is essentially due to the power boosting effect, which occurs in the pseudo-inverse computation of ill-conditioned channel matrices.

An approach to improve the performance of the ZF precoding is to take advantage of multi-user diversity by applying user selection when $K \gg N_b$. We write $\mathcal{A} \subset \{1, 2, \ldots, K\}$ to denote a set of the selected users. Different selections result in different sum rate, and therefore the maximal rate of the system is obtained by considering all the possible sets, i.e.

$$R_{\max} = \max_{\mathcal{A} \subset \{1,2,\ldots,K\}} R_{ZF}. \tag{9.43}$$

A greedy user-selection ZF algorithm is depicted as follows [Sanguinetti and Poor, 2009]:

1. Initialization:
   Set $n = 1$ and find the best user such that

   $$k_1 = \arg \max_{k=1,2,\ldots,K} \left(\|\mathbf{h}_k\|^2 \right) \tag{9.44}$$

   where $\mathbf{h}_k$ is the kth row of $\mathbf{H}$.
   Set $\mathcal{A}_1 = \{k_1\}$, and denote its achieved sum rate $R_{\max}(\mathcal{A}_1)$.
2. While $n \leqslant K$
   Find a user k_n from the unselected users such that

   $$k_n = \arg \max_{k \in \{1,2,\ldots,K\} - \mathcal{A}_{n-1}} R_{\max}(\mathcal{A}_{n-1} \cup \{k\}). \tag{9.45}$$

   Set $\mathcal{A}_n = \mathcal{A}_{n-1} + \{k_n\}$ and denote the achieved rate by $R_{\max}(\mathcal{A}_n)$.
   If $R_{\max}(\mathcal{A}_n) < R_{\max}(\mathcal{A}_{n-1})$ stop and set $n = n - 1$;
3. Determine the precoding matrix $\mathbf{W}$ for the set of selected users.

9.1.5 Block Diagonalization

In the case of multi-antenna receivers, the ZF precoding can be applied straightforward as long as treating the multiple receive antennas of each user as individual single-antenna receivers without cooperation. However, albeit a simple receiver architecture, it does not allow exploitation of the cooperation gain of multiple receive antennas in the signal detection processing. One approach to overcoming this drawback is represented by the BD scheme proposed independently by Choi and Murch [2004] and Spencer et al. [2004]. Its principle is that the transmitter completely suppresses the inter-user interference while each receiver mitigates the inter-stream interference among its respective data streams.

Using a transmit precoding technique based on a decomposition approach at the base station, a multi-user MIMO broadcast channel is transformed into multiple parallel SU-MIMO channels. Each equivalent SU-MIMO channel has the same properties as a conventional SU-MIMO channel. Therefore, any SU-MIMO technique, such as Vertical Bell Laboratories Layer Space-Time (V-BLAST), maximum likelihood detection, linear detection (e.g. ZF, MF, and MMSE), and singular value decomposition-based precoding, can be applied for each user of the multi-user MIMO system. Meanwhile, increasing the number of transmit antennas of the multi-user system by one increases the number of spatial channels to each user by one.

According to Eq. (9.14), such a system can be modeled into

$$
\begin{aligned}
\mathbf{r}_k &= \mathbf{H}^{(k)} \sum_{k=1}^{K} \mathbf{B}_k \mathbf{u}_k + \mathbf{n}_k \\
&= \underbrace{\mathbf{H}^{(k)} \mathbf{B}_k \mathbf{u}_k}_{\text{Desired signal}} + \underbrace{\mathbf{H}^{(k')} \sum_{k'=1,k'\neq k}^{K} \mathbf{B}_{k'} \mathbf{u}_{k'} + \mathbf{n}_k}_{\text{Multi-user Interference}}.
\end{aligned}
\tag{9.46}
$$

The second item represents the interference to user k due to other $K-1$ users, and therefore the primary objective is to completely null this interference. This objective can be mathematically depicted as

$$
\mathbf{H}^{(k')} \sum_{k'=1,k'\neq k}^{K} \mathbf{B}_{k'} \mathbf{u}_{k'} = \mathbf{0},
\tag{9.47}
$$

when $\mathbf{B}_{k'} \neq \mathbf{0}$. It is equivalent to

$$
\mathbf{H}^{(k')} \sum_{k'=1,k'\neq k}^{K} \mathbf{B}_{k'} = \mathbf{0},
\tag{9.48}
$$

since $\mathbf{u}_{k'} \neq \mathbf{0}$.

As presented in Choi and Murch [2004], this problem can be solved through the Singular Value Decomposition (SVD) of a sub-matrix of $\mathbf{H}$. That is

$$
\tilde{\mathbf{H}}_k =
\begin{bmatrix}
\mathbf{H}^{(1)} \\
\vdots \\
\mathbf{H}^{(k-1)} \\
\mathbf{H}^{(k+1)} \\
\vdots \\
\mathbf{H}^{(K)}
\end{bmatrix}
= \mathbf{U}_k \mathbf{\Sigma} \mathbf{V}_k^H = \mathbf{U}_k
\begin{bmatrix}
\mathbf{\Sigma}' & \mathbf{0} \\
\mathbf{0} & \mathbf{0}
\end{bmatrix}
\begin{bmatrix}
\mathbf{V}_k^{\varnothing} & \mathbf{V}_k^0
\end{bmatrix}^H,
\tag{9.49}
$$

where $\tilde{\mathbf{H}}_k \in \mathbb{C}^{\tilde{N}_k \times N_b}$ with

$$
\tilde{N}_k = \sum_{k'=1,k'\neq k}^{K} N_k
$$

is a sub-matrix of $\mathbf{H}$ removing the rows for user k. The number of zero columns $\mathbf{0}$ at the least columns of the diagonal matrix is $\tilde{N}_x$, satisfying $\tilde{N}_x \geq N_b - \tilde{N}_k$. The matrix $\mathbf{V}_k^0 \in \mathbb{C}^{N_b \times \tilde{N}_x}$ corresponds to these zero columns. Then, the precoding matrix for user k is

$$\mathbf{B}_k = \mathbf{V}_k^0 \mathbf{A}_k, \tag{9.50}$$

where $\mathbf{A}_k$ is a nonzero $\tilde{N}_k \times L_k$ matrix, which can be designed alone by some criteria or can be jointly designed with the structure of the receiver.

To provide an insight into this technique, we provide a concrete example to demonstrate the process of achieving BD. For a $([2, 1], 4)$ MU-MIMO system consisting of a four-antenna base station and two users, where user 1 has 2 antennas and user 1 has a single antenna. A channel realization is

$$\mathbf{H} = \begin{bmatrix} 2.0563 - 1.1151i & -0.7482 + 0.0237i & 0.7767 + 0.2476i & -1.4509 - 0.1853i \\ 0.5835 + 0.3592i & -0.3314 - 0.9431i & -0.1965 - 0.2115i & -0.2502 - 1.2376i \\ 0.9751 + 0.1994i & -0.1927 + 0.7973i & 0.4961 + 0.0162i & -0.5824 - 0.2020i \end{bmatrix}. \tag{9.51}$$

Applying the SVD over

$$\tilde{\mathbf{H}}_1 = \begin{bmatrix} 0.9751 + 0.1994i & -0.1927 + 0.7973i & 0.4961 + 0.0162i & -0.5824 - 0.2020i \end{bmatrix}, \tag{9.52}$$

yields

$$\mathbf{V}_1 = \begin{bmatrix} -0.6444 + 0.1318i & 0.0418 - 0.5404i & -0.3254 + 0.0416i & 0.4012 + 0.0705i \\ 0.1273 + 0.5269i & 0.8225 + 0.0142i & 0.0302 + 0.1034i & 0.0022 - 0.1338i \\ -0.3278 + 0.0107i & 0.0134 - 0.1069i & 0.9350 + 0.0052i & 0.0790 + 0.0178i \\ 0.3849 - 0.1335i & 0.0235 + 0.1318i & 0.0752 - 0.0301i & 0.8997 + 0.0080i \end{bmatrix}, \tag{9.53}$$

and then derives out

$$\mathbf{V}_1^0 = \begin{bmatrix} 0.0418 - 0.5404i & -0.3254 + 0.0416i & 0.4012 + 0.0705i \\ 0.8225 + 0.0142i & 0.0302 + 0.1034i & 0.0022 - 0.1338i \\ 0.0134 - 0.1069i & 0.9350 + 0.0052i & 0.0790 + 0.0178i \\ 0.0235 + 0.1318i & 0.0752 - 0.0301i & 0.8997 + 0.0080i \end{bmatrix}. \tag{9.54}$$

Similarly, we can figure out

$$\mathbf{V}_2^0 = \begin{bmatrix} -0.3605 - 0.2350i & 0.2311 + 0.1576i \\ -0.4085 - 0.1008i & -0.5849 - 0.3009i \\ 0.7751 + 0.0742i & -0.1731 + 0.0567i \\ -0.0554 + 0.1684i & 0.6666 + 0.1070i \end{bmatrix}. \tag{9.55}$$

Letting $\mathbf{V}^0 = [\mathbf{V}_1^0, \mathbf{V}_2^0]$, we have

$$\mathbf{HV}^0 = \begin{bmatrix} -1.11 - 1.42i & -0.04 + 0.64i & -0.34 - 0.35i & 0 & 0 \\ 0.09 - 1.12i & -0.36 - 0.46i & -0.15 - 0.91i & 0 & 0 \\ 0 & 0 & 0 & 0.30 - 0.64i & 0.09 - 0.38i \end{bmatrix},$$

(9.56)

exhibiting that the channel matrix is successfully block diagonalized. Along with $\mathbf{A}_1 \in \mathbb{C}^{3 \times L_1}$, where $L_1 = 1$ or 2, and $\mathbf{A}_2 \in \mathbb{C}^{2 \times 1}$, two parallel SU-MIMO channels are formed. The solution provided by Choi and Murch [2004] is not the unique method to achieve BD, there exists other ways such as the solution presented in Chen et al. [2007].

9.2 Massive MIMO

From a cellular network perspective, a base station needs to simultaneously support a reasonable number of active terminals. SU-MIMO is a point-to-point MIMO system where multiple transmit and receive antennas are dedicated to a single user for spatial multiplexing. However, it does not mean the system has only one user. Instead, different users are accommodated in orthogonal time-frequency resource units employing *time-division multiplexing* and *frequency-division multiplexing*. In theory, its channel capacity grows linearly by simultaneously increasing the number of transmit antennae and the number of receive antennae. However, SU-MIMO is not scalable to high-order spatial multiplexing due to three practical factors. First, it is challenging to embed too many antennas at a terminal and employ advanced signal processing algorithms to separate high-dimensional data streams due to the constraints of hardware size, power supply, and equipment cost. Second, a compact antenna array is hard to support a large number of independent sub-channels in a point-to-point link, even in a rich-scattering environment. Particularly, the channel matrix has the minimal rank of one in line-of-sight conditions. Third, the channel capacity scales slowly at the low SNR regime, e.g. at the cell edge, where most terminals usually locate, exhibiting high path loss and strong inter-cell interference [Marzetta, 2015].

By breaking up spatial-multiplexed streams among multiple terminals, MU-MIMO gains superiority over SU-MIMO with two fundamental advantages. First, MU-MIMO requires only single-antenna terminals, facilitating the use of low-complexity, low-cost, power-saving equipment. Second, it is less vulnerable to the propagation environment due to the spatial distribution of terminals. It can function well even under line-of-sight conditions if the typical angular separation among terminals is greater than the angular resolution of the base station array.

Nevertheless, the conventional MU-MIMO is still hard to scale up for high-order spatial multiplexing since the capacity-achieving precoding and decoding impose exponentially growing complexity. Most seriously, the transmitter requires the knowledge of the downlink channel, and the resources spent acquiring CSI rise with the number of service antennas and the number of users.

Massive MIMO proposed by Marzetta [2010] breaks this scalability barrier by not attempting to achieve the full Shannon limit and paradoxically by increasing the size of the system. It departs from Shannon-theoretic practice in three ways:

- Only the base station learns the channel knowledge for precoding in the downlink and decoding in the uplink, whereas the terminals do not need to know it. Taking advantage of the channel reciprocity of a TDD system, the overhead required to acquire CSI depends on the number of terminals, irrespective of the number of base station antennas.
- The base station side installs a large-scale antenna array so that the number of service antennas is typically increased to several times the number of active users. The scale of users keeps small such that the implementation complexity is low.
- A simple linear precoding multiplexing is employed in the downlink, coupled with linear decoding in the uplink. As the number of base station antennas increases, the performance of linear precoding and decoding can approach the Shannon limit.

9.2.1 CSI Acquisition

We can use a *coherence block* to define a time-frequency plane during which the channel is regarded as time-invariant and frequency-flat. In the temporal domain, the duration of a coherence block equals the channel coherence time T_c, and its width in the frequency domain is the same as the channel coherence bandwidth B_c. The number of time-frequency resource units is $\tau_c = T_c B_c$, which can be used for carrying τ_c complex-valued symbols. Massive MIMO relies on measuring the actual responses of propagation channels. To this end, a unique reference signal is assigned to each terminal (per coherence block), and these reference signals need to be mutually orthogonal. Without loss of generality, we only focus on a single coherence block, where the signal transmission is divided into three phases: uplink data transmission, uplink training, and downlink data transmission. The first part of this subsection will study how to acquire CSI in a massive MIMO system.

A single-cell massive MIMO system is generally comprised of a base station with M antennas and K single-antenna terminals, where $M \gg K$. A typical terminal k,

$k = 1, 2, \ldots, K$ is assigned to a reference signal with length τ_p, denoted by a vector $\boldsymbol{\phi}_k \in \mathbb{C}^{\tau_p \times 1}$, where $\tau_c \geq \tau_p \geq K$. In order to form K orthogonal reference signals, the condition

$$\boldsymbol{\Phi}^H \boldsymbol{\Phi} = \mathbf{I}_K \tag{9.57}$$

should be satisfied, where $\boldsymbol{\Phi} \in \mathbb{C}^{\tau_p \times K}$ is given by

$$\boldsymbol{\Phi} = [\boldsymbol{\phi}_1, \boldsymbol{\phi}_2, \ldots, \boldsymbol{\phi}_K]. \tag{9.58}$$

These terminals simultaneously transmit their reference signals over τ_p time-frequency resource units. The transmitted signals can be denoted by

$$\mathbf{X}_p = \sqrt{p_u \tau_p} \boldsymbol{\Phi}^H, \tag{9.59}$$

where the normalization is applied so that each terminal expends a total power that equals to the length of reference signals, and p_u denotes the uplink power constraint. The reference signals are always transmitted at maximum possible power without power control.

Then, the base station observes the $M \times \tau_p$ received symbols

$$\mathbf{Y}_p = \mathbf{G}_u \mathbf{X}_p + \mathbf{Z}_p, \tag{9.60}$$

where $\mathbf{Z}_p$ corresponds to $M \times \tau_p$ independent complex Gaussian noise with each entry $z \sim \mathcal{CN}(0, \sigma_n^2)$, $\mathbf{G}_u \in \mathbb{C}^{M \times K}$ models the uplink channel matrix from K terminals to M antennas of the base station. The complex-valued channel coefficient between terminal k and antenna m is denoted by

$$g_{mk} = \sqrt{\beta_{mk}} h_{mk} \tag{9.61}$$

with the large-scale fading coefficient β_{mk}, and the small-scale fading gain, which is generally *i.i.d.* Rayleigh faded, i.e. $h_{mk} \sim \mathcal{CN}(0, 1)$. A reasonable assumption is that β_{mk} is known (by measurement, for example) by the system. We can further assume that all β_{mk} for a typical terminal is the same since the large-scale fading depends on the propagation distance and shadowing, resulting in $\beta_{mk} = \beta_k, \forall m = 1, 2, \ldots, M$. Then, we get the prior distribution of $g_{mk} \sim \mathcal{CN}(0, \beta_k)$. Equation (9.61) is further simplified to

$$g_{mk} = \sqrt{\beta_k} h_{mk}. \tag{9.62}$$

The base station decorrelates the received signals with the known reference signals

$$\begin{aligned} \tilde{\mathbf{Y}}_p = \mathbf{Y}_p \boldsymbol{\Phi} &= \mathbf{G}_u \mathbf{X}_p \boldsymbol{\Phi} + \mathbf{Z}_p \boldsymbol{\Phi} \\ &= \sqrt{p_u \tau_p} \mathbf{G}_u \boldsymbol{\Phi}^H \boldsymbol{\Phi} + \mathbf{Z}_p \boldsymbol{\Phi} \\ &= \sqrt{p_u \tau_p} \mathbf{G}_u + \tilde{\mathbf{Z}}_p, \end{aligned} \tag{9.63}$$

where each entry of $\tilde{\mathbf{Z}}_p \in \mathbb{C}^{M \times K}$ is also independent complex Gaussian noise $\tilde{z} \sim \mathcal{CN}(0, \sigma_n^2)$ due to the multiplication with a unitary matrix [Marzetta et al., 2016]. Due to the independence of g and z, Eq. (9.63) can be decomposed into

$$\tilde{y}_{mk,p} = \sqrt{p_u \tau_p} g_{mk} + \tilde{z}_{mk}. \tag{9.64}$$

Conducting channel estimation with linear MMSE, the estimate [Tse and Viswanath, 2005] is obtained by

$$\hat{g}_{mk} = \mathbb{E}\left[g_{mk} | \tilde{y}_{mk,p}\right] = \frac{\mathbb{E}\left[\tilde{y}_{mk,p}^* g_{mk}\right] \tilde{y}_{mk,p}}{\mathbb{E}\left[\left|\tilde{y}_{mk,p}\right|^2\right]} = \left(\frac{\sqrt{p_u \tau_p} \beta_k}{p_u \tau_p \beta_k + \sigma_n^2}\right) \tilde{y}_{mk,p}.$$

Let $\hat{g}_{mk}$ be an estimate of g_{mk} and $\tilde{g}_{mk}$ be the estimation error raised by additive noise, we have

$$\hat{g}_{mk} = g_{mk} - \tilde{g}_{mk}. \tag{9.65}$$

The variance of $\hat{g}_{mk}$ is computed by

$$\mathbb{E}\left[\left|\hat{g}_{mk}\right|^2\right] = \frac{p_u \tau_p \beta_k^2}{p_u \tau_p \beta_k + \sigma_n^2}. \tag{9.66}$$

Then, we can write $\hat{g}_{mk} \sim \mathcal{CN}(0, \alpha_k)$ with $\alpha_k = \frac{p_u \tau_p \beta_k^2}{p_u \tau_p \beta_k + \sigma_n^2}$, and the MSE is thus

$$\mathbb{E}\left[\left|\tilde{g}_{mk}\right|^2\right] = \beta_k - \alpha_k = \frac{\sigma_n^2 \beta_k}{p_u \tau_p \beta_k + \sigma_n^2}. \tag{9.67}$$

9.2.2 Linear Detection in Uplink

In the uplink, K terminals simultaneously transmit their respective symbols toward the base station. There is no explicit cooperation among the terminals for performing joint precoding. The only thing these terminals can do is to weight their respective symbols independently. We write η_k to denote the power-control coefficient for a typical terminal k, satisfying $0 \leqslant \eta_k \leqslant 1$. The transmitted symbols $u_k, k = 1, 2, \ldots, K$ are uncorrelated with the power constraint p_u.

Consequently, the covariance matrix for the vector of the transmitted symbols $\mathbf{u} = [u_1, u_2, \ldots, u_K]^T$ is

$$\mathbb{E}\left[\mathbf{u}\mathbf{u}^H\right] = p_u \mathbf{I}_K. \tag{9.68}$$

The transmitted symbol for a typical terminal k is $\sqrt{\eta_k} u_k$. Thus, the base station observes a $M \times 1$ vector of the received symbols

$$\mathbf{r} = \mathbf{G}_u \mathbf{D}_\eta \mathbf{u} + \mathbf{n} \tag{9.69}$$

applying a diagonal matrix $\mathbf{D}_\eta \in \mathbb{C}^{K \times K}$ formed by power-control coefficients η_k, $k = 1, 2, \ldots, K$, namely

$$\mathbf{D}_\eta = \begin{bmatrix} \sqrt{\eta_1} & 0 & \cdots & 0 \\ 0 & \sqrt{\eta_2} & \cdots & 0 \\ \vdots & \vdots & \ddots & \vdots \\ 0 & 0 & \cdots & \sqrt{\eta_K} \end{bmatrix}. \tag{9.70}$$

The base station performs detection processing in terms of the observation and channel knowledge to recover the transmitted symbols. This process can be mathematically denoted by

$$\hat{\mathbf{u}} = f(\mathbf{r}, \mathbf{G}_u), \tag{9.71}$$

where we assume the channel knowledge at the base station is perfect (the impact of channel estimation error on performance can be found in the literature such as [Marzetta et al., 2016]). Linear detection is attractive from the practical perspective due to its low complexity while achieving good performance. Three linear algorithms are typically applied for the uplink detection of a massive MIMO system.

9.2.2.1 Matched Filtering

The philosophy behind the matched filtering, as also known as maximum-ratio combining, is to amplify the desired signal as much as possible, whereas disregarding the inter-user interference. For the case of single-user transmission, it would be optimal. The decoding matrix can be given by $\mathbf{G}_u^H$, resulting in the post-processing output of

$$\tilde{\mathbf{u}} = \mathbf{G}_u^H \mathbf{r} = \mathbf{G}_u^H \mathbf{G}_u \mathbf{D}_\eta \mathbf{u} + \mathbf{G}_u^H \mathbf{n}.$$

Decomposing this equation yields the kth soft estimate

$$\tilde{u}_k = \underbrace{\|\mathbf{g}_k\|^2 \sqrt{\eta_k} u_k}_{\text{Desired signal}} + \underbrace{\sum_{i=1, i \neq k}^{K} \mathbf{g}_k^H \mathbf{g}_i \sqrt{\eta_i} u_i}_{\text{Inter-user interference}} + \underbrace{\mathbf{g}_k^H \mathbf{n}}_{\text{Noise}}, \tag{9.72}$$

where $\mathbf{g}_k \in \mathbb{C}^{M \times 1}$ is the kth column of $\mathbf{G}_u$, or $\mathbf{G}_u = [\mathbf{g}_1, \mathbf{g}_2, \ldots, \mathbf{g}_K]$. Treating the inter-user interference as a colored noise, the base station can get a hard estimate $\hat{u}_k$ for each transmitted symbol u_k.

9.2.2.2 ZF Detection

Instead of maximizing the strength of the desired signal, better performance can be achieved by cancelling the inter-user interference completely. The decoding matrix is the pseudo inverse of the channel matrix, i.e. $(\mathbf{G}_u^H \mathbf{G}_u)^{-1} \mathbf{G}_u^H$. Thus, the output of post-processing is

$$\begin{aligned} \tilde{\mathbf{u}} &= (\mathbf{G}_u^H \mathbf{G}_u)^{-1} \mathbf{G}_u^H \mathbf{r} \\ &= (\mathbf{G}_u^H \mathbf{G}_u)^{-1} \mathbf{G}_u^H \mathbf{G}_u \mathbf{D}_\eta \mathbf{u} + (\mathbf{G}_u^H \mathbf{G}_u)^{-1} \mathbf{G}_u^H \mathbf{n} \\ &= \mathbf{D}_\eta \mathbf{u} + (\mathbf{G}_u^H \mathbf{G}_u)^{-1} \mathbf{G}_u^H \mathbf{n}. \end{aligned} \tag{9.73}$$

Similarly, decomposing the aforementioned equation yields the kth soft estimate

$$\tilde{u}_k = \underbrace{\sqrt{\eta_k}u_k}_{\text{Desired signal}} + \underbrace{\mathfrak{g}_k\mathbf{n}}_{\text{Noise}}, \tag{9.74}$$

where $\mathfrak{g}_k \in \mathbb{C}^{1 \times M}$ denotes the kth row of $(\mathbf{G}_u^H \mathbf{G}_u)^{-1} \mathbf{G}_u^H$. The inter-user interference is now completely eliminated, but it is possible that the noise is amplified if the decoding matrix is ill-conditioned.

9.2.2.3 MMSE Detection

To mitigate the effect of noise amplification in zero forcing, there is a regularized version with the decoding matrix $(\mathbf{G}_u^H \mathbf{G}_u + \sigma_n^2 \mathbf{I})^{-1} \mathbf{G}_u^H$. It can minimize the MSE of the estimated symbols, and therefore is called MMSE detection or regularized ZF detection. The output of post-processing becomes

$$\begin{aligned} \tilde{\mathbf{u}} &= (\mathbf{G}_u^H \mathbf{G}_u + \sigma_n^2 \mathbf{I})^{-1} \mathbf{G}_u^H \mathbf{r} \\ &= (\mathbf{G}_u^H \mathbf{G}_u + \sigma_n^2 \mathbf{I})^{-1} \mathbf{G}_u^H \mathbf{G}_u \mathbf{u} + (\mathbf{G}_u^H \mathbf{G}_u + \sigma_n^2 \mathbf{I})^{-1} \mathbf{G}_u^H \mathbf{n}. \end{aligned} \tag{9.75}$$

At the low SNR regime $\sigma_n^2 \to 0$, the regularized ZF detection achieves a comparable performance with the ZF detection. At the high SNR regime $\sigma_n^2 \gg 0$, it behaves similar to the matched filtering. Therefore, the regularized ZF detection performs well in the whole range of SNRs.

9.2.3 Linear Precoding in Downlink

In the downlink of a massive MIMO system, the base station spatially multiplexes the information-bearing symbols intended for K terminals, denoted by $\mathbf{u} = [u_1, u_2, \ldots, u_K]^T$, through precoding or transmit beamforming. Then, it sends the spatially multiplexed signals over the same time-frequency resource unit. A significant distinction with the uplink transmission is that joint processing can be performed among M transmit antennas. We write η_k to denote the power-control coefficient for the kth information symbol, and η_k, $k = 1, 2, \ldots, K$ are also jointly determined, subject to

$$\sum_{k=1}^{K} \eta_k \leqslant 1. \tag{9.76}$$

Then, the vector of the transmitted symbols $\mathbf{s} = [s_1, s_2, \ldots, s_M]^T$ can be formed by

$$\mathbf{s} = \mathbf{P} \mathbf{D}_\eta \mathbf{u}, \tag{9.77}$$

where $\mathbf{P}$ denotes a $M \times K$ precoding matrix, and $\mathbf{D}_\eta$ is specified in Eq. (9.70). The choice of non-negative power-control coefficients and the scaling of the precoding matrix ensure that the total transmit power satisfies

$$\mathbb{E}\left[\mathbf{s}^H \mathbf{s}\right] = tr\left(\mathbb{E}\left[\mathbf{s}\mathbf{s}^H\right]\right) \leqslant P. \tag{9.78}$$

Collectively, the $K \times 1$ vector of the received symbols for all terminals is given by

$$\mathbf{r} = \mathbf{G}_d \mathbf{s} + \mathbf{n} = \mathbf{G}_u^T \mathbf{s} + \mathbf{n}, \tag{9.79}$$

where $\mathbf{G}_d \in \mathbb{C}^{K \times M}$ denotes the channel matrix from the base station to the terminals, and we assume $\mathbf{G}_d = \mathbf{G}_u^T$ due to the channel reciprocity in a TDD system. Similar to the linear detection, there are three typical linear precoding methods, i.e. maximum-ratio precoding or also called conjugate beamforming, ZF precoding, and regularized ZF precoding.

9.2.3.1 Conjugate Beamforming

To maximize the array gain of the transmission, the precoding matrix is given by $\mathbf{P}_{cb} = \alpha_{cb} \mathbf{G}_d^H$ with the normalizing scalar α_{cb}. It results in a received symbol vector

$$\mathbf{r} = \mathbf{G}_d \mathbf{s} + \mathbf{n} = \mathbf{G}_d \mathbf{P}_{cb} \mathbf{D}_\eta \mathbf{u} + \mathbf{n} = \alpha_{cb} \mathbf{G}_d \mathbf{G}_d^H \mathbf{D}_\eta \mathbf{u} + \mathbf{n}. \tag{9.80}$$

Equivalently, the kth terminal has the observation of

$$r_k = \underbrace{\alpha_{cb} \|\mathbf{g}_k\|^2 \sqrt{\eta_k} u_k}_{\text{Desired signal}} + \underbrace{\alpha_{cb} \sum_{i=1, i \neq k}^{K} \mathbf{g}_k \mathbf{g}_i^H \sqrt{\eta_i} u_i}_{\text{Inter-user interference}} + \underbrace{\mathbf{n}}_{\text{Noise}}. \tag{9.81}$$

where $\mathbf{g}_k \in \mathbb{C}^{1 \times M}$ is the kth row of $\mathbf{G}_d$. Compared with Eq. (9.72), it is known that conjugate beamforming can form a received signal equaling to a soft estimate after post-processing [Yang and Marzetta, 2013]. Hence, the signal detection of the terminal side is simplified.

9.2.3.2 ZF Precoding

Through zero-forcing precoding at the transmitter, the inter-user interference of the received signals can be completely suppressed. The precoding matrix is the pseudo inverse of the channel matrix, i.e. $\mathbf{P}_{ZF} = \alpha_{ZF} \mathbf{G}_d^H (\mathbf{G}_d \mathbf{G}_d^H)^{-1}$. The received symbol vector then becomes

$$\begin{aligned} \mathbf{r} = \mathbf{G}_d \mathbf{s} + \mathbf{n} &= \mathbf{G}_d \mathbf{P}_{ZF} \mathbf{D}_\eta \mathbf{u} + \mathbf{n} \\ &= \alpha_{ZF} \mathbf{G}_d \mathbf{G}_d^H (\mathbf{G}_d \mathbf{G}_d^H)^{-1} \mathbf{D}_\eta \mathbf{u} + \mathbf{n} \\ &= \alpha_{ZF} \mathbf{D}_\eta \mathbf{u} + \mathbf{n}. \end{aligned} \tag{9.82}$$

Accordingly, the observation of the kth terminal is expressed by

$$r_k = \underbrace{\alpha_{ZF} \sqrt{\eta_k} u_k}_{\text{Desired signal}} + \underbrace{n_k}_{\text{Noise}}, \tag{9.83}$$

which is equivalent to an AWGN channel

$$\tilde{r}_k = u_k + \tilde{n}_k \tag{9.84}$$

for the reason that the factor $\alpha_{ZF} \sqrt{\eta_k}$ is deterministic and easy to know.

9.2.3.3 Regularized ZF Precoding

It is possible to form a linear combination of the ZF precoding and conjugate beamforming by means of regularization. A diagonal loading factor is added prior to the inversion of the matrix $\mathbf{G}_d \mathbf{G}_d^H$. In consequence, the precoding matrix becomes

$$\mathbf{P}_{\mathrm{rZF}} = \alpha_{\mathrm{rZF}} \mathbf{G}_d^H (\mathbf{G}_d \mathbf{G}_d^H + \delta \mathbf{I})^{-1}. \tag{9.85}$$

Then, the transmitted symbol vector is

$$\mathbf{s} = \alpha_{\mathrm{rZF}} \mathbf{G}_d^H (\mathbf{G}_d \mathbf{G}_d^H + \delta \mathbf{I})^{-1} \mathbf{D}_\eta \mathbf{u}, \tag{9.86}$$

where $\delta > 0$ is the regularization factor, and can be optimized based on the design requirements. A regularized ZF precoding becomes the ZF precoding as $\delta \to 0$, and becomes conjugate beamforming when $\delta \gg 0$.

9.3 Multi-Cell Massive MIMO

From the perspective of a cellular network, the downlink and uplink transmission, especially the uplink training, of a cell are affected by its neighboring cells. Consider a cellular system with a network of non-overlapping cells. Two adjacent cells are generally assigned to orthogonal frequency bands to eliminate inter-cell interference. Figure 9.3 demonstrates a network consisting of hexagonal cells with frequency reuse factor of seven. Suppose a total of L cells, indexed by $l = 1, 2, \ldots, L$, share the same frequency band, referring to as co-channel cells, whereas ignoring other co-channel cells with negligible mutual interference due to significant separation distance. Each cell consists of one base station with M antennas and K single-antenna users.

We write $g_{mk}^{l\mu}$ to model the channel between the mth service antenna at the base station of cell l and the kth user of cell μ, where our focus is on cell l, and μ denotes one of its co-channel cells nearby. A general channel gain is

$$g_{mk}^{l\mu} = \sqrt{\beta_{mk}^{l\mu}} \, h_{mk}^{l\mu}, \tag{9.87}$$

where large-scale fading coefficient $\beta_{mk}^{l\mu}$ is a non-negative constant and assumed to be known to everybody, and small-scale fading gain $h_{mk}^{l\mu}$ is $i.i.d.$ zero-mean, circularly symmetric complex Gaussian random variables, i.e. $h_{mk}^{l\mu} \in \mathcal{CN}(0, 1)$. The $\beta_{mk}^{l\mu}$ values model path-loss and shadowing that change slowly and can be learned over long period of time, while the $h_{mk}^{l\mu}$ values model fading that change relatively fast and must be learned and used very quickly. Since the cell layout and shadowing are captured using the constant $\beta_{mk}^{l\mu}$ values, the specific details of the cell layout and shadowing model are irrelevant. In a TDD system, we assume channel reciprocity for the forward and reverse links, i.e. $g_{mk}^{l\mu} = g_{km}^{l\mu}$, and block fading, namely, $h_{mk}^{l\mu}$ remains constant for a number of symbol periods.

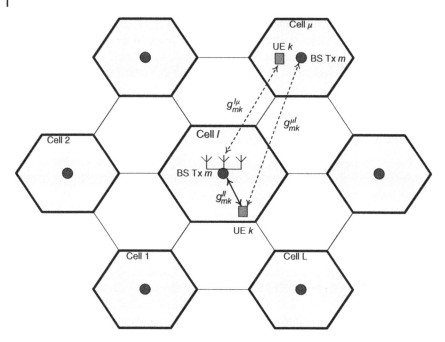

Figure 9.3 The model of a multi-cell massive MIMO system, where cell l is centered by $L - 1$ co-channel cells indexed by $\mu = 1, \ldots, L$ and $\mu \neq l$ that utilize the same frequency band with a frequency-reuse factor of seven. Denote by $g_{mk}^{l\mu}$ the channel gain between the mth service antenna of cell l and the kth user of cell μ, while $g_{mk}^{\mu l}$ points to the channel gain between the mth service antenna of cell μ and the kth user of cell l.

The small-scale fading coefficients are usually different for different user-antenna pairs due to multi-path signals' constructive and destructive addition. In contrast, the large-scale fading coefficients are identical for all antennas at the same base station but only user-dependent because it is related to propagation distance and shadowing. Thus, $\beta_{mk}^{l\mu}$ in Eq. (9.87) can be replaced with $\beta_{k}^{l\mu}$, namely

$$g_{mk}^{l\mu} = \sqrt{\beta_{k}^{l\mu}} h_{mk}^{l\mu}. \tag{9.88}$$

The uplink channel matrix from all K users in the μth cell to the base station of the lth cell can be expressed as

$$\mathbf{G}^{l\mu} = \begin{pmatrix} g_{11}^{l\mu} & g_{12}^{l\mu} & \cdots & g_{1K}^{l\mu} \\ g_{21}^{l\mu} & g_{22}^{l\mu} & \cdots & g_{2K}^{l\mu} \\ \vdots & \vdots & \ddots & \vdots \\ g_{M1}^{l\mu} & g_{M2}^{l\mu} & \cdots & g_{MK}^{l\mu} \end{pmatrix} = \mathbf{H}^{l\mu} \left(\mathbf{B}^{l\mu} \right)^{1/2}, \tag{9.89}$$

where

$$\mathbf{H}^{l\mu} = \begin{pmatrix} h_{11}^{l\mu} & h_{12}^{l\mu} & \cdots & h_{1K}^{l\mu} \\ h_{21}^{l\mu} & h_{22}^{l\mu} & \cdots & h_{2K}^{l\mu} \\ \vdots & \vdots & \ddots & \vdots \\ h_{M1}^{l\mu} & h_{M2}^{l\mu} & \cdots & h_{MK}^{l\mu} \end{pmatrix}, \tag{9.90}$$

and

$$\mathbf{B}^{l\mu} = \begin{pmatrix} \beta_1^{l\mu} & & & \\ & \beta_2^{l\mu} & & \\ & & \ddots & \\ & & & \beta_K^{l\mu} \end{pmatrix}. \tag{9.91}$$

The downlink channel matrix from the base station of the lth cell to all K users in the μth cell is the transpose of $\mathbf{G}^{l\mu}$, i.e.

$$\left(\mathbf{G}^{l\mu}\right)^T = \begin{pmatrix} g_{11}^{l\mu} & g_{21}^{l\mu} & \cdots & g_{M1}^{l\mu} \\ g_{12}^{l\mu} & g_{22}^{l\mu} & \cdots & g_{M2}^{l\mu} \\ \vdots & \vdots & \ddots & \vdots \\ g_{1K}^{l\mu} & g_{2K}^{l\mu} & \cdots & g_{MK}^{l\mu} \end{pmatrix}. \tag{9.92}$$

It is noted that the uplink channel matrix from all K users in the lth cell to the base station of the μth cell is expressed as

$$\mathbf{G}^{\mu l} = \begin{pmatrix} g_{11}^{\mu l} & g_{12}^{\mu l} & \cdots & g_{1K}^{\mu l} \\ g_{21}^{\mu l} & g_{22}^{\mu l} & \cdots & g_{2K}^{\mu l} \\ \vdots & \vdots & \ddots & \vdots \\ g_{M1}^{\mu l} & g_{M2}^{\mu l} & \cdots & g_{MK}^{\mu l} \end{pmatrix}, \tag{9.93}$$

where the first letter of the superscripts is used for the index of the cell where the base station locates, and the second letter of the superscripts is for the index of the cell where the users locate.

9.3.1 Pilot Contamination

In a multi-cell massive MIMO system, a typical terminal k, $k = 1, 2, \ldots, K$ in cell l, $l = 1, 2, \ldots, L$ is assigned to a reference signal $\boldsymbol{\phi}_{k,l} \in \mathbb{C}^{\tau_p \times 1}$. Ideally, the reference signals employed by users within the same cell and in the adjacent co-channel cells are orthogonal, that is

$$\boldsymbol{\phi}_{k,l}^H \boldsymbol{\phi}_{i,\mu} = \delta[k - i]\delta[l - \mu], \tag{9.94}$$

where $\delta[\cdot]$ is the delta function

$$\delta[n] = \begin{cases} 1 & n = 0 \\ 0 & \text{otherwise} \end{cases}. \tag{9.95}$$

In vector form, it satisfies

$$\mathbf{\Phi}_l^H \mathbf{\Phi}_\mu = \delta[l - \mu]\mathbf{I}_K, \tag{9.96}$$

where $\tau_p \geq K$, and $\mathbf{\Phi}_l \in \mathbb{C}^{\tau_p \times K}$ is given by

$$\mathbf{\Phi}_l = [\boldsymbol{\phi}_{1,l}, \boldsymbol{\phi}_{2,l}, \dots, \boldsymbol{\phi}_{K,l}]. \tag{9.97}$$

However, the number of orthogonal reference sequences with a given period and bandwidth is limited, which in turn limits the number of users that can be served. In order to handle more users, non-orthogonal reference sequences are used in neighboring cells. As a result, the estimate of the channel vector to a user becomes correlated with the channel vectors of the users with non-orthogonal reference sequences. There are many schemes for assigning reference sequences to users in different cells. One simple scheme is to reuse the same set of orthogonal reference sequences in all co-channel cells. This means that the kth user in a cell will be assigned the reference sequence $\boldsymbol{\phi}_k$. Identical reference sequences assigned to users in adjacent co-channel cells will interfere with each other, leading to *pilot contamination* [Lu et al., 2014].

Consider a system with L co-channel cells, and the remaining cells operated in other frequency bands are assumed to be ideally isolated. All L cells use the same set of K reference sequences $\mathbf{\Phi} = [\boldsymbol{\phi}_1, \boldsymbol{\phi}_2, \dots, \boldsymbol{\phi}_K]$, and the kth user in each cell is assigned to an identical reference sequence $\boldsymbol{\phi}_k$. The terminals in all L cells simultaneously transmit their reference signals, and assume further that the transmission from different cells is synchronized (from the standpoint of pilot contamination, this constitutes the worst possible case). The transmitted signals can be denoted by

$$\mathbf{X}_p = \sqrt{P_r \tau_p} \mathbf{\Phi}^H, \tag{9.98}$$

where the normalization is applied so that each terminal expends a total power that equals to the length of reference signals, and P_r denotes the power constraint in the reverse link. Then, the base station at cell l observes the $M \times \tau_p$ received symbols

$$\mathbf{Y}_p^l = \sum_{\mu=1}^{L} \mathbf{G}^{l\mu} \mathbf{X}_p + \mathbf{Z}_p, \tag{9.99}$$

where $\mathbf{Z}_p$ corresponds to $M \times \tau_p$ independent complex Gaussian noise with each entry $z \sim \mathcal{CN}(0, \sigma_n^2)$, $\mathbf{G}^{l\mu} \in \mathbb{C}^{M \times K}$ models the uplink channel matrix from K terminals of cell μ to M base station antennas of cell l, as defined in Eq. (9.89).

The base station decorrelates the received signals with the known reference signals

$$
\begin{aligned}
\tilde{\mathbf{Y}}_p^l &= \frac{1}{\sqrt{P_r \tau_p}} \mathbf{Y}_p^l \boldsymbol{\Phi} = \frac{1}{\sqrt{P_f \tau_p}} \sum_{\mu=1}^{L} \mathbf{G}^{l\mu} \mathbf{X}_p \boldsymbol{\Phi} + \frac{1}{\sqrt{P_r \tau_p}} \mathbf{Z}_p \boldsymbol{\Phi} \\
&= \sum_{\mu=1}^{L} \mathbf{G}^{l\mu} \boldsymbol{\Phi}^H \boldsymbol{\Phi} + \frac{1}{\sqrt{P_r \tau_p}} \mathbf{Z}_p \boldsymbol{\Phi} \\
&= \underbrace{\mathbf{G}^{ll}}_{\text{Desired CSI}} + \underbrace{\sum_{\mu=1,\mu\neq l}^{L} \mathbf{G}^{l\mu}}_{\text{Pilot contamination}} + \frac{1}{\sqrt{P_r \tau_p}} \tilde{\mathbf{Z}}_p,
\end{aligned} \tag{9.100}
$$

where each entry of $\tilde{\mathbf{Z}}_p = \mathbf{Z}_p \boldsymbol{\Phi}$ is also independent complex Gaussian noise $\tilde{z} \sim \mathcal{CN}(0, \sigma_n^2)$ due to the multiplication with a unitary matrix. The aforementioned equation can be decomposed into

$$
\tilde{y}_{mk,p}^l = g_{mk}^{ll} + \sum_{\mu=1,\mu\neq l}^{L} g_{mk}^{l\mu} + \frac{1}{\sqrt{P_r \tau_p}} \tilde{z}_{mk}. \tag{9.101}
$$

Conducting channel estimation with linear MMSE, the estimate [Tse and Viswanath, 2005] is obtained by

$$
\begin{aligned}
\hat{g}_{mk}^{ll} = \mathbb{E}\left[g_{mk}^{ll} | \tilde{y}_{mk,p}^l\right] &= \frac{\mathbb{E}\left[\left(\tilde{y}_{mk,p}^l\right)^* g_{mk}^{ll}\right] \tilde{y}_{mk,p}^l}{\mathbb{E}\left[\left|\tilde{y}_{mk,p}^l\right|^2\right]} \\
&= \left(\frac{\beta_k^{ll}}{\sum_{\mu=1}^{L} \beta_k^{l\mu} + \frac{\sigma_n^2}{P_r \tau_p}}\right) \tilde{y}_{mk,p}^l.
\end{aligned} \tag{9.102}
$$

The variance of $\hat{g}_{mk}^{ll}$ is computed by

$$
\mathbb{E}\left[\left|\hat{g}_{mk}^{ll}\right|^2\right] = \frac{P_r \tau_p \left(\beta_k^{ll}\right)^2}{P_r \tau_p \sum_{\mu=1}^{L} \beta_k^{l\mu} + \sigma_n^2} = \alpha_k^{ll}. \tag{9.103}
$$

Then, we can write $\hat{g}_{mk}^{ll} \sim \mathcal{CN}(0, \alpha_k^{ll})$. Let $\tilde{g}_{mk}^{ll}$ be the estimation error raised by additive noise and pilot contamination, we have

$$
\tilde{g}_{mk}^{ll} = g_{mk}^{ll} - \hat{g}_{mk}^{ll}. \tag{9.104}
$$

The MSE of this channel estimate is thus computed by

$$
\mathbb{E}\left[\left|\tilde{g}_{mk}\right|^2\right] = \beta_k^{ll} - \alpha_k^{ll}. \tag{9.105}
$$

Equivalently, Eq. (9.100) can also be expressed as

$$
\begin{aligned}
\tilde{\mathbf{Y}}_p^l &= \mathbf{G}^{ll} + \sum_{\mu=1,\mu\neq l}^{L} \mathbf{G}^{l\mu} + \frac{1}{\sqrt{P_r\tau_p}}\tilde{\mathbf{Z}}_p \\
&= \mathbf{H}^{ll}\left(\mathbf{B}^{ll}\right)^{1/2} + \sum_{\mu=1,\mu\neq l}^{L} \mathbf{H}^{l\mu}\left(\mathbf{B}^{l\mu}\right)^{1/2} + \frac{1}{\sqrt{P_r\tau_p}}\tilde{\mathbf{Z}}_p.
\end{aligned}
\tag{9.106}
$$

Given that $\mathbf{B}^{l\mu}$ is known, the MMSE estimate of $\mathbf{H}^{ll}$ is [Jose et al., 2011]:

$$
\hat{\mathbf{H}}^{ll} = \sqrt{P_r\tau_p}\left(\mathbf{B}^{ll}\right)^{1/2}\left(\sigma_n^2\mathbf{I} + P_r\tau_p\sum_{\mu=1}^{L}\mathbf{B}^{l\mu}\right)^{-1}\tilde{\mathbf{Y}}_p^l.
\tag{9.107}
$$

9.3.2 Uplink Data Transmission

In the uplink of a multi-cell massive MIMO system, the K terminals in each cell independently transmit signals toward their respective base station. Let u_k^μ, $\forall k = 1, 2, \ldots, K$ and $\mu = 1, 2, \ldots, L$ denote a zero-mean, unit-variance symbol from user k in the μth cell. The vector of information-bearing symbols from all K users in cell μ is expressed as $\mathbf{u}_\mu = [u_1^\mu, u_2^\mu, \ldots, u_K^\mu]^T$.

Ignoring power control, the base station in the lth cell receives a $M \times 1$ vector comprising the received signals from all terminals in the L cells, i.e.

$$
\begin{aligned}
\mathbf{r}_l &= \sqrt{P_r}\sum_{\mu=1}^{L}\mathbf{G}^{l\mu}\mathbf{u}_\mu + \mathbf{n}_l \\
&= \underbrace{\sqrt{P_r}\mathbf{G}^{ll}\mathbf{u}_l}_{\text{Desired signal}} + \underbrace{\sqrt{P_r}\sum_{\mu=1,\mu\neq l}^{L}\mathbf{G}^{l\mu}\mathbf{u}_\mu + \mathbf{n}_l}_{\text{Inter-cell interference}}
\end{aligned}
\tag{9.108}
$$

with the reverse-link power constraint P_r, and the uplink channel matrix $\mathbf{G}^{l\mu} \in \mathbb{C}^{M\times K}$ from K users in cell μ to the base station in cell l.

Equivalently, the received signal at the mth base station antenna of cell l is expressed as

$$
\begin{aligned}
r_m^l &= \sqrt{P_r}\sum_{\mu=1}^{L}\sum_{k=1}^{K}g_{mk}^{l\mu}u_k^\mu + n_m^l \\
&= \underbrace{\sqrt{P_r}\sum_{k=1}^{K}g_{mk}^{ll}u_k^l}_{\text{Desired signal}} + \underbrace{\sqrt{P_r}\sum_{\mu=1,\mu\neq l}^{L}\sum_{k=1}^{K}g_{mk}^{l\mu}u_k^\mu + n_m^l}_{\text{Inter-cell interference}}.
\end{aligned}
\tag{9.109}
$$

Suppose we use matched filtering to detect $\mathbf{u}_l$, the lth base station processes the received signal by multiplying it with the conjugate of its estimated CSI $\hat{\mathbf{G}}_{ll}$, see Eq. (9.100), which can be rewritten as

$$
\hat{\mathbf{G}}^{ll} = \sum_{\mu=1}^{L} \mathbf{G}^{l\mu} + \mathbf{W}_l
$$

$$
= \sum_{\mu=1}^{L} \mathbf{H}^{l\mu} \left(\mathbf{B}^{l\mu} \right)^{1/2} + \mathbf{W}_l. \tag{9.110}
$$

It results in

$$
\mathbf{y}_l = \left(\hat{\mathbf{G}}^{ll} \right)^H \mathbf{r}_l
$$

$$
= \left[\sum_{\mu=1}^{L} \mathbf{G}^{l\mu} + \mathbf{W}_l \right]^H \left[\sqrt{P_r} \sum_{\mu'=1}^{L} \mathbf{G}^{l\mu'} \mathbf{u}_{\mu'} + \mathbf{n}_l \right]. \tag{9.111}
$$

According to Marzetta [2010],

$$
\frac{1}{M} \left[\mathbf{G}^{l\mu} \right]^H \mathbf{G}^{l\mu'} = (\mathbf{B}^{l\mu})^{1/2} \left(\frac{\left[\mathbf{H}^{l\mu} \right]^H \mathbf{H}^{l\mu'}}{M} \right) (\mathbf{B}^{l\mu'})^{1/2}. \tag{9.112}
$$

As the number of base station antennas grows without limit $M \to +\infty$, we have

$$
\frac{\left[\mathbf{H}^{l\mu} \right]^H \mathbf{H}^{l\mu'}}{M} \longrightarrow \mathbf{I}_K \delta[\mu - \mu']. \tag{9.113}
$$

Substituting Eqs. (9.112) and (9.113) into Eq. (9.111) yields

$$
\frac{1}{\sqrt{P_r} M} \mathbf{y}_l \longrightarrow \sum_{\mu=1}^{L} \mathbf{B}^{l\mu} \mathbf{u}_{\mu}. \tag{9.114}
$$

The kth entry of the processed signal becomes

$$
\frac{1}{\sqrt{P_r} M} y_k^l \longrightarrow \beta_k^{ll} u_k^l + \sum_{\mu=1,\mu\neq l}^{L} \beta_k^{l\mu} u_k^{\mu}. \tag{9.115}
$$

The salutary effect of using an unlimited number of base station antennas is that the effects of uncorrelated receiver noise and fast fading are eliminated completely, and transmissions from terminals within one's own cell do not interfere. However transmission from terminals in other cells that use the same pilot sequence constitute a residual interference. The effective Signal-to-Interference Ratio (SIR) is

$$
\gamma_k^l = \frac{\left(\beta_k^{ll} \right)^2}{\sum_{\mu=1,\mu\neq l}^{L} \left(\beta_k^{l\mu} \right)^2}, \tag{9.116}
$$

which is a random quantity depending on the random positions of the terminals and the shadow fading.

9.3.3 Downlink Data Transmission

In the downlink of a multi-cell massive MIMO system, the μth base station transmits a vector of message-bearing symbols $\mathbf{u}_\mu = [u_1^\mu, u_2^\mu, \ldots, u_K^\mu]^T$ through a precoding matrix toward K terminals in its respective cell independently, where $u_k^\mu, \forall k = 1, 2, \ldots, K$ and $\mu = 1, 2, \ldots, L$ denotes a zero-mean, unit-variance symbol intended for user k in the μth cell. Applying conjugate beamforming, $\mathbf{u}_\mu$ is multiplied with the conjugate of its estimate for the channel matrix. Thus, the vector of transmitted symbols in the μth base station is computed by

$$\mathbf{s}_\mu = \left(\hat{\mathbf{G}}^{\mu\mu} \right)^* \mathbf{u}_\mu, \tag{9.117}$$

where $\hat{\mathbf{G}}^{\mu\mu} \in \mathbb{C}^{M \times K}$ denotes the estimate of the uplink channel matrix between K users in cell μ and the base station in cell μ.

The K users in the lth cell obtain a $K \times 1$ vector comprising the received signals from all L base stations, i.e.

$$
\begin{aligned}
\mathbf{r}_l &= \sqrt{P_f} \sum_{\mu=1}^{L} \left(\mathbf{G}^{l\mu} \right)^T \mathbf{s}_\mu + \mathbf{n}_l \\
&= \sqrt{P_f} \sum_{\mu=1}^{L} \left(\mathbf{G}^{l\mu} \right)^T \left(\hat{\mathbf{G}}^{\mu\mu} \right)^* \mathbf{u}_\mu + \mathbf{n}_l \\
&= \sqrt{P_f} \sum_{\mu=1}^{L} \left(\mathbf{G}^{l\mu} \right)^T \left(\sum_{\mu'=1}^{L} \mathbf{G}^{\mu\mu'} + \mathbf{W}_\mu \right)^* \mathbf{u}_\mu + \mathbf{n}_l
\end{aligned}
\tag{9.118}
$$

where P_f is the forward-link power constraint, $\mathbf{G}^{l\mu} \in \mathbb{C}^{M \times K}$ denotes the uplink channel matrix from K users in cell μ to the base station in cell l, and the downlink channel matrix equals to $\left(\mathbf{G}^{l\mu} \right)^T$ because of channel reciprocity.

As the number of base station antennas grows without limit $M \to +\infty$, like Eq. (9.114), we have

$$\frac{1}{\sqrt{P_f M}} \mathbf{r}_l \longrightarrow \sum_{\mu=1}^{L} \mathbf{B}^{\mu l} \mathbf{u}_\mu. \tag{9.119}$$

The kth entry of the processed signal becomes

$$\frac{1}{\sqrt{P_f M}} r_k^l \longrightarrow \sum_{\mu=1}^{L} \beta_k^{\mu l} u_k^\mu = \beta_k^{ll} u_k^l + \sum_{\mu=1, \mu \neq l}^{L} \beta_k^{\mu l} u_k^\mu. \tag{9.120}$$

The effective SIR is then

$$\gamma_k^l = \frac{\left(\beta_k^{ll}\right)^2}{\sum_{\mu=1,\mu\neq l}^{L}\left(\beta_k^{\mu l}\right)^2}.$$

(9.121)

9.4 Cell-Free Massive MIMO

A base station with a large-scale antenna array simultaneously serves many users within a cell of a network of cells over the same time-frequency resource is a promising wireless access technology. With simple signal processing, it can provide high throughput, reliability, and energy efficiency. The massive number of service antennas in a cell can be deployed in collocated or distributed setups. Collocated massive MIMO architectures, where all service antennas are located in a compact area, have low backhaul requirements and joint processing. Nonetheless, high performance is primarily achieved by users that stay near the cell centers. Meanwhile, most users at the cell edges restrict to substantially worse quality of services due to inter-cell interference and handover issues that are inherent to the cellular architecture. The network densification for a high system capacity also leads to severe inter-cell interference and more frequent handovers [Zhang et al., 2020].

Consequently, most traffic congestion in cellular networks nowadays happens at the cell edge. The so-called 95%-likely user data rates, which can be guaranteed to 95% of the users and thus define the user-experienced performance, remain mediocre in 5G networks. The solution to these issues might be to connect each user with a multitude of distributed antennas. If there is only one huge cell in the network, no inter-cell interference appears and no handover is needed. This solution has been explored in the past, using names such as network MIMO, distributed MIMO, distributed antenna array, and Coordinated Multi-Point (CoMP) transmission and reception. Thanks to their ability to exploit spatial diversity against shadow fading more efficiently, a distributed system can offer a much higher probability of coverage than a collocated system at the expense of increased backhaul requirements.

In Ngo et al. [2017], a distributed massive MIMO system, where a large number of service antennas serve a much smaller number of autonomous users spread over a wide area, has been proposed. All antennas cooperate phase-coherently via a fronthaul network and serve all users in the same time-frequency resource. To avoid the huge amount of overhead to acquiring CSI, the system operates in the TDD mode and exploit the channel reciprocity. There are no cells or cell boundaries. Therefore, this system is referred to as *cell-free massive MIMO*. Since this setup combines the distributed MIMO and massive MIMO concepts, it is expected to reap all benefits from these two systems.

9.4.1 Cell-Free Network Layout

The initial setup of a cell-free massive MIMO system consists of M Access Points (APs) and K users, where $M \gg K$. All APs and user terminals are equipped with a single antenna, and they randomly spread over a geographical area. Furthermore, all APs connect to a Central Processing Unit (CPU) via a backhaul network, as shown in Figure 9.4, and the bandwidth of backhauling is assumed to be unlimited and the transmission is error-free to focus on the precoding and detection. All M APs simultaneously serve all K users in the same time-frequency resource. The downlink transmission from the APs to the users and the uplink transmission from the users to the APs are separated by the TDD operation. Each coherence interval is divided into three phases: uplink training, downlink payload data transmission, and uplink payload data transmission. In the uplink training phase, the users send reference signals to the APs and each AP estimates the channel to all users independently. Exploiting the channel reciprocity of a TDD system, the base station knows the downlink channel knowledge from the estimated uplink CSI. The acquired channel estimates are employed to precode the transmitted signals in the downlink, and to detect the signals transmitted from the users in the uplink.

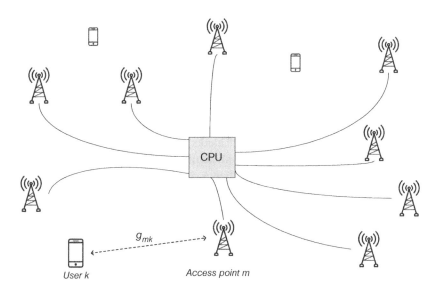

Figure 9.4 Schematic diagram of a cell-free massive MIMO system where M single-antenna access points controlled by a centralized processing unit serves K terminals.

We can write

$$g_{mk} = \sqrt{\beta_{mk}} h_{mk} \tag{9.122}$$

to model the fading channel between a generic AP $m = 1, \dots, M$ and a typical User Equipment (UE) $k = 1, \dots, K$, where β_{mk} and h_{mk} represent large-scale and small-scale fading, respectively. Small-scale fading is assumed to be frequency flat [Jiang and Schotten, 2021b] and is modeled by a circularly symmetric complex Gaussian random variable with zero mean and unit variance, i.e. $h_{mk} \sim \mathcal{CN}(0, 1)$. Large-scale fading is frequency independent and keeps constant for a relatively long period. It is computed by

$$\beta_{mk} = 10^{\frac{PL_{mk} + X_{mk}}{10}} \tag{9.123}$$

with shadowing fading $X_{mk} \sim \mathcal{N}(0, \sigma_{sd}^2)$ and path loss PL_{mk}. Ngo et al. [2017] applies the COST-Hata model, i.e.

$$PL_{mk} = \begin{cases} -L - 35 \log_{10}(d_{mk}), & d_{mk} > d_1 \\ -L - 15 \log_{10}(d_1) - 20 \log_{10}(d_{mk}), & d_0 < d_{mk} \leq d_1 \ , \\ -L - 15 \log_{10}(d_1) - 20 \log_{10}(d_0), & d_{mk} \leq d_0 \end{cases} \tag{9.124}$$

where d_{mk} represents the distance between AP m and UE k, d_0 and d_1 are the break points of the three-slope model, and

$$L = 46.3 + 33.9 \log_{10}\left(f_c\right) - 13.82 \log_{10}\left(h_{AP}\right) - \left[1.1 \log_{10}(f_c) - 0.7\right] h_{UE}$$
$$+ 1.56 \log_{10}\left(f_c\right) - 0.8 \tag{9.125}$$

with carrier frequency f_c, the antenna height of AP h_{AP}, and the antenna height of UE h_{UE}. Unlike collocated massive MIMO where the large-scale fading from a typical terminal k to all base station antennas is identical, denoted by $\beta_{mk} = \beta_k$, $\forall m = 1, 2, \dots, M$, each antenna pair between terminal k and AP m has a unique β_{mk}. An example cell-free massive MIMO system consisting of $M = 128$ APs and $K = 20$ UEs is demonstrated as Figure 9.5.

9.4.2 Uplink Training

As a massive MIMO system, a typical terminal k is assigned to an orthogonal reference sequence $\boldsymbol{\phi}_k \in \mathbb{C}^{\tau_p \times 1}$, where $\tau_p \geq K$. These terminals simultaneously transmit their reference signals over τ_p time-frequency resource units, resulting in the transmitted signals

$$\mathbf{X}_p = \sqrt{p_u \tau_p} \boldsymbol{\Phi}^H, \tag{9.126}$$

where

$$\boldsymbol{\Phi} = [\boldsymbol{\phi}_1, \boldsymbol{\phi}_2, \dots, \boldsymbol{\phi}_K]. \tag{9.127}$$

Then, the mth AP observes a $1 \times \tau_p$ received symbol vector

$$\mathbf{y}_m^p = \mathbf{g}_m \mathbf{X}_p + \mathbf{z}_m^p, \tag{9.128}$$

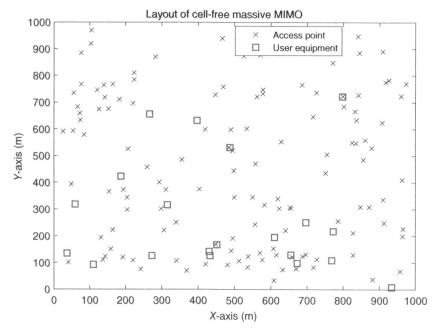

Figure 9.5 The layout of a cell-free massive MIMO system where $M = 128$ access points and $K = 20$ users are distributed over a square area with the length of 1 km.

where $\mathbf{z}_m^p$ corresponds to $1 \times \tau_p$ independent complex Gaussian noise with each entry $z \sim C\mathcal{N}(0, \sigma_n^2)$, $\mathbf{g}_m \in \mathbb{C}^{1 \times K}$ models the spatial signature from K terminals to AP m, which can be expressed by

$$\mathbf{g}_m = [g_{m1}, g_{m2}, \dots, g_{mK}], \tag{9.129}$$

corresponding to the mth row of the uplink channel matrix $\mathbf{G}_u \in \mathbb{C}^{M \times K}$.

Unlike a massive MIMO system, where the base station performs joint channel estimation, each AP in a cell-free system estimates its own spatial signature $\mathbf{g}_m$ independently

$$
\begin{aligned}
\tilde{\mathbf{y}}_m^p = \mathbf{y}_m^p \mathbf{\Phi} &= \mathbf{g}_m \mathbf{X}_p \mathbf{\Phi} + \mathbf{z}_m^p \mathbf{\Phi} \\
&= \sqrt{p_u \tau_p} \mathbf{g}_m \mathbf{\Phi}^H \mathbf{\Phi} + \mathbf{z}_m^p \mathbf{\Phi} \\
&= \sqrt{p_u \tau_p} \mathbf{g}_m + \tilde{\mathbf{z}}_m^p,
\end{aligned}
\tag{9.130}
$$

where each entry of $\tilde{\mathbf{z}}_m^p \in \mathbb{C}^{1 \times K}$ is also independent complex Gaussian noise $\tilde{z} \sim C\mathcal{N}(0, \sigma_n^2)$ due to the multiplication with a unitary matrix. Equivalently, we have

$$\tilde{y}_{mk,p} = \sqrt{p_u \tau_p} g_{mk} + \tilde{z}_{mk}, \quad \forall k = 1, 2, \dots, K. \tag{9.131}$$

Employing the linear MMSE estimation, AP m gets the estimates

$$\hat{g}_{mk} = \mathbb{E}\left[g_{mk}|\tilde{y}_{mk,p}\right] = \left(\frac{\sqrt{p_u\tau_p}\beta_{mk}}{p_u\tau_p\beta_{mk} + \sigma_n^2}\right)\tilde{y}_{mk,p}, \quad \forall k = 1, 2, \dots, K.$$

The variance of $\hat{g}_{mk}$ is computed by

$$\mathbb{E}\left[|\hat{g}_{mk}|^2\right] = \frac{p_u\tau_p\beta_{mk}^2}{p_u\tau_p\beta_{mk} + \sigma_n^2} = \alpha_{mk}, \tag{9.132}$$

resulting in $\hat{g}_{mk} \sim \mathcal{CN}(0, \alpha_{mk})$. The MSE of the channel estimate equals to

$$\mathbb{E}\left[|\tilde{g}_{mk}|^2\right] = \epsilon_{mk} = \beta_{mk} - \alpha_{mk} = \frac{\sigma_n^2\beta_{mk}}{p_u\tau_p\beta_{mk} + \sigma_n^2}, \tag{9.133}$$

corresponding to $\tilde{g}_{mk} \sim \mathcal{CN}(0, \epsilon_{mk})$. In the subsequent precoding and decoding, AP m is assumed to know the estimate of its own spatial signature

$$\hat{\mathbf{g}}_m = [\hat{g}_{m1}, \hat{g}_{m2}, \dots, \hat{g}_{mK}]. \tag{9.134}$$

9.4.3 Uplink Signal Detection

Similar to the uplink of a cellular massive MIMO system, K terminals in a cell-free network simultaneously transmit their respective symbols toward M APs. There is no direct cooperation among the terminals to perform joint precoding. The only thing these terminals can do is to weight their respective symbols independently. The transmitted symbols u_k, $k = 1, 2, \dots, K$ are zero-mean, unit-variance, and mutually uncorrelated, and therefore the covariance matrix for the vector $\mathbf{u} = [u_1, u_2, \dots, u_K]^T$ is

$$\mathbb{E}\left[\mathbf{u}\mathbf{u}^H\right] = \mathbf{I}_K. \tag{9.135}$$

The transmitted symbol for a typical terminal k is $\sqrt{\eta_k}u_k$, where η_k denotes a power-control coefficient, satisfying $0 \leqslant \eta_k \leqslant 1$. Thus, the vector of received symbols is computed by

$$\mathbf{r} = \sqrt{P_r}\mathbf{G}\mathbf{D}_\eta\mathbf{u} + \mathbf{n} \tag{9.136}$$

with the power constraint P_r, the uplink channel matrix $\mathbf{G} \in \mathbb{C}^{M \times K}$, and a diagonal matrix $\mathbf{D}_\eta = \text{diag}\{\sqrt{\eta_1}, \dots, \sqrt{\eta_K}\}$.

Equivalently, the observation at the mth AP is expressed by

$$r_m = \sqrt{P_r}\mathbf{g}_m\mathbf{D}_\eta\mathbf{u} + n_m$$

$$= \sqrt{P_r}\sum_{k=1}^{K}g_{mk}\sqrt{\eta_k}u_k + n_m, \tag{9.137}$$

where $\mathbf{g}_m = [g_{m1}, g_{m2}, \dots, g_{mK}]$ is the spatial signature of AP m, equivalent to the mth row of $\mathbf{G}$.

Similarly, three typical algorithms can be applied to recover the information symbols, i.e. matched filtering, zero-forcing detection, and minimum mean-squared error detection.

9.4.3.1 Matched Filtering

To detect u_k, the mth AP pre-processes the received signal by multiplying it with the conjugate of its local CSI $\hat{g}_{mk}$, and sends the result $\hat{g}_{mk}^* r_m$ to the CPU, resulting in

$$
\begin{aligned}
r_i &= \sum_{m=1}^{M} \hat{g}_{mi}^* r_m \\
&= \sum_{m=1}^{M} \hat{g}_{mi}^* \left(\sqrt{P_r} \sum_{k=1}^{K} g_{mk} \sqrt{\eta_k} u_k + n_m \right) \\
&= \underbrace{\sqrt{P_r} \sum_{m=1}^{M} \hat{g}_{mi}^* g_{mi} \sqrt{\eta_i} u_i}_{\text{Desired signal}} + \underbrace{\sqrt{P_r} \sum_{m=1}^{M} \hat{g}_{mi}^* \sum_{k=1, k\neq i}^{K} g_{mk} \sqrt{\eta_k} u_k}_{\text{Inter-user interference}} + \underbrace{\sum_{m=1}^{M} \hat{g}_{mi}^* n_m}_{\text{Noise}}.
\end{aligned}
\tag{9.138}
$$

Treating the inter-user interference as a colored noise, the CPU can get a hard estimate $\hat{u}_i$ for u_i, $\forall i = 1, 2, \ldots, K$.

9.4.3.2 ZF Detection

In this case, each AP needs to send its observation r_m and local CSI $\hat{g}_m$ to the CPU via the fronthaul network. Consequently, the CPU knows $\mathbf{r}$ and builds a decoding matrix as $(\hat{\mathbf{G}}^H \hat{\mathbf{G}})^{-1} \hat{\mathbf{G}}^H$. Thus, the output of post-processing is

$$
\begin{aligned}
\tilde{\mathbf{u}} &= (\hat{\mathbf{G}}^H \hat{\mathbf{G}})^{-1} \hat{\mathbf{G}}^H \mathbf{r} \\
&= \sqrt{P_r} (\hat{\mathbf{G}}^H \hat{\mathbf{G}})^{-1} \hat{\mathbf{G}}^H \mathbf{G} \mathbf{D}_\eta \mathbf{u} + (\hat{\mathbf{G}}^H \hat{\mathbf{G}})^{-1} \hat{\mathbf{G}}^H \mathbf{n}.
\end{aligned}
$$

Ignoring the estimation error, i.e. $\hat{\mathbf{G}} = \mathbf{G}$, we have

$$
\tilde{\mathbf{u}} = \sqrt{P_r} \mathbf{D}_\eta \mathbf{u} + (\hat{\mathbf{G}}^H \hat{\mathbf{G}})^{-1} \hat{\mathbf{G}}^H \mathbf{n}.
\tag{9.139}
$$

Decomposing this equation yields the kth soft estimate

$$
\tilde{u}_k = \underbrace{\sqrt{P_r \eta_k} u_k}_{\text{Desired signal}} + \underbrace{\mathfrak{g}_k \mathbf{n}}_{\text{Noise}},
\tag{9.140}
$$

where $\mathfrak{g}_k \in \mathbb{C}^{1 \times M}$ denotes the kth row of $(\hat{\mathbf{G}}^H \hat{\mathbf{G}})^{-1} \hat{\mathbf{G}}^H$. The inter-user interference is completely eliminated, and a hard estimate $\hat{u}_k$ can be derived from $\tilde{u}_k$.

9.4.3.3 MMSE Detection

To mitigate the effect of noise amplification in zero-forcing detection, the CPU can employ the MMSE method to minimize the MSE of the estimated symbols. Each AP needs to send its observation r_m and local CSI $\hat{\mathbf{g}}_m$ to the CPU via the fronthaul network. Consequently, the CPU knows $\mathbf{r}$ and builds a decoding matrix as $(\hat{\mathbf{G}}^H \hat{\mathbf{G}} + \sigma_n^2 \mathbf{I})^{-1} \hat{\mathbf{G}}^H$. The output of post-processing becomes

$$\tilde{\mathbf{u}} = (\hat{\mathbf{G}}^H \hat{\mathbf{G}} + \sigma_n^2 \mathbf{I})^{-1} \hat{\mathbf{G}}^H \mathbf{r}$$

$$= \sqrt{P_r} (\hat{\mathbf{G}}^H \hat{\mathbf{G}} + \sigma_n^2 \mathbf{I})^{-1} \hat{\mathbf{G}}^H \mathbf{G} \mathbf{D}_\eta \mathbf{u} + (\hat{\mathbf{G}}^H \hat{\mathbf{G}} + \sigma_n^2 \mathbf{I})^{-1} \hat{\mathbf{G}}^H \mathbf{n}. \tag{9.141}$$

9.4.4 Conjugate Beamforming

In the downlink of a cell-free massive MIMO system, two typical transmission methods, i.e. conjugate beamforming and zero-forcing precoding, are applied to spatially multiplex the information-bearing symbols intended for K terminals, denoted by $\mathbf{u} = [u_1, u_2, \dots, u_K]^T$, where the symbols are normalized

$$\mathbb{E}[|u_k|^2] = 1, \quad k = 1, 2, \dots, K. \tag{9.142}$$

A cell-free massive MIMO system applying conjugate beamforming operates as follows:

- A typical AP m measures β_{mk}, $k = 1, 2, \dots, K$ and reports them to the CPU. Usually, the large-scale fading keeps constant for a relative long period with respective to the channel coherence time. Consequently, the knowledge of β_{mk} can be regarded as perfect, and the overhead of measurement and distribution is small.
- The CPU computes power-control coefficients η_{mk}, $\forall m$, $\forall k$ as a function of β_{mk}, and sends them to corresponding APs. Meanwhile, the CPU distributes the information-bearing symbols $\mathbf{u}$ to all APs. Note that β_{mk} might be the same for tens of symbol periods, and therefore only $\mathbf{u}$ is needed to be transmitted.
- Users synchronously transmit their pilot sequences $\boldsymbol{\phi}_k$, $k = 1, \dots, K$ with a duration of τ_p.
- The mth AP, where $m = 1, \dots, M$, acquires the estimate of its own spatial signature $\hat{\mathbf{g}}_m = [\hat{g}_{m1}, \hat{g}_{m2}, \dots, \hat{g}_{mK}]^T$.
- The APs treat the channel estimates as the true channels, and use conjugate beamforming to generate the transmitted signals. The mth AP sends the signal

$$s_m = \sqrt{\eta_{mk} P_m} \hat{\mathbf{g}}_m^H \mathbf{u} = \sqrt{P_m} \sum_{k=1}^K \sqrt{\eta_{mk}} \hat{g}_{mk}^* u_k, \tag{9.143}$$

where P_m is the transmit power limit of AP m. The choice of power-control coefficients is subjected to

$$\mathbb{E}[|s_m|^2] \leqslant P_m, \quad \forall m = 1, 2, \dots, M, \tag{9.144}$$

which can be interpreted as

$$\sum_{k=1}^K \eta_{mk} \mathbb{E}[|\hat{g}_{mk}|^2] \leqslant 1 \tag{9.145}$$

or

$$\sum_{k=1}^{K} \eta_{mk} \leqslant \frac{1}{\mathbb{E}[|\hat{g}_{mk}|^2]} = \frac{p_u \tau_p \beta_{mk} + \sigma_n^2}{p_u \tau_p \beta_{mk}^2} \tag{9.146}$$

applying Eqs. (9.132) and (9.142).

Then, the observation of the ith user, $\forall i = 1, 2, \dots, K$ is

$$
\begin{aligned}
r_i &= \sum_{m=1}^{M} g_{mi} s_m + n_i \\
&= \sum_{m=1}^{M} g_{mi} \left(\sqrt{P_m} \sum_{k=1}^{K} \sqrt{\eta_{mk}} \hat{g}_{mk}^* u_k \right) + n_i \\
&= \underbrace{\sum_{m=1}^{M} \sqrt{P_m \eta_{mi}} g_{mi} \hat{g}_{mi}^* u_i}_{\text{Desired signal}} + \underbrace{\sum_{m=1}^{M} \sqrt{P_m} \sum_{k=1, k \neq i}^{K} \sqrt{\eta_{mk}} g_{mi} \hat{g}_{mk}^* u_k + n_i}_{\text{Multi-user interference}}.
\end{aligned}
$$

The salutary effect of using an unlimited number of base station antennas is that the effects of uncorrelated noise and fast channel fading disappear. It is a result of **channel hardening** in massive MIMO systems [Marzetta, 2010]. Consequently, the detection of received signals is done conditioned on that a generic user k is only aware of the *statistics* of the estimated channel coefficients, i.e. $\mathbb{E}\left[\left|\hat{g}_{mk}\right|^2\right] = \alpha_{mk}$, $\forall m = 1, 2, \dots, M$, since there is no reference signals and channel estimation in the downlink. Thus, the observation of the ith user can be rewritten as (assuming the power constraint of each AP is identical, i.e. $P_m = P_f, \forall m = 1, 2, \dots, M$ for brevity)

$$
\begin{aligned}
r_i &= \sum_{m=1}^{M} \sqrt{P_f \eta_{mi}} [\hat{g}_{mi} + \tilde{g}_{mi}] \hat{g}_{mi}^* u_i + \sum_{m=1}^{M} \sum_{k=1, k \neq i}^{K} \sqrt{P_f \eta_{mk}} [\hat{g}_{mi} + \tilde{g}_{mi}] \hat{g}_{mk}^* u_k + n_i \\
&= \sum_{m=1}^{M} \sqrt{P_f \eta_{mi}} |\hat{g}_{mi}|^2 u_i + \sum_{m=1}^{M} \sum_{k=1, k \neq i}^{K} \sqrt{P_f \eta_{mk}} \hat{g}_{mi} \hat{g}_{mk}^* u_k \\
&\quad + \sum_{m=1}^{M} \sum_{k=1}^{K} \sqrt{P_f \eta_{mk}} \tilde{g}_{mi} \hat{g}_{mk}^* u_k + n_i \\
&= \underbrace{\sum_{m=1}^{M} \sqrt{P_f \eta_{mi}} \alpha_{mi} u_i}_{S_0 : \text{ Useful signal}} + \underbrace{\sum_{m=1}^{M} \sqrt{P_f \eta_{mi}} (|\hat{g}_{mi}|^2 - \alpha_{mi}) u_i}_{\mathcal{I}_1 : \text{ No CSI at user}} \\
&\quad + \underbrace{\sum_{m=1}^{M} \sum_{k=1, k \neq i}^{K} \sqrt{P_f \eta_{mk}} \hat{g}_{mi} \hat{g}_{mk}^* u_k}_{\mathcal{I}_2 : \text{ Multi-user interference}} + \underbrace{\sum_{m=1}^{M} \sum_{k=1}^{K} \sqrt{P_f \eta_{mk}} \tilde{g}_{mi} \hat{g}_{mk}^* u_k}_{\mathcal{I}_3 : \text{ CSI estimate error}} + \underbrace{n_i}_{\mathcal{N}_4}
\end{aligned}
\tag{9.147}
$$

Since information symbols intended for different users are independent, and additive Gaussian noise is uncorrelated with information symbols and channel realizations, the terms S_0, I_1, I_2, I_3, and N_4 are mutually uncorrelated [Nayebi et al., 2017]. According to Hassibi and Hochwald [2003], the worst-case noise for mutual information is Gaussian additive noise with the variance equaling to the variance of $I_1 + I_2 + I_3 + N_4$.

Thus, the downlink achievable rate for user k is lower bounded by

$$R_i = \log(1 + \gamma_i), \tag{9.148}$$

where

$$\gamma_i = \frac{\mathbb{E}\left[|S_0|^2\right]}{\mathbb{E}\left[|I_1 + I_2 + I_3 + N_4|^2\right]}$$

$$= \frac{\mathbb{E}\left[|S_0|^2\right]}{\mathbb{E}\left[|I_1|^2\right] + \mathbb{E}\left[|I_2|^2\right] + \mathbb{E}\left[|I_3|^2\right] + \mathbb{E}\left[|N_4|^2\right]} \tag{9.149}$$

with

$$\mathbb{E}\left[|S_0|^2\right] = P_f \left(\sum_{m=1}^{M} \sqrt{\eta_{mi}}\alpha_{mi}\right)^2 \tag{9.150}$$

$$\mathbb{E}\left[|I_1|^2\right] = P_f \sum_{m=1}^{M} \eta_{mi}\alpha_{mi}^2 \tag{9.151}$$

$$\mathbb{E}\left[|I_2|^2\right] = P_f \sum_{m=1}^{M} \sum_{k=1,k\neq i}^{K} \eta_{mk}\alpha_{mi}\alpha_{mk} \tag{9.152}$$

$$\mathbb{E}\left[|I_3|^2\right] = P_f \sum_{m=1}^{M} \sum_{k=1}^{K} \eta_{mk}\epsilon_{mi}\alpha_{mk} \tag{9.153}$$

Substituting Eqs. from (9.149) to (9.153) into Eq. (9.148), yields

$$R_i = \log\left(1 + \frac{P_f \left(\sum_{m=1}^{M} \sqrt{\eta_{mi}}\alpha_{mi}\right)^2}{\sigma_n^2 + P_f \sum_{m=1}^{M} \sum_{k=1}^{K} \eta_{mk}\beta_{mi}\alpha_{mk}}\right). \tag{9.154}$$

9.4.5 Zero-Forcing Precoding

The philosophy behind the ZF precoding is to completely suppress the interference among different users given the knowledge of downlink channels. A cell-free massive MIMO system applying the ZF precoding operates as follows:

- AP m measures β_{mk}, $k = 1, 2, \ldots, K$ and reports them to the CPU.
- As conjugate beamforming, the CPU computes power-control coefficients in terms of β_{mk}. It is necessary to have $\eta_{1k} = \cdots = \eta_{Mk}$, $\forall k$, and therefore power coefficients should be only the functions of k, i.e. $\eta_{mk} = \eta_k$.

- Users synchronously transmit their pilot sequences $\boldsymbol{\phi}_k$, $k = 1, \ldots, K$.
- The mth AP, where $m = 1, \ldots, M$, acquires the estimate of its own spatial signature $\hat{\mathbf{g}}_m = [\hat{g}_{m1}, \hat{g}_{m2}, \ldots, \hat{g}_{mK}]^T$.
- Each AP sends its local CSI to the CPU, and therefore the CPU gets the global CSI $\hat{\mathbf{G}} = [\hat{\mathbf{g}}_1, \hat{\mathbf{g}}_2, \ldots, \hat{\mathbf{g}}_M] \in \mathbb{C}^{K \times M}$.
- The CPU jointly encodes the information-bearing symbols in terms of

$$\mathbf{s} = \hat{\mathbf{G}}^H \left(\hat{\mathbf{G}} \hat{\mathbf{G}}^H \right)^{-1} \mathbf{D}_\eta \mathbf{u}, \tag{9.155}$$

where $\mathbf{D}_\eta \in \mathbb{C}^{K \times K}$ is a diagonal matrix consisting of power-control coefficients, i.e. $\mathbf{D}_\eta = \text{diag}\{ \sqrt{\eta_1}, \ldots, \sqrt{\eta_K} \}$.

- The CPU distributes the precoded symbol s_m to AP m, and these APs synchronously send their respective transmitted symbols toward the users.

Then, the vector of received symbols can be rewritten as

$$
\begin{aligned}
\mathbf{r} &= \sqrt{P_f} \mathbf{G} \mathbf{s} + \mathbf{n} \\
&= \sqrt{P_f} \mathbf{G} \hat{\mathbf{G}}^H \left(\hat{\mathbf{G}} \hat{\mathbf{G}}^H \right)^{-1} \mathbf{D}_\eta \mathbf{u} + \mathbf{n}.
\end{aligned} \tag{9.156}
$$

Equivalently, the ith user observes

$$
\begin{aligned}
r_i &= \sqrt{P_f} \mathbf{g}_i \mathbf{s} + n_i \\
&= \sqrt{P_f} \mathbf{g}_i \hat{\mathbf{G}}^H \left(\hat{\mathbf{G}} \hat{\mathbf{G}}^H \right)^{-1} \mathbf{D}_\eta \mathbf{u} + n_i \\
&= \sqrt{P_f} \left(\hat{\mathbf{g}}_i + \tilde{\mathbf{g}}_i \right) \hat{\mathbf{G}}^H \left(\hat{\mathbf{G}} \hat{\mathbf{G}}^H \right)^{-1} \mathbf{D}_\eta \mathbf{u} + n_i \\
&= \underbrace{\sqrt{P_f \eta_i} u_i}_{S_0: \text{ Useful signal}} + \underbrace{\sqrt{P_f} \tilde{\mathbf{g}}_i \hat{\mathbf{G}}^H \left(\hat{\mathbf{G}} \hat{\mathbf{G}}^H \right)^{-1} \mathbf{D}_\eta \mathbf{u}}_{I_1: \text{ CSI estimate error}} + \underbrace{n_i}_{I_2: \text{ Noise}},
\end{aligned} \tag{9.157}
$$

where $\mathbf{g}_i \in \mathbb{C}^{1 \times M} = [g_{1i}, g_{2i}, \ldots, g_{Mi}]$ stands for the actual channel signature for user i, which is the ith row of the channel matrix $\mathbf{G}$, $\hat{\mathbf{g}}_i \in \mathbb{C}^{1 \times M} = [\hat{g}_{1i}, \hat{g}_{2i}, \ldots, \hat{g}_{Mi}]$ expresses an estimate of $\mathbf{g}_i$, and the corresponding estimation error $\tilde{\mathbf{g}}_i = \mathbf{g}_i - \hat{\mathbf{g}}_i$.

Because of the independence of the transmitted symbols, additive noise, and channel realizations, the terms S_0, I_1, and I_2 are mutually uncorrelated. Based on the worst-case uncorrelated additive noise [Hassibi and Hochwald, 2003], the achievable rate of user i with the ZF precoding is lower bounded by

$$R_i^{\text{ZF}} = \log \left(1 + \gamma_i^{\text{ZF}} \right), \tag{9.158}$$

where

$$\gamma_i^{\text{ZF}} = \frac{\mathbb{E}\left[|S_0|^2 \right]}{\mathbb{E}\left[|I_1|^2 \right] + \mathbb{E}\left[|I_2|^2 \right]}. \tag{9.159}$$

According to Nayebi et al. [2017], the variance of $\mathcal{I}_1$ can be figured out as

$$\mathbb{E}\left[|\mathcal{I}_1|^2\right] = P_f \mathbb{E}\left[\left|\tilde{\mathbf{g}}_i \hat{\mathbf{G}}^H \left(\hat{\mathbf{G}}\hat{\mathbf{G}}^H\right)^{-1} \mathbf{D}_\eta \mathbf{u}\right|^2\right]$$

$$= P_f tr \left(\mathbf{D}_\eta^2 \mathbb{E}[\left(\hat{\mathbf{G}}\hat{\mathbf{G}}^H\right)^{-1} \hat{\mathbf{G}} \mathbb{E}\left[\tilde{\mathbf{g}}_i^H \tilde{\mathbf{g}}_i\right] \hat{\mathbf{G}}^H \left(\hat{\mathbf{G}}\hat{\mathbf{G}}^H\right)^{-1}]\right). \tag{9.160}$$

We write $\chi_k^i, k = 1, 2, \dots, K$ to denote the kth diagonal element of the $K \times K$ matrix dedicated to user i:

$$\mathbb{E}\left[\left(\hat{\mathbf{G}}\hat{\mathbf{G}}^H\right)^{-1} \hat{\mathbf{G}} \mathbb{E}\left[\tilde{\mathbf{g}}_i^H \tilde{\mathbf{g}}_i\right] \hat{\mathbf{G}}^H \left(\hat{\mathbf{G}}\hat{\mathbf{G}}^H\right)^{-1}\right], \tag{9.161}$$

where $\mathbb{E}\left[\tilde{\mathbf{g}}_i^H \tilde{\mathbf{g}}_i\right]$ is a diagonal matrix with ϵ_{mi} on its mth diagonal element, i.e.

$$\mathbb{E}\left[\tilde{\mathbf{g}}_i^H \tilde{\mathbf{g}}_i\right] = \begin{bmatrix} \epsilon_{1i} & 0 & \dots & 0 \\ 0 & \epsilon_{2i} & \dots & 0 \\ \vdots & \vdots & \ddots & \vdots \\ 0 & 0 & \dots & \epsilon_{Mi} \end{bmatrix}. \tag{9.162}$$

Then, Eq. (9.159) can be further expressed by

$$\gamma_i^{ZF} = \frac{P_f \eta_i}{\sigma_n^2 + P_f \sum_{k=1}^{K} \eta_k \chi_k^i}. \tag{9.163}$$

9.4.6 Impact of Channel Aging

In cell-free massive MIMO systems, linear precoding is mainly implemented through conjugate beamforming and zero-forcing precoding. The former utilizes the local CSI to independently produce transmitted signals at each AP. It is simple with a low requirement on backhauling but suffers from inter-user interference. Hence, conjugate beamforming is inferior to zero-forcing precoding in terms of spectral and power efficiency. In the cell-free architecture, however, zero-forcing precoding requires the exchange of instantaneous CSI and precoded data between the CPU and APs via a fronthaul network. In addition to high implementation complexity and a significant backhaul burden, it causes considerable propagation and processing delays. In practice, the system performance is vulnerable to such a delay since the knowledge of CSI is outdated quickly, referred to as *channel aging*, subjected to the channel fading and imperfect hardware.

9.4.6.1 Channel Aging

Because of the processing and propagation delays, there exists a time gap between the instant when reference signals sound the uplink channels and the instant when the downlink data transmission based on the measured CSI happens. The acquired CSI may be outdated under the fluctuation of channels raised by *user mobility* and *phase noise*.

User Mobility The relative movement between an AP and a UE as well as their surrounding reflectors leads to a time-varying channel. Given the moving speed v_k of a typical UE k, its maximal Doppler shift is obtained by $f_d^k = v_k/\lambda$, where λ represents the wavelength of carrier frequency. The higher the mobility, the faster the channel varies. To quantify the aging of CSI raised by the Doppler effect, a metric known as correlation coefficient is applied, as defined by Jiang and Schotten [2021c]:

$$\rho_k = \frac{\mathbb{E}\left[h_{mk,d}h_{mk,p}^*\right]}{\sqrt{\mathbb{E}[|h_{mk,p}|^2]\mathbb{E}[|h_{mk,d}|^2]}}, \tag{9.164}$$

where $h_{mk,p}$ and $h_{mk,d}$ denote the small-scale channel fading between AP m and UE k at the instants of the uplink training (notated by p) and the actual downlink data transmission (notated by d), respectively. Under the classical Doppler spectrum of the Jakes model, it takes the value

$$\rho_k = J_0(2\pi f_d^k \,\Delta\tau), \tag{9.165}$$

where $\Delta\tau$ stands for the overall delay, and $J_0(\cdot)$ denotes the *zero*th order Bessel function of the first kind. According to Jiang et al. [2016], we have

$$h_{mk,d} = \left(\rho_k h_{mk,p} + \kappa_{mk}\sqrt{1-\rho_k^2}\right) \tag{9.166}$$

with an innovation component κ_{mk} that is a random variable with standard normal distribution $\kappa_{mk} \sim \mathcal{CN}(0,1)$.

Phase Noise It is attractive for cost-efficient implementation of massive MIMO systems with low-cost transceivers, whereas raising the problem of hardware impairments. Meanwhile, each distributed AP in a cell-free massive MIMO system has to operate a local oscillator, in contrast to a common oscillator in a collocated massive MIMO setup. Due to imperfect oscillators at the transmitter, the transmitted signals suffer from phase noise during the up-conversion processing from baseband to passband signals, and *vice versa* at the receiver. Such phase noise is not only random but also time-varying, leading to the outdated CSI that is equivalent to that of user mobility.

Utilizing a well-established Wiener process [Krishnan et al., 2016], the phase noise of the mth AP and the kth user at *discrete-time* instant t can be modeled as

$$\begin{cases} \phi_{m,t} = \phi_{m,t-1} + \Delta\phi_t, & \Delta\phi_t \sim \mathcal{CN}(0,\sigma_\phi^2) \\ \varphi_{k,t} = \varphi_{k,t-1} + \Delta\varphi_t, & \Delta\varphi_t \sim \mathcal{CN}(0,\sigma_\varphi^2), \end{cases} \tag{9.167}$$

where the increment variances are given by $\sigma_i^2 = 4\pi^2 f_c c_i T_s$, $\forall i = \phi, \varphi$ with symbol period T_s and oscillator-dependent constant c_i.

Until now, we can write

$$g_{mk,t} = \sqrt{\beta_{mk}} h_{mk,t} e^{j(\phi_{m,t} + \varphi_{k,t})} \tag{9.168}$$

to denote the overall channel gain between AP m and UE k at instant t combining the effects of path loss, shadowing, small-scale fading, and phase noise. In particular, the acquired CSI $g_{mk,p} = \sqrt{\beta_{mk}} h_{mk,p} e^{j(\phi_{m,p} + \varphi_{k,p})}$ is an outdated version of its actual value $g_{mk,d} = \sqrt{\beta_{mk}} h_{mk,d} e^{j(\phi_{m,d} + \varphi_{k,d})}$. Under good condition where the channels exhibit slow fading under low mobility and the quality of oscillators is high, the effect of channel aging is not explicit and the performance loss is possible to be small. Otherwise, the impact should be serious either in fast fading environments or low-cost hardware utilization.

Propagation and Processing Delays Assume that the APs and the UEs are well-synchronized, the knowledge of β_{mk} is perfectly available, and the fronthaul network provides error-free and infinite capacity. As illustrated in Figure 9.6, the propagation and processing delays can be modeled as follows:

- The users simultaneously transmit their reference signals i_k, $k = 1, \ldots, K$ toward the APs with a duration of T_p. The propagation delay is τ_{ul}.
- The mth AP estimates its own channel signature $\hat{g}_{mk,p}$, $\forall k$ with a processing time of τ_{ce}.
- AP m sends its local CSI $\hat{\mathbf{g}}_m = \left[\hat{g}_{m1,p}, \ldots, \hat{g}_{mK,p}\right]^T \in \mathbb{C}^{K \times 1}$ to the CPU, leading to a propagation delay of τ_{fh}^u.
- Using $\hat{\mathbf{G}} = [\hat{\mathbf{g}}_1, \ldots, \hat{\mathbf{g}}_M] \in \mathbb{C}^{K \times M}$, the CPU precodes a block of information-bearing symbols $\mathbf{U} \in \mathbb{C}^{K \times N_T}$, where N_T denotes the number of symbols per user. The transmitted symbol block is given by $\mathbf{X} = \hat{\mathbf{G}}^H \left(\hat{\mathbf{G}} \hat{\mathbf{G}}^H\right)^{-1} \mathbf{D}_\eta \mathbf{U}$. The precoding costs a time τ_{ZF}.
- The CPU distributes the precoded symbol vector $\mathbf{x}_m \in \mathbb{C}^{1 \times N_T}$ to AP m, using the time of τ_{fh}^d.
- The transmitter of AP m needs a preparation time of τ_{tx} to start the transmission after the reception of $\mathbf{x}_m$, and the signal propagation takes τ_{dl}.

In particular, let $\Delta\tau$ denote the gap between the time when the reference signals probe the channels and the instant that all APs synchronously transmit the precoded symbols. As depicted in Figure 9.6, we get

$$\Delta\tau = T_p + \tau_{ce} + \tau_{fh}^u + \tau_{ZF} + \tau_{fh}^d + \tau_{tx}, \tag{9.169}$$

which is normalized by the sampling period to $n_{\Delta\tau} = \left\lceil \frac{\Delta\tau}{T_s} \right\rceil$.

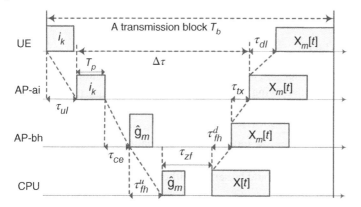

Figure 9.6 Schematic diagram for time alignment of zero-forcing precoding in a cell-free massive MIMO system where M APs serve K users. Based on reference signals $\mathbf{i}_k$, AP m gets $\hat{\mathbf{g}}_m$ and transfers it to the CPU. Due to the delay $\triangle \tau$, the CSI during the downlink data transmission changes, namely, $g_{mk,d} \neq g_{mk,p}$. Note that *AP-ai* stands for the air interface of an AP while *AP-bh* points to the part that interacts with the fronthaul network. Source: Jiang and Schotten [2021a]/With permission of IEEE.

9.4.6.2 Performance Degradation

According to Eq. (9.168), the overall CSI during the downlink data transmission is given by

$$g_{mk,d} = \sqrt{\beta_{mk}} h_{mk,d} e^{j(\phi_{m,d}+\varphi_{k,d})}. \tag{9.170}$$

We write $\hat{g}_{mk,p}$ to denote an estimate of $g_{mk,p}$, and then the estimation error raised by additive noise is computed by

$$\tilde{g}_{mk,p} = g_{mk,p} - \hat{g}_{mk,p}. \tag{9.171}$$

The innovation component in Eq. (9.166) corresponds to a compound element in the overall CSI, which is written as $e_{mk} = \sqrt{\beta_{mk}} \kappa_{mk} e^{j(\phi_{m,p}+\varphi_{k,p})}$. Substituting Eq. (9.166) into Eq. (9.170) and applying Eq. (9.171), yields

$$
\begin{aligned}
g_{mk,d} &= \sqrt{\beta_{mk}} \left(\rho_k h_{mk,p} + \kappa_{mk} \sqrt{1 - \rho_k^2} \right) e^{j(\phi_{m,d}+\varphi_{k,d}+\phi_{m,p}+\varphi_{k,p}-\phi_{m,p}-\varphi_{k,p})} \\
&= \left(\rho_k g_{mk,p} + e_{mk} \sqrt{1 - \rho_k^2} \right) e^{j(\phi_{m,d}+\varphi_{k,d}-\phi_{m,p}-\varphi_{k,p})} \\
&= \left(\rho_k \hat{g}_{mk,p} + \rho_k \tilde{g}_{mk,p} + e_{mk} \sqrt{1 - \rho_k^2} \right) e^{j(\phi_{m,d}-\phi_{m,p})} e^{j(\varphi_{k,d}-\varphi_{k,p})}. \tag{9.172}
\end{aligned}
$$

Denote the kth row of $\hat{\mathbf{G}}$ as $\hat{\mathbf{g}}_k = \left[\hat{g}_{1k,p}, \dots, \hat{g}_{Mk,p} \right]$, $\tilde{\mathbf{g}}_k = \left[\tilde{g}_{1k,p}, \tilde{g}_{2k,p}, \dots, \tilde{g}_{Mk,p} \right]$, $\mathbf{e}_k = \left[e_{1k}, \dots, e_{Mk} \right]$, and a diagonal matrix

$$\triangle \mathbf{\Phi} = \text{diag}\{ e^{j(\phi_{1,d}-\phi_{1,p})}, \dots, e^{j(\phi_{M,d}-\phi_{M,p})} \}. \tag{9.173}$$

Building a channel vector $\mathbf{g}_{k,d} = [g_{1k,d}, \ldots, g_{Mk,d}] \in \mathbb{C}^{1 \times M}$ and substituting Eq. (9.172) into it, yields

$$\mathbf{g}_{k,d} = e^{j(\varphi_{k,d} - \varphi_{k,p})} \left(\rho_k \hat{\mathbf{g}}_k + \rho_k \tilde{\mathbf{g}}_k + \sqrt{1 - \rho_k^2} \mathbf{e}_k \right) \triangle \mathbf{\Phi}. \tag{9.174}$$

In the presence of outdated CSI, the received signal of user k given in Eq. (9.157) can be rewritten as

$$
\begin{aligned}
r_k &= \sqrt{P_f} \mathbf{g}_{k,d} \mathbf{s} + n_k \\
&= \sqrt{P_f} \mathbf{g}_{k,d} \hat{\mathbf{G}}^H \left(\hat{\mathbf{G}} \hat{\mathbf{G}}^H \right)^{-1} \mathbf{D}_\eta \mathbf{u} + n_k \\
&= \sqrt{P_f} e^{j(\varphi_{k,d} - \varphi_{k,p})} \left(\rho_k \hat{\mathbf{g}}_k + \rho_k \tilde{\mathbf{g}}_k + \sqrt{1 - \rho_k^2} \mathbf{e}_k \right) \\
&\quad \times \triangle \mathbf{\Phi} \hat{\mathbf{G}}^H \left(\hat{\mathbf{G}} \hat{\mathbf{G}}^H \right)^{-1} \mathbf{D}_\eta \mathbf{u} + n_k \\
&= \underbrace{\sqrt{P_f} \eta_k e^{j(\varphi_{k,d} - \varphi_{k,p})} e^{-\frac{n_{\Delta \tau} \sigma_\phi^2}{2}} \rho_k u_k}_{\mathcal{D}_0 : \text{ desired signal}}
\end{aligned}
$$

$$
+ \underbrace{\sqrt{P_f} e^{j(\varphi_{k,d} - \varphi_{k,p})} \rho_k \tilde{\mathbf{g}}_k \triangle \mathbf{\Phi} \hat{\mathbf{G}}^H \left(\hat{\mathbf{G}} \hat{\mathbf{G}}^H \right)^{-1} \mathbf{D}_\eta \mathbf{u}}_{\mathcal{I}_1 : \text{ effective noise}}
$$

$$
+ \underbrace{\sqrt{P_f (1 - \rho_k^2)} e^{j(\varphi_{k,d} - \varphi_{k,p})} \mathbf{e}_k \triangle \mathbf{\Phi} \hat{\mathbf{G}}^H \left(\hat{\mathbf{G}} \hat{\mathbf{G}}^H \right)^{-1} \mathbf{D}_\eta \mathbf{u}}_{\mathcal{I}_2 : \text{ effective noise}} + \underbrace{n_k}_{\mathcal{I}_3}. \tag{9.175}
$$

During the derivation, $T_{\text{PN}} = \lim_{M \to \infty} \frac{1}{M} \text{tr} \{ \triangle \mathbf{\Phi} \} = e^{-n_{\Delta \tau} \sigma_\phi^2 / 2}$ is employed according to Krishnan et al. [2016], implying that phase noise hardens to a deterministic value when $M \to \infty$.

The information symbols, estimation errors, innovation components, and additive noise are independent, such that the terms $\mathcal{D}_0$, $\mathcal{I}_1$, $\mathcal{I}_2$, and $\mathcal{I}_3$ in Eq. (9.175) are mutually uncorrelated. By applying the fact that uncorrected Gaussian noise represents the worst case [Ngo et al., 2017], the achievable rate for user k is lower bounded by $\log_2 (1 + \gamma_k)$ with the effective signal-to-interference-plus-noise ratio (SINR)

$$\gamma_k = \frac{\mathbb{E}\left[|\mathcal{D}_0|^2\right]}{\mathbb{E}\left[|\mathcal{I}_1|^2\right] + \mathbb{E}\left[|\mathcal{I}_2|^2\right] + \mathbb{E}\left[|\mathcal{I}_3|^2\right]}. \tag{9.176}$$

It is straightforward to figure out

$$\mathbb{E}\left[|\mathcal{D}_0|^2\right] = P_f \eta_k \rho_k^2 e^{-n_{\Delta \tau} \sigma_\phi^2} \tag{9.177}$$

and $\mathbb{E}[|\mathcal{I}_3|^2] = \sigma_n^2$. Similar to Eq. (9.160), the variance of $\mathcal{I}_1$ is computed by

$$
\begin{aligned}
\mathbb{E}\left[|\mathcal{I}_1|^2\right] &= \mathbb{E}\left[|\sqrt{P_f}e^{j(\varphi_{k,d}-\varphi_{k,p})}\rho_k \tilde{\mathbf{g}}_k \vartriangle \boldsymbol{\Phi}\hat{\mathbf{G}}^H\left(\hat{\mathbf{G}}\hat{\mathbf{G}}^H\right)^{-1}\mathbf{D}_\eta \mathbf{u}|^2\right] \\
&= P_f \rho_k^2 \mathbb{E}\left[\left\|\tilde{\mathbf{g}}_k \vartriangle \boldsymbol{\Phi}\hat{\mathbf{G}}^H\left(\hat{\mathbf{G}}\hat{\mathbf{G}}^H\right)^{-1}\mathbf{D}_\eta \mathbf{u}\right\|^2\right] \\
&= P_f \rho_k^2 e^{-n_{\Delta\tau}\sigma_\phi^2} tr\left\{\mathbf{D}_\eta^2 \mathbb{E}\left[\left(\hat{\mathbf{G}}\hat{\mathbf{G}}^H\right)^{-1}\hat{\mathbf{G}}\mathbb{E}\left[\tilde{\mathbf{g}}_k^H \tilde{\mathbf{g}}_k\right]\hat{\mathbf{G}}^H\left(\hat{\mathbf{G}}\hat{\mathbf{G}}^H\right)^{-1}\right]\right\} \\
&= P_f \rho_k^2 e^{-n_{\Delta\tau}\sigma_\phi^2}\sum_{i=1}^K \eta_i \chi_{ki},
\end{aligned}
\tag{9.178}
$$

where χ_{ki} denotes the ith diagonal element of

$$
\mathbb{E}\left[\left(\hat{\mathbf{G}}\hat{\mathbf{G}}^H\right)^{-1}\hat{\mathbf{G}}\mathbb{E}\left[\tilde{\mathbf{g}}_k^H \tilde{\mathbf{g}}_k\right]\hat{\mathbf{G}}^H\left(\hat{\mathbf{G}}\hat{\mathbf{G}}^H\right)^{-1}\right].
\tag{9.179}
$$

Likewise, the variance of $\mathcal{I}_2$ is given by

$$
\begin{aligned}
\mathbb{E}\left[|\mathcal{I}_2|^2\right] &= \mathbb{E}\left[|\sqrt{P_f(1-\rho_k^2)}e^{j(\varphi_{k,d}-\varphi_{k,p})}\mathbf{e}_k \vartriangle \boldsymbol{\Phi}\hat{\mathbf{G}}^H\left(\hat{\mathbf{G}}\hat{\mathbf{G}}^H\right)^{-1}\mathbf{D}_\eta \mathbf{u}|^2\right] \\
&= P_f\left(1-\rho_k^2\right)\mathbb{E}\left[\left\|\mathbf{e}_k \vartriangle \boldsymbol{\Phi}\hat{\mathbf{G}}^H\left(\hat{\mathbf{G}}\hat{\mathbf{G}}^H\right)^{-1}\mathbf{D}_\eta \mathbf{u}\right\|^2\right] \\
&= P_f\left(1-\rho_k^2\right)e^{-n_{\Delta\tau}\sigma_\phi^2} tr\left\{\mathbf{D}_\eta^2 \mathbb{E}\left[\left(\hat{\mathbf{G}}\hat{\mathbf{G}}^H\right)^{-1}\hat{\mathbf{G}}\mathbf{E}_k\hat{\mathbf{G}}^H\left(\hat{\mathbf{G}}\hat{\mathbf{G}}^H\right)^{-1}\right]\right\} \\
&= P_f\left(1-\rho_k^2\right)e^{-n_{\Delta\tau}\sigma_\phi^2}\sum_{i=1}^K \eta_i \xi_{ki}
\end{aligned}
\tag{9.180}
$$

where $\mathbf{E}_k = \mathbb{E}\left[\mathbf{e}_k^H \mathbf{e}_k\right] = \text{diag}\left\{\beta_{1k}, \beta_{2k}, \dots, \beta_{Mk}\right\} \in C^{M\times M}$, and ξ_{ki} represents the ith diagonal element of $\mathbb{E}\left[\left(\hat{\mathbf{G}}\hat{\mathbf{G}}^H\right)^{-1}\hat{\mathbf{G}}\mathbf{E}_k\hat{\mathbf{G}}^H\left(\hat{\mathbf{G}}\hat{\mathbf{G}}^H\right)^{-1}\right]$.

Substituting Eqs. (9.178) and (9.180) into Eq. (9.176), yields

$$
\gamma_k = \frac{\rho_k^2 \eta_k}{\rho_k^2 \sum_{i=1}^K \eta_i \chi_{ki} + \left(1-\rho_k^2\right)\sum_{i=1}^K \eta_i \xi_{ki} + \dfrac{\sigma_n^2}{P_f e^{-n_{\Delta\tau}\sigma_\phi^2}}}.
\tag{9.181}
$$

Taking into account the propagation delay over the air interface $n_{ai} = \left\lceil \frac{\tau_{ul}+\tau_{dl}}{T_s} \right\rceil$ and the delay $n_{\Delta\tau}$, the achievable spectral efficiency of the kth user is given by:

$$
R_k = \left(1 - \frac{n_{ai}+n_{\Delta\tau}}{T_b}\right)\log_2\left(1+\gamma_k\right).
\tag{9.182}
$$

Figure 9.7 provides a comparison with respect to cumulative distribution functions (CDFs) of per-user Spectral Efficiency (SE) by varying the velocity v or accumulative phase noise T_{PN}. The performance curve of Zero-Forcing Precoding

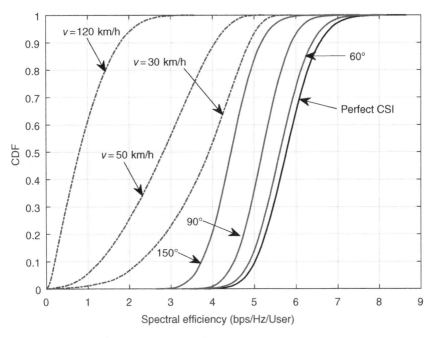

Figure 9.7 The CDF comparison under different conditions of user mobility and phase noise. Source: Jiang and Schotten [2021a]/With permission of IEEE.

(ZFP) using perfect CSI is applied as a benchmark, where the UEs are stationary ($v = 0$ km/h) and the transceivers have perfect local oscillators ($T_{PN} = 0°$). To observe the effect of user mobility, we first set $T_{PN} = 0°$ and select three typical values: $v = 30$, 50, and 120 km/h. Without loss of generality, the overall delay is simply set to $\triangle \tau = 1$ ms since the aging effect of user mobility is decided by the combination of velocity and delay. Even at low mobility of $v = 30$ km/h, which is equivalent to very high correlation $\rho = 0.97$, the performance deterioration is already remarkable. To be specific, the 5%-likely per-user SE reduces to 1.8 bps/Hz, in comparison with 4.8 bps/Hz of the benchmark, amounting to a loss of 62.5%. The 50%-likely (median) per-user SE degrades 32%, dropping from 5.7 to 3.9 bps/Hz. With the increase of v, the performance loss becomes more substantial. At high mobility of $v = 120$ km/h, the 5%-likely and median SE further decrease to 0.13 and 0.79 bps/Hz, amounting to a very high loss of 97% and 86%, respectively. In addition, the impact of phase noise is investigated by using the selected phase noise of $T_{PN}^2 = 60°$, $90°$, and $150°$, where the UEs are set to be stationary $v = 0$ km/h. With small phase noise of 60°, as shown in the figure, the performance loss is marginal. Increased to 150°, the 5%-likely and median SE degrade to 3.5 and 4.5 bps/Hz, equivalent to a loss of 27% and 23%, respectively.

9.5 Opportunistic Cell-Free Communications

Exploiting the degree of freedom in the frequency domain enabled by the OFDM transmission in wideband systems and the near-far effect among different APs, a cell-free massive MIMO system can implement opportunistic communications to improve its power efficiency and spectral efficiency. The key idea is to assign orthogonal frequency-domain resources to different users so that each subcarrier or Resource Block (RB) carries only a single user. This setup not only avoids multi-user interference but also simplifies the system design. Then, a number of access points with strong large-scale fading (defined as the near APs) are opportunistically selected to serve this user. At the same time, the far APs with weak large-scale fading are deactivated over the assigned subcarriers or RBs for this user. As a side effect, the number of active APs per subcarrier becomes small, which enables the use of downlink pilots so that the user can perform coherent detection. The major technical benefits of Opportunistic AP Selection (OAS) scheme are twofold:

- *Opportunistic Gain*: From the point of view of a typical user, a near AP has a favorable channel with a small path loss. In contrast, the energy radiated from a far AP is wasted over its long propagation distance. In other words, the same amount of power transmitted from a near AP generates a much stronger received power than a far AP, resulting in high power and spectral efficiencies.
- *Coherent Gain*: From the perspective of each subcarrier or resource block, only a few APs serve a single user. Then, a high-dimensional massive MIMO system is transformed into a low-dimensional Multiple-Input Single-Output (MISO) system. As a result, the prohibitive overhead of inserting downlink pilots, which is proportional to the massive number of base-station antennas, can be alleviated. The user can obtain the instantaneous CSI by estimating downlink pilots, rather than only knowing statistical CSI, and then perform coherent detection. Hence, a fundamental problem restricting the downlink performance of massive MIMO can be solved thanks to opportunistic AP selection.

9.5.1 Cell-free Massive Wideband Systems

Consider a cell-free massive MIMO system where M randomly distributed APs that are connected to a CPU serve K users in a geographical area. Without losing generality, assume that each AP and UE is equipped with a single antenna for simple analysis. Consider frequency-selective fading in wideband systems, where the channel between AP m and user k can be modeled as a linear time-varying filter in a baseband equivalent basis, i.e.

$$\mathbf{h}_{mk}[t] = [h_{mk,0}[t], h_{mk,1}[t], \dots, h_{mk,L_{mk}-1}[t]]^T, \tag{9.183}$$

where the filter length L_{mk} depends on the delay spread and the sampling interval. Taking into account large-scale fading β_{mk}, the channel filter between AP m and user k can be modeled by

$$
\begin{aligned}
\mathbf{g}_{mk}[t] &= [g_{mk,0}[t], g_{mk,1}[t], \dots, g_{mk,L_{mk}-1}[t]]^T \\
&= \sqrt{\beta_{mk}} \mathbf{h}_{mk}[t],
\end{aligned}
\tag{9.184}
$$

with $g_{mk,l}[t] = \sqrt{\beta_{mk}} h_{mk,l}[t]$, $\forall l = 0, 1, \dots, L_{mk} - 1$. It is observed that the cell-free structure results in the *near-far* effect among different APs from the perspective of a typical user. The APs can be therefore divided into two categories: the near APs and the far APs, similar to the near and far users from the perspective of a base station in the conventional cellular systems.

The signal transmission in an OFDM system is organized in block-wise. Denote the frequency-domain symbol block of AP m on the tth OFDM symbol by

$$
\tilde{\mathbf{x}}_m[t] = \left[\tilde{x}_{m,0}[t], \tilde{x}_{m,1}[t], \dots, \tilde{x}_{m,N-1}[t] \right]^T.
\tag{9.185}
$$

Performing an N-point inverse discrete Fourier transform (IDFT), $\tilde{\mathbf{x}}_m[t]$ is converted into a time-domain sequence

$$
\mathbf{x}_m[t] = \left[x_{m,0}[t], x_{m,1}[t], \dots, x_{m,N-1}[t] \right]^T
\tag{9.186}
$$

in terms of

$$
x_{m,n'}[t] = \frac{1}{N} \sum_{n=0}^{N-1} \tilde{x}_{m,n}[t] e^{\frac{2\pi j n' n}{N}}, \quad \forall n'.
\tag{9.187}
$$

Defining the discrete Fourier transform (DFT) matrix

$$
\mathbf{D} = \begin{bmatrix} \Omega_N^{00} & \cdots & \Omega_N^{0(N-1)} \\ \vdots & \ddots & \vdots \\ \Omega_N^{(N-1)0} & \cdots & \Omega_N^{(N-1)(N-1)} \end{bmatrix}
\tag{9.188}
$$

with a primitive Nth root of unity $\Omega_N^{nn'} = e^{-\frac{2\pi j n' n}{N}}$, the OFDM modulation is expressed in matrix form as

$$
\mathbf{x}_m[t] = \mathbf{D}^{-1} \tilde{\mathbf{x}}_m[t] = \frac{1}{N} \mathbf{D}^* \tilde{\mathbf{x}}_m[t].
\tag{9.189}
$$

Cyclic prefix (CP) is inserted between two transmission blocks to preserve subcarrier orthogonality and absorb inter-symbol interference. The transmitted baseband signal with CP is denoted by $\mathbf{x}_m^{cp}[t]$. Going through the wireless channel, it results in a received signal component $\mathbf{x}_m^{cp}[t] * \mathbf{g}_{mk}[t]$ at the typical user k, where $*$ denotes *the linear convolution*. Consequently, the received signal at user k is given by

$$
\mathbf{y}_k^{cp}[t] = \sum_{m=1}^{M} \mathbf{g}_{mk}[t] * \mathbf{x}_m^{cp}[t] + \mathbf{z}_k[t],
\tag{9.190}
$$

where $\mathbf{z}_k[t]$ denotes a vector of additive white Gaussian noise with zero mean and variance σ_z^2, i.e. $\mathbf{z}_k \sim \mathcal{CN}(\mathbf{0}, \sigma_z^2 \mathbf{I})$. Removing the CP, we get

$$\mathbf{y}_k[t] = \sum_{m=1}^{M} \mathbf{g}_{mk}^{N}[t] \otimes \mathbf{x}_m[t] + \mathbf{z}_k[t], \qquad (9.191)$$

where $\otimes$ stands for *the cyclic convolution*, and $\mathbf{g}_{mk}^{N}[t]$ is an N-length zero-padded vector of $\mathbf{g}_{mk}[t]$. Then, the frequency-domain received signal is computed by

$$\tilde{\mathbf{y}}_k[t] = \mathbf{D}\mathbf{y}_k[t]. \qquad (9.192)$$

Substituting Eq. (9.191) into Eq. (9.192), and applying *the convolution theorem* [Jiang and Kaiser, 2016], yields

$$\tilde{\mathbf{y}}_k[t] = \sum_{m=1}^{M} \mathbf{D}\left(\mathbf{g}_{mk}^{N}[t] \otimes \mathbf{x}_m[t]\right) + \mathbf{D}\mathbf{z}_k[t]$$

$$= \sum_{m=1}^{M} \tilde{\mathbf{g}}_{mk}[t] \odot \tilde{\mathbf{x}}_m[t] + \tilde{\mathbf{z}}_k[t], \qquad (9.193)$$

where $\odot$ represents the Hadamard product [Figure 9.8]. For a typical subcarrier n, the downlink signal model is expressed by

$$\tilde{y}_{k,n}[t] = \sum_{m=1}^{M} \tilde{g}_{mk,n}[t]\tilde{x}_{m,n}[t] + \tilde{z}_{k,n}[t], \quad k \in \{1, \dots, K\}. \qquad (9.194)$$

9.5.2 Opportunistic AP Selection

The downlink transmission from the APs to the users and the uplink transmission from the users to the APs are separated by time-division multiplexing (TDD) with the assumption of perfect channel reciprocity. The length of a radio frame is generally less than the channel coherence time, and therefore the channel condition is regarded as constant within a frame. Without losing generality, the time indexing of signals is ignored for simple analysis.

The communication process of the OAS scheme in a cell-free massive MIMO-OFDM system is depicted as follows:

- AP m, $\forall m$ measures the large-scale fading β_{mk}, $k = 1, 2, \dots, K$, in a long-term basis, and reports this information periodically to the CPU. Thus, the CPU has a global knowledge of large-scale CSI $\mathbf{B} \in \mathbb{C}^{M \times K}$, where $[\mathbf{B}]_{m,k} = \beta_{mk}$ with $[\cdot]_{m,k}$ denotes the (m, k)th entry of a matrix. Since β_{mk} is frequency independent and varies slowly, this measurement is practically easy to implement.
- UE k, $\forall k$ periodically reports its data-rate request with a scalar $r_{q,k}$ through the uplink signaling. Then, the CPU knows $\mathbf{r}_q = \{r_{q,1}, r_{q,2}, \dots, r_{q,K}\}$.
- *Frequency-Domain Resource Allocation*: The CPU makes a decision of resource allocation as a function of the users' requests, i.e. $\{\mathbb{B}_1, \dots, \mathbb{B}_K\} = f(\mathbf{r}_q)$, where

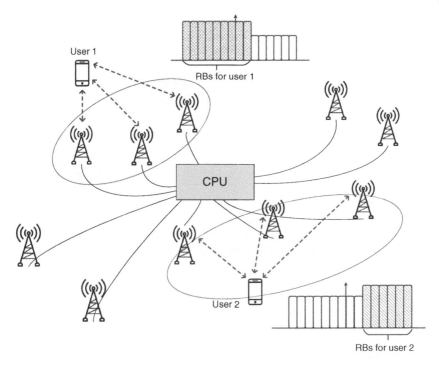

Figure 9.8 Illustration of opportunistic AP selection in a cell-free massive MIMO-OFDM system.

the specific implementation of $f(\cdot)$ is based on some particular criteria, e.g. fairness, priority, and performance. The resource pool consists of N OFDM subcarriers, denoted by the set of subcarrier indices $\mathbb{B} = \{0, 1, 2 \ldots, N-1\}$. Using $\mathbb{B}_k$ to denote the indices of subcarriers assigned to user k, we have $\bigcup_{k=1}^{K} \mathbb{B}_k \in \mathbb{B}$ (when all subcarrier are allocated, $\bigcup_{k=1}^{K} \mathbb{B}_k = \mathbb{B}$). The subcarriers are allocated orthogonally, satisfying $\mathbb{B}_k \cap \mathbb{B}_{k'} = \varnothing, \forall k' \neq k$. The time interval of resource allocation depends on the system design.

- *Opportunistic Selection*: The CPU selects the opportunistic APs for each user in terms of large-scale fading. Assume the number of selected APs is M_s, where $1 \leqslant M_s \leqslant M$. Order the indices of the APs in terms of their large-scale fading in a descending order, and then select the first M_s APs. We denote the set of opportunistic APs for user k by $\mathbb{M}_k = \{\dot{m}_1^k, \ldots, \dot{m}_{M_s}^k\}$. If $M_s = M$, all APs participate into the transmission, without any selection. If $M_s = 1$, only a single AP with the largest large-scale fading is determined, i.e.

$$\dot{m}_1^k = \arg \max_{m=1,\ldots,M} (\mathbf{b}_k),$$ (9.195)

where $\mathbf{b}_k$ denotes the kth row of $\mathbf{B}$.

- *Uplink Transmission*: User k, $\forall k$ transmits its data and particular pilot sequence over its assigned subcarriers $\mathbb{B}_k$. The opportunistic APs within $\mathbb{M}_k$ estimate the uplink CSI $\hat{g}_{mk,n}$, where $m \in \mathbb{M}_k$ and $n \in \mathbb{B}_k$. The APs detect the uplink data coherently with the knowledge of uplink CSI.
- *Conjugate Beamforming*: Afterward, AP m, $\forall m \in \mathbb{M}_k$ knows the downlink CSI $\hat{g}_{mk,n}$ according to channel reciprocity. Then, it transmits the modulated symbol $s_{k,n}$, $n \in \mathbb{B}_k$ with $\mathbb{E}[|s_{k,n}|^2] = 1$, and downlink pilot sequences over the assigned subcarriers $\mathbb{B}_k$. Applying conjugate beamforming in the frequency domain, the transmitted symbol at the mth AP is

$$\tilde{x}_{m,n} = \sqrt{\eta_{mk} P_d} \hat{g}_{mk,n}^* s_{k,n}, \tag{9.196}$$

where $\sqrt{\eta_{mk}}$, $0 \le \eta_{mk} \le 1$ denotes the power-control coefficient, and P_d is a unified power constraint of each AP.
- *Coherent Detection*: User k estimates the downlink CSI $\hat{g}_{mk,n}$, where $m \in \mathbb{M}_k$ and $n \in \mathbb{B}_k$, and detects the downlink data coherently with the aid of $\hat{g}_{mk,n}$.

9.5.3   Spectral Efficiency Analysis

The performance of three different schemes are investigated to shed light on the gains of opportunistic selection and downlink channel estimation. First, the performance in terms of SE for the conventional CFmMIMO-OFDM system without opportunistic AP selection, denoted by *Full AP*, is analyzed, as a benchmark for comparison. Second, the performance of the system that selects M_s opportunistic APs among M APs but does not insert downlink pilots is also derived. Finally, the SE of opportunistic AP selection with the insertion of downlink pilots, denoted by *OAS-DP*, is analyzed.

As a baseline, the conventional conjugate beamforming in CFmMIMO [Nayebi et al., 2017] is directed applied over each subcarrier in CFmMIMO-OFDM. To do so, each AP multiplexes a total of K symbols, i.e. $s_{k,n}$ intended to user k, $k = 1, \ldots, K$, before transmission. With a power-control coefficient $\sqrt{\eta_{mk}}$, $0 \le \eta_{mk} \le 1$, the transmitted signal of the mth AP at subcarrier n is

$$\tilde{x}_{m,n} = \sqrt{P_d} \sum_{k=1}^{K} \sqrt{\eta_{mk}} \hat{g}_{mk,n}^* s_{k,n}, \tag{9.197}$$

where $\hat{g}_{mk,n}$ denotes an estimate of $\tilde{g}_{mk,n}$, and $\hat{g}_{mk,n} = \tilde{g}_{mk,n} - \xi_{mk,n}$ with estimation error $\xi_{mk,n}$ raised by additive noise. Applying the MMSE estimation, we gets $\hat{g}_{mk,n} \in \mathcal{CN}(0, \alpha_{mk})$ with $\alpha_{mk} = \frac{P_u \beta_{mk}^2}{P_u \beta_{mk} + \sigma_z^2}$, where P_u is the uplink power constraint, in comparison with $\tilde{g}_{mk,n} \in \mathcal{CN}(0, \beta_{mk})$ [Jiang and Schotten, 2021a].

In the conventional CFmMIMO, there is no downlink pilot and channel estimation due to the prohibitive overhead of inserting pilots over a massive number of antennae. Consequently, each user is assumed to only have the knowledge of channel statistics $\mathbb{E}\left[\left|\hat{g}_{mk,n}\right|^2\right] = \alpha_{mk}$, a.k.a. channel hardening, rather than the channel realization $\hat{g}_{mk,n}$. Substituting Eq. (9.197) into Eq. (9.194), yields the received signal at user k:

$$\tilde{y}_{k,n} = \sqrt{P_d} \sum_{m=1}^{M} \tilde{g}_{mk,n} \sum_{k'=1}^{K} \sqrt{\eta_{mk'}} \hat{g}_{mk',n}^* s_{k',n} + \tilde{z}_{k,n}$$

$$= \underbrace{\sqrt{P_d} \sum_{m=1}^{M} \sqrt{\eta_{mk}} \mathbb{E}\left[\left|\hat{g}_{mk,n}\right|^2\right] s_{k,n}}_{\text{Desired signal}} + \underbrace{\sqrt{P_d} \sum_{m=1}^{M} \hat{g}_{mk,n} \sum_{k' \neq k}^{K} \sqrt{\eta_{mk'}} \hat{g}_{mk',n}^* s_{k',n}}_{\text{Inter-user interference}}$$

$$+ \underbrace{\sqrt{P_d} \sum_{m=1}^{M} \sqrt{\eta_{mk}} \left(\left|\hat{g}_{mk,n}\right|^2 - \mathbb{E}\left[\left|\hat{g}_{mk,n}\right|^2\right]\right) s_{k,n}}_{\text{Error due to CSI statistics}}$$

$$+ \underbrace{\sqrt{P_d} \sum_{m=1}^{M} \xi_{mk,n} \sum_{k'=1}^{K} \sqrt{\eta_{mk'}} \hat{g}_{mk',n}^* s_{k',n}}_{\text{Channel-estimation error}} + \underbrace{\tilde{z}_{k,n}}_{\text{Noise}} \ .$$

The spectral efficiency of user k on subcarrier n, $\forall n = 0, 1, \dots, N-1$ is lower bounded by $\log_2\left(1 + \gamma_k^{\langle n \rangle}\right)$ with

$$\gamma_k^{\langle n \rangle} = \frac{\left(\sum_{m=1}^{M} \sqrt{\eta_{mk}} \alpha_{mk}\right)^2}{\sum_{m=1}^{M} \beta_{mk} \sum_{k'=1}^{K} \eta_{mk'} \alpha_{mk'} + \frac{1}{\gamma_t}}, \tag{9.198}$$

with the transmit SNR $\gamma_t = P_d / \sigma_z^2$.

The OAS scheme exploits the degree of freedom enabled by the frequency domain to assign different users to orthogonal resources. As a result, the inter-user interference vanishes since each OFDM subcarrier accommodates a single user. Therefore, substituting $K = 1$ into Eq. (9.198) yields the performance of the first scheme with *Full AP* transmission, i.e.

$$\gamma_k^{\langle n \rangle} = \frac{\left(\sum_{m=1}^{M} \sqrt{\eta_{mk}} \alpha_{mk}\right)^2}{\sum_{m=1}^{M} \beta_{mk} \eta_{mk} \alpha_{mk} + \frac{1}{\gamma_t}}. \tag{9.199}$$

To shed light on the effect of opportunistic selection, the performance of selecting M_s APs without adding downlink pilots is then investigated. That is, each user only have the knowledge of channel statistics rather than the channel realization.

Substituting Eq. (9.196) into Eq. (9.194) to get the received signal at user k on subcarrier $n \in \mathbb{B}_k$ as

$$
\begin{aligned}
\tilde{y}_{k,n} = & \sqrt{P_d} \sum_{m \in \mathbb{M}_k} \tilde{g}_{mk,n} \sqrt{\eta_{mk}} \hat{g}^*_{mk,n} s_{k,n} + \tilde{z}_{k,n} \\
= & \underbrace{\sqrt{P_d} \sum_{m \in \mathbb{M}_k} \sqrt{\eta_{mk}} \mathbb{E}\left[\left|\hat{g}_{mk,n}\right|^2\right] s_{k,n}}_{\text{Desired signal}} \\
& + \underbrace{\sqrt{P_d} \sum_{m \in \mathbb{M}_k} \sqrt{\eta_{mk}} \left(\left|\hat{g}_{mk,n}\right|^2 - \mathbb{E}\left[\left|\hat{g}_{mk,n}\right|^2\right]\right) s_{k,n}}_{\text{Error due to CSI statistics}} \\
& + \underbrace{\sqrt{P_d} \sum_{m \in \mathbb{M}_k} \xi_{mk,n} \sqrt{\eta_{mk}} \hat{g}^*_{mk,n} s_{k,n}}_{\text{Channel-estimation error}} + \underbrace{\tilde{z}_{k,n}}_{\text{Noise}} \ .
\end{aligned}
\tag{9.16}
$$

Note that the received signal of user k on subcarrier $n \in \{\mathbb{B} - \mathbb{B}_k\}$ is $\tilde{y}_{k,n} = 0$. Likewise, the spectral efficiency of user k on subcarrier $n \in \mathbb{B}_k$ is lower bounded by $\log_2\left(1 + \gamma_k^{\langle n \rangle}\right)$ with

$$
\gamma_k^{\langle n \rangle} = \frac{\left(\sum_{m \in \mathbb{M}_k} \sqrt{\eta_{mk}} \alpha_{mk}\right)^2}{\sum_{m \in \mathbb{M}_k} \eta_{mk} \beta_{mk} \alpha_{mk} + \frac{1}{\gamma_t}}.
\tag{9.17}
$$

Thanks to the opportunistic AP selection, there are only a few number of active APs over each subcarrier in the proposed scheme, while other far APs are turned off. From the perspective of a typical subcarrier, it is a low-dimensional MISO system, where the overhead of inserting downlink pilots is acceptable. As a result, user k obtains the estimated CSI $\hat{g}_{mk,n}$ rather than the channel statistics α_{mk}. Thus, the received signal at user k in Eq. (9.16) can be reformed as

$$
\begin{aligned}
\tilde{y}_{k,n} = & \sqrt{P_d} \sum_{m \in \mathbb{M}_k} \tilde{g}_{mk,n} \sqrt{\eta_{mk}} \hat{g}^*_{mk,n} s_{k,n} + \tilde{z}_{k,n} \\
= & \underbrace{\sqrt{P_d} \sum_{m \in \mathbb{M}_k} \sqrt{\eta_{mk}} \left|\hat{g}_{mk,n}\right|^2 s_{k,n}}_{\text{Desired signal}} + \underbrace{\sqrt{P_d} \sum_{m \in \mathbb{M}_k} \xi_{mk,n} \sqrt{\eta_{mk}} \hat{g}^*_{mk,n} s_{k,n}}_{\text{Channel-Estimation error}} + \underbrace{\tilde{z}_{k,n}}_{\text{Noise}} \ .
\end{aligned}
$$

$$
\tag{9.18}
$$

Applying the coherent detection, the spectral efficiency of user k on subcarrier $n \in \mathbb{B}_k$ is expressed by $\log_2\left(1 + \gamma_k^{\langle n \rangle}\right)$ with

$$\gamma_k^{\langle n \rangle} = \frac{\left(\sum_{m \in \mathbb{M}_k} \sqrt{\eta_{mk}} \left|\hat{g}_{mk,n}\right|^2\right)^2}{\sum_{m \in \mathbb{M}_k} \eta_{mk}(\beta_{mk} - \alpha_{mk})\alpha_{mk} + \frac{1}{\gamma_t}}. \tag{9.19}$$

Figure 9.9 demonstrates the CDFs of different schemes. First, the curve of *Full AP* stands for the conventional CFmMIMO-OFDM system, where all $M = 128$ APs serve an assigned user on a typical subcarrier without opportunistic AP selection. The achieved 95%-likely spectral efficiency is around 2.2 bps/Hz and the 50%-likely or median spectral efficiency is approximately 3.2 bps/Hz. If $M_s = 10$ active APs are selected in terms of large-scale fading, and the power constraint of each AP is the same as that of the *Full AP*, the achieved SE of *OAS Power Saving* is slightly inferior to the *Full AP*. It has 95%-likely SE of around 1.9 bps/Hz and the median SE of approximately 2.7 bps/Hz. However, it significantly outperforms in terms of power efficiency since only $M_s = 10$ APs are active, compared with $M = 128$ APs in the *Full AP*, amounting to a power saving of 92.19%. That is because that the power of the far APs cannot effectively transfer to the received power due to severe propagation loss. Turning off the far APs does not affect the total received power at the user.

As a fair comparison, assume the selected APs has the same total power as the *Full AP*, i.e. each opportunistic AP uses a power $M/M_s = 12.8$ times higher. As shown by the CDF of *OAS Equal Total Power*, the 95%-likely SE substantially

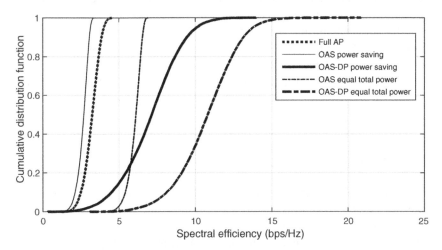

Figure 9.9 CDFs of the achievable spectral efficiency for different schemes in a cell-free massive MIMO-OFDM system.

increases to 5.2 bps/Hz and the median SE reaches 6.1 bps/Hz. Next, we can observe the significant performance gain of downlink pilots that enable the coherent detection at the user. Even if the total transmit power is less than 10% of the *Full AP*, *OAS-DP Power Saving* achieves a 95%-likely SE of around 3.8 bps/Hz and the median SE of 7.2 bps/Hz. Compared with the *Full AP*, it realizes the performance gain of about 70% and 125% in 95%-likely and median SE, respectively, while achieving a 10-fold power efficiency. Under the same total power, the superiority of opportunistic AP selection with the aid of downlink pilot is more significant. In this case, the 95%-likely SE substantially increases to 7.4 bps/Hz and the median SE reaches 10.8 bps/Hz. In a nutshell, the numerical results corroborate the great advantage of opportunistic AP selection, as well as the empowered downlink CSI, to boost both the power and spectral efficiencies in a cell-free massive MIMO system.

9.6 Summary

This chapter first introduced the critical issues of multi-user MIMO techniques, including the principle of the well-known capacity-achieving method called dirty-paper coding. MU-MIMO facilitates the use of low-complexity, low-cost terminals and is less vulnerable to propagation environments. Most importantly, it achieves a sum throughput that is higher than the channel capacity of SU-MIMO. Nevertheless, the conventional MU-MIMO is still hard to scale up for high-order spatial multiplexing. This chapter then studied a revolutionary technique called massive MIMO that breaks this scalability barrier by not attempting to achieve the full Shannon limit and paradoxically by increasing the size of the system. Finally, a distributed massive MIMO system called cell-free massive MIMO was presented, where a large number of service antennas randomly spread over a wide area. Cell-free setup is particularly attractive for some of 5G and upcoming 6G deployment scenarios, such as a campus or private network dedicated to an industrial site.

References

Caire, G. and Shamai, S. [2001], On achievable rates in a multi-antenna Gaussian broadcast channel, *in* 'Proceedings of 2001 IEEE International Symposium on Information Theory', Washington, DC, USA, p. 147.

Caire, G. and Shamai, S. [2003], 'On the achievable throughput of a multiantenna Gaussian broadcast channel', *IEEE Transactions on Information Theory* **49**(3), 1691–1706.

Chen, R., Heath, R. W. and Andrews, J. G. [2007], 'Transmit selection diversity for unitary precoded multiuser spatial multiplexing systems with linear receivers', *IEEE Transactions on Signal Processing* **55**(3), 1159–1171.

Choi, L.-U. and Murch, R. [2004], 'A transmit preprocessing technique for multiuser MIMO systems using a decomposition approach', *IEEE Transactions on Wireless Communications* **3**(1), 20–24.

Costa, M. [1983], 'Writing on dirty paper', *IEEE Transactions on Information Theory* **29**(3), 439–441.

Hassibi, B. and Hochwald, B. [2003], 'How much training is needed in multiple-antenna wireless links?', *IEEE Transactions on Information Theory* **49**(4), 951–963.

Jiang, W. and Kaiser, T. [2016], From OFDM to FBMC: Principles and Comparisons, *in* F. L. Luo and C. Zhang, eds, '*Signal Processing for 5G: Algorithms and Implementations*', John Wiley & Sons and IEEE Press, United Kingdom, Chapter 3.

Jiang, W. and Schotten, H. [2021a], 'Impact of channel aging on zero-forcing precoding in cell-free massive MIMO systems', *IEEE Communications Letters* **25**(9), 3114–3118.

Jiang, W. and Schotten, H. D. [2021b], 'Cell-free massive MIMO-OFDM transmission over frequency-selective fading channels', *IEEE Communications Letters* **25**(8), 2718–2722.

Jiang, W. and Schotten, H. D. [2021c], 'A simple cooperative diversity method based on deep-learning-aided relay selection', *IEEE Transactions on Vehicular Technology* **70**(5), 4485–4500.

Jiang, W., Kaiser, T. and Vinck, A. J. H. [2016], 'A robust opportunistic relaying strategy for co-operative wireless communications', *IEEE Transactions on Wireless Communications* **15**(4), 2642–2655.

Jose, J., Ashikhmin, A., Marzetta, T. L. and Vishwanath, S. [2011], 'Pilot contamination and precoding in multi-cell TDD systems', *IEEE Transactions on Wireless Communications* **10**(8), 2640–2651.

Krishnan, R., Khanzadi, M. R., Krishnan, N., Wu, Y., Amat, A. G., Eriksson, T. and Schober, R. [2016], 'Linear massive MIMO precoders in the presence of phase noise – a large-scale analysis', *IEEE Transactions on Vehicular Technology* **65**(5), 3057–3071.

Lu, L., Li, G. Y., Swindlehurst, A. L., Ashikhmin, A. and Zhang, R. [2014], 'An overview of massive MIMO: Benefits and challenges', *IEEE Journal of Selected Topics in Signal Processing* **8**(5), 742–758.

Marzetta, T. L. [2010], 'Noncooperative cellular wireless with unlimited numbers of base station antennas', *IEEE Transactions on Wireless Communications* **9**(11), 3590–3600.

Marzetta, T. L. [2015], 'Massive MIMO: An introduction', *Bell Labs Technical Journal* **20**, 11–22.

Marzetta, T. L., Larsson, E. G., Yang, H. and Ngo, H. Q. [2016], *Fundamentals of Massive MIMO*, Cambridge University Press, Cambridge, United Kingdom.

Nayebi, E., Ashikhmin, A., Marzetta, T. L., Yang, H. and Rao, B. D. [2017], 'Precoding and power optimization in cell-free massive MIMO systems', *IEEE Transactions on Wireless Communications* **16**(7), 4445–4459.

Ngo, H. Q., Ashikhmin, A., Yang, H., Larsson, E. G. and Marzetta, T. L. [2017], 'Cell-free massive MIMO versus small cells', *IEEE Transactions on Wireless Communications* **16**(3), 1834–1850.

Sanguinetti, L. and Poor, H. V. [2009], Fundamentals of multi-user MIMO communications, *in* V. Tarokh, ed., '*New Directions in Wireless Communications Research*', Springer, Boston, USA, Chapter 6, pp. 139–173.

Spencer, Q., Swindlehurst, A. and Haardt, M. [2004], 'Zero-forcing methods for downlink spatial multiplexing in multiuser MIMO channels', *IEEE Transactions on Signal Processing* **52**(2), 461–471.

Tse, D. and Viswanath, P. [2005], *Fundamentals of Wireless Communication*, Cambridge University Press, Cambridge, United Kingdom.

Yang, H. and Marzetta, T. L. [2013], 'Performance of conjugate and zero-forcing beamforming in large-scale antenna systems', *IEEE Journal on Selected Areas in Communications* **31**(2), 172–179.

Zhang, J., Bjornson, E., Matthaiou, M., Ng, D. W. K., Yang, H. and Love, D. J. [2020], 'Prospective multiple antenna technologies for beyond 5G', *IEEE Journal on Selected Areas in Communications* **38**(8), 1637–1660.

10

Adaptive and Non-Orthogonal Multiple Access Systems in 6G

One of the technological trends in wireless communications is that the signal bandwidth becomes increasingly wide, aiming to support a higher transmission rate. However, with the decrease of the symbol period, the delay spread in a multipath fading channel raises severe inter-symbol interference and substantially constrains the achievable transmission rate. In traditional single-carrier transmission, an equalizer with several hundred taps might be necessary to effectively mitigate the inter-symbol interference in a wideband system, which is too complex to implement in a practical system. In this regard, multi-carrier modulation provides an efficient alternative by splitting a wideband signal into a set of orthogonal narrowband signals. Due to its ability to cope with multipath frequency-selective fading without the need for complex equalization and simple implementation through the use of the digital Fourier transform, Orthogonal Frequency-Division Multiplexing (OFDM), as a kind of multi-carrier modulation, has become the most dominant waveform design technique for wired and wireless communication systems over the past two decades since its first proposal by Chang [1966]. As a result, it has been extensively applied in many well-known standards, e.g. Digital Subscriber Line (DSL), Digital Video Broadcasting-Terrestrial (DVB-T), Wi-Fi, WiMAX, LTE, LTE-Advanced, and 5G NR. On the other hand, a cellular network needs to accommodate a lot of active subscribers simultaneously over a finite amount of time-frequency resources. Therefore, efficient allocation of radio resources among users is a critical design aspect of both uplink and downlink channels since the bandwidth is usually scarce and expensive. The share of a communications channel among multiple users that are geographically distributed is referred to as *multiple access*. The conventional multiple-access techniques orthogonally split the signaling dimensions over time-domain, frequency-domain, or code-domain. As an extension of OFDM transmission to implementing a multi-user system, Orthogonal Frequency-Division Multiple Access (OFDMA) provides an efficient and flexible multi-access technology by exploiting time-frequency resources simultaneously.

6G Key Technologies: A Comprehensive Guide, First Edition. Wei Jiang and Fa-Long Luo.
© 2023 The Institute of Electrical and Electronics Engineers, Inc. Published 2023 by John Wiley & Sons, Inc.

Historically, most of the mobile systems were based on orthogonal multiple-access techniques for simple system design and low-complexity receiver implementation. To meet the heterogeneous requirements on massive connectivity, high spectral efficiency, low latency, and improved fairness, the 5G system has accepted a new technique called Non-Orthogonal Multiple Access (NOMA) as one of its multi-access methods. In contrast to the conventional orthogonal schemes, the key distinguishing feature of NOMA is to serve a higher number of users than the number of orthogonal resource units with the aid of non-orthogonal resource sharing, with the price of the sophisticated inter-user interference cancelation at the receiver. Both orthogonal and NOMA are envisioned to be evolved further and play a critical role in the upcoming 6G system design.

This chapter will introduce both orthogonal and NOMA techniques, consisting of

- Modeling a frequency-selective fading channel in wireless broadband communications.
- The fundamentals of multi-carrier modulation, including the synthesis and analysis filters, polyphase implementation, and filter-bank multi-carrier.
- A thorough introduction of the OFDM technique, consisting of its basic principle, the efficient implementation through digital Fourier transform, the insertion of cyclic prefix, frequency-domain signal processing, and out-of-band emission suppression.
- OFDMA transmission in the downlink and single-carrier FDMA in the uplink.
- The basic principle and advantage of cyclic delay diversity.
- Multi-cell OFDMA and cell-free massive MIMO-OFDM.
- The fundamentals of NOMA, including the principles of power-domain NOMA and code-domain NOMA, multi-user superposition transmission, and grant-free transmission.

10.1 Frequency-Selective Fading Channel

A multipath fading channel can be described by the response at time t to an input impulse at time $t - \tau$, namely

$$h(\tau, t) = \sum_{l=1}^{L} a_l(t)\delta\left(\tau - \tau_l(t)\right), \tag{10.1}$$

where $a_l(t)$ and $\tau_l(t)$ denote the attenuation and propagation delay of the lth path at time t, respectively, and L is the total number of resolvable paths. This expression is quite nice. It means that the effect of mobile users, arbitrarily moving reflectors and absorbers, and all of the complexities of solving Maxwell's equations, finally reduce to an input–output relation, which is represented as the impulse response of a linear time-varying channel filter. In a particular situation where the transmitter, receiver, and the environment are all stationary, the attenuation and propagation delays do not change over time, and we have the linear time-invariant channel with an impulse response [Tse and Viswanath, 2005]

$$h(\tau) = \sum_{l=1}^{L} a_l \delta \left(\tau - \tau_l \right). \tag{10.2}$$

Practical wireless communications are passband transmission that is carried out in a bandwidth at carrier frequency f_c. However, most of the signal processing in wireless communications, such as channel coding, modulation, detection, synchronization, and estimation, are usually implemented at the baseband. Hence, it makes sense to obtain a complex baseband equivalent model:

$$h_b(\tau, t) = \sum_{l=1}^{L} a_l(t) \delta \left(\tau - \tau_l(t) \right) e^{-2\pi j f_c \tau_l(t)}. \tag{10.3}$$

The next step is to convert the continuous-time channel to a discrete-time channel. Following the sampling theorem, we can create a more useful discrete-time channel model by figuring out the ζth tap of the channel filter at (discrete) time n, i.e.

$$h_\zeta[n] = \sum_{l=1}^{L} a_l(nT_s) e^{-2\pi j f_c \tau_l(nT_s)} \mathrm{sinc} \left(\zeta - \frac{\tau_l(nT_s)}{T_s} \right), \quad \zeta = 0, 1, \ldots, Z-1, \tag{10.4}$$

where $T_s = 1/B_w$ stands for the sampling period with the bandwidth of the transmitted signal B_w, and the sinc function is defined as

$$\mathrm{sinc}(t) := \frac{\sin(\pi t)}{\pi t}. \tag{10.5}$$

In a special case, where the gains and delays of the paths are time-invariant, Eq. (10.4) is simplified to

$$h_\zeta = \sum_{l=1}^{L} a_l e^{-2\pi j f_c \tau_l} \mathrm{sinc} \left(\zeta - \frac{\tau_l}{T_s} \right). \tag{10.6}$$

Then, the discrete-time input–output relationship of a baseband equivalent system can be expressed by

$$y[n] = \sum_{\zeta=0}^{Z-1} h_\zeta[n] x[n - \zeta], \tag{10.7}$$

assuming that the channel filter has an infinite length Z, which is determined by the delay spread and the sampling rate of a system. Figure 10.1 illustrates some examples of the filter taps generated from 3GPP Extended Typical Urban (ETU)

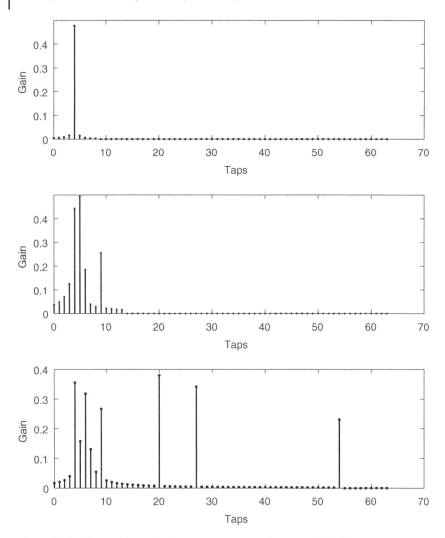

Figure 10.1 Illustration of the filter taps generated from the 3GPP ETU channel model with the delay spread of 5 ms. The sampling rates corresponding to the uppermost, middle, and lowermost figures are 0.1, 1.0, 10 MHz, respectively. Source: Adapted from Jiang and Schotten [2019].

channel model under different sampling rates. With a higher sampling rate, the resolution becomes better, resulting in a more accurate channel filter described by more taps.

When a signal pulse goes through a multipath channel, the received signal will appear as a pulse train, with each pulse corresponding to the direct path or an NLOS path. An important feature of radio propagation is *multipath delay spread* or *time dispersion* raised from the distinct arrival time of various propagation paths. Assuming τ_1 in the multipath channel model given by Eq. (10.3) denotes the propagation time of the first arriving multipath component, the *minimum excess delay* is equal to τ_1, as a reference delay of zero. Meanwhile, the propagation time of the last arriving multipath component is τ_L. Delay spread can be simply measured by the difference of the arrival time between the shortest and longest resolvable paths, also referred to as the *maximum excess delay*, in terms of

$$T_d := \tau_L - \tau_1. \tag{10.8}$$

The impulse response of a wireless channel varies both in time and frequency, and the delay spread determines how quickly it changes in frequency. There exists a phase difference $2\pi f(\tau_i - \tau_j)$ between two arbitrary multipath components i and j. Given the maximal phase difference among all paths as $2\pi f T_d$, the magnitude of the overall frequency response changes significantly when the phase difference increases or decreases with a value of π. Thus, the *coherence bandwidth*, which indicates the fading rate of a wireless channel in the frequency domain, is defined as

$$B_c := \frac{1}{2T_d}. \tag{10.9}$$

In narrowband transmission, the bandwidth of a transmitted signal is usually far less than the coherence bandwidth, i.e. $B_w \ll B_c$. Thus, the fading across the entire bandwidth is highly correlated, called *frequency-flat fading*. In this case, the delay spread is considerably less than the symbol period T_s, and therefore a single tap is sufficient to represent the channel filter, e.g.

$$h = -1.3162 + 0.3671i, \tag{10.10}$$

as demonstrated in the uppermost plot of Figure 10.1. Accordingly, the input–output relation in Eq. (10.7) is simplified to

$$y[n] = h[n]x[n]. \tag{10.11}$$

On the contrary, if the signal bandwidth $B_w \gg B_c$, two frequency points separated by more than the coherence bandwidth exhibit roughly independent response. Thus, wideband communications suffer from *frequency-selective fading* and Inter-Symbol Interference (ISI). The multipath delay spreads across multiple symbols, and a channel filter can only be represented by a series of taps, e.g.

$$\begin{aligned}
\mathbf{h} = [\, &- 1.316 + 0.367i, -0.144 - 0.08161i, 0.0772 + 0.0243i, -0.0515 - 0.014i \\
&0.0386 + 0.0097i, -0.0308 - 0.0074i, 0.0257 + 0.0060i, -0.0220 - 0.0051i \\
&0.0192 + 0.0044i, -0.0171 - 0.0038i, 0.0154 + 0.0034i, -0.0140 - 0.0031i \\
&0.0128 + 0.0028i, -0.0118 - 0.0026i, 0.0110 + 0.0024i, -0.0102 - 0.0022i\,].
\end{aligned} \tag{10.12}$$

In wireless communications, the mechanisms of mitigating ISI play a vital role in designing wideband signal formatting and receiver structure. Several techniques can be applied to mitigate the distortion due to multipath delay spread, including single-carrier equalization, spread spectrum, and multi-carrier modulation. The first two techniques are classical and can refer to the literature such as [Goldsmith, 2005]. The following section will briefly introduce the principle of multi-carrier modulation, aiming to provide readers with a self-contained illustration.

10.2 Multi-Carrier Modulation

One of the technological trends in wireless communications is that the signal bandwidth becomes increasingly wider to support a higher transmission rate [Jiang et al., 2021]. In the conventional single-carrier transmission, a higher bandwidth in the frequency domain corresponds to a shorter symbol period in the time domain. With the decrease of the symbol period, the delay spread in a multi-path fading channel raises severe ISI and substantially constrains the achievable transmission rate. Traditionally, a digital filter, referred to as the equalizer, is applied at the receiver to reverse the distortion incurred in a channel. The number of filter taps required for the equalizer is proportional to the signal bandwidth. An equalizer with several hundred taps might be necessary to effectively mitigate the ISI in an extremely large bandwidth signal. It is too complex to implement in a practical system. Therefore, the wireless community has to find an alternative to replace the single-carrier transmission. Multi-Carrier Modulation (MCM) is a broadband communication technique where a wideband signal is split into a set of orthogonal narrowband signals. The symbol period of a narrowband signal is substantially extended and is far longer than that of a wideband signal. Thus, the effect of ISI can be alleviated in an MCM system if the delay spread becomes negligible compared with the extended symbol period.

10.2.1 The Synthesis and Analysis Filters

In multi-carrier modulation, an array of filters, also known as a filter bank, is applied to synthesize multi-carrier signals at the transmitter. Accordingly, another filter bank is used for analyzing received multi-carrier signals at the receiver. When a signal $x(t)$ goes through a channel with an impulse response of $h(t)$, the resultant signal is given by

$$s(t) = h(t) * x(t), \tag{10.13}$$

where $*$ stands for the linear convolution. As illustrated in Figure 10.2, a *synthesis filter bank* is comprised of an array of filters denoted by $h_n(t)$, $n = 1, 2, \ldots, N$.

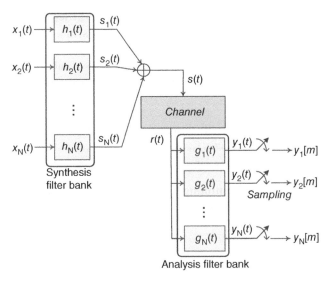

Figure 10.2 Block diagram of a multi-carrier modulation system consisting of a synthesis filter bank at the transmitter and an analysis filter bank at the receiver.

Each filter responses its impulse independently, following

$$s_n(t) = h_n(t) * x_n(t). \tag{10.14}$$

Their output signals are summed up, synthesizing a composite signal, which is given by

$$s(t) = \sum_{n=1}^{N} s_n(t) = \sum_{n=1}^{N} h_n(t) * x_n(t). \tag{10.15}$$

Although $h_n(t)$ can be any possible filtering in principle, the impulse response of a synthesis filter is specifically designed for processing the input signal for each subcarrier in an MCM system. The transmitted signal on the nth subcarrier is expressed by

$$x_n(t) = \sum_{m=-\infty}^{\infty} u_{m,n} \delta(t - mT), \quad n = 1, \ldots, N, \tag{10.16}$$

where $u_{m,n}$ denotes an information-bearing symbol corresponding to the time-frequency resource unit over the nth subcarrier during the mth symbol period, N denotes the total number of subcarriers, and T is the symbol period.

Substituting Eq. (10.16) into Eq. (10.15), yields a continuous-time baseband multi-carrier signal

$$s(t) = \sum_{m=-\infty}^{+\infty} \sum_{n=1}^{N} u_{m,n} h_n(t - mT). \tag{10.17}$$

To achieve the orthogonality among the subcarriers, the inter-subcarrier spacing denoted by $\triangle f$ needs to be an integer multiple of the inverse of the symbol period. In general, the spacing is set to $\triangle f = 1/T$ so as to maximize spectral efficiency [Jiang and Kaiser, 2016]. Without loss of generality, the frequencies of subcarriers in an equivalent baseband model can be written as

$$f_n = n \triangle f = \frac{n}{T}, \quad n = 1, 2, \ldots, N. \tag{10.18}$$

The synthesis filters are based on a specially designed prototype filter $p_T(t)$, and are modulated by the subcarrier frequencies f_n, as follows:

$$h_n(t) = p_T(t)e^{2\pi j f_n t + j\phi_n} = p_T(t)e^{2\pi j n \triangle f t + j\phi_n}, \tag{10.19}$$

where ϕ_n stands for a phase shift. Substituting Eq. (10.19) into Eq. (10.17), the baseband multi-carrier signal can be rewritten as

$$
\begin{aligned}
s(t) &= \sum_{m=-\infty}^{\infty} \sum_{n=1}^{N} u_{m,n} p_T(t - mT)e^{2\pi j n \triangle f(t-mT)+j\phi_n} \\
&= \sum_{m=-\infty}^{\infty} \sum_{n=1}^{N} u_{m,n} p_T(t - mT)e^{2\pi j n \triangle f t + j\phi_n} e^{-2\pi j n m \triangle f T} \\
&= \sum_{m=-\infty}^{\infty} \sum_{n=1}^{N} u_{m,n} p_T(t - mT)e^{2\pi j n \triangle f t + j\phi_n},
\end{aligned}
\tag{10.20}
$$

applying

$$e^{-2\pi j n m \triangle f T} = e^{-2\pi j n m} = e^{j0} = 1. \tag{10.21}$$

Correspondingly, the receiver is equipped with an *analysis filter bank* consisting of an array of filters, which have a common incoming multi-carrier signal $r(t)$. Although any possible filtering can be carried out, each analysis filter processes a different subcarrier of the received signal $r(t)$ in multi-carrier communications. Neglecting channel impairments and additive noise for simplicity, the input signal for the analysis filter bank equals to the generated signal of the synthesis filter bank, i.e. $r(t) = s(t)$. Similar to the synthesis filters, the analysis filters are based on a specifically designed prototype filter $p_R(t)$. Like Eq. (10.19), the impulse response of a typical analysis filter can be expressed by

$$g_k(t) = p_R(t)e^{-(2\pi j f_k t + j\phi_k)} = p_R(t)e^{-(2\pi j k \triangle f t + j\phi_k)}, \quad k = 1, 2, \ldots, N. \tag{10.22}$$

Feeding $r(t)$ into a typical analysis filter $g_k(t)$, the resultant signal is calculated by

$$
\begin{aligned}
y_k(t) &= g_k(t) * r(t) \\
&= \sum_{m=-\infty}^{\infty} \sum_{n=1}^{N} u_{m,n} p_R(t) * p_T(t - mT)e^{2\pi j n \triangle f t + j\phi_n} e^{-2\pi j k \triangle f t - j\phi_k} \\
&= \sum_{m=-\infty}^{\infty} \sum_{n=1}^{N} u_{m,n} p_R(t) * p_T(t - mT)e^{2\pi j(n-k)\triangle f t + j(\phi_n - \phi_k)}.
\end{aligned}
\tag{10.23}
$$

To properly recover the information-bearing symbol at each time-frequency resource unit, two major criteria need to be satisfied:

- No Inter-Carrier Interference (ICI) in the frequency domain
- No ISI in the time domain

First, the subcarriers need to constitute an orthogonal basis set within a symbol period to avoid the generation of ICI, i.e.

$$\frac{1}{T}\int_0^T e^{2\pi j(n-k)\Delta ft + j(\phi_n - \phi_k)}dt = \delta[n-k]. \tag{10.24}$$

To achieve the orthogonality, the subcarrier spacing needs to equal an integer multiple of the inverse of the symbol period. As mentioned previously, the spacing is usually selected as $\Delta f = 1/T$ to maximize spectral efficiency. The subcarriers denoted in an exponential form $e^{2\pi jn\Delta ft}$, $t \in [0, T)$, each of which has two branches: the In-phase (I) and Quadrature (Q) components. The information-bearing symbol $u_{m,n}$ is complex-valued, i.e. $u_{m,n} = a_{m,n} + jb_{m,n}$, where $a_{m,n}$ and $b_{m,n}$ are real-valued numbers. The modulated signal on a typical subcarrier n is denoted by $\Re[u_{m,n}e^{2\pi jn\Delta ft}]$, which can be transformed into I- and Q-branch form as:

$$a_{m,n}\cos(2\pi n \Delta ft) - b_{m,n}\sin(2\pi n \Delta ft). \tag{10.25}$$

That is to say, the real part of the information-bearing symbol is modulated on the I-branch signal of the subcarrier, while the imaginary part is carried by the Q-branch signal. As demonstrated in Figure 10.3, the sinusoidal waves of $\cos(2\pi n\Delta ft)$ and $\sin(2\pi n\Delta ft)$, $n = 1, 2, 3$ are mutually orthogonal during one symbol period T, satisfying the requirement of Eq. (10.24).

Second, the two prototype filters should satisfy the condition that the output signal does not cause ISI in the time domain. It does not necessarily mean that no any overlapping exists among consecutive symbols, but requiring at least no interference at the sampling points, namely

$$p_T(t) * p_R(t)\Big|_{t_s=iT} = \begin{cases} 1, & i = 0 \\ 0, & i \neq 0 \end{cases},$$

where $t_s = iT$ stands for the sampling points at the time axis. This condition can be satisfied by, for example, employing the *sinc* function in the pulse shaping.

10.2.2 Polyphase Implementation

The multi-carrier signal contains N subcarriers with the inter-subcarrier spacing Δf, resulting in a signal bandwidth of $B_w = N \Delta f$. According to the sampling

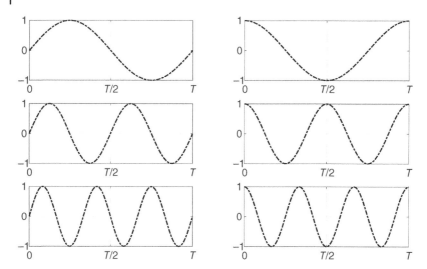

Figure 10.3 Illustration of a group of orthogonal subcarriers [Jiang and Kaiser, 2016]. The sinusoidal waves in the first column correspond to $\sin(2\pi n\Delta ft)$, $n = 1, 2, 3$, while the second column corresponds to $\cos(2\pi n\Delta ft)$.

theorem [Oppenheim et al., 1998], the sampling interval T_s can equal to the inverse of the signal bandwidth, i.e.

$$T_s = \frac{1}{B_w} = \frac{1}{N\,\Delta f} = \frac{T}{N}. \tag{10.26}$$

The length of symbol period is T, so the corresponding number of signal samples within each symbol period is $T/T_s = N$. The discrete-time prototype filter can be obtained by sampling the continuous-time prototype filter $p_T(t)$ with the sampling rate of T_s, resulting in

$$p_T[l] = p_T(lT_s), \quad l = 0, 1, \dots, \mathfrak{L} - 1. \tag{10.27}$$

It is possible that the length of the discrete-time prototype filter is larger than the symbol period, and $\mathfrak{L}$ can be selected as an integer multiple of N. Assuming the overlapping factor is K, which means the length of the prototype filter is K times of the symbol period, i.e.

$$\mathfrak{L} = KN. \tag{10.28}$$

In the field of signal processing [Oppenheim et al., 1998], the Z-transform is an important analysis tool, which converts a discrete-time sequence into a complex frequency-domain representation. The Z-transform of $p_T[l]$ is calculated by

$$P_T(z) = \sum_{l=0}^{\mathfrak{L}-1} p_T[l]z^{-l} = \sum_{l=0}^{KN-1} p_T[l]z^{-l}. \tag{10.29}$$

Letting $l = k'N + n'$, where $k' = 0, 1, \dots, (K-1)$ and $n' = 0, 1, \dots, (N-1)$, Eq. (10.29) can be further transformed into

$$
\begin{aligned}
P_T(z) &= \sum_{n'=0}^{N-1} \sum_{k'=0}^{K-1} p_T[k'N+n'] z^{-(k'N+n')} \\
&= \sum_{n'=0}^{N-1} z^{-n'} \sum_{k'=0}^{K-1} p_T[k'N+n'] z^{-k'N}.
\end{aligned}
\tag{10.30}
$$

Constituting a series of N subsequences $p_T^{n'}[k'] = k'N + n'$, $n' = 0, 1, \dots,$ $(N-1)$, where $p_T^{n'}[k']$ has a length of K and is called the $(n')^{\text{th}}$ polyphase component of the prototype filter $p_T[l]$. The Z-transform of $p_T^{n'}[k']$, namely

$$
P_T^{n'}(z^N) = \sum_{k'=0}^{K-1} p_T[k'N+n'] z^{-k'N}
\tag{10.31}
$$

is referred to as the (n')th polyphase decomposition of $P_T(z)$. Substituting Eq. (10.31) into Eq. (10.30) yields

$$
P_T(z) = \sum_{n'=0}^{N-1} P_T^{n'}(z^N) z^{-n'},
\tag{10.32}
$$

which is the polyphase decomposition of the prototype filter.

Similarly, sampling $h_n(t)$ with an interval of T_s gets

$$
\begin{aligned}
h_n[l] = h_n(lT_s) &= p_T[l] e^{2\pi j n \Delta f l T_s + j\phi_n} \\
&= p_T[l] e^{2\pi j n l/N + j\phi_n}, \quad l = 0, 1, \dots, \mathscr{L}-1.
\end{aligned}
\tag{10.33}
$$

The Z-transform of the nth synthesis filter is computed by Jiang and Kaiser [2016]

$$
\begin{aligned}
H_n(z) &= \sum_{l=0}^{\mathscr{L}-1} h_n[l] z^{-l} \\
&= \sum_{l=0}^{\mathscr{L}-1} p_T[l] e^{2\pi j n l/N + j\phi_n} z^{-l} \\
&= \sum_{n'=0}^{N-1} \sum_{k'=0}^{K-1} p_T[k'N+n'] e^{2\pi j n (k'N+n')/N + j\phi_n} z^{-(k'N+n')} \\
&= e^{j\phi_n} \sum_{n'=0}^{N-1} e^{2\pi j n n'/N} z^{-n'} \sum_{k'=0}^{K-1} p_T[k'N+n'] e^{2\pi j n k'} z^{-k'N} \\
&= e^{j\phi_n} \sum_{n'=0}^{N-1} e^{2\pi j n n'/N} P_T^{n'}(z^N) z^{-n'}.
\end{aligned}
\tag{10.34}
$$

The synthesis filter bank consists of N filters, and, consequently, its Z-transform can be expressed in matrix form as

$$\begin{bmatrix} H_1(z) \\ H_2(z) \\ \vdots \\ H_N(z) \end{bmatrix} = \begin{bmatrix} e^{j\phi_1} \\ e^{j\phi_2} \\ \vdots \\ e^{j\phi_N} \end{bmatrix} \begin{bmatrix} 1 & 1 & \cdots & 1 \\ 1 & W^{-1} & \cdots & W^{-N+1} \\ \vdots & \cdots & \ddots & \vdots \\ 1 & W^{-N+1} & \cdots & W^{(-N+1)^2} \end{bmatrix} \begin{bmatrix} P_T^0(z^N) \\ P_T^1(z^N)z^{-1} \\ \vdots \\ P_T^N(z^N)z^{-N+1} \end{bmatrix}, \tag{10.35}$$

applying a primitive Nth root of unity $W = e^{-j2\pi/N}$. The left vector stands for the phase rotations, the matrix in the middle implies the inverse discrete Fourier transform (DFT), and the right vector is a polyphase decomposition of the prototype filter.

10.2.3 Filter Bank Multi-Carrier

In principle, a prototype filter can be designed to achieve a side lobe as small as possible by means of the filter bank. This form of multi-carrier transmission is known as Filter Bank Multi-Carrier (FBMC), exhibiting an excellent out-of-band (OOB) emission property [Jiang and Schellmann, 2012]. The following frequency-domain coefficients can be applied to constitute a desired prototype filter with the overlapping factor of $K = 4$

$$p = \begin{bmatrix} 1, 0.97196, 0.707, 0.235147 \end{bmatrix} \tag{10.36}$$

Based on these coefficients, the frequency response of the prototype filter is obtained through an interpolation operation, which is expressed as

$$P(f) = \sum_{k=-K+1}^{K-1} p_k \frac{\sin\left(\pi NK \left[f - \frac{k}{NK}\right]\right)}{NK \sin\left(\pi \left[f - \frac{k}{NK}\right]\right)}, \tag{10.37}$$

where N is the total number of subcarriers, K is the overlapping factor, and p_k is mapped from the aforementioned coefficients, i.e. $p_0 = 1, p_{\pm 1} = 0.97196, p_{\pm 2} = 0.707$, and $p_{\pm 3} = 0.235147$. Then, its impulse response $p_T(t)$ can be obtained by an inverse Fourier transform, as follows:

$$p_T(t) = 1 + \sum_{k=1}^{K-1} p_k \cos\left(\frac{2\pi kt}{KT}\right). \tag{10.38}$$

The frequency response of a FBMC subcarrier is very compact. The ripples of a FBMC subcarrier can be neglected, and there is even no ICI between two non-neighboring subcarriers. However, this frequency feature comes at the price in the temporal domain, where the prototype filter spans over $K = 4$ symbols, rather than a rectangular prototype filter occupying only a single symbol. By sampling $p_T(t)$, a discrete-time prototype filter with a length of KN is obtained:

$$
\begin{aligned}
p_T[s] &= p_T(sT_s) \\
&= 1 + \sum_{k=1}^{K-1} p_k \cos\left(\frac{2\pi ksT_s}{KNT_s}\right) \\
&= 1 + \sum_{k=1}^{K-1} p_k \cos\left(\frac{2\pi ks}{KN}\right), \quad s = 0, 1, \dots, KN - 1.
\end{aligned}
\tag{10.39}
$$

Then, the $(n')^{\text{th}}$ polyphase decomposition can be obtained by

$$
p_T^{n'}[k'] = p_T[k'N + n'], \quad k' = 0, 1, \dots, K - 1,
\tag{10.40}
$$

which is actually an Finite Impulse Response (FIR) filter with a length of K. This filter is applied for the (n')th FBMC subcarrier, and a number of N FIR filters constitute the polyphase network to generate FBMC signals. Thanks to the utilization of DFT, the difference between FBMC and Orthogonal Frequency-Division Multiplexing (OFDM) transmission is only on the polyphase implementation. Hence, the OFDM technology, which has been widely applied in mobile and wireless communication systems, such as Wireless Fidelity (Wi-Fi), 4G LTE, and 5G NR, can be regarded as a special case of FBMC. The next section will discuss the principle and key issues of the OFDM technology, as a promising modulation technique for the upcoming 6G system.

10.3 Orthogonal Frequency-Division Multiplexing

Due to its ability to cope with multipath frequency-selective fading without the need for complex equalization and simple implementation through the use of the digital Fourier transform, OFDM has become the most dominant modulation technique for wired and wireless communication systems over the past two decades. It has been extensively applied in many well-known standards, e.g. Digital Subscriber Line (DSL), Digital Video Broadcasting-Terrestrial (DVB-T), Wi-Fi, Worldwide Inter-operability for Microwave Access (WiMAX), LTE, and LTE-Advanced. After an extensive comparison among all possible techniques, OFDM has been adopted as one of the critical elements of 5G NR, as a comprehensive trade-off among the performance, complexity, comparability, and robustness. It is envisioned that it will serve as a key technology in the forthcoming 6G transmission [Jiang and Schotten, 2021c] either in the conventional sub-6 GHz band and high frequency bands.

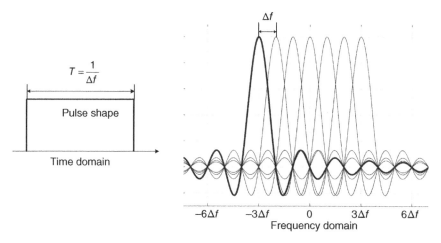

Figure 10.4 The rectangular prototype filter of the OFDM signal in the time domain, and the packing of orthogonal OFDM subcarriers in the frequency domain.

As a kind of multi-carrier modulation, the major features of OFDM transmission, which distinguish it from frequency-division multiplexing of multiple narrow-band channels, are

- The use of a typically very large number of orthogonal subcarriers, rather than only a few non-overlapping carriers.
- The use of simple rectangular pulse shaping, and a *sinc*-shaped subcarrier spectrum, as shown in Figure 10.4.
- Tight frequency-domain packing of the subcarriers with an inter-subcarrier spacing of $\Delta f = 1/T$, where T is the duration of a symbol period.

The OFDM scheme can be regarded as a special case of FMBC, where a rectangular prototype filter modeled by

$$p_0(t) = \begin{cases} 1, & -\frac{T}{2} \leqslant t < \frac{T}{2} \\ 0, & \text{others} \end{cases} \tag{10.41}$$

is applied [Jiang and Zhao, 2012], as shown in Figure 10.4. Sampling with an interval of $T_s = T/N$, the discrete-time rectangular prototype filter can be given by:

$$p_0[l] = \begin{cases} 1, & 0 \leqslant l < N - 1 \\ 0, & \text{others} \end{cases} \tag{10.42}$$

The Fourier transform of the rectangular prototype filter is

$$\begin{aligned} P_0(f) &= \int_{-\infty}^{\infty} p_0(t)e^{-2\pi jft} dt \\ &= \int_{-T/2}^{T/2} e^{-2\pi jft} dt = \frac{\sin \pi fT}{\pi f} = T\text{sinc}\left(\frac{f}{\Delta f}\right), \end{aligned} \tag{10.43}$$

implying a *sinc*-shaped spectrum that achieves the subcarrier orthogonality in the frequency domain.

Accordingly, the synthesis filters in Eq. (10.19) and the analysis filters in Eq. (10.22) become

$$
h_n(t) = \begin{cases} e^{2\pi jn\Delta ft+j\phi_n}, & -\frac{T}{2} \leqslant t < \frac{T}{2} \\ 0, & \text{others} \end{cases}
\tag{10.44}
$$

and

$$
g_k(t) = \begin{cases} e^{-(2\pi jk\Delta ft+j\phi_k)}, & -\frac{T}{2} \leqslant t < \frac{T}{2} \\ 0, & \text{others} \end{cases},
\tag{10.45}
$$

respectively. Since a phase rotation does not affect the orthogonality of OFDM subcarriers, we can neglect the phase and use $\phi_n = 0$, $\forall n = 1, 2, \ldots, N$. In complex baseband notation, a basic OFDM signal $s(t)$ during the symbol period $mT \leqslant t < (m+1)T$ can be thus given by

$$
s(t) = \sum_{n=0}^{N-1} s_n(t) = \sum_{n=0}^{N-1} u_{m,n} e^{2\pi jn\Delta ft},
\tag{10.46}
$$

where $s_n(t)$ is the nth modulated subcarrier with frequency $f_n = n\Delta f$. The Fourier transform of the nth *unmodulated* subcarrier can be calculated by

$$
\begin{aligned}
P_n(f) &= \int_{-\infty}^{\infty} p_0(t) e^{2\pi jn\Delta ft} e^{-2\pi jft} dt \\
&= \int_{-T/2}^{T/2} e^{2\pi jn\Delta ft} e^{-2\pi jft} dt \\
&= T\text{sinc}\left(\frac{f}{\Delta f} - n\right).
\end{aligned}
\tag{10.47}
$$

That is to say, the spectrum of the nth subcarrier can be obtained simply by shifting $P_0(f)$ in the frequency axis with a shift of $n\Delta f$, i.e. $P_n(f) = P_0(f - n\Delta f)$, as shown in Figure 10.4.

The basic principles of OFDM modulation and demodulation can be depicted as Figure 10.5. OFDM demodulation uses a bank of correlators, one for each subcarrier. With the orthogonality among subcarriers, two arbitrary OFDM subcarriers do not cause any interference in the ideal case, even though the spectrum of neighbor subcarriers overlaps. Thus, the avoidance of interference between OFDM subcarriers is not simply due to the separation in frequency, which is the case for frequency-division multiplexing. Instead, the subcarrier orthogonality is due to the specific frequency-domain structure of each subcarrier in combination with the particular choice of a subcarrier spacing equal to the symbol rate, i.e. $\Delta f = 1/T$ [Dahlman et al., 2011].

Neglecting channel impairments and additive noise, i.e. $r(t) = s(t)$, and going through the corresponding analysis filter $g_k(t)$, the resultant signal at the kth

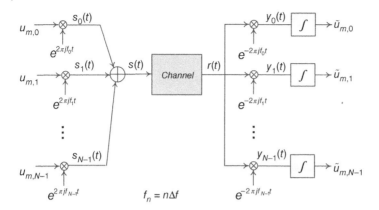

Figure 10.5 The principle of OFDM modulation and demodulation over a set of orthogonal subcarriers.

subcarrier during the mth symbol period is

$$y_k(t) = \sum_{n=0}^{N-1} u_{m,n} e^{2\pi j n \Delta f t} e^{-2\pi j k \Delta f t}$$

$$= \sum_{n=0}^{N-1} u_{m,n} e^{2\pi j (n-k) \Delta f t}.$$

Integrating over one symbol period, yields

$$\tilde{u}_{m,k} = \frac{1}{T} \int_{mT}^{(m+1)T} y_k(t) dt$$

$$= \frac{1}{T} \int_{mT}^{(m+1)T} \sum_{n=0}^{N-1} u_{m,n} e^{2\pi j (n-k) \Delta f t} dt$$

$$= \frac{1}{T} \int_{mT}^{(m+1)T} u_{m,n} dt + \underbrace{\frac{1}{T} \sum_{n=0, n \neq k}^{N-1} u_{m,n} \int_{mT}^{(m+1)T} e^{2\pi j (n-k) \Delta f t} dt}_{\text{Inter-carrier interference}}.$$

Due to the orthogonality among subcarriers, the ICI can be in principle eliminated, and we have

$$\tilde{u}_{m,k} = u_{m,n}, \quad \forall n = k. \tag{10.48}$$

Consequently, the information-bearing symbol $u_{m,n}$ can be delivered over the nth subcarrier during the mth symbol period, and a total of N information-bearing symbols are transmitted in parallel over a single symbol period.

10.3.1 DFT Implementation

Although a bank of modulator-correlator pairs shown in Figure 10.5 can be used for describing the basic principles of OFDM modulation and demodulation, it is not the most appropriate structure for actual implementation. Actually, due to its specific design and the particular selection of the inter-subcarrier spacing, OFDM allows for low-complexity implementation employing the DFT processing. It can be further implemented by a computationally efficient algorithm named Fast Fourier Transform (FFT) if the number of subcarriers is a power of 2.

Because of the rectangular pulse shaping, the OFDM symbols are non-overlapping in the time domain (i.e. the overlapping factor $K = 1$). As a result, the polyphase decomposition defined in Eq. (10.31) is simplified to

$$
\begin{aligned}
P_0^{n'}(z^N) &= \sum_{k'=0}^{K-1} p_0[k'N + n']z^{-k'N} \\
&= \sum_{k'=0}^{0} p_0[n']z^0 \\
&= 1, \quad n' = 0, 1, \ldots, N-1.
\end{aligned}
\tag{10.49}
$$

Thus, Eq. (10.35) can be rewritten as

$$
\begin{bmatrix} H_1(z) \\ H_2(z) \\ \vdots \\ H_N(z) \end{bmatrix} = \begin{bmatrix} 1 & 1 & \cdots & 1 \\ 1 & W^{-1} & \cdots & W^{-N+1} \\ \vdots & \cdots & \ddots & \vdots \\ 1 & W^{-N+1} & \cdots & W^{(-N+1)^2} \end{bmatrix},
\tag{10.50}
$$

which is equivalent to an Inverse Discrete Fourier Transform (IDFT), implying that an OFDM signal is possible to be generated by an IDFT modulator.

As $B_w = N \Delta f$ can be seen as the nominal bandwidth of OFDM transmission, a discrete-time OFDM signal can be obtained by sampling $s(t)$ in Eq. (10.46) with a rate of $f_s \geq N \Delta f$ to comply with the sampling theorem. Assuming that the sampling rate is a multiple of the inter-subcarrier spacing, i.e. $f_s = N_s \Delta f$, we know that $N_s \geq N$, which is called *over-sampling* in OFDM. For example, the LTE transmission supports approximately $N = 1200$ subcarriers, whereas the DFT size is selected as $N_s = 2048$. This corresponds to a sampling rate $f_s = N_s \Delta f = 30.72 \, \text{MHz}$, given $\Delta f = 15 \, \text{kHz}$ in LTE. The remaining 848 subcarriers do not carry anything by assigning a null symbol '0', and are generally called *virtual OFDM subcarriers* [Jiang, 2016].

With these assumptions, the sampled OFDM signal during the mth symbol, in other words, the discrete-time OFDM sequence can be expressed as

$$
\begin{aligned}
s^m[k] = s(kT_s) &= \sum_{n=0}^{N-1} s_n(kT_s) \\
&= \sum_{n=0}^{N-1} u_{m,n} e^{2\pi jn\Delta fkT_s} \\
&= \sum_{n=0}^{N-1} u_{m,n} e^{2\pi jnk/N_s}, \quad k = 0, 1, \ldots, N_s - 1.
\end{aligned}
\tag{10.51}
$$

Adding some virtual subcarriers at the end, we can form an alternative expression of the transmitted symbols

$$
a_{m,n} = \begin{cases} u_{m,n}, & 0 \leqslant n < N \\ 0, & N \leqslant n < N_s \end{cases}.
\tag{10.52}
$$

Then, Eq. (10.51) can be rewritten as

$$
\begin{aligned}
s^m[k] &= \sum_{n=0}^{N-1} u_{m,n} e^{2\pi jnk/N_s} \\
&= \sum_{n=0}^{N-1} u_{m,n} e^{2\pi jnk/N_s} + \sum_{n=N}^{N_s-1} 0 \cdot e^{2\pi jnk/N_s} \\
&= \sum_{n=0}^{N_s-1} a_{m,n} e^{2\pi jnk/N_s},
\end{aligned}
\tag{10.53}
$$

which is equivalent to an N_s-point IDFT. It reveals that a discrete-time OFDM sequence equals exactly to the DFT of the modulation symbols, followed by digital-to-analog conversion.

OFDM is a kind of block-wise transmission. A block of modulation symbols $\mathbf{u}_m = [u_{m,0}, u_{m,1}, \ldots, u_{m,N-1}]^T$ is first extended with zeros to length N_s, yielding

$$
\mathbf{a}_m = [u_{m,0}, u_{m,1}, \ldots, u_{m,N-1}, 0, \ldots, 0]^T.
\tag{10.54}
$$

Perform the IDFT processing over $\mathbf{a}_m$ to obtain the OFDM sequence

$$
\mathbf{s}_m = [s_m[0], s_m[1], \ldots, s_m[N_s - 1]]^T.
\tag{10.55}
$$

Defining a matrix

$$
\mathbf{F} = \begin{pmatrix}
\omega_{N_s}^{0 \cdot 0} & \omega_{N_s}^{0 \cdot 1} & \cdots & \omega_{N_s}^{0 \cdot (N_s-1)} \\
\omega_{N_s}^{1 \cdot 0} & \omega_{N_s}^{1 \cdot 1} & \cdots & \omega_{N_s}^{1 \cdot (N_s-1)} \\
\vdots & \vdots & \ddots & \vdots \\
\omega_{N_s}^{(N_s-1) \cdot 0} & \omega_{N_s}^{(N_s-1) \cdot 1} & \cdots & \omega_{N_s}^{(N_s-1) \cdot (N_s-1)}
\end{pmatrix}
\tag{10.56}
$$

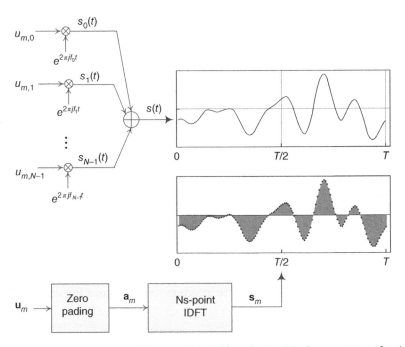

Figure 10.6 The equivalence between multi-carrier modulation over a set of orthogonal subcarriers and the direct DFT conversion of the modulation symbols. The OFDM signal in the upper part is obtained by summing up $N = 8$ modulated subcarriers, while the OFDM sequence in the lower part is generated through a 64-point IDFT over a 64×1 symbol vector consisting of 8 modulation symbols padded with 56 zeros at the tail.

with a primitive N_s^{th} root of unity $\omega_{N_s} = e^{-2\pi j/N_s}$, the OFDM modulation (i.e. an IDFT) can be written in matrix form as

$$\mathbf{s}_m = \mathbf{F}^*\mathbf{a}_m = N_s\mathbf{F}^{-1}\mathbf{a}_m, \tag{10.57}$$

where the superscript $(\cdot)^*$ denotes the complex conjugate, and the superscript $(\cdot)^{-1}$ stands for the inverse of a square matrix.

As shown in Figure 10.6, an OFDM signal, consisting of N modulated subcarriers, has the same envelope as that of an OFDM sequence obtained through the N_s-point DFT of the modulation symbols. Similarly, the DFT processing can be used for OFDM demodulation, replacing the bank of N decorrelators with sampling at a rate of $f_s = N_s \,\Delta f$, followed by an N_s-point DFT/FFT operation.

10.3.2 Cyclic Prefix

The previous subsection illustrated that an uncorrupted OFDM signal could be demodulated at the receiver without interference among subcarriers.

The orthogonality is achieved since each subcarrier has an integer number of periods of complex exponentials during one OFDM symbol period, mathematically denoted by

$$e^{2\pi j f_n t} = e^{2\pi j n \Delta f t} = e^{2\pi j \frac{nt}{T}}, \quad -\frac{T}{2} \leqslant t < \frac{T}{2}. \tag{10.58}$$

In practice, however, this orthogonality can be lost in a time-dispersive channel. That is because the correlation interval for one path will overlap with the symbol boundary of a different path. The integration interval will not necessarily correspond to an integer number of periods of complex exponentials since the modulation symbols may differ between two consecutive symbols. As a result, a time-dispersive channel causes not only inter-symbol interference but also inter-carrier interference in OFDM transmission, as illustrated in Figure 10.7.

To solve this problem and make an OFDM signal robust to the time dispersion, the Cyclic Prefix (CP), or called cyclic extension originally, referring to a prefixing of an OFDM symbol, was proposed by Peled and Ruiz [1980]. Cyclic-prefix insertion implies that the last part of an OFDM symbol is copied and inserted at the beginning of the OFDM symbol. The insertion thus increases the length of

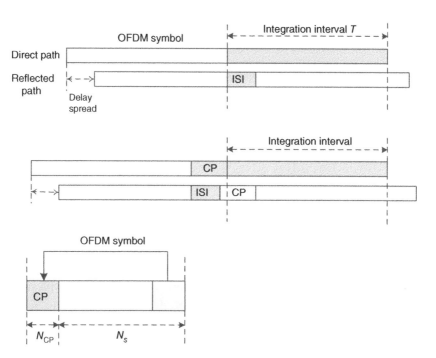

Figure 10.7 The cyclic prefix can alleviate the inter-symbol interference and inter-carrier interference due to the time dispersion.

the OFDM symbol from T to $T_o = T + T_{CP}$, where T_{CP} is the length of the cyclic prefix, leading to a reduction in the OFDM symbol rate. If the correlation interval at the demodulator is still carried out over a symbol period T, the subcarrier orthogonality can be preserved as long as the CP duration is greater than the delay spread.

In practice, the insertion of CP is carried out over the time-discrete OFDM sequence, where the last N_{CP} samples are copied and inserted at the beginning of the block, increasing the block length from N_s to $N_s + N_{CP}$. At the demodulator, the corresponding samples are discarded before the DFT demodulation. The discrete-time impulse response of a frequency-selective fading channel at time m is assumed to be

$$\mathbf{h}_m = \left[h_m[0], h_m[1], \ldots, h_m[\mathfrak{L} - 1]\right]^T, \tag{10.59}$$

where $\mathfrak{L}$ denotes the length of the channel filter. When the mth OFDM symbol $\mathbf{s}_m$ goes through the channel, the resultant signal in discrete-time baseband notation is computed by Oppenheim et al. [1998]

$$\mathbf{r}_m = \mathbf{h}_m * \mathbf{s}_m, \tag{10.60}$$

where $*$ means a linear convolution. The length of the output signal is $\mathfrak{L} + N_s - 1$, which is longer than the input signal since the length of the channel filter is greater than one in a frequency-selective fading channel. Accordingly, the received signal can be denoted by

$$\mathbf{r}_m = \left[r_m[0], r_m[1], \cdots, r_m[Q - 1]\right]^T, \tag{10.61}$$

where $Q = \mathfrak{L} + N_s - 1$, and

$$r_m[q] = \sum_{l=0}^{\mathfrak{L}-1} h_m[l] s_m[q - l], \quad q = 0, 1, \ldots, Q - 1. \tag{10.62}$$

Without a CP, the residual samples with a length of $\mathfrak{L} - 1$ at the end of each OFDM symbol overlaps with its subsequent OFDM symbol, causing ISI and destroying the subcarrier orthogonality. Intuitively, inserting a guard interval between two consecutive OFDM symbols can absorb the residual of the previous OFDM symbol. Given Eq. (10.55), the OFDM sequence with the insertion of CP can be expressed by

$$\mathbf{x}_m = \big[\underbrace{s_m[N_s - N_{CP}], \ldots, s_m[N_s - 1]}_{\text{CP Insertion}}, s_m[0], s_m[1], \ldots, s_m[N_s - 1]\big]^T. \tag{10.63}$$

Convolving $\mathbf{x}_m$ with $\mathbf{h}_m$ yields

$$\mathbf{y}_m = \mathbf{h}_m * \mathbf{x}_m. \tag{10.64}$$

The length of $\mathbf{y}_m$ is $S = \mathfrak{L} + N_s + N_{CP} - 1$, and then we can write the received signal by

$$\mathbf{y}_m = \left[y_m[0], y_m[1], \ldots, y_m[S-1] \right], \tag{10.65}$$

whose entry equals to

$$y_m[s] = \sum_{l=0}^{\mathfrak{L}-1} h_m[l] x_m[s-l]. \tag{10.66}$$

As long as the span of the delay spread does not exceed the length of the CP, the inter-symbol interference can be absorbed, while the inter-carrier interference can also be avoided since the subcarrier orthogonality is preserved during an integration interval. The drawback of cyclic-prefix insertion is the loss of power and bandwidth as the OFDM symbol rate reduces. One way to minimize such a loss is to reduce the inter-subcarrier spacing Δf, with a corresponding increase in the symbol period T as a consequence. However, this will increase the sensitivity of the OFDM transmission to fast channel fluctuation due to high Doppler spread. It is also essential to understand that the CP does not necessarily have to cover the entire length of the channel time dispersion. In general, there is a trade-off between the power loss and the signal corruption (inter-symbol and inter-subcarrier interference) due to the residual time dispersion not covered by the cyclic prefix.

10.3.3 Frequency-Domain Signal Processing

In addition to

- eliminating the ISI from the previous symbol, and
- preserving the subcarrier orthogonality,

the employment of cyclic prefix can

- convert the linear convolution with a channel filter into *circular convolution*, also known as *cyclic convolution*, allowing for simple frequency-domain signal processing.

Assuming a sufficiently large cyclic prefix, the linear convolution of a time-dispersive radio channelwill appear as circular convolution during the demodulator integration interval. The combination of OFDM modulation, a time-dispersive radio channel, and OFDM demodulation can then be seen as a set of parallel frequency-domain sub-channels, as illustrated in Figure 10.8.

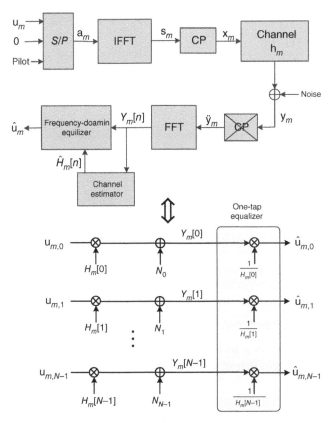

Figure 10.8 Block diagram of an end-to-end OFDM transmission system, and its equivalent frequency-domain representation.

At the receiver, the CP is discarded and only the samples within the integration interval from N_{CP} to $N_{CP} + N_s - 1$ are used for demodulation. Mathematically, the input to the DFT demodulator is

$$\ddot{\mathbf{y}}_m = \left[\ddot{y}_m[0], \ddot{y}_m[1], \ldots, \ddot{y}_m[N_s - 1]\right]^T$$
$$= \left[y_m[N_{CP}], y_m[N_{CP} + 1], \ldots, y_m[N_{CP} + N_s - 1]\right]^T, \tag{10.67}$$

extracted from $\mathbf{y}_m$ in Eq. (10.65). It is amazing that the extracted sequence is exactly equivalent to the circular convolution of $\mathbf{s}_m$ and an extended channel filter padding with zeros to length N_s, i.e.

$$\ddot{\mathbf{y}}_m = \mathbf{s}_m \otimes \mathbf{h}_{N_s}, \tag{10.68}$$

and alternatively

$$\ddot{y}_m[k] = y_m[N_{CP} + k]$$

$$= \sum_{i=0}^{N_s-1} s_m[i] h_{N_s}\left[(k-i)_{N_s}\right], \quad \forall k = 0, \ldots, N_s - 1, \tag{10.69}$$

where $\otimes$ stands for the circular convolution, $(\cdot)_{N_s}$ denotes a periodic shift with N_s, and

$$\mathbf{h}_{N_s} = \left[h_{N_s}[0], h_{N_s}[1], \ldots, h_{N_s}[N_s - 1]\right]^T$$

$$= [h_m[0], h_m[1], \ldots, h_m[\mathfrak{L} - 1], \underbrace{0, \ldots, 0}_{N_s - \mathfrak{L}}]^T. \tag{10.70}$$

A particular advantage of forming circular convolution in the OFDM transmission is to enable simple frequency-domain processing and *one-tap equalization* [Jiang and Schotten, 2019]. According to the theory of signal processing, the circular convolution of two finite sequences with the same length in the time domain, rather than the linear convolution, corresponds to the multiplication of their DFTs in the frequency domain. Mathematically, if $\ddot{y}_m = s_m \otimes \mathbf{h}_{N_s}$, we have

$$Y_m[n] = H_m[n] X_m[n], \quad \forall n = 0, 1, \ldots, N_s - 1, \tag{10.71}$$

where $Y_m[n]$ and $X_m[n]$ are the nth entry of frequency-domain received and transmitted sequences, which are the DFTs of $\ddot{y}_m$ and s_m, respectively, giving by

$$Y_m[n] = \sum_{k=0}^{N_s-1} \ddot{y}_m[k] e^{-2\pi jnk/N_s}$$

$$= \sum_{k=0}^{N_s-1} y_m[k + N_{CP}] e^{-2\pi jnk/N_s}, \tag{10.72}$$

and

$$X_m[n] = \sum_{k=0}^{N_s-1} s_m[k] e^{-2\pi jnk/N_s}. \tag{10.73}$$

Recalling Eq. (10.57), the OFDM sequence s_m is the IDFT of the information-bearing symbols

$$\mathbf{a}_m = [u_{m,0}, u_{m,1}, \ldots, u_{m,N-1}, 0, \ldots, 0]^T, \tag{10.74}$$

and therefore we have

$$X_m[n] = a_m[n], \quad \forall n = 0, 1, \ldots, N_s - 1, \tag{10.75}$$

and

$$X_m[n] = u_{m,n}, \quad \forall n = 0, 1, \ldots, N - 1. \tag{10.76}$$

Taking into account frequency-domain noise denoted by $N_m[n]$, Eq. (10.71) can be rewritten as

$$Y_m[n] = H_m[n]u_{m,n} + N_m[n], \quad \forall n = 0, 1, \ldots, N - 1, \tag{10.77}$$

implying that the OFDM transmission can be regarded as a set of parallel frequency-domain sub-channels, as illustrated in Figure 10.8. The complex frequency-domain channel taps $H_m[n]$, $n = 0, 1, \ldots, N_s - 1$ can be obtained by performing the DFT over $\mathbf{h}_{N_s}$, i.e.

$$H_m[n] = \sum_{l=0}^{N_s-1} h_{N_s}[l]e^{-2\pi jnl/N_s} = \sum_{l=0}^{\mathfrak{L}-1} h_m[l]e^{-2\pi jnl/N_s}. \tag{10.78}$$

Since the modulation and demodulation are independently carried out on each subcarrier, the channel estimation and equalization become simpler in the frequency domain. Suppose $P[n]$ is a known transmit symbol at the receiver, referred to as pilot, we can estimate the channel response at the insertion point of the pilot as

$$\hat{H}_p[n] = \frac{Y_p[n]}{P[n]}, \tag{10.79}$$

where $Y_p[n]$ is the received signal corresponding to $P[n]$. Based on the channel estimate at the pilot $\hat{H}_p[n]$, the channel responses at all subcarriers $\hat{H}[n]$ can be estimated through an interpolation operation. Thus, the recovery of the transmitted symbol can be realized by a simple *one-tap* equalizer

$$\hat{X}[n] = \frac{Y[n]}{\hat{H}[n]}. \tag{10.80}$$

Advanced algorithms can be used for the channel estimation, ranging from simple averaging in combination with linear interpolation to MMSE estimation relying on more detailed knowledge of the channel time-domain and frequency-domain characteristics.

10.3.4 Out-of-Band Emission

Due to the large side lobes that decays asymptotically with f^{-2}, the OOB power leakage of OFDM signals is unacceptably high for practical systems. The interference of OFDM signals on its adjacent channels is around −20 dB, much higher than the requirement of the Adjacent Channel Interference power Ratio (ACIR) of −45 dB specified in 3GPP LTE. Moreover, for some particular deployment scenarios such as cognitive radio-based networks over white TV bands [Jiang et al., 2013; Jiang, 2012], the ACIR defined by Federal Communications Commission (FCC) is much lowered to −72 dB. Due to its low-complexity implementation, the insertion of *guard bands* is often utilized in OFDM systems to

Table 10.1 Transmission bandwidth configuration of 3GPP LTE.

Channel bandwidth (MHz)	1.4	3	5	10	15	20
Number of RBs	6	15	25	50	75	100
Transmission bandwidth (MHz)	1.08	2.7	4.5	9	13.5	18
Guard band (MHz)	0.32	0.3	0.5	1	1.5	2
Spectral loss	22.85%	10%	10%	10%	10%	10%

Source: Jiang and Kaiser [2016]/With permission of IEEE.

minimize the power leakage. Although the whole channel bandwidth is allocated to a dedicated LTE channel, this spectrum cannot be fully used to transmit signals. The guard bands are inserted by deactivating the subcarriers lying at the edges of the spectrum band, applying virtual OFDM subcarriers. The utilization of guard bands somehow alleviates the amount of OOB power leakage but inevitably comes at the cost of spectral-efficiency loss. The channel bandwidth is the amount of spectral resource allocated to a dedicated system, while the transmission bandwidth is the width of spectrum that is actually occupied by the transmit signals. Obviously, the transmission bandwidth isn't allowed to be larger than the channel bandwidth. In the LTE, the term of Resource Block (RB) is defined as a set of OFDM subcarriers, which equals to 12 subcarriers spanning over a signal bandwidth of 180 kHz. The transmission bandwidth in the LTE is parameterized by the number of RBs. For example, the channel of 1.4 MHz is able to transmit up to 6 RBs, which is equivalent to a signal bandwidth of 1.08 MHz. The difference between the channel bandwidth and transmission bandwidth is exactly the width of guard bands. To give a quantitative evaluation of the loss of spectral efficiency, the parameters related to guard bands specified in 3GPP LTE is summarized in Table 10.1 as an example. We can see from this table that the loss of spectral efficiency due to the utilization of guard bands is more than 10% in the LTE system.

In addition to inserting guard bands, advanced signal-processing algorithms, such as time-domain windowing, active interference cancelation, subcarrier weighting, spectral precoding, and low-pass filtering, have been designed to suppress the power leakage of OFDM signals [Jiang and Schellmann, 2012].

Time-Domain Windowing This scheme applies appropriate windows, such as *half-sine* or *Hanning* window, to the transmitted signal. By smoothing the signal's amplitude to zero at the symbol boundaries, the sidelobes can be remarkably confined. Different windowing functions can be uniformly formulated by

$$w(t) = R(t/T) * g(t), \tag{10.81}$$

where $R(t)$ denotes a normalized rectangular pulse

$$R(t) = \begin{cases} 1, & 0 \leqslant t < 1 \\ 0, & \text{otherwise} \end{cases}.$$ (10.82)

The commonly used half-sine pulse is defined as

$$g(t) = \frac{\pi}{2\beta T} \sin\left(\frac{\pi t}{\beta T}\right) R(t/\beta T),$$ (10.83)

where β represents the roll-off factor. Note that $\beta > 0$ effectively broadens the duration of an OFDM symbol, thus increasing the overhead and lowering spectral efficiency. Even if a small value of β is used, the windowing scheme can achieve a considerable sidelobe suppression. Moreover, the computational complexity caused by the windowing of each OFDM symbol is negligible compared with that of the DFT processing.

Active Interference Cancellation Another solution to lower the sidelobes of OFDM signals is to insert additional cancelation subcarriers at the edges of the OFDM spectrum. The information symbols transmitted during the nth symbol is $\mathbf{u}_m = [u_{m,0}, u_{m,1}, \ldots, u_{m,N-1}]^T$. Inserting N_c additional complex symbols $\mathbf{g}_m = [g_{m,1}, g_{m,2}, \ldots, g_{m,N_c}]^T$ results in

$$\mathbf{a}_m = \left[g_{m,1}, \ldots, g_{m,\frac{N_c}{2}}, u_{m,0}, \ldots, u_{m,N-1}, g_{m,\frac{N_c}{2}+1}, \ldots, g_{m,N_c}\right]^T.$$ (10.84)

To measure the power leakage, L selected frequency observation points are chosen in the OOB spectrum range. Defining f_{l,n_c} as the contribution of n_c^{th} cancelation subcarrier to the lth observation point, an $L \times N_c$ matrix $\mathbf{C}$ with entries f_{l,n_c} can be formed. Denote by $\mathbf{f}_m$ an L-dimensional vector containing the contributions of the subcarriers bearing $\mathbf{u}_m$ to the L observation points. Then, the optimization problem can be formulated as a linear least squares problem [Brandes et al., 2006], i.e.

$$\mathbf{g}_m = \arg\min_{\tilde{\mathbf{g}}_m} \left\| \mathbf{f}_m + \mathbf{C}\tilde{\mathbf{g}}_m \right\|.$$ (10.85)

Solving this problem translates to finding the optimal weighting vector $\mathbf{g}_m$ that minimizes the OOB power leakage. To simplify the optimization process, the symbols g_{m,n_c} can be limited to a pre-defined set of quantified symbols. The optimization process in Eq. (10.85) could then be realized by using an exhaustive search.

Subcarrier Weighting Subcarrier weighting multiplies each information symbol with an optimized weight as [Cosovic et al., 2006]

$$\hat{u}_{m,n} = g_{m,n} u_{m,n}, \quad n = 0, \ldots, N-1.$$ (10.86)

The weighting vector can be obtained by solving the optimization equation

$$\mathbf{g}_m = \arg \min_{\tilde{\mathbf{g}}_m} \|\mathbf{S}\tilde{\mathbf{g}}_m\|^2, \tag{10.87}$$

where $\mathbf{S}$ is a $L \times N$ matrix, whose elements $S_{l,n}$ reflect the spectral component of the nth subcarrier to the lth frequency observation point, as defined in active interference cancelation.

Both the weights for the cancelation subcarriers in active interference cancelation and for the information symbols in this scheme are data-dependent and therefore must be determined by solving constrained optimization problems per OFDM symbol. The price to pay for both methods is an increased peak-to-average power ratio as well as a decreased average signal-to-noise-plus-interference ratio, resulting in a degraded Bit Error Rate (BER) performance.

Spectral Precoding To avoid iterative computation of weights on a per-OFDM-symbol basis, a data-independent approach called *spectral precoding* has been proposed in Jiang and Zhao [2012], where the transmitted symbols is precoded prior to the OFDM modulation as

$$\tilde{\mathbf{a}}_k = \mathbf{G}\mathbf{a}_k. \tag{10.88}$$

A precoding matrix $\mathbf{G}$ aims at rendering the OFDM signal and its first n derivatives continuous in phase and amplitude. Denoting by $x_m(t)$, $mT \leqslant t < (m+1)T$ the mth OFDM symbol with the CP, the sidelobes can be drastically decreased by rendering two consecutive OFDM symbols continuous in their first n derivatives, which is expressed by Chung [2008]

$$\left.\frac{d^n}{dt^n}x_{m-1}(t)\right|_{t=mT} = \left.\frac{d^n}{dt^n}x_m(t)\right|_{t=mT}. \tag{10.89}$$

Increasing the order of derivatives n, the BER degrades since the precoding will result in an uneven distribution of the signal over the subcarriers. To compensate for that, iterative decoding could be applied, which would come at the cost of additional computational complexity. The selection of an appropriate n will be a question of the desired trade-off between the OOB power leakage reduction and complexity.

Low-Pass Filtering Another effective method in practice is low-pass filtering, where the transmitted signal is filtered before digital-to-analog conversion

$$\tilde{s}[n] = s[n] * f[n] \tag{10.90}$$

where $s[n]$ denotes the OFDM sequence and $f[n]$ represents a FIR filter. The sidelobe of OFDM signals can be remarkably suppressed by designing a filter with large OOB attenuation. For example, a low-pass filter with 88 taps can achieve attenuation of 50 dB [Jiang and Schellmann, 2012].

10.4   Orthogonal Frequency-Division Multiple Access

A cellular network needs to accommodate a lot of active subscribers simultaneously over a finite amount of time-frequency resources. Therefore, efficient allocation of radio resources among users is a critical design aspect of both uplink (UL) and downlink (DL) channels since the bandwidth is usually scarce and expensive. The share of a communications channel among multiple users that are geographically distributed is referred to as *multiple access*. The most common multiple-access techniques, which orthogonally or non-orthogonally split the signaling dimensions into channels and then assign these channels to different users, include Time-Division Multiple Access (TDMA), Frequency-Division Multiple Access (FDMA), Code-Division Multiple Access (CDMA), and Space-Division Multiple Access (SDMA). As an extension of OFDM transmission to implementing a multi-user system, Orthogonal Frequency-Division Multiple Access (OFDMA) provides an efficient and flexible multi-access technology over a time-frequency resource grid.

10.4.1   Orthogonal Frequency-Division Multiple Access

The discussion in the previous section has implicitly assumed that the OFDM transmission is carried out in a point-to-point communications link for simplicity. In this context, all OFDM subcarriers are used to multiplex the data intended for a single user in the DL transmission, and a single user is assigned to all subcarriers in the UL transmission. Because of the independence among subcarriers, OFDM transmission can also be used as a user-multiplexing or multiple-access scheme, allowing for simultaneous frequency-separated transmissions with multiple users. In the DL, a subset of OFDM subcarriers is used for transmission to one user, while another subset of OFDM subcarriers is used for another user. Similarly, in the UL, one user transmits its data over a subset of OFDM subcarriers, while another user can simultaneously transmit its data over another subset of OFDM subcarriers.

As illustrated in Figure 10.9, three spatially distributed users are assigned to three different portions of OFDM subcarriers. The simplest way is to allocate a group of consecutive subcarriers for transmission and reception of a user. Meanwhile, the subcarriers for a user can be distributed across the whole bandwidth to exploit frequency diversity, with a price of a bit higher implementation complexity and higher vulnerability to hardware impairments.

When OFDMA is employed as an UL multiple-access scheme, the transmitted signals from different terminals must arrive at the base station approximately at the same time. More specifically, the time arrival difference should be less than the length of the cyclic prefix so as to preserve the subcarrier orthogonality and thus avoid inter-carrier interference among different users. Due to the difference

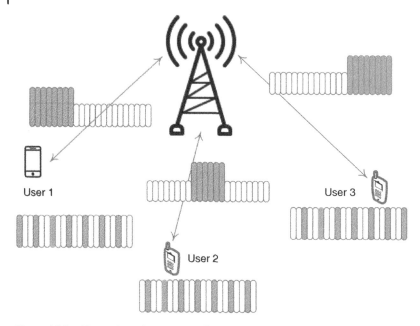

Figure 10.9 Illustration of orthogonal frequency-division multiple access using continuous and distributed subcarriers.

in propagation delays, it is necessary to control the UL transmission timing of each terminal, e.g. letting the terminal far away from the base station send its signal in advance. Such transmission-timing control should adjust the transmit timing of each terminal to ensure that UL transmissions arrive approximately time-aligned at the base station. Furthermore, as the propagation time changes as the terminals move within the cell, the transmission-timing control should be a dynamic process, continuously adjusting the exact timing of each terminal.

Additionally, there exists inter-carrier interference even in the case of perfect transmission-timing control due to frequency errors. Such interference is typically low with reasonable frequency errors and Doppler spread. However, it assumes that different subcarriers are received with at least approximately the same power level. The propagation distances and corresponding path losses may differ significantly in the UL. Hence, the received-signal strengths may thus differ significantly, implying a potentially significant interference from a stronger subcarrier to its weaker adjacent subcarrier unless the subcarrier orthogonality is perfectly retained. To avoid this, at least some degree of UL power control may need to be applied in OFDMA, reducing the transmission power of user terminals close to the base station and ensuring that all received signals will be approximately at the same power level [Dahlman et al., 2011].

10.4.2 Single-Carrier Frequency-Division Multiple Access

A major technical challenge of OFDM transmission, as any multi-carrier modulation, is the significant variations in the instantaneous power of its transmitted signals. Such power variations are generally measured by the Peak-to-Average Power Ratio (PAPR), which is defined by

$$\text{PAPR} := \frac{\max_{n=0,\ldots,N_s}\left(|s_m[n]|^2\right)}{\mathbb{E}\left[|s_m[n]|^2\right]}, \tag{10.91}$$

or

$$\text{PAPR} := \frac{\max_{mT\leqslant t<(m+1)T}\left(|s_m(t)|^2\right)}{\frac{1}{T}\int_{t=mT}^{(m+1)T}|s_m(t)|^2 dt}. \tag{10.92}$$

High PAPR implies a reduced power-amplifier efficiency and higher power-amplifier cost. It is a particular design constraint for the UL transmission due to the requirements of low-power consumption and low cost for mobile terminals. Several methods, such as *tone reservation*, where a subset of OFDM subcarriers are not used for data transmission and instead are modulated to suppress the largest peak, and *selective scrambling*, which selects a transmitted signal with the lowest PAPR from a number of scrambled signals using different codes, have been proposed. However, most of these methods have limitations to what extent they can lower power variations. Thus, it is attractive to consider wider-band single-carrier transmission, exhibiting a constant envelope with very low PAPR, as an alternative to multi-carrier transmission, especially in the UL for mobile terminals.

The single-carrier property is realized by a transmission scheme called DFT-spread OFDM or DFT-s-OFDM, which has minor variations in the instantaneous power of the transmitted signals and enables flexible bandwidth assignment. The basic principle of DFT-s-OFDM is to perform DFT-based precoding in the normal OFDM transmission. A block of M information-bearing symbols is first applied to a DFT with a size of M. Its output is then assigned to a subset of M consecutive or distributed subcarriers of an OFDM modulator implemented by a size-N_s IFFT (generally assume that N_s is set to a power of 2 whereas M is more flexible). If the DFT size M equals the IFFT size N_s, the cascaded DFT-IFFT processing will cancel each other, and the transmitted signal belongs to single-carrier transmission. However, if M is smaller than N_s and the remaining inputs to the IFFT are set to zero, the output of the OFDM modulation will be a signal with low power variations, exhibiting single-carrier property. The major advantage of DFT-s-OFDM compared with the normal OFDM transmission is the low PAPR in the instantaneous transmission power, resulting in an improved power-amplifier efficiency, which can lower power consumption and enable low-cost mobile terminals.

The nominal bandwidth of the transmitted signal for a user is $M \triangle f$. Thus, by dynamically adjusting the block size M, the instantaneous bandwidth of the transmitted signal can be varied, allowing for flexible bandwidth assignment. Furthermore, by assigning the DFT output to different subsets of OFDM sub-carriers, the transmitted signal can be shifted in the frequency domain. Multiple users can simultaneously transmit their data by using the DFT-s-OFDM transmission, enabling not only low power variations like single-carrier transmission but also OFDMA. Therefore, this technique is referred to as Single-Carrier Frequency-Division Multiple Access (SC-FDMA), which has been adopted as the UL transmission scheme in 3GPP Long-Term Evolution (LTE), as well as 5G NR. Without any loss of generality, as shown in Figure 10.10, we assume two users transmitting

$$\mathbf{u}_1 = \left[u_1[0], u_1[1], \dots, u_1[M_1 - 1] \right]^T \tag{10.93}$$

and

$$\mathbf{u}_2 = \left[u_2[0], u_2[1], \dots, u_2[M_2 - 1] \right]^T, \tag{10.94}$$

respectively, toward the base station. Each user makes the DFT precoding over its transmitted symbol blocks before the OFDM modulation, e.g.

$$\tilde{u}_1[n] = \sum_{m=0}^{M_1 - 1} u_1[m] e^{-2\pi j n m / M_1}, \quad n = 0, 1, \dots, M_1 - 1 \tag{10.95}$$

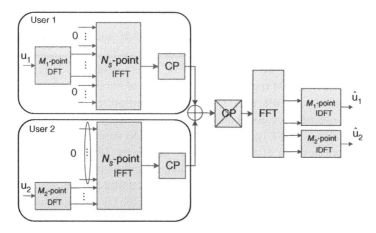

Figure 10.10 Block diagram of single-carrier frequency-division multiple access implemented by DFT-s-OFDM, where two users transmit $\mathbf{u}_1$ and $\mathbf{u}_2$, respectively, toward the base station.

for user 1. Then, the precoded symbols for users 1 and 2 are assigned to different portions of OFDM subcarriers, resulting in

$$\mathbf{a}_1 = \left[\ \underbrace{0, \dots, 0}_{N_s - M_1 - M_2}, \tilde{u}_1[0], \dots, \tilde{u}_1[M_1 - 1], \underbrace{0, \dots, 0}_{M_2}\ \right]^T \tag{10.96}$$

and

$$\mathbf{a}_2 = \left[\underbrace{0, \dots, 0}_{N_s - M_2}, \tilde{u}_2[0], \dots, \tilde{u}_2[M_2 - 1]\right]^T. \tag{10.97}$$

Neglecting channel impairments and additive noise for simplicity, the receiver observes the combined signal from users 1 and 2. Perform the N_s-point FFT demodulation, it gets

$$\mathbf{a}_{rx} = \left[\ \underbrace{0, \dots, 0}_{N_s - M_1 - M_2}, \tilde{u}_1[0], \dots, \tilde{u}_1[M_1 - 1], \tilde{u}_2[0], \dots, \tilde{u}_2[M_2 - 1]\right]^T. \tag{10.98}$$

Separating user-dependent symbols and then making size-M_1 and M_2 IDFT, respectively, the information symbols from two users, i.e. $\mathbf{u}_1$ and $\mathbf{u}_2$, are successfully delivered.

10.4.3 Cyclic Delay Diversity

In the single-carrier wideband transmission, such as wideband CDMA, each modulation symbol is spread over the entire signal bandwidth. Since the channel is highly frequency-selective, a transmitted signal experiences frequency portions with relatively high gains and frequency portions with high attenuation. Such transmission of information over multiple frequency portions with different instantaneous channel quality gets a gain of frequency diversity. On the other hand, in the case of OFDM transmission, each modulation symbol is confined to a narrow-band subcarrier. Thus, some modulation symbols may be fully confined to a frequency portion with low instantaneous channel gain. Therefore, the individual modulation symbols cannot exploit frequency diversity even if the channel is highly frequency-selective over the overall OFDM transmission bandwidth. Consequently, the error probability of OFDM transmission is poor and significantly worse than the error rate in a single-carrier wideband system.

A radio channel subject to time dispersion, with the transmitted signal propagating to the receiver via multiple, independently fading paths with different delays, provides the possibility for multi-path diversity or, equivalently, frequency diversity. Recall that the impulse response for the multipath fading channel is

$$h(\tau, t) = \sum_l a_l(t)\delta(\tau - \tau_l(t)), \tag{10.99}$$

where $a_l(t)$ and $\tau_l(t)$ denote the time-varying attenuation and propagation delay of path l, respectively. Its frequency response is computed by Tse and Viswanath [2005]

$$H(f; t) = \int_{-\infty}^{\infty} h(\tau, t)e^{-2\pi jf\tau}d\tau = \sum_l a_l(t)e^{-2\pi jf\tau_l(t)}. \tag{10.100}$$

There is a differential phase $2\pi f[\tau_{l_1}(t) - \tau_{l_2}(t)]$ among different paths, causing selective fading in frequency. This says that the frequency response changes significantly when f changes by $B_c = \frac{1}{2T_d}$, where B_c is the coherence bandwidth and T_d stands for the delay spread. It implies that a larger delay spread corresponds to a faster variation in the frequency response, or equivalently, more severe frequency selectivity.

If the channel itself is not time dispersive or frequency selectivity is insufficient, a technique called *delay diversity* can be used to create *artificial* time dispersion or, equivalently, artificial frequency selectivity by transmitting identical signals with delays over multiple transmit antennas. The induced delay should be determined to ensure a suitable amount of frequency selectivity over the signal bandwidth. Delay diversity is transparent to the terminal side, which observes a single radio channel subject to additional time dispersion. Delay diversity can thus straightforwardly be employed in an existing mobile-communication system without any compatibility issue in a legacy air-interface standard.

Cyclic-Delay Diversity (CDD) is similar to delay diversity with the main difference that it operates block-wise and applies cyclic shifts, rather than linear delays, to the different antennas. Thus, cyclic-delay diversity is applicable to block-based transmission schemes such as OFDM and DFT-s-OFDM. In the case of OFDM transmission, a cyclic shift of the time-domain signal corresponds to a frequency-dependent phase shift before OFDM modulation, as illustrated in Figure 10.11. Similar to delay diversity, this will create artificial frequency selectivity as seen by the receiver. To avoid the limit of delay length, CDD circularly shifts the samples in an OFDM symbol instead of adding a linear delay to the whole symbol. Assume that the transmitter is equipped with an antenna array of N_t elements, indexed by $n_t = 1, 2, \ldots, N_t$, corresponding to a cyclic delay σ_{n_t}, the receiver has a single antenna. According to the *shift theorem* of the DFT [Oppenheim et al., 1998], a circular shift of a finite-length sequence corresponds to a multiplication of a phase factor in the frequency domain, and this phase

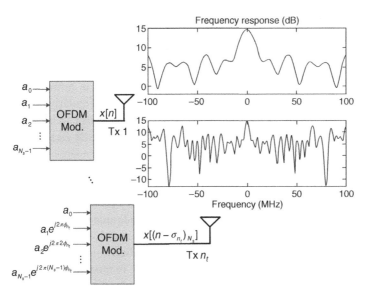

Figure 10.11 Schematic diagram of cyclic delay diversity in an OFDM system, and a comparison of the frequency responses of a time-dispersive fading channel with [the lower] and without [the upper] the CDD.

linearly increases with the index. Referring to Eq. (10.53), we have the original OFDM modulation

$$s[k] = \sum_{n=0}^{N_s-1} a_n e^{2\pi jnk/N_s},\tag{10.101}$$

and its circular shift is given by

$$\begin{aligned}s\left[(k - \sigma_{n_t})_{N_s}\right] &= \sum_{n=0}^{N_s-1} a_n e^{2\pi jn(k-\sigma_{n_t})/N_s} \\ &= \sum_{n=0}^{N_s-1} a_n e^{2\pi jnk/N_s} e^{-2\pi jn\sigma_{n_t}/N_s} \\ &= \sum_{n=0}^{N_s-1} \left(a_n e^{jn\phi_{n_t}}\right) e^{2\pi jnk/N_s}\end{aligned}\tag{10.102}$$

with $\phi_{n_t} = -\frac{2\pi j\sigma_{n_t}}{N_s}$.

In summary, the delay spread due to multipath propagation in a wireless channel raises frequency selectivity. By artificially inducing a larger delay through the use of multiple transmit antennas, delay diversity can increase the variation rate of the channel frequency response, facilitating the exploit of frequency diversity. CDD is a particular implementation of delay diversity in an OFDM system to replace the linear delay with a circular shift. Since a phase rotation in the frequency domain corresponds to a cyclic shift of the time-domain signal. Adding an additional phase shift to each transmitted symbol before the DFT modulation as $\tilde{a}_n = a_n e^{jn\phi_{n_l}}$, where this phase linearly increases in terms of the index of subcarriers, the generated OFDM sequence becomes a circularly shifted version of the original sequence with a shift of σ_{n_l}. Then, a receiver treats these signals the same as multipath components from a single transmitter except for a larger delay spread, which boosts frequency selectivity.

10.4.4 Multi-Cell OFDMA

In early mobile systems, a base station generally covers a wide area with a diameter of tens of kilometers by mounting its antenna at a high elevation and transmitting radio signals with high power. All mobile users within this coverage area share the allocated spectrum, leading to a limited capacity. To meet the increasing demand for a high capacity, the concept of cellular networks through frequency reuse emerged. In 1947, William R. Young first reported his idea about the hexagonal cell layout throughout a wide area so that every mobile telephone can connect to at least one cell. Then, Douglas H. Ring expanded on Young's concept and sketched out the basic design for a standard cellular network as a technical memorandum entitled *Mobile Telephony - Wide Area Coverage* on 11 December 1947 [Ring, 1947].

Because signal power decays drastically with the propagation distance, the same frequency spectrum is possible to reuse at spatially separated locations. If the distance is sufficiently large, co-channel interference is not objectionable. Divide the available spectrum into β non-overlapping portions, and each cell within a cluster of β adjacent cells is assigned to a different portion. Therefore, the same channel is not used in neighboring cells to lower co-channel interference. The ratio β denotes how often a channel can be reused and is termed the *frequency reuse factor*. Through frequency reuse, every β neighboring cells, also known as a cluster, share the whole spectrum. Depending on the geometry of the cellular arrangement and the interference avoidance pattern, the reuse factor can be different.

The classical interference avoidance scheme divides the frequency band into, e.g. three equal subbands, allocated to cells so that adjacent cells always use different frequencies. This scheme is called *hard frequency reuse* and leads to low adjacent-cell interference, with a price to a large capacity loss because only one-third of the resources are used in each cell. For instance, the well-known GSM standard, which is based on the multi-access technique of TDMA, adopted a reuse factor of 3. The simplest scheme to allocate frequencies in a cellular network is to use a reuse factor of 1, i.e. to allocate all chunks to each cell, maximizing spectrum utilization. For example, the CDMA system can use a reuse factor of 1, referred to as *universal frequency reuse*, with the aid of advanced interference suppression techniques. However, in this case, high inter-cell interference is observed, especially at cell edges, where the desired signal is at the weakest level whereas the interference is at the strongest level.

Due to the flexibility of multi-carrier transmission, the frequency reuse can be achieved on the granularity of one subcarrier. That is to say, different subcarriers can be assigned to different cells, in addition to the traditional frequency reuse method that can only allocate different carriers in different cells. For effective Inter-Cell Interference Coordination (ICIC) and the maximization of spectrum usage, the design of OFDMA-based cellular systems should consider seriously the frequency allocation schemes in a multi-cell basis. Fractional Frequency Reuse (FFR) and Soft Frequency Reuse (SFR) [Yang, 2014] are two wide-recognized methods that have been proposed to improve spectrum efficiency and reduce ICI in a multi-carrier communication system. In both FFR and SFR, the subcarriers are divided into different groups, which are treated differently in terms of a cell center and cell edge.

In the FFR scheme, the bandwidth is split into two subbands, and each cell is correspondingly divided into the inner part and the outer part. One sub-band is dedicated to the inner part and reused at all cell centers. The other subband is further divided into three non-overlapping portions, which are assigned to three adjacent cells, respectively, so that the inter-cell interference is minimized at the cell edge. The SFR scheme, proposed by Yang in 3GPP technical proposal R1-050507 [Yang, 2005], employs the frequency reuse factor of 1 in the inner part of the cell, and the frequency reuse factor of 3 at the outward cell region close to the cell edge. For the inner part of the cell, through the limitation of the transmission power, some isolated islands are formed and do not interfere with each other. As demonstrated by Figure 10.12, mobile station 11 and 12 are linked to base station 1, mobile station 21 and 22 are linked to base station 2, and mobile station 31 and 32 are linked to base station 3. Mobile stations 11, 21, and 31 are located at the intersection of 3 cells, mobile stations 12, 22, and 32 are at the inner part of their respective cells. For the mobile station at the cell edge, different subcarriers are allocated to them to avoid the inter-cell interference. For the mobile stations near the base station, all the sub-carriers are available, in comparison with only a fraction of all subcarriers in FFR. When the power ratio between the subcarriers in the cell center and the subcarriers at the cell edge is 0, SFR is equivalent to the hard frequency reuse with a factor of 3. When the power ratio reaches to 1, SFR is equivalent to the universal frequency reuse with a factor of 1. Through the adjustment of power ratio in the range from 0 to 1, the frequency reuse factor from 3 to 1 can be implemented. This is the reason why it can be called soft frequency reuse scheme.

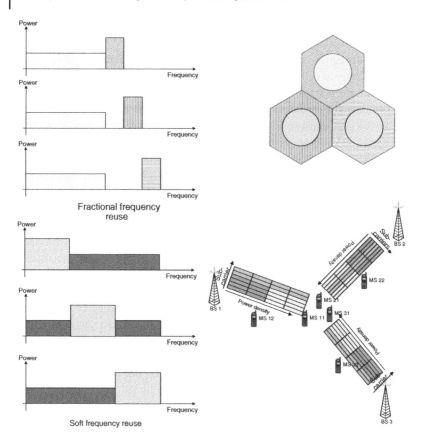

Figure 10.12 The principles of fractional frequency reuse and soft frequency reuse. Source: Adapted from Yang [2005].

10.5 Cell-Free Massive MIMO-OFDMA

Most traffic congestion in cellular networks nowadays happens at the cell edge. The so-called 95%-likely user data rates, which can be guaranteed to 95% of the users and thus define the user-experienced performance, remain mediocre in 5G networks. The solution to these issues might be to connect each user with a multitude of distributed antennas. If there is only one huge cell in the network, no inter-cell interference appears. A distributed massive multiple-input multiple-output (MIMO) system, where a large number of service antennas serve a much smaller number of autonomous users spread over a wide area, has been proposed. All antennas cooperate phase-coherently via a fronthaul network and serve all users in the same time-frequency resource. There are no cells or cell boundaries. Therefore, this system is referred to as *cell-free massive MIMO*. Due to the ability to exploit spatial diversity against shadow fading more efficiently, a distributed system can offer a much higher probability of coverage than a collocated system at the expense of increased backhaul requirements. On the other hand, the signal bandwidth of mobile communications becomes increasingly wider to meet the demand for higher throughput, especially in high-frequency bands such as millimeter-wave and terahertz communications [Jiang and Schotten, 2022a,b]. Nevertheless, wide-band communications suffer from highly frequency selectivity of fading channels. Consequently, the combination of cell-free massive MIMO with OFDM transmission, coined cell-free massive MIMO-OFDM or cell-free massive MIMO-OFDMA [Jiang and Schotten, 2021b], is promising in the forthcoming 6G systems.

10.5.1 The System Model

Consider a geographical area where M randomly distributed Access Points (APs) are connected to a Central Processing Unit (CPU) via a fronthaul network and serve K users. Without loss of generation, assume that each AP and user is equipped with a single antenna but its adaptability to multi-antenna APs is straightforward. In contrast to the conventional Cell-Free massive MIMO (CFmMIMO) that requires $K \ll M$, the number of users in Cell-Free massive MIMO-OFDM (CFmMIMO-OFDM) is scalable, ranging from small $K \ll M$ to very large $K \gg M$. Users are divided into groups and each group is assigned to different RBs. Thus, the constraint that the number of users is far smaller than the number of APs is still satisfied on each subcarrier or RB. In CFmMIMO transmission [Jiang and Schotten, 2021a], the small-scale fading is assumed to be frequency-flat, modeled by a circularly symmetric complex Gaussian random variable with zero mean and unit variance, i.e. $h[t] \sim \mathcal{CN}(0,1)$. This assumption is only valid for narrow-band communications. Nevertheless, most of the current and future mobile communications are broadband, suffering from severe frequency selectivity. A frequency-selective fading channel is modeled as a time-varying linear filter $\mathbf{h}[t] = \begin{bmatrix} h_0[t], \dots, h_{L-1}[t] \end{bmatrix}^T$, where the filter length L is related to the multi-path delay spread T_d and the sampling interval T_s. The tap gain is computed by

$$h_l[t] = \sum_i a_i(tT_s)e^{-j2\pi f_c \tau_i(tT_s)}\text{sinc}\left[l - \frac{\tau_i(tT_s)}{T_s}\right] \qquad (10.103)$$

for $l = 0, \ldots, L - 1$, with carrier frequency f_c, and time-varying attenuation $a_i(t)$ and delay $\tau_i(t)$ of the ith signal path. The fading channel between AP m and user k is given by

$$\mathbf{g}_{mk}[t] = \left[g_{mk,0}[t], \ldots, g_{mk,L_{mk}-1}[t] \right]^T$$

$$= \sqrt{\beta_{mk}[t]} \left[h_{mk,0}[t], \ldots, h_{mk,L_{mk}-1}[t] \right]^T = \sqrt{\beta_{mk}[t]} \mathbf{h}_{mk}[t], \qquad (10.104)$$

where $g_{mk,l}[t] = \sqrt{\beta_{mk}[t]} h_{mk,l}[t]$ and $\beta_{mk}[t]$ indicates large-scale fading, which is frequency independent and varies slowly, and L_{mk} denotes the channel length.

The data transmission in an OFDM system is organized in block-wise, as shown in Figure 10.13. We write

$$\tilde{\mathbf{x}}_m[t] = \left[\tilde{x}_{m,0}[t], \ldots, \tilde{x}_{m,n}[t], \ldots, \tilde{x}_{m,N-1}[t] \right]^T \qquad (10.105)$$

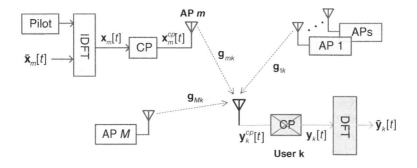

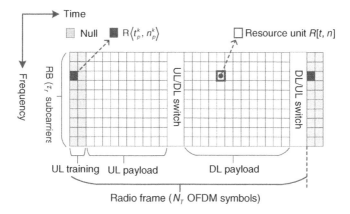

Figure 10.13 Schematic diagram of a cell-free massive MIMO-OFDM system where M APs serve K users, as well as a time-frequency resource grid of a single RB.

to denote the frequency-domain transmission block of AP m on the tth OFDM symbol. Transform $\tilde{\mathbf{x}}_m[t]$ into a time-domain sequence

$$\mathbf{x}_m[t] = \left[x_{m,0}[t], \ldots, x_{m,k}[t], \ldots, x_{m,N-1}[t]\right]^T. \tag{10.106}$$

through an N-point IDFT, i.e.

$$x_{m,k}[t] = \frac{1}{N}\sum_{n=0}^{N-1}\tilde{x}_{m,n}[t]e^{2\pi jkn/N} \tag{10.107}$$

for $k = 0, 1, \ldots, N-1$. Defining the DFT matrix

$$\mathbf{F} = \begin{bmatrix} \omega_N^{0 \cdot 0} & \cdots & \omega_N^{0 \cdot (N-1)} \\ \vdots & \ddots & \vdots \\ \omega_N^{(N-1) \cdot 0} & \cdots & \omega_N^{(N-1) \cdot (N-1)} \end{bmatrix} \tag{10.108}$$

with a primitive Nth root of unity $\omega_N^{n \cdot k} = e^{2\pi jnk/N}$, the OFDM modulation can be written in matrix form as

$$\mathbf{x}_m[t] = \mathbf{F}^{-1}\tilde{\mathbf{x}}_m[t] = \frac{1}{N}\mathbf{F}^*\tilde{\mathbf{x}}_m[t]. \tag{10.109}$$

A cyclic prefix with a length of N_{CP} is added between two consecutive blocks to avoid inter-symbol interference and preserve the orthogonality of subcarriers. Thus, the transmitted signal is expressed by

$$\mathbf{x}_m^{\mathrm{CP}}[t] = \left[\underbrace{x_{m,N-N_{\mathrm{CP}}}[t], \ldots, x_{m,N-1}[t]}_{\text{Cyclic prefix}}, x_{m,0}[t], \ldots, x_{m,N-1}[t]\right]^T. \tag{10.110}$$

The signal $\mathbf{x}_m^{\mathrm{CP}}[t]$ goes through the channel $\mathbf{g}_{mk}[t]$ to reach a typical user k, resulting in $\mathbf{x}_m^{\mathrm{CP}}[t] * \mathbf{g}_{mk}[t]$, where $*$ denotes *the linear convolution*. Thus, the overall received signal at user k is $\mathbf{y}_k^{\mathrm{CP}}[t] = \sum_{m=1}^{M}\mathbf{x}_m^{\mathrm{CP}}[t] * \mathbf{g}_{mk}[t] + \mathbf{z}_k[t]$, where $\mathbf{z}_k[t]$ is a vector of additive noise. Removing the CP, yields

$$\mathbf{y}_k[t] = \sum_{m=1}^{M}\mathbf{g}_{mk}^N[t] \otimes \mathbf{x}_m[t] + \mathbf{z}_k[t], \tag{10.111}$$

where $\otimes$ stands for *the cyclic convolution* [Jiang and Kaiser, 2016] and $\mathbf{g}_{mk}^N[t]$ is an N-point channel filter formed by padding zeros at the tail of $\mathbf{g}_{mk}[t]$, i.e.

$$\mathbf{g}_{mk}^N[t] = \left[g_{mk,0}[t], \ldots, g_{mk,L_{mk}-1}[t], 0, \ldots, 0\right]^T. \tag{10.112}$$

The DFT demodulator outputs the frequency-domain received signal

$$\tilde{\mathbf{y}}_k[t] = \mathbf{F}\mathbf{y}_k[t]. \tag{10.113}$$

Substituting Eqs. (10.109) and (10.111) into Eq. (10.113), and applying *the convolution theorem* for DFT, we know

$$\tilde{\mathbf{y}}_k[t] = \sum_{m=1}^{M} \mathbf{F}\left(\mathbf{g}_{mk}^N[t] \otimes \mathbf{x}_m[t]\right) + \mathbf{F}\mathbf{z}_k[t]$$

$$= \sum_{m=1}^{M} \tilde{\mathbf{g}}_{mk}[t] \odot \tilde{\mathbf{x}}_m[t] + \tilde{\mathbf{z}}_k[t], \tag{10.114}$$

where $\odot$ represents the Hadamard product (element-wise multiplication), the frequency-domain channel response and noise are given by

$$\tilde{\mathbf{g}}_{mk}[t] = \mathbf{F}\mathbf{g}_{mk}^N[t] \tag{10.115}$$

and

$$\tilde{\mathbf{z}}_k[t] = \mathbf{F}\mathbf{z}_k[t], \tag{10.116}$$

respectively. In final, a frequency-selective channel is transformed into a set of N independent frequency-flat subcarriers. The signal transmission in the DL on the nth subcarrier is given by

$$\tilde{y}_{k,n}[t] = \sum_{m=1}^{M} \tilde{g}_{mk,n}[t]\tilde{x}_{m,n}[t] + \tilde{z}_{k,n}[t], \quad k \in \{1, \dots, K\}, \tag{10.117}$$

where $\tilde{g}_{mk,n}[t]$ is the nth element of $\tilde{\mathbf{g}}_{mk}[t]$. Similarly, the UL transmission is expressed by

$$\tilde{y}_{m,n}[t] = \sum_{k=1}^{K} \tilde{g}_{mk,n}[t]\tilde{x}_{k,n}[t] + \tilde{z}_{m,n}[t], \quad m \in \{1, \dots, M\}. \tag{10.118}$$

10.5.2 The Communication Process

The DL transmission from the APs to the users and the UL from the users to the APs are separated by Time-Division Duplex (TDD) with the assumption of perfect channel reciprocity. A radio frame is mainly divided into three phases: UL training, UL payload transmission, and DL payload transmission.

10.5.2.1 Uplink Training

We write $\mathscr{R}\langle t, n \rangle$ to denote a Resource Unit (RU) on the nth subcarrier of the tth OFDM symbol. The time-frequency resource of a radio frame is divided into N_{RB} RBs, each of which contains $\lambda_{\text{RB}} = N/N_{\text{RB}}$ (assumed to be an integer) consecutive subcarriers, as shown in Figure 10.13. The rth RB is defined as

$$B_r \triangleq \{\mathscr{R}\langle t, n \rangle | 1 \leq t \leq N_T \text{ and } (r-1)\lambda_{\text{RB}} \leq n < r\lambda_{\text{RB}}\}, \tag{10.119}$$

for any $r \in \{1, \dots, N_{RB}\}$. The transmission of a radio frame in CFmMIMO is carried out within the *coherence time* and the width of one RB is smaller than the *coherence bandwidth*. Taking advantage of time and frequency correlation, the channel coefficient of any RU can be obtained by interpolating the channel estimates of pilots. Without loss of generality, the *block fading* model is adopted where the channel coefficients for all RUs within one RB are assumed to be identical, i.e.

$$\tilde{g}_{mk,n}[t] = \tilde{g}^r_{mk} \impliedby \mathcal{R}\langle t, n \rangle \in \mathcal{B}_r. \tag{10.120}$$

The channel estimation of a conventional CFmMIMO system relies on time-domain pilot sequences, where the maximal number of orthogonal sequences is τ_p by using τ_p pilot symbols. If $K \leqslant \tau_p$, pilot contamination can be avoided. Owing to the limitation of the frame length, however, some users need to share the same sequence when $K > \tau_p$, leading to pilot contamination. In contrast, CFmMIMO-OFDM is able to provide more orthogonal pilots by means of frequency-division multiplexing thanks to the extra degree of freedom gained from the frequency domain. To estimate $\tilde{g}^r_{mk}$, each user on $\mathcal{B}_r$ needs only one pilot symbol. Suppose the first τ_p OFDM symbols are dedicated to the UL training, one RB has $N_p = \tau_p \lambda_{RB}$ orthogonal pilots. The number of users allocated to $\mathcal{B}_r$ is denoted by K_r, there is no pilot contamination if $K_r \leqslant N_p$, which is a very relaxed condition. We write $\mathcal{R}\langle t^k_p, n^k_p \rangle$ with $1 \leqslant t^k_p \leqslant \tau_p$ and $(r-1)\lambda_{RB} \leqslant n^k_p < r\lambda_{RB}$ to denote the RU reserved for the pilot symbol of user k, $k \in \{1, \dots, K_r\}$. Other users keep silence (null) on this RU to achieve orthogonality. Mathematically, the pilot assignment is specified by

$$\tilde{x}_{k,n}[t] = \begin{cases} \sqrt{p_u}\, \mathbb{P}_k, & \text{if } t = t^k_p \wedge n = n^k_p \\ 0, & \text{otherwise} \end{cases}, \quad 1 \leqslant t \leqslant \tau_p, \tag{10.121}$$

where $\wedge$ represents logical AND, $\mathbb{P}_k$ is the known pilot symbol with $\mathbb{E}[|\mathbb{P}_k|^2] = 1$, and p_u denotes the UL transmit power limit.

Substituting Eqs. (10.120) and (10.121) into Eq. (10.118), yields the received signal of the mth AP on $\mathcal{R}\langle t^k_p, n^k_p \rangle$

$$\begin{aligned}
\tilde{y}_{m,n^k_p}[t^k_p] &= \sum_{k=1}^{K_r} \tilde{g}_{mk,n^k_p}[t^k_p] \tilde{x}_{k,n^k_p}[t^k_p] + \tilde{z}_{m,n^k_p}[t^k_p] \\
&= \tilde{g}_{mk,n^k_p}[t^k_p] \tilde{x}_{k,n^k_p}[t^k_p] + \sum_{k' \neq k}^{K_r} \tilde{g}_{mk',n^k_p}[t^k_p] \tilde{x}_{k',n^k_p}[t^k_p] + \tilde{z}_{m,n^k_p}[t^k_p] \\
&= \sqrt{p_u}\,\tilde{g}_{mk,n^k_p}[t^k_p] \mathbb{P}_k + \tilde{z}_{m,n^k_p}[t^k_p] \\
&= \sqrt{p_u}\,\tilde{g}^r_{mk} \mathbb{P}_k + \tilde{z}_{m,n^k_p}[t^k_p].
\end{aligned} \tag{10.122}$$

Let $\hat{g}^r_{mk}$ be an estimate of $\tilde{g}^r_{mk}$, we have $\hat{g}^r_{mk} = \tilde{g}^r_{mk} - \xi^r_{mk}$ with estimation error ξ^r_{mk} raised by additive noise. Applying the MMSE estimation gets

$$\hat{g}^r_{mk} = \left(\frac{R_{gg}\mathbb{P}^*_k}{R_{gg}|\mathbb{P}_k|^2 + R_{nn}} \right) \tilde{y}_{m,n^k_p}[t^k_p] = \left(\frac{\beta_{mk}\mathbb{P}^*_k}{\beta_{mk}|\mathbb{P}_k|^2 + \sigma^2_z} \right) \tilde{y}_{m,n^k_p}[t^k_p], \quad (10.123)$$

which applies $R_{gg} = \mathbb{E}\left[\left| \tilde{g}^r_{mk} \right|^2 \right] = \beta_{mk}$ and $R_{nn} = \mathbb{E}\left[\left| \tilde{z}_{m,n}[t] \right|^2 \right] = \sigma^2_z$. Compute the variance of $\hat{g}^r_{mk}$ as

$$\mathbb{E}\left[\hat{g}^r_{mk}(\hat{g}^r_{mk})^* \right] = \mathbb{E}\left[\frac{\beta^2_{mk}|\mathbb{P}_k|^2 \left| \sqrt{P_u}\tilde{g}^r_{mk}\mathbb{P}_k + \tilde{z}_{m,n^k_p}[t^k_p] \right|^2}{(\beta_{mk}|\mathbb{P}_k|^2 + \sigma^2_z)^2} \right]$$

$$= \frac{\beta^2_{mk}\mathbb{E}\left[\left| \sqrt{P_u}\tilde{g}^r_{mk}\mathbb{P}_k + \tilde{z}_{m,n^k_p}[t^k_p] \right|^2 \right]}{(\beta_{mk} + \sigma^2_z)^2}$$

$$= \frac{P_u\beta^2_{mk}}{P_u\beta_{mk} + \sigma^2_z}. \quad (10.124)$$

Consequently, we know that $\hat{g}^r_{mk} \in \mathcal{CN}(0, \alpha_{mk})$ with $\alpha_{mk} = \frac{P_u\beta^2_{mk}}{P_u\beta_{mk}+\sigma^2_z}$, in comparison with $\tilde{g}^r_{mk} \in \mathcal{CN}(0, \beta_{mk})$.

10.5.2.2 Uplink Payload Data Transmission

Suppose τ_u OFDM symbols are used for UL transmission, on the RU $\mathcal{R}\langle t, n \rangle \in \mathcal{B}_r$, $\tau_p < t \le \tau_p + \tau_u$, all K_r users simultaneously transmit their signals to the APs. The kth user weights its transmit symbol $q_{k,n}[t]$, satisfying

$$\mathbb{E}\left[|q_{k,n}[t]|^2 \right] = 1, \quad (10.125)$$

by a power-control coefficient $\sqrt{\psi_k}, 0 \le \psi_k \le 1$. Substituting $\tilde{x}_{k,n}[t] = \sqrt{\psi_k P_u}q_{k,n}[t]$ into Eq. (10.118) yields

$$\tilde{y}_{m,n}[t] = \sqrt{P_u}\sum_{k=1}^{K_r}\tilde{g}_{mk,n}[t]\sqrt{\psi_k}q_{k,n}[t] + \tilde{z}_{m,n}[t]. \quad (10.126)$$

10.5.2.3 Downlink Payload Data Transmission

As CFmMIMO applying conjugate beamforming in the DL, CFmMIMO-OFDM employs frequency-domain conjugate beamforming. On $\mathcal{R}\langle t, n \rangle \in \mathcal{B}_r$, $\tau_p + \tau_u < t \le N_T$, each AP multiplexes a total of K_r symbols, i.e. $s_{k,n}[t]$ intended to user k, $k = 1, \ldots, K_r$, before transmission. With a power-control coefficient $\sqrt{\eta_{mk}}, 0 \le \eta_{mk} \le 1$, the transmitted signal of the mth AP is

$$\tilde{x}_{m,n}[t] = \sqrt{P_d}\sum_{k=1}^{K_r}\sqrt{\eta_{mk}}(\hat{g}_{mk,n}[t])^* s_{k,n}[t]. \quad (10.127)$$

Substituting Eq. (10.127) into Eq. (10.117) to get the received signal at user k

$$\tilde{y}_{k,n}[t] = \sqrt{P_d} \sum_{m=1}^{M} \tilde{g}_{mk,n}[t] \sum_{k'=1}^{K_r} \sqrt{\eta_{mk'}} \left(\hat{g}_{mk',n}[t]\right)^* s_{k',n}[t] + \tilde{z}_{k,n}[t]$$

$$= \underbrace{\sqrt{P_d} \sum_{m=1}^{M} \sqrt{\eta_{mk}} \left|\hat{g}_{mk,n}[t]\right|^2 s_{k,n}[t]}_{\text{Desired signal}}$$

$$+ \underbrace{\sqrt{P_d} \sum_{m=1}^{M} \hat{g}_{mk,n}[t] \sum_{k'\neq k}^{K_r} \sqrt{\eta_{mk'}} \left(\hat{g}_{mk',n}[t]\right)^* s_{k',n}[t]}_{\text{Multi-user interference}}$$

$$+ \underbrace{\sqrt{P_d} \sum_{m=1}^{M} \xi_{mk,n}[t] \sum_{k'=1}^{K_r} \sqrt{\eta_{mk'}} \left(\hat{g}_{mk',n}[t]\right)^* s_{k',n}[t]}_{\text{Channel-estimate error}} + \underbrace{\tilde{z}_{k,n}[t]}_{\text{Noise}}. \qquad (10.128)$$

Each user is assumed to have the knowledge of channel statistics

$$\mathbb{E}\left[\sum_{m=1}^{M} \sqrt{\eta_{mk}} \left|\hat{g}_{mk,n}[t]\right|^2\right] \qquad (10.129)$$

rather than channel realizations $\hat{g}_{mk,n}[t]$ since there is no pilot and channel estimation in the DL. We can derive that the spectral efficiency of user k on $\mathscr{R}\langle t,n\rangle \in B_r$ is lower bounded by $\log_2\left(1 + \gamma_k^{\langle t,n\rangle}\right)$ with

$$\gamma_k^{\langle t,n\rangle} = \frac{P_d \left(\sum_{m=1}^{M} \sqrt{\eta_{mk}} \alpha_{mk}\right)^2}{\sigma_z^2 + P_d \sum_{m=1}^{M} \beta_{mk} \sum_{k'=1}^{K_r} \eta_{mk'} \alpha_{mk'}}, \qquad (10.130)$$

implying that the effect of small-scale fading is vanished (a.k.a. channel hardening), such that $\gamma_k^{\langle t,n\rangle} = \gamma_k^r$, for all $\mathscr{R}\langle t,n\rangle \in B_r$.

10.5.3 User-Specific Resource Allocation

The conventional CFmMIMO systems can support only very few $K \ll M$ users with uniform quality of service. By exploiting the frequency domain, CFmMIMO-OFDM is adaptive to different numbers of users from a few $K \ll M$ to massive $K \gg M$ and is flexible to offer diverse data rates for heterogeneous users. As demonstrated in Figure 10.14, classify all users

$$\mathcal{U} = \{u_1, u_2, \ldots, u_K\} \qquad (10.131)$$

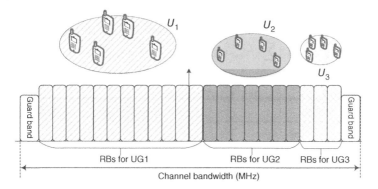

Figure 10.14 Illustration of user-specific resource allocation in a cell-free massive MIMO-OFDM system.

into different groups $\mathcal{U}_s, s = 1, 2, \ldots, S$ in terms of their demands on data throughput, subjecting to

$$\bigcup_{s=1}^{S} \mathcal{U}_s = \mathcal{U} \tag{10.132}$$

and

$$\mathcal{U}_s \bigcap \mathcal{U}_{s'} = \varnothing, \quad \forall s' \neq s. \tag{10.133}$$

The number of users in $\mathcal{U}_s$ satisfies $|\mathcal{U}_s| \ll M$ and $\sum_{s=1}^{S} |\mathcal{U}_s| = K$, where $|\cdot|$ stands for the cardinality of a set. The resource pool is $\mathbb{B} = \{B_r | 1 \leq r \leq N_{\mathrm{RB}}\}$ where the granularity for allocation is one RB. Using $\mathbb{B}_s$ to denote the RBs allocated to $\mathcal{U}_s$, we have $\bigcup_{s=1}^{S} \mathbb{B}_s \in \mathbb{B}$ (when all RBs are used, $\bigcup_{s=1}^{S} \mathbb{B}_s = \mathbb{B}$) and $\mathbb{B}_s \bigcap \mathbb{B}_{s'} = \varnothing$, $\forall s' \neq s$. If $B_r \in \mathbb{B}_s$, the number of users served by this RB is $K_r = |\mathcal{U}_s|$. For any user $k \in \mathcal{U}_s$, its per-user data rate in the DL is

$$R_k = \left(1 - \frac{\tau_p + \tau_u}{N_T}\right) \sum_{B_r \in \mathbb{B}_s} \lambda_{\mathrm{RB}} \, \triangle f \log_2 \left(1 + \gamma_k^r\right), \tag{10.134}$$

where $\triangle f$ is the inter-subcarrier spacing. Then, the DL sum data throughput of the system is $R_d = \sum_{s=1}^{S} \sum_{k \in \mathcal{U}_s} R_k$.

10.6 Non-Orthogonal Multiple Access

The design of an efficient multiple-access technique is a critical aspect of a cellular system, which has been a dominant way to distinguish different wireless communications from the first generation to the fifth generation. For instance, FDMA has been used for 1G, time-division multiple access in most of 2G, and code-division multiple access for 3G. In 4G, OFDMA and SC-FDMA have been adopted for the DL and UL transmission, respectively. Most of these techniques follow the philosophy of Orthogonal Multiple Access (OMA), where an orthogonal resource unit, e.g. a time slot, a subcarrier, and an orthogonal spread code, is dedicated to a single user. In such a way, the multiplexing gain is achieved with reasonable complexity while alleviating multi-user interference.

In order to meet the heterogeneous requirements on massive connectivity, high spectral efficiency, low latency, and improved fairness, the 5G system has accepted a brand new technique called Non-Orthogonal Multiple Access (NOMA) as one of its multi-access methods. In contrast to the conventional OMA schemes, the key distinguishing feature of NOMA is to serve a higher number of users than the number of orthogonal resource units with the aid of non-orthogonal resource sharing, with the price of the sophisticated inter-user interference cancelation at the receiver. Ding et al. [2017] provides a simple example to illustrate the superiority of NOMA in comparison with OMA. Consider a scenario where a user with very poor channel conditions needs to be served for fairness purposes, e.g. this user has high priority data or has not been served for a long time. In this case, the use of OMA means that it is inevitable that one of the scarce bandwidth resources is solely occupied by this user, despite its poor channel conditions. This design harms the spectrum efficiency and the capacity of the overall system. In such a situation, the use of NOMA ensures that the user with poor channel conditions gets served and that users with better channel conditions can concurrently access the same resources. Consequently, if user fairness has to be guaranteed, the system capacity of NOMA can be significantly larger than that of OMA. In addition to its spectral efficiency gain, NOMA can effectively support more users, ensuring massive connectivity in the deployment scenario of the Internet of Things.

10.6.1 Fundamentals of NOMA

Although the inter-user interference among orthogonally multiplexed users enables low-complexity Multi-User Detection (MUD) at the receiver, it is widely recognized that OMA cannot achieve the sum-rate capacity of a multi-user wireless system. Superposition coding at the transmitter and Successive Interference Cancellation (SIC) at the receiver make it possible to reuse each orthogonal resource unit by more than one user. At the transmitter side, all the individual information symbols are superimposed into a single waveform, while SIC at the receiver side decodes the signals iteratively until it gets the desired signal. This scheme is sometimes called power-domain NOMA. The basic NOMA principle and its sum-rate capacity comparison with OMA will be introduced in this section in terms of the DL and UL transmission, respectively.

10.6.1.1 Downlink Non-Orthogonal Multiplexing

In the DL, a single-antenna base station superimposes the information-bearing symbols intended for K single-antenna users. The multiplexed signal at the base station can be expressed by

$$s = \sum_{k=1}^{K} \sqrt{\alpha_k P_d} s_k, \tag{10.135}$$

where s_k is the information-bearing symbol for a general user k, $k = 1, 2, \ldots, K$, satisfying $\mathbb{E}\left[|s_k|^2\right] = 1$, α_k represents the power allocation coefficient subjecting to $\sum_{k=1}^{K} \alpha_k \leqslant 1$, and P_d denotes the total transmit power of the base station. The challenge is to decide how to allocate the power among the users, which is critical for interference cancelation at the receiver. That is why NOMA is regarded as a kind of power-domain multiple access. Generally, more power is allocated to a user with a smaller channel gain, e.g. located farther from the base station, to improve their received SNR, so that a high detection reliability can be guaranteed. Despite of less power assigned to a user with a stronger channel gain, e.g. close to the base station, it is capable of detecting its signal correctly with reasonable SNR. User i observes the received signal

$$
\begin{aligned}
r_i &= g_i s + n_i \\
&= g_i \sum_{k=1}^{K} \sqrt{\alpha_k P_d} s_k + n_i \\
&= \underbrace{g_i s_i}_{\text{Desired signal}} + \underbrace{g_i \sum_{k=1, k\neq i}^{K} \sqrt{\alpha_k P_d} s_k + n_i}_{\text{Multi-user interference}},
\end{aligned}
\tag{10.136}
$$

where g_k stands for the complex channel gain between the base station and user k, n_i is the additive white Gaussian noise with zero mean and power spectral density N_0 (W/Hz). Without loss of generality, we can assume that user 1 has the largest channel gain, and user K is the weakest, i.e.

$$|g_1| \geqslant |g_2| \geqslant \cdots \geqslant |g_K|. \tag{10.137}$$

The same signal s that contains all information symbols is delivered to all users. The optimal order of interference cancelation is detecting the user with the most power allocation (the weakest channel gain) to the user with the least power allocation (the strongest channel gain). With this order, each user decodes s_K first, and then subtracts its component from the received signal. As a result, a typical user i after the first SIC iteration gets

$$\tilde{r}_i = r_i - \sqrt{\alpha_K P_d} g_i s_K = g_i \sum_{k=1}^{K-1} \sqrt{\alpha_k P_d} s_k + n_i, \tag{10.138}$$

assuming error-free detection and perfect channel knowledge. In the second iteration, each user decodes s_{K-1} using the remaining signal $\tilde{r}_i$ without the interference
of the weakest user. The cancelation iterates until each user gets its own signal.
Particularly, the weakest user decodes its own signal directly without successive
interference cancelation since it is allocated the most power. Therefore, the SNR
for user K can be written as

$$\gamma_K = \frac{|g_K|^2 \alpha_K P_d}{|g_K|^2 \sum_{k=1}^{K-1} \alpha_k P_d + N_0 B_w}, \tag{10.139}$$

where B_w denotes the signal bandwidth. In general, the SNR for user i is

$$\gamma_i = \frac{|g_i|^2 \alpha_i P_d}{|g_i|^2 \sum_{k=1}^{i-1} \alpha_k P_d + N_0 B_w}, \tag{10.140}$$

resulting in the achievable rate of

$$R_i = B_w \log\left(1 + \frac{|g_i|^2 \alpha_i P_d}{|g_i|^2 \sum_{k=1}^{i-1} \alpha_k P_d + N_0 B_w}\right). \tag{10.141}$$

The sum rate of DL NOMA transmission is computed by

$$R = \sum_{i=1}^{K} R_i. \tag{10.142}$$

In the case of two users, as illustrated in Figure 10.15, a far user at the cell edge is
allocated to more power and a near user at the cell center gets less power. Regardless of the difference of g_1 and g_2, this power ratio will be kept in the received

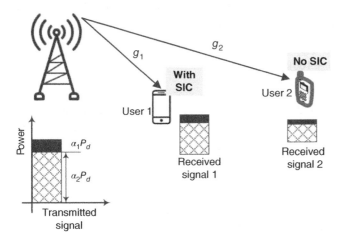

Figure 10.15 Illustration of downlink NOMA consisting of a base station, a far user, and
a near user.

signal of any user. The near user firstly detects the symbol of the far user s_2, and then subtracts its regenerated component from the received signal, resulting in the available rate of

$$R_1 = \log\left(1 + \frac{|g_1|^2 \alpha_1 P_d}{N_0}\right), \tag{10.143}$$

assuming the signal bandwidth is normalized to $B_w = 1$ Hz. The far user detects its signal directly by treating the signal of the near user as a colored noise, getting

$$R_2 = \log\left(1 + \frac{|g_2|^2 \alpha_2 P_d}{|g_2|^2 \alpha_1 P_d + N_0}\right). \tag{10.144}$$

Consequently, two-user DL transmission using NOMA gets a sum rate of

$$R = R_1 + R_2 = \log\left(1 + \frac{|g_1|^2 \alpha_1 P_d}{N_0}\right) + \log\left(1 + \frac{|g_2|^2 \alpha_2 P_d}{|g_2|^2 \alpha_1 P_d + N_0}\right). \tag{10.145}$$

Using the OMA scheme as a comparison, the near user is assigned to the bandwidth of η Hz ($0 < \eta < 1$), leaving the rest of the bandwidth $(1 - \eta)$ Hz to the far user. Their achievable rates can be computed by

$$R_1 = \eta \log\left(1 + \frac{|g_1|^2 \alpha_1 P_d}{\eta N_0}\right)$$
$$R_2 = (1 - \eta) \log\left(1 + \frac{|g_2|^2 \alpha_2 P_d}{(1 - \eta) N_0}\right). \tag{10.146}$$

Suppose the geometries of the near user and the far user are $\frac{|g_1|^2 P_d}{N_0} = 20$ dB and $\frac{|g_2|^2 P_d}{N_0} = 0$ dB, respectively. When equal bandwidth (i.e. $\eta = 0.5$) and equal power (i.e. $\alpha_1 = \alpha_2 = 0.5$) are allocated to either user with the proportional fairness criteria, the per-user rates of OMA are calculated according to Eq. (10.146) as

$$R_1 = 0.5\log_2\left(1 + 100\right) = 3.3291 \text{ bps/Hz}$$
$$R_2 = 0.5\log_2\left(1 + 1\right) = 0.5 \text{ bps/Hz}. \tag{10.147}$$

On the other hand, when the power allocation in NOMA is conducted as $\alpha_1 = 0.2$ and $\alpha_2 = 0.8$, the per-user rates are calculated according to Eqs. (10.143) and (10.144) as

$$R_1 = \log_2\left(1 + 20\right) = 4.3923 \text{ bps/Hz}$$
$$R_2 = \log_2\left(1 + 0.8/1.2\right) = 0.7370 \text{ bps/Hz}, \tag{10.148}$$

corresponding to a spectral-efficiency gain of approximately 32% and 47% in terms of the OMA scheme [Saito et al., 2013].

Such a gain is harvested by making full use of the difference of channel gains among the users. The larger difference usually corresponds to a higher spectral-efficiency gain of NOMA in comparison with OMA, and vice versa. If two users have identical channel conditions, i.e. $|g_1| = |g_2|$, the sum rate of NOMA in Eq. (10.145) can be rewritten as

$$R = \log\left(1 + \frac{|g_1|^2 \alpha_1 P_d}{N_0}\right) + \log\left(1 + \frac{|g_2|^2 \alpha_2 P_d}{|g_2|^2 \alpha_1 P_d + N_0}\right)$$

$$= \log\left(1 + \frac{|g_1|^2 P_d}{N_0}\right) = \log\left(1 + \frac{|g_2|^2 P_d}{N_0}\right), \tag{10.149}$$

which is exactly identical to the sum rate of OMA. It implies that the performance gain of NOMA vanishes if the users have the same or similar channel conditions.

10.6.1.2 Uplink Non-Orthogonal Multiple Access

UL transmission of NOMA slightly differs from its DL counterpart, where K spatially distributed users equipped with a single antenna simultaneously transmit their information-bearing symbols toward a single-antenna base station over the same resource unit. The received signal at the base station can be expressed by

$$r = \sum_{k=1}^{K} g_k \sqrt{P_k} s_k + n, \tag{10.150}$$

where s_k is the information-bearing symbol of user k, $k = 1, 2, \dots, K$, satisfying $\mathbb{E}\left[|s_k|^2\right] = 1$, and P_k denotes the power constraint of user k, g_k stands for the complex channel gain from user k to the base station, and n is the additive white Gaussian noise with zero mean and power spectral density N_0 (W/Hz). Similarly, we can also assume that user 1 has the largest channel gain, and user K is the weakest, i.e.

$$|g_1| \geqslant |g_2| \geqslant \cdots \geqslant |g_K|. \tag{10.151}$$

As a kind of power-domain technique, the power allocation plays a critical role in the DL of NOMA, which has at least two major functions: artificially creating sufficient power difference at the received signal to facilitate successive interference cancelation, and guaranteeing a reasonable received power for cell-edge users. In the UL, the users may again optimize their transmit powers according to their locations as in the DL. However, the users are possible to be well distributed in the cell coverage, and the received power levels from different users are already well separated.

The base station can firstly decode the symbol of the user with the largest received power, treating all other users' signals as a colored noise. Then, it cancels the corresponding interference from the received signal and continues to decode the symbol of another user with the second largest received power. This SIC process iterates at the base station until all symbols are detected.

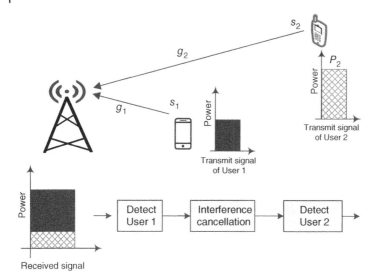

Figure 10.16 Illustration of uplink NOMA consisting of a base station, a far user, and a near user.

Following the decoding order of $1 \rightarrow 2 \rightarrow \cdots \rightarrow K$, the achievable rate of a typical user k can be computed by

$$R_k = B_w \log \left(1 + \frac{|g_k|^2 P_k}{\sum_{i=k+1}^{K} |g_i|^2 P_i + N_0 B_w} \right). \tag{10.152}$$

In the case of two users, as illustrated in Figure 10.16, a near user at the cell center transmits s_1 with power P_1 toward the base station while a far user at the cell edge simultaneously transmits s_2 with power P_2. Their power can be identical or the far user has higher power to guarantee a reasonable received signal strength. In practice, the difference of g_1 and g_2 raises a sufficient power difference between the received signal components of these two users, regardless of their transmit power gap. The base station can first detect the near user directly without SIC, resulting in

$$R_1 = \log \left(1 + \frac{|g_1|^2 P_1}{|g_2|^2 P_2 + N_0} \right), \tag{10.153}$$

assuming the signal bandwidth is normalized to $B_w = 1$ Hz. Then, the base station cancels the interference of the first user and detects the signal of the second user, getting

$$R_2 = \log \left(1 + \frac{|g_2|^2 P_2}{N_0} \right). \tag{10.154}$$

The sum rate of two-user UL NOMA transmission is computed by

$$
\begin{aligned}
R &= R_1 + R_2 \\
&= \log\left(1 + \frac{|g_1|^2 P_1}{|g_2|^2 P_2 + N_0}\right) + \log\left(1 + \frac{|g_2|^2 P_2}{N_0}\right) \\
&= \log\left(1 + \frac{|g_1|^2 P_1 + |g_2|^2 P_2}{N_0}\right).
\end{aligned}
\tag{10.155}
$$

If the base station first detects the far user directly and then detects the near user without inter-user interference, their achievable rates become

$$
\begin{aligned}
R_1 &= \log\left(1 + \frac{|g_1|^2 P_1}{N_0}\right) \\
R_2 &= \log\left(1 + \frac{|g_2|^2 P_2}{|g_1|^2 P_1 + N_0}\right).
\end{aligned}
\tag{10.156}
$$

Interestingly, if we do not consider a higher error probability due to the low SNR of detecting the far user first, the sum rate of two-user UL NOMA transmission is the same regardless of the SIC order, because

$$
\begin{aligned}
R &= R_1 + R_2 \\
&= \log\left(1 + \frac{|g_1|^2 P_1}{N_0}\right) + \log\left(1 + \frac{|g_2|^2 P_2}{|g_1|^2 P_1 + N_0}\right) \\
&= \log\left(1 + \frac{|g_1|^2 P_1 + |g_2|^2 P_2}{N_0}\right)
\end{aligned}
\tag{10.157}
$$

exactly equals to Eq. (10.155). Consequently, the two-user UL transmission with NOMA forms the following capacity region

$$
\mathfrak{C} = \left\{ (R_1, R_2) \in \mathbb{R}_+^2 \;\middle|\;
\begin{array}{l}
R_1 \leqslant \log\left(1 + \dfrac{|g_1|^2 P_1}{N_0}\right) \\[2mm]
R_2 \leqslant \log\left(1 + \dfrac{|g_2|^2 P_2}{N_0}\right) \\[2mm]
R_1 + R_2 \leqslant \log\left(1 + \dfrac{|g_1|^2 P_1 + |g_2|^2 P_2}{N_0}\right)
\end{array}
\right\}.
\tag{10.158}
$$

In an OMA scheme, the first user occupies η of the total time-frequency resources, leaving the rest of $(1 - \eta)$ time-frequency resources to the second user. Their achievable rates can be computed by

$$
\begin{aligned}
R_1 &= \eta \log\left(1 + \frac{|g_1|^2 P_1}{\eta N_0}\right) \\
R_2 &= (1 - \eta) \log\left(1 + \frac{|g_2|^2 P_2}{(1 - \eta)N_0}\right).
\end{aligned}
\tag{10.159}
$$

Suppose the geometries of the near user and the far user are $\frac{|g_1|^2 P_1}{N_0} = 20\,\mathrm{dB}$ and $\frac{|g_2|^2 P_2}{N_0} = 0\,\mathrm{dB}$, respectively. When equal bandwidth (i.e. $\eta = 0.5$) is allocated to either user with the proportional fairness criteria, the per-user rates of OMA are calculated according to Eq. (10.159) as

$$R_1 = 0.5\,\log_2\,(1 + 200) = 3.8255\ \mathrm{bps/Hz}$$
$$R_2 = 0.5\,\log_2\,(1 + 2) = 0.7925\ \mathrm{bps/Hz}. \tag{10.160}$$

In contrast, the per-user rates of UL NOMA transmission are calculated according to Eqs. (10.153) and (10.154) (decoding the near user at first) as

$$R_1 = \log_2\,(1 + 50) = 5.6724\ \mathrm{bps/Hz}$$
$$R_2 = \log_2\,(1 + 1) = 1\ \mathrm{bps/Hz}, \tag{10.161}$$

corresponding to a spectral-efficiency gain of approximately 48% and 26% in terms of the OMA scheme.

Reversing the SIC order, the per-user rates of UL NOMA transmission are calculated according to Eq. (10.156) (decoding the far user at first) as

$$R_1 = \log_2\,(1 + 100) = 6.6582\ \mathrm{bps/Hz}$$
$$R_2 = \log_2\,(1 + 1/101) = 0.0142\ \mathrm{bps/Hz}. \tag{10.162}$$

These results correspond to a spectral-efficiency gain of approximately 74% for the near user but a loss of -98% for the far user, implying the significance of SIC ordering. It can be seen that different SIR ordering obtains the same sum rate of $R = 6.6724\ \mathrm{bps/Hz}$, verifying the equivalence of Eqs. (10.155) and (10.157).

10.6.2 Multi-User Superposition Coding

With the advancement of receiver implementation and hardware capability, interference cancelation becomes more affordable in mobile terminals. It makes non-orthogonal transmission more feasible. In the DL, NOMA is a promising technique to boost system capacity and improve user experience, e.g. Multi-User Superposition Transmission (MUST) specified in 3GPP for DL mobile broadband services. In general, multiple users can be multiplexed over each orthogonal resource unit. However, considering the diminishing gains when superposing more users and the signaling overhead, the two-user superposition is generally focused.

The simplest method is a linear superposition, independently mapping the coded bits of two or more co-scheduled users into component constellation symbols that are superposed with adaptive power ratio. Figure 10.17 demonstrates the direct superposition coding over two users, where n and m coded bits of the near and far users, denoted as $b_1, b_2, \ldots, b_n$ and $c_1, c_2, \ldots, c_n$, respectively,

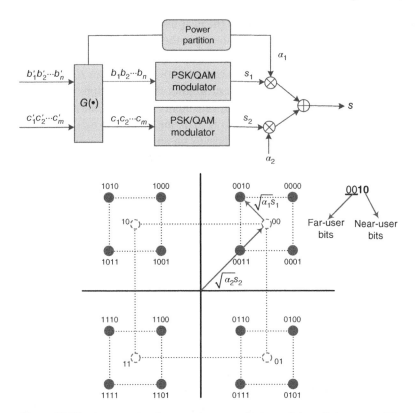

Figure 10.17 An example of composite constellations of downlink superposition transmission for a far user and a near user.

are simultaneously transmitted. The bits are separately modulated with legacy Phase-Shift Keying (PSK) or Quadrature Amplitude Modulation (QAM) modulators, each having 2^n and 2^m constellation points. Flexible power partition between the near and far user, denoted as $\alpha_1 P_d$ and $\alpha_2 P_d$, is employed. The power ratio α_1/α_2 is chosen to maximize the sum rate under a certain fairness criterion, while taking into account channel gains, and so on. After being scaled by appropriate power coefficients, the two modulation symbols, s_1 and s_2, are superposed, resulting in

$$s = \sqrt{\alpha_1 P_d}s_1 + \sqrt{\alpha_2 P_d}s_2. \tag{10.163}$$

The composite symbol s has a constellation with 2^{n+m} points. The 16-point constellation in the figure is an example when QPSK is employed for both the near and far users, where $\alpha_1 < 0.5$ since the near user is assigned to a small portion of the power, and its symbol amplitude is therefore small. The symbol of the far user has more power, determining in which quadrant the composite symbol would lie.

In other words, the constellation point of the near user depends on the constellation point of the far user. As an example shown in the figure, the coded bits of the far user *00* first decide the first quadrant, and then one of the four points in the first quadrant is selected according to the codes bits of the near user *10*.

Although the constellation of either user follows the Gray mapping (i.e. two adjacent constellation points differ in only one binary digit), the constellation of the direct superposition is not a kind of Gray mapping, leading to a capacity loss. A non-Gray mapped constellation relies heavily on an advanced receiver. In this case, symbol-level interference cancelation cannot sufficiently resolve the blurred constellation due to inter-user interference. More complicated receivers, e.g. codeword-level interference cancelation, are needed to achieve acceptable performance. The Gray mapping ensures robust performance even when interference cancelation is performed at the symbol level without channel decoding, which is therefore significantly simpler than codeword-level interference cancelation. As illustrated in Figure 10.17, inputting the coded bits into a bit converter denoted as $G(\cdot)$ can ensure Gray property in the composite constellation. Alternatively, a bit-partition-based scheme, rather than power partition, can be employed to implement the Gray mapping, where the constellation is legacy QAM mapper with a uniformly spaced rectangular grid and the coded bits of two or more users are directly superposed onto the symbols of a composite constellation [Yifei, 2016].

A 3GPP study item, "Study on DL Multiuser Superposition Transmission," has been carried out at 3GPP to evaluate the system performance of potential LTE enhancements enabling DL MUST. The objectives include defining target deployment scenarios and an evaluation methodology for MUST, identifying potential MUST schemes and corresponding LTE enhancements, and assessing the feasibility and system-level performance of the possible MUST schemes. As a result, 3GPP recommended three different categories of MUST in 3GPP TR36.859 [2015]:

- *MUST Category 1: Superposition transmission with adaptive power ratio on component constellations and non-Gray-mapped composite constellation*, where the coded bits of two or more co-scheduled users are independently mapped to component constellation symbols, but the composite constellation does not have Gray mapping.
- *MUST Category 2: Superposition transmission with adaptive power ratio on component constellations and Gray-mapped composite constellation*, where the coded bits of two or more co-scheduled users are jointly mapped to component constellations, and then the composite constellation has Gray mapping.
- *MUST Category 3: Superposition transmission with label-bit assignment on Gray-mapped composite constellation*, where the coded bits of two or more co-scheduled users are directly mapped onto the symbols of a composite constellation.

Accordingly, several candidate receiver schemes to enable various MUST schemes for both the far and near users have been suggested by 3GPP. The promising receiver schemes for the far user include

- Linear MMSE with interference rejection combining
- Maximum-likelihood detection
- Reduced-complexity maximum-likelihood detection
- Symbol-level interference cancelation

The receiver schemes for the near user could be

- Maximum-likelihood detection
- Reduced-complexity maximum-likelihood detection
- Symbol-level interference cancelation
- Linear codeword-level successive interference cancelation
- Maximum-likelihood codeword-level successive interference cancelation

10.6.3 Uplink Grant-Free Transmission

The majority of UL transmission in traditional mobile systems schedules users orthogonally, with a dedicated time-frequency resource unit per user. Such user scheduling leads to heavy signaling overhead between the network and terminals. On the other hand, the data payload is typically patchy and small in the deployment scenarios of the modern Internet of Things, but the number of connections is massive. High power consumption of devices due to significant overhead per user and the tight control mechanism make the low-cost, low-power terminal design more challenging. NOMA allows more users to be served simultaneously, and it facilitates grant-free transmission so that the system is not strictly constricted by the number of orthogonal resources and their scheduling granularity.

To alleviate the effect of resource collision in non-orthogonal transmission, spreading can be applied. Historically, non-orthogonal transmission has been used in the UL transmission of IS-95, CDMA2000, and Wideband Code Division Multiple Access (WCDMA), to name a few, which primarily serve circuit-switched voice users with continuous but small data streaming. These systems are based on direct sequence spread-spectrum, allowing for multiple users to share a common time-frequency resource by means of a set of spreading codes. Similarly, the concept of spreading can be used in the UL for massive connectivity with more advanced techniques. It raises a new philosophy of designing non-orthogonal transmission referred to as *code-domain NOMA*, in comparison with the conventional design called *power-domain NOMA*. UL grant-free transmission is comprised of the following critical ingredients [Yifei, 2016]:

- *Code Spreading*: As evidenced in the 3G technology based on direct sequence spread-spectrum, spreading can improve the system robustness against co-channel interference and inter-user interference. A factor graph is an effective tool for design optimization. In a factor graph, a number of variable nodes are connected to a number of factor nodes. The connections between variable nodes and factor nodes define the key property of non-orthogonal access schemes. Additionally, the map of connections provides guidance for receiver implementation, e.g. low density signature-based spreading, which can lower the detection complexity by avoiding the metric calculation of the full connections between the variable nodes and factor nodes. Moreover, non-binary complex sequences have lower cross-correlation between different sequences, compared with binary sequences, even when they are very short. These features can facilitate the accommodation of significantly more active users in shared resources when those users randomly choose spreading sequences.
- *Operation Mode*: In grant-free transmission, link adaptation is performed in a long-term style. Long-term means that the selection of Modulation and Coding Scheme (MSC) depends only on the large-scale fading and open-loop power control. Modulation Coding Scheme (MCS) is not frequently adjusted since the fluctuation of large-scale fading varies relatively slow, and the focus of grant-free scheduling is not on maximizing the system capacity.
- *Receiver Design*: Two kinds of receivers have gained much attention:
  - Bit-level successive interference cancelation
  - Message Passing Algorithm (MPA)

  The bit-level interference cancelation is similar to codeword-level successive interference cancelation. For the UL, bit-level interference cancelation becomes more affordable than the DL since the base station needs to decode the bits of all active devices. MPA is a sub-optimal algorithm of maximum-likelihood detection of a factor graph. The detection process is iterative, similar to that of a LDPC (Low Density Parity Check) decoder where belief metric or extrinsic information flows back and forth between the variable nodes and factor nodes.

Different schemes have been designed for UL NOMA to support massive connectivity and enable grant-free transmission with low latency. For example, a total of 15 proposals have been submitted to 3GPP during the design of New Radio UL transmission [Chen et al., 2018], including both code-domain and power-domain implementation, i.e. Sparse Code Multiple Access (SCMA), Multi-User Shared Access (MUSA), Low Code Rate Spreading, Frequency-Domain Spreading, Non-orthogonal Coded Multiple Access (NCMA), NOMA, Pattern Division Multiple Access (PDMA), Resource Spread Multiple Access (RSMA), Interleave-Grid Multiple Access (IGMA), Low Density Spreading with Signature Vector Extension (LDS-SVE), Low code rate and Signature-based Shared Access

(LSSA), Non-Orthogonal Coded Access (NOCA), Interleave-Division Multiple Access (IDMA), Repetition-Division Multiple Access (RDMA), and Group Orthogonal Coded Access (GOCA). These schemes have a common basis and many similarities regardless of their particular properties. To provide insight, we will study the principles and design aspects of several typical code-domain NOMA schemes in the subsequent part.

10.6.4 Code-Domain NOMA

In addition to differentiating different users' signals in the power domain, as introduced in Section 10.6.1, the practical implementation of NOMA can also be realized in the code domain. The most representative code-domain NOMA schemes include Low-Density Signature (LDS)-based CDMA or OFDM, and SCMA. This part will briefly present their basic principles and major features to give readers an insight into these spreading-based non-orthogonal transmission schemes.

10.6.4.1 Low-Density Signature-CDMA/OFDM

In the conventional CDMA, it is impossible to achieve orthogonal channelization when the number of active users is larger than the processing gain, namely, an overloaded condition. Using a dense density structure, each chip of the received signal contains the contribution from all co-scheduled users in the system. In other words, each user suffers from multiple-access interference from all other users at every chip. If the cross-correlation matrix of the signature follows a certain format, the complexity of optimal MUD algorithms can be lowered, but its complexity is still too high to become affordable. Inspired by the success of the low-density structure in LDPC codes, Hoshyar et al. [2008] proposed a novel LDS -based CDMA transmission. Along with an iterative chip-level message-passing algorithm or MPA-based detection, its achievable performance approaches that of the single-user system with the overloading factor of 200% under an affordable computational complexity, in comparison with the conventional structure employing optimal MUD.

Consider an UL CDMA system where K synchronous users simultaneously transmit their symbols x_k, $k = 1, 2, \ldots, K$ toward a base station with the aid of a set of N-length spreading sequences. The modulated symbol x_k is formed by mapping a sequence of independent information bits to a constellation alphabet. Then, the modulated symbol is multiplexed with a spreading sequence $\mathbf{s}_k = [s_{1,k}, s_{2,k}, \ldots, s_{N,k}]^T$, assigned uniquely to each user. In the conventional CDMA structure, each component of a spreading sequence usually takes non-zero values, which is optimized to certain constraints, e.g. high auto-correlation and low cross-correlation. Then, the signal received at chip n, $n = 1, 2, \ldots, N$ can be expressed by

$$y_n = \sum_{k=1}^{K} g_k s_{n,k} x_k + z_n, \tag{10.164}$$

where g_k stands for the channel gain between user k and the base station, $s_{n,k}$ is the nth component of the spreading sequence $\mathbf{s}_k$, and z_n denotes additive Gaussian noise.

Stacking N successive chips together on symbol-by-symbol basis, the received signal vector $\mathbf{y} = [y_1, y_2, \ldots, y_N]^T$ is the superposition of the transmitted signals form all users, which can be formulated as

$$\begin{aligned} \mathbf{y} &= \sum_{k=1}^{K} g_k \mathbf{s}_k x_k + \mathbf{z} \\ &= \mathbf{H}\mathbf{x} + \mathbf{z}, \end{aligned} \tag{10.165}$$

where $\mathbf{x} = [x_1, x_2, \ldots, x_K]^T$, $\mathbf{z} = [z_1, z_2, \ldots, z_N]^T$, and the effective receive signature $\mathbf{H} = [g_1 \mathbf{s}_1, g_2 \mathbf{s}_2, \ldots, g_K \mathbf{s}_K]$.

Instead of optimizing the N-chip signatures, the LDS structure intentionally arranges each user to spread its modulated symbol over a small number of d_v chips, followed by a zero-padding process such that the processing gain is still N, as shown in Figure 10.18. Additionally, let d_c be the maximum number of users that is allowed to interfere within a single chip. The spread and padded sequences are then interleaved uniquely for each user such that the resultant

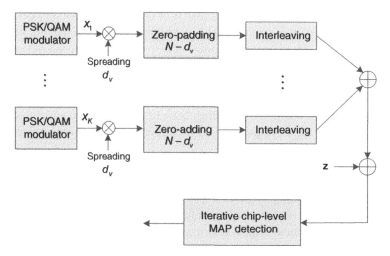

Figure 10.18 Block diagram of an LDS-CDMA system, where multiple users simultaneously transmit their symbols toward a base station.

signatures matrix becomes sparse. This low-density structure can be expressed by an indicator matrix, where the entry 1 on the nth row and kth column means user k spreads its signal at chip n, where 0 means user k turns off at this chip. The set of the position of 1's in the nth row of the indicator matrix denotes the set of users that contribute their data at the nth chip, while its kth column represents the set of chips over which user k spreads its data. Let ξ_n and ζ_k be the set of 1's position in the nth row and kth column, respectively, Eq. (10.164) can be rewritten as

$$y_n = \sum_{k \in \xi_n} g_k s_{n,k} x_k + z_n. \tag{10.166}$$

As a result, the number of the superimposed signals at each chip will be less than the number of active users so as to lower the multi-user interference. For example, an indicator matrix of

$$\begin{bmatrix} 1 & 1 & 1 & 0 & 0 & 0 \\ 1 & 0 & 0 & 1 & 1 & 0 \\ 0 & 1 & 0 & 1 & 0 & 1 \\ 0 & 0 & 1 & 0 & 1 & 1 \end{bmatrix}, \tag{10.167}$$

standing for the setting where $K = 6$ users superimpose their symbols x_k, $k = 1, \ldots, 6$ over $N = 4$ chips. Since the number of users is larger than the number of orthogonal resource units, NOMA is achieved. Each user spreads its symbol using a unique 4-chip spreading sequence consisting of only two non-zero components, i.e. $d_v = 2$. Each chip accommodates only three users, namely, $d_c = 3$, rather than all six users, so that multiple-access interference is reduced. Meanwhile, the spreading operation can be expressed by a factor graph that contains a number of variable nodes and factor nodes, as shown in Figure 10.19. The variable nodes usually represent the modulated symbols or coded bits,

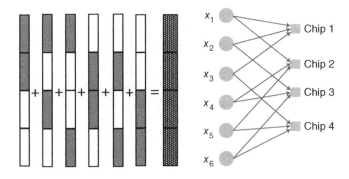

Figure 10.19 An example of the indication matrix and its corresponding factor graph.

and the factor nodes stand for orthogonal time-frequency resource units. The connection between the variable and factor nodes can distinguish non-orthogonal access schemes. A variable node is able to connect to multiple factor nodes, which is actually the spreading process. A factor node can connect to multiple variable nodes, implying non-orthogonal resource allocation [Yifei, 2016].

Apply the low-density spreading over OFDM is straightforward if replacing each chip with an OFDM subcarrier, forming a new technique called Low-Density Structure-OFDM or LDS-OFDM [Hoshyar et al., 2010]. It is noted that the design constraint of LDS-CDMA and LDS-OFDM is that the number of users is larger than the number of chips (or subcarriers), namely, $K > N$, to achieve NOMA. Meanwhile, the number of users corresponding to each chip or subcarrier should be less than the number of chips or subcarriers, e.g. $d_c < N$ to guarantee that the multi-access interference is lowered, compared with the conventional dense-density structure.

10.6.4.2 Sparse Code Multiple Access

SCMA is an enhanced version of the basic LDS-CDMA. The fundamental of LDS-CDMA and SCMA is the same: to use low density (sparse non-zero component sequence) to reduce the complexity of MPA detection at the receiver. The key idea of SCMA is to merge the constellation mapping and the spreading to map coded bits to a codeword directly. The whole process can be interpreted as a coding procedure from the binary domain to a complex multidimensional domain. SCMA has been proposed in Nikopour and Baligh [2013] with the following properties:

- Binary-domain data are directly encoded to multidimensional complex-domain codewords selected from a predefined codebook set
- Multiple access is achieved by designing multiple codebooks, one for each layer or user
- Codewords are sparse such that the MPA multi-user detection is applicable to detect the multiplexed codewords with affordable complexity
- Non-orthogonal transmission is implemented by multiplexing a number of layers or users that is larger than the spreading factor

Without losing generality, we can assume there are K codebooks available dedicated to K users or K spatial layers. Each codebook is comprised of M codewords of length N, and the number of non-zero multidimensional elements in each codeword is d_v. All codewords of a particular codebook contain zeros in the same $N - d_v$ dimensions, and the positions of zeros in different codebooks are distinct to facilitate the collision avoidance of any pair of users. Consequently, the maximum number of codebooks is limited by the selection of N and d_v, equaling to $\binom{N}{d_v}$. Non-zero values in the codewords can take various complex values. Each user or

Codebooks

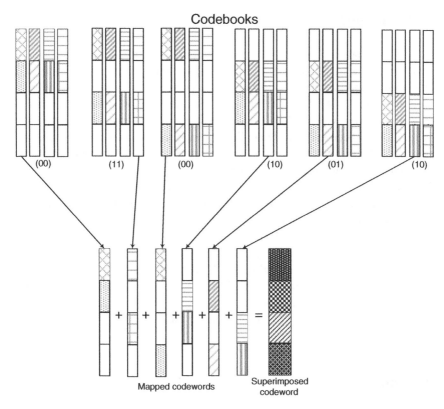

Figure 10.20 Illustration of an SCMA system with six users and four subcarriers.

layer maps $\log_2 M$ coded bits directly to a multidimensional complex codeword. The selected codewords of all users or layers are superimposed into a multiplexed codeword, which is transmitted over N shared orthogonal resource units, such as CDMA chips or OFDM subcarriers.

For example, as illustrated in Figure 10.20, there are $K = 6$ users in an SCMA system, and each user has a unique codebook. The length of codewords is $N = 4$, and all codewords of a particular codebook contain $d_v = 2$ non-zero complex values in the same two dimensions. Therefore, it can support up to $\binom{4}{2} = 6$ different codebooks, and the positions of non-zeros in different codebooks are distinct to facilitate collision avoidance. For each user, a pair of coded bits, e.g. 00 for user 1, 11 for user 2, are mapped to a complex codeword in each codebook. The selected codewords for six users are superimposed into a multiplexed codeword, which is then transmitted over $N = 4$ shared orthogonal resources.

10.7 Summary

After the original concept of OFDM proposed by Chang [1966], two fundamental breakthroughs were followed: the efficient implementation using FFT proposed by Weinstein and Ebert [1971], and the employment of cyclic prefix proposed by Peled and Ruiz [1980]. Then, OFDM has become the most dominant modulation technique for wired and wireless communication systems over the past two decades. It has been extensively applied in many well-known standards, xDSL, DVB-T, Wi-Fi, WiMAX, LTE, and LTE-Advanced, to name a few. After an extensive comparison among all possible techniques, OFDMA has been adopted as one of the critical elements of 5G NR, as a comprehensive trade-off among the performance, complexity, comparability, and robustness. It is the first time in the history of mobile communications where the same multiple access technique becomes the basis for two generations of mobile systems. It is envisioned that OFDM, OFDMA, and SC-FDMA will also serve as critical technologies in the forthcoming 6G transmission either in the conventional sub-6 GHz band and high-frequency bands. In addition, the 5G standard has integrated NOMA due to its advantages in spectral efficiency and massive connections. It is also envisioned that more advanced and efficient NOMA schemes will be designed and more tightly integrated into the next-generation system.

References

3GPP TR36.859 (2015), Study on downlink multiuser superposition transmission (MUST) for LTE (release 13), Technical Report 36.859, The 3rd Generation Partnership Project.

Brandes, S., Cosovic, I. and Schnell, M. (2006), 'Reduction of out-of-band radiation in OFDM systems by insertion of cancellation carriers', *IEEE Communications Letters* **10**(6), 420–422.

Chang, R. W. (1966), 'Synthesis of band-limited orthogonal signals for multichannel data transmission', *Bell System Technology Journal* **45**, 1775–1796.

Chen, Y., Bayesteh, A., Wu, Y., Ren, B., Kang, S., Sun, S., Xiong, Q., Qian, C., Yu, B., Ding, Z., Wang, S., Han, S., Hou, X., Lin, H., Visoz, R. and Razavi, R. (2018), 'Toward the standardization of non-orthogonal multiple access for next generation wireless networks', *IEEE Communications Magazine* **56**(3), 19–27.

Chung, C.-D. (2008), 'Spectral precoding for rectangularly pulsed OFDM', *IEEE Transactions on Communications* **56**(9), 1498–1510.

Cosovic, I., Brandes, S. and Schnell, M. (2006), 'Subcarrier weighting: A method for sidelobe supression in OFDM systems', *IEEE Communications Letters* **10**(6), 444–446.

Dahlman, E., Parkvall, S. and Sköld, J. (2011), *4G LTE/LTE-Advanced for Mobile Broadband*, Academic Press, Elsevier, Oxford, The United Kingdom.

Ding, Z., Lei, X., Karagiannidis, G. K., Schober, R., Yuan, J. and Bhargava, V. K. (2017), 'A survey on non-orthogonal multiple access for 5G networks: Research challenges and future trends', *IEEE Journal on Selected Areas in Communications* **35**(10), 2181–2195.

Goldsmith, A. (2005), *Wireless Communications*, Cambridge University Press, Stanford University, California.

Hoshyar, R., Wathan, F. P. and Tafazolli, R. (2008), 'Novel low-density signature for synchronous CDMA systems over AWGN channel', *IEEE Transactions on Signal Processing* **56**(4), 1616–1626.

Hoshyar, R., Razavi, R. and Al-Imari, M. (2010), LDS-OFDM an efficient multiple access technique, *in* 'Proceedings of 2010 IEEE 71st Vehicular Technology Conference (VTC)', Taipei, Taiwan.

Jiang, W. (2012), Multicarrier transmission schemes in cognitive radio, *in* 'Proceedings of 2012 International Symposium on Signals, Systems, and Electronics (ISSSE)', Potsdam, Germany.

Jiang, W. (2016), 'Method for dynamically setting virtual subcarriers, receiving method, apparatus and system'. U.S. Patent, US9,319,885, 19 April 2016.

Jiang, W. and Kaiser, T. (2016), From OFDM to FBMC: Principles and comparisons, *in* F. L. Luo and C. Zhang, eds, '*Signal Processing for 5G: Algorithms and Implementations*', John Wiley & Sons and IEEE Press, United Kingdom, Chapter 3.

Jiang, W. and Schellmann, M. (2012), Suppressing the out-of-band power radiation in multi-carrier systems: A comparative study, *in* 'Proceedings of 2012 IEEE Global Communications Conference (GLOBECOM)', Anaheim, CA, USA, pp. 1477–1482.

Jiang, W. and Schotten, H. D. (2019), Recurrent neural network-based frequency-domain channel prediction for wideband communications, *in* 'Proceedings of IEEE 89th Vehicular Technology Conference (VTC2019-Spring)', Kuala Lumpur, Malaysia.

Jiang, W. and Schotten, H. (2021a), 'Impact of channel aging on zero-forcing precoding in cell-free massive MIMO systems', *IEEE Communications Letters* **25**(9), 3114–3118.

Jiang, W. and Schotten, H. D. (2021b), 'Cell-free massive MIMO-OFDM transmission over frequency-selective fading channels', *IEEE Communications Letters* **25**(8), 2718–2722.

Jiang, W. and Schotten, H. D. (2021c), The kick-off of 6G research worldwide: An overview, *in* 'Proceedings of 2021 IEEE Seventh International Conference on Computer and Communications (ICCC)', Chengdu, China.

Jiang, W. and Schotten, H. D. (2022a), Initial access for millimeter-wave and terahertz communications with hybrid beamforming, *in* 'Proceedings of 2022 IEEE International Communications Conference (ICC)', Seoul, South Korea.

Jiang, W. and Schotten, H. D. (2022b), Initial beamforming for millimeter-wave and terahertz communications in 6G mobile systems, *in* 'Proceedings of 2022 IEEE Wireless Communications and Networking Conference (WCNC)', Austin, USA.

Jiang, W. and Zhao, Z. (2012), Low-complexity spectral precoding for rectangularly pulsed OFDM, *in* 'Proceedings of 2012 IEEE Vehicular Technology Conference (VTC Fall)', Quebec City, QC, Canada.

Jiang, W., Cao, H., Nguyen, T. T., Güven, A. B., Wang, Y., Gao, Y., Kabbani, A., Wiemeler, M., Kreul, T., Zheng, F. and Kaiser, T. (2013), Key issues towards beyond LTE-advanced systems with cognitive radio, *in* 'Proceedings of 2013 14th Workshop on Signal Processing Advances in Wireless Communications (SPAWC)', Darmstadt, Germany, pp. 510–514.

Jiang, W., Han, B., Habibi, M. A. and Schotten, H. D. (2021), 'The road towards 6G: A comprehensive survey', *IEEE Open Journal of the Communications Society* **2**, 334–366.

Nikopour, H. and Baligh, H. (2013), Sparse code multiple access, *in* 'Proceedings of 2013 IEEE 24th Annual International Symposium on Personal, Indoor, and Mobile Radio Communications (PIMRC)', London, UK, pp. 332–336.

Oppenheim, A. V., Willsky, A. S. and Nawab, S. H. (1998), *Signals and Systems*, second edn, Pearson Education Inc., Prentice-Hall, New Jersey, The United States.

Peled, A. and Ruiz, A. (1980), Frequency domain data transmission using reduced computational complexity algorithms, *in* 'Proceedings of 1980 IEEE International Conference on Acoustics, Speech, and Signal Processing (ICASSP)', Denver, CO, USA, pp. 964–967.

Ring, D. H. (1947), 'Mobile telephony - wide area coverage', *Bell Telephone Laboratories.*

Saito, Y., Kishiyama, Y., Benjebbour, A., Nakamura, T., Li, A. and Higuchi, K. (2013), Non-orthogonal multiple access (NOMA) for cellular future radio access, *in* 'Proceedings of 2013 IEEE 77th Vehicular Technology Conference (VTC Spring)', Dresden, Germany.

Tse, D. and Viswanath, P. (2005), *Fundamentals of Wireless Communication*, Cambridge University Press, Cambridge, United Kingdom.

Weinstein, S. and Ebert, P. (1971), 'Data transmission by frequency-division multiplexing using the discrete Fourier transform', *IEEE Transactions on Communication Technology* **19**(5), 628–634.

Yang, X. (2005), Soft frequency reuse scheme for UTRAN LTE, *in* '3GPP TSG RAN WG1 Meeting #41, R1-050507', Athens, Greece.

Yang, X. (2014), 'A multilevel soft frequency reuse technique for wireless communication systems', *IEEE Communications Letters* **18**(11), 1983–1986.

Yifei, Y. (2016), Non-orthogonal multi-user superposition and shared access, *in* F. L. Luo and C. Zhang, eds, '*Signal Processing for 5G: Algorithms and Implementations*', John Wiley & Sons and IEEE Press, United Kingdom, Chapter 6.

Index

6G Key Technologies: A Comprehensive Guide, First Edition. Wei Jiang and Fa-Long Luo.
© 2023 The Institute of Electrical and Electronics Engineers, Inc. Published 2023 by John Wiley & Sons, Inc.

Printed and bound by CPI Group (UK) Ltd, Croydon, CR0 4YY

27/10/2024

14580670-0004